T0213340

Igor Kriz • Sophie Kriz

Introduction to Algebraic Geometry

 Birkhäuser

Igor Kriz
Department of Mathematics
University of Michigan
Ann Arbor, MI, USA

Sophie Kriz
Department of Mathematics
University of Michigan
Ann Arbor, MI, USA

ISBN 978-3-030-62643-3   ISBN 978-3-030-62644-0   (eBook)
https://doi.org/10.1007/978-3-030-62644-0

Mathematics Subject Classification: 14-01

This book is published under the imprint Birkhäuser, www.birkhauser-science.com, by the registered company
Springer Nature Switzerland AG.
The registered company address is: Gewerbestrasse 11, 6330 Cham, Switzerland

To Professor Aleš Pultr

Preface

Algebraic geometry is an important subject in many fields of mathematics. Its modern foundations are over 50 years old, formulated by Grothendieck and his co-authors, and recorded in the iconic EGA and SGA volumes. It should therefore now be a classical subject. Yet, algebraic geometry is still not easily accessible to the beginning student. No single book seems to allow the student to learn all the basic concepts together with their motivations and foundations. We hope, in the present text, to give an introductory approach following such a philosophy, without unnecessarily limiting our attention to elementary concepts only.

An amazing feature of algebraic geometry is how little it, at least initially, requires in the way of prerequisites. Basic concepts of algebra, for example, a *commutative ring*, suggest a geometric view: we want to think of a ring as a ring of functions, of the prime ideals as the functions' vanishing points. To make these visualizations, we do not need analysis.

The role of analysis, however, is replaced by the theory of commutative rings, known as *commutative algebra*, a deep subject in its own right. In addition, analysis and topology eventually must reappear. Many concepts of algebraic geometry imitate the notions of analysis and topology. We make an effort to introduce all the motivating concepts as they arise, together with other basic notions of higher mathematics.

Perhaps the greatest excursion into another field occurs in the subject of *cohomology*. This concept originated in algebraic topology, and to understand it properly, one must include the topological, categorical, and analytical points of view. We dedicate an entire chapter to the foundations of cohomology, and then another chapter to the various manifestations of the idea of cohomology within the subject of algebraic geometry. Some of these discussions touch on more advanced topics and suggest possible directions of further reading.

Algebraic geometry is an involved subject, and we hope to help the reader get a good start on it.

Ann Arbor, MI, USA
Ann Arbor, MI, USA

Igor Kriz
Sophie Kriz

Introduction

Algebraic geometry introduces an appealing way of thinking about geometry solely based on algebra, and its basic concepts can be introduced in an "elementary" fashion, without assuming, for example, any familiarity with advanced calculus. Saying this, of course, hides the fact that one must use a lot of abstraction early on. In addition, the algebra involved, when treated in detail, is not always easy or elementary, and the development of the theory often imitates concepts of topology and analysis. Without having seen those concepts, the right intuition may be hard to form.

What is the right set of prerequisites, then, for studying this book? Roughly speaking, we assume that the typical reader will have seen undergraduate analysis, algebra, and point set topology. However, we try to review those subjects as they arise. We think that for many concepts, their use in algebraic geometry serves as additional motivation, and vice versa. This is why we hope to make this text as self-contained as possible. Somewhat special is the role of *category theory*, an abstract language central to a modern treatment of algebraic geometry, which is typically not a part of undergraduate curricula. For this reason, we introduce the concepts of category theory in more detail.

The book is divided into six chapters. In Chap. 1, we focus on algebraic varieties over the field of complex numbers $\mathbb{C}$. In the first section, we introduce very quickly affine and projective varieties and regular functions. Even at this early stage, we will get a first glimpse of the concept of a *sheaf*. The notion of a *category* naturally arises from looking at these structures, so we describe it next, along with basic strategies of how categories are used in mathematics, and, in particular, how a category of the algebraic varieties we introduced is formed. We then go on to describe a few other geometric notions, such as rational maps of varieties (i.e., maps defined "generically"), as well as smooth maps and dimension.

At this point, it becomes clear that for a deeper understanding of the concepts introduced, and to be able to calculate with them, we need to revisit algebra. The type of algebra needed is called *commutative algebra*. The aim of this book is to be completely self-contained in commutative algebra; this means that essentially all necessary concepts of commutative algebra are treated. This will not be possible without some forward references (the structure of Chap. 1 being the first example) and without confining some material to the Exercises (this does not occur in Chap. 1, however).

The commutative algebra foundations of Chap. 1 are divided into two sections. In Sect. 4, we deal with polynomials. We cover the basic properties of polynomial rings, the basic algorithms (including finding a Gröbner basis), and the Nullstellensatz, which we have used without proof in the very beginning. In Sect. 5, we begin our treatment of commutative algebra more systematically, essentially with the goal of proving the elementary facts about dimension and regular local rings. Much of this material will be useful in subsequent chapters.

By the end of Chap. 1, the level of abstraction achieved in algebra has exceeded the abstraction introduced in geometry. It becomes clear that the "patchwork" notion of variety, as introduced, neither is satisfactory nor exhausts fully the possibilities of the methods commutative algebra has to offer. Thus, it is time to introduce Grothendieck's concept of a *scheme*. This is done in Chap. 2. On this occasion, we need to strengthen our foundations in category theory also. We need to discuss sheaves in more detail and explore the methods of category theory in their full strength, introducing, for example, limits, colimits, and adjoint functors. Abelian categories, axiomatizing the basic properties of the category of abelian groups, also become a necessary notion. This is because abelian sheaves are so intrinsically embedded in the foundations of schemes, and many of their formal properties are contained in the fact that abelian sheaves form an abelian category.

In Chap. 2, we discuss the basic properties of the category of schemes, explore which limits and colimits exist, and detail the basic construction methods of schemes. This includes affine schemes (which form the opposite category of commutative rings), projective schemes, and gluing. Basic finiteness properties of schemes, such as quasicompact and Noetherian schemes, and morphisms of finite type, as well as finite morphisms, are also formulated.

Chapter 3 discusses more advanced, as well as more geometrical, properties of schemes. We encounter some more striking examples of algebraic geometry modeling the concepts of point set topology. In particular, we introduce separated morphisms of schemes, which are an analogue of the concept of a Hausdorff space, and universally closed morphisms of schemes, which are an analogue of the concept of a compact space. Schemes that are both separated and universally closed are called *proper*.

We also introduce the concept of regular schemes and smooth morphisms of schemes, thus generalizing the corresponding concepts introduced in Chap. 1 for varieties. And on that occasion, it is now time to introduce a more abstract concept of varieties. Using these concepts, we prove the classification theorem of curves over algebraically closed fields with respect to birational equivalence. In the process, new commutative algebra is needed as well. For example, we introduce the techniques of valuation rings and also give another proof (more abstract and general) of the Nullstellensatz.

It also becomes apparent at this point that working over an algebraically closed field when discussing varieties restricts the point of view unduly. We therefore discuss the Galois group (a review of Galois theory is presented in the Exercises) and explain how varieties over a perfect field can be viewed as varieties over the algebraic closure, with

Galois group action. To be able to make these statements, we cover the basics of group actions, and also profinite groups.

At this point, for the sake of an analogy, we make our first excursion into algebraic topology, describing the fundamental group of a space, and also showing how, at the cost of working profinitely, it can be unified with the Galois group in the concept of the *étale fundamental group of a scheme*.

Finally, we briefly state the famous Weil conjectures on zeta functions of varieties over finite fields (solved by Deligne, based on earlier work of Grothendieck and others) to provide an illustration of the deep arithmetic information we encounter when leaving the world of algebraically closed fields.

In Chap. 4, our pursuit of geometric concepts continues. We begin with the notion of algebraic vector bundles and then move on to quasicoherent and coherent sheaves. Once again, we need to revisit our foundations of sheaves, covering in more detail such topics as sheaves of universal algebras, and abstract sheafification, as well as functors on categories of sheaves associated with maps of the underlying spaces. We also discuss the zeroth algebraic K-group.

Perhaps the largest part of the chapter is dedicated to divisors. We start with line bundles and the Picard group, and then discuss Cartier divisors and finally Weil divisors and divisibility. We also illustrate these concepts with many examples. In the process, we discuss the first cohomology group of an abelian sheaf. Chapter 4 is concluded by a treatment of very ample divisors, and blow-ups.

An important point needs to be made here. Many concrete computations are done using Weil divisors on regular schemes. However, this requires yet another layer of foundations from commutative algebra, for example, the Auslander-Buchsbaum theorem asserting that a regular local ring is a unique factorization domain, or the theorem that a localization of a regular ring is regular, or that a polynomial ring in one generator over a regular ring is regular. This already came up in Chap. 3 and arises even much more strongly in the context of divisors. Many texts hide the fact that none of this is rigorously proved without the cohomological characterization of regular local rings, which does not fit neatly the narrative of algebraic geometry in the most narrow sense. Because of the way the present book is structured, we have a unique opportunity to prove all the necessary foundational facts, but we postpone them to Chap. 5, where a general introduction to cohomology is given.

The purpose of Chap. 5 then is to give an adequately broad introduction to cohomology. This cannot be just cohomology of sheaves, which is perhaps the most native part in algebraic geometry. As we already saw, we also need to discuss *Ext* groups of modules, to fill a gap in our foundations. The fact is that cohomology and homology are really concepts of algebraic topology, and it is impossible to give a good introduction without venturing into that subject in some depth.

To motivate things, we start Chap. 5 with de Rham cohomology of manifolds, which was not historically the first approach but is geometrically self-motivating, at the cost of making a brief excursion into *analytic* geometry. We introduce manifolds using

Grothendieck's general approach to geometry using ringed spaces, thereby demonstrating the fact that the scope of the method extends beyond schemes. Covering this concept has the added benefit of distinguishing it from a purely algebraic de Rham cohomology, which is covered (and then compared) in Chap. 6.

The real reason why analytical de Rham cohomology works is that it is an example of sheaf cohomology. Sheaf cohomology and Ext are both derived functors. To understand them properly, we give an introduction to derived categories in general and derived categories of abelian categories in particular. After introducing these methods, we are in position to prove the cohomological characterization of regular local rings.

As already mentioned, the origin of homology and cohomology is in algebraic topology. To have a full picture (and computational tools), we briefly cover singular homology and cohomology of topological spaces as well as CW-homology. We also introduce spectral sequences and give some useful examples, in particular the Grothendieck and Leray spectral sequences. We prove the coincidence between singular cohomology with coefficients $\mathbb{R}$ and de Rham cohomology, thus proving the topological version of the de Rham theorem.

The aim of Chap. 6 is to bring the discussion of cohomology closer to its applications in algebraic geometry and to introduce some more advanced cohomological methods. We begin by discussing the rich additional structure present on the de Rham cohomology of a smooth projective variety over $\mathbb{C}$ (in its analytic topology). This is known as *Hodge theory*. Then, we move on to constructing a purely algebraic analog of de Rham cohomology of a variety over any field. In the process, we explore cohomology of schemes with coefficients in quasicoherent sheaves. Using these methods, we then prove the results of Serre and Grothendieck, comparing the algebraic and geometric versions of de Rham cohomology, and prove the algebraic version of the de Rham theorem over $\mathbb{C}$. We also explain why we do not have a natural algebraic de Rham theorem over $\mathbb{R}$.

While de Rham cohomology can be discussed over any field (or even ring), its good properties deteriorate in positive characteristic. To begin with, we only get cohomology groups that are modules over a finite field. We show how this can be remedied using the tool of *crystalline cohomology*, which gives p-adic information for varieties over fields of characteristic p. We also show how using *étale cohomology*, a technique based on étale morphisms enabled by Grothendieck's generalization of the concept of topology, lets us obtain ℓ-adic information for varieties over a field of characteristic p with $\ell \neq p$. Finally, we briefly introduce Voevodsky's *motivic cohomology* that gives *integral* (i.e., $\mathbb{Z}$-valued) cohomological information for smooth schemes over any field.

In conclusion, we would like to say a few words about the exercises. In this book, there are usually about 30–40 exercises per chapter. We found it more convenient to group the exercises by chapter: they seemed easier to find and cross-reference than if divided between the individual sections. The exercises are of several main types: There are routine computational exercises, and exercises asking to fill in details of proofs. There are also deeper larger-scale exercises on foundational topics that do not naturally fit into the narrative of the text. This includes review topics (such as Galois theory), as well as new topics (such as the Auslander-Buchsbaum theorem or the Zariski Main Theorem). These exercises contain detailed hints, basically describing the main stages of the proof of what is claimed. The student is expected to follow the proof along and write it up, justifying every step. Finally, there are exercises describing applications of the material covered. Again, there are more or less routine applications (such as, say, the Riemann-Roch theorem for curves or the cohomology of smooth hypersurfaces), as well as deeper applications, some of which may not be ordinarily mentioned in this context (such as the lack of finite étale morphisms over $Spec(\mathbb{Z})$, the Quillen-Suslin theorem, or the Artin-Mumford birational invariant for smooth projective varieties over $\mathbb{C}$). This diversity of exercises is intentional. Algebraic geometry is an indispensable and central part of many fields of mathematics. This is the main reason mathematicians interested in many fields should study it.

Contents

Beginning Concepts

The purpose of this chapter is to give a quick introduction to algebraic varieties, which will give us an idea of what algebraic geometry studies. In contrast with, say, differential geometry, almost no prerequisites are needed to start with. In particular, no analysis is required: All the foundations come from algebra. We begin this chapter by assuming the key results from algebra as given, and define the first concepts of algebraic geometry in Sect. 1. After that, some concepts from category theory and universal algebra will become necessary, which are introduced in Sect. 2. In Sect. 3, some additional geometrical concepts are introduced.

By that time, it will become clear that we should really explore the underlying algebraic concepts in more detail. In Sect. 4, we introduce the main facts about polynomials of several variables over a field, and treat some of the basic methods of computing with polynomials. The reader should consult this section before attempting the calculational exercises at the end of the chapter.

Section 5 delves deeper into abstract *commutative algebra*, which contains the algebraic foundations of algebraic geometry. One of the goals of this section is to prove the main theorems about *dimension*, which are very useful in working with algebraic varieties and with the concepts introduced in subsequent chapters.

1   The Definition of Algebraic Varieties

1.1   Affine Algebraic Sets

The starting point of algebraic geometry is studying solutions of systems of polynomial equations in several variables over a field. (A *field* is an algebraic structure with operations of addition and multiplication which satisfy all the formal properties of the real numbers,

© The Author(s), under exclusive license to Springer Nature Switzerland AG 2021
I. Kriz, S. Kriz, *Introduction to Algebraic Geometry*,
https://doi.org/10.1007/978-3-030-62644-0_1

i.e. commutativity and associativity of both operations, the existence of $0 \neq 1$ with their usual properties, distributivity, and the existence of an additive inverse—or minus sign— as well as multiplicative inverses of non-zero elements). Systems of polynomial equations in variables $x_1, \ldots, x_n$ can always be written in the form

$$p_1(x_1, \ldots, x_n) = 0,$$
$$\ldots$$
$$p_m(x_1, \ldots, x_n) = 0. \tag{1.1.1}$$

where $p_1, \ldots, p_m$ are polynomials. Solutions of the Eqs. (1.1.1) are n-tuples of elements $(x_1, \ldots, x_n)$ of the given field which satisfy the equations. Such n-tuples are also called *zeros of the polynomials* $p_1, \ldots, p_n$. Sets of zeros of sets of polynomials are called *affine algebraic sets*.

The set of all polynomials in n variables over a field forms a *commutative ring*, which means that it has operations of addition and multiplication satisfying all the formal properties of integers, i.e. commutativity and associativity of both operations, the existence of 0 and 1, distributivity, and the existence of an additive inverse. One defines not necessarily commutative rings by dropping the assumption that multiplication be commutative (we then must require that 1 be a left and right unit and that left and right distributivities hold). In this book, by a ring, we shall mean a commutative ring, unless specified otherwise.

Solutions of (1.1.1), or zeros of the polynomials $p_1, \ldots, p_m$, are also zeros of all *linear combinations*

$$a_1 p_1 + \cdots + a_m p_m \tag{1.1.2}$$

where $a_1, \ldots, a_m$ are arbitrary polynomials.

The elements (1.1.2) form *the ideal generated by* $p_1, \ldots, p_m$, which is denoted by

$$(p_1, \ldots, p_m).$$

An *ideal* in a commutative ring is a subset which contains 0, is closed under $+$, and multiples by elements of the ring. By the *Hilbert basis theorem*, which we prove in Sect. 4 (Theorem 4.2.1), the ring of polynomials in n variables over a field is *Noetherian*, which means that every ideal is finitely generated (i.e. generated by finitely many elements). Because of this, it is sufficient to consider systems of finitely many polynomial equations (1.1.1).

Note that a commutative ring R is Noetherian if and only if it satisfies the ascending chain condition (ACC) with respect to ideals. To satisfy the ACC with respect to subsets of a certain kind means that there does not exist an infinite chain

$$I_1 \subsetneq I_2 \subsetneq \cdots \subsetneq I_n \subsetneq \cdots \tag{1.1.3}$$

of such sets. Thus, we claim that a ring R is Noetherian if and only if (1.1.3) does not occur in R where I_n are ideals. To see this, if R is not Noetherian, it has an ideal I which is not finitely generated, so having picked, by induction, elements $r_1 \ldots, r_n \in I$, they cannot generate I, so we can pick $r_{n+1} \in I \setminus (r_1, \ldots, r_n)$. Thus, R fails the ACC for ideals. On the other hand, if R fails the ACC for ideals, then we have ideals (1.1.3) in R. Assume, for contradiction, that R is Noetherian. Let

$$I = \bigcup_n I_n.$$

Then the ideal is finitely generated, say, by elements $r_1, \ldots, r_k$. Thus, there exists an n such that $r_1, \ldots, r_k \in I_n$, which implies $I_n = I$, which is a contradiction.

1.1.1 Complex Numbers

Zeros of polynomials behave better when we work in the field $\mathbb{C}$ of complex numbers than in the field $\mathbb{R}$ of real numbers. The field $\mathbb{C}$ contains the number i which has the property

$$i^2 = -1,$$

and more generally, a complex number can be uniquely written as $a + bi$ where a, b are real numbers. Addition and multiplication are then determined by the properties of a field. Division is possible because $(a + bi)(a - bi) = a^2 + b^2$, and we can thus make the denominator real.

Thus, the polynomial equation

$$x^2 + 1 = 0$$

has solutions in $\mathbb{C}$, namely i and $-i$, while it has no solution over the field of real numbers $\mathbb{R}$.

It turns out that more generally, *every* non-constant polynomial in one variable with coefficients in $\mathbb{C}$ has at least one zero (we also say *root*). A field which satisfies this property is called *algebraically closed*. The fact that $\mathbb{C}$ is algebraically closed is known as the *fundamental theorem of algebra*. In the first three sections of this chapter, we will assume from now on that we are working over the field $\mathbb{C}$. More generally, in much of what we say (excluding connections with analysis), we could work over any algebraically closed field.

1.1.2 Nullstellensatz

The fact that a non-constant polynomial over $\mathbb{C}$ always has a root can be generalized to several variables as follows: Let I be an ideal in the ring $\mathbb{C}[x_1, \ldots, x_n]$ of polynomials in n variables over $\mathbb{C}$. Let $X = Z(I)$ be the affine algebraic set which is the set of zeros of the ideal I. If $I = (f_1, \ldots, f_m)$, we also write

$$Z(f_1, \ldots, f_m) = Z(I).$$

Let, on the other hand, $I(X)$ be the ideal of all polynomials which are zero on X (i.e. $p(x_1, \ldots, x_n) = 0$ for every $(x_1, \ldots, x_n) \in X$). Then

$$I(X) = \sqrt{I} \tag{1.1.4}$$

where the right hand side of (1.1.4) is called the *radical* of I and consists of all polynomials p for which $p^k \in I$ for some non-negative integer k. Equation (1.1.4) is called the *Nullstellensatz*, and is due to Hilbert. In German, Nullstelle means zero, literally "zero place," a point at which a polynomial is zero. In English, as we already remarked, such a point is called just a "zero," which can be confusing.

As many facts in algebraic geometry, a proof of the Nullstellensatz requires certain methods from algebra. The kind of algebra relevant to the foundations of algebraic geometry is known as *commutative algebra*, to which we will keep returning throughout this book. The Nullstellensatz will be restated and proved in Sect. 4.3 below.

1.2 Zariski Topology

1.2.1 Topology

Algebraic geometry builds fundamental concepts of geometry out of pure algebra (rings and polynomials). A very basic concept of geometry is *topology*. A *topology* on a set X is specified by *open* (and/or *closed*) sets. An open set containing a point $x \in X$ is also called an *open neighborhood* of x. A set X with a topology is called a *topological space*. An open set is the same thing as a complement of a closed set, and vice versa, so it suffices to specify either open sets or closed sets. Open sets in a topology are required to satisfy the following properties (or *axioms*):

1. $\emptyset$, X are open.
2. A union of arbitrarily (possibly infinitely) many open sets is open.
3. An intersection of two (hence finitely many) open sets is open.

One can equivalently formulate the axioms for closed sets by swapping union and intersection.

For any set $S \subseteq X$, we then have a smallest closed set (with respect to inclusion) $\overline{S}$ containing S (namely, the intersection of all closed sets containing S). It is called the *closure* of S. Symmetrically, the *interior* S° is the largest open set (i.e. the union of all open sets) contained in S.

1.2.2 Zariski and Analytic Topology

In algebraic geometry, the set of all n-tuples of complex numbers is called the *affine space* $\mathbb{A}_{\mathbb{C}}^n$. For the purposes of algebraic geometry, we consider the *Zariski topology* on $\mathbb{A}_{\mathbb{C}}^n$, in which closed sets are affine algebraic sets (see Sect. 1.1). Similarly, in the Zariski topology on any affine algebraic set X, the closed sets are affine algebraic sets in $\mathbb{A}_{\mathbb{C}}^n$ which are subsets of X. To verify the axioms of topology, one notes that for sets of n-variable polynomials S_i, we have

$$Z\left(\bigcup_i S_i\right) = \bigcap_i Z(S_i)$$

and for sets of n-variable polynomials S, T, we have

$$Z(\{p \cdot q \mid p \in S, \ q \in T\}) = Z(S) \cup Z(T).$$

The Zariski topology is not the most typical kind of topology one considers outside of algebraic geometry. In analysis, the key example of a topology is the *analytic topology*. In the analytic topology on $\mathbb{A}_{\mathbb{C}}^n = \mathbb{C}^n$ (or on $\mathbb{R}^n$), a set U is open when with any point $x \in U$, the set U also contains all points of distance $< \epsilon$ for some $\epsilon > 0$ (where ϵ can depend on x). As the name suggests, the analytic topology is very important in mathematical analysis. The Zariski topology has "far fewer" closed (and open) sets than the analytic topology. For example, in $\mathbb{R}^n$ (or $\mathbb{C}^n$), any open ball is open and any closed ball is closed in the analytic topology. On the other hand, the only Zariski closed sets in $\mathbb{A}_{\mathbb{C}}^1$ are itself and finite subsets.

Still, we can use the analytic topology for intuition about the Zariski topology on algebraic sets. For example, a single point is closed (in both analytic and Zariski topology), and is not open, unless we are in $\mathbb{A}_{\mathbb{C}}^0$.

1.3   Affine and Projective Varieties

1.3.1   Affine and Quasi-Affine Varieties

In a topological space X, a non-empty closed set Z is called *irreducible* if there do not exist closed subsets $Z_1 \neq Z$, $Z_2 \neq Z$ of Z such that $Z = Z_1 \cup Z_2$ (i.e. Z is not a union of two closed subsets other than itself). Z is called *connected* if it is not a union of two *disjoint* closed subsets other than itself.

An *affine variety* is an affine algebraic set which is irreducible in the Zariski topology. A *quasi-affine variety* is a Zariski open subset U of an affine variety X. (Caution: U is open in X, not necessarily in $\mathbb{A}_{\mathbb{C}}^n$.) For any topological space X and any subset $S \subseteq X$, we have a topology on S where, by definition, open (resp. closed) sets in S are of the form $V \cap S$ where V is an open (resp. closed) set in X. This topology is called the *induced topology*.

The Zariski topology on an affine algebraic set is induced from the Zariski topology on $\mathbb{A}_{\mathbb{C}}^n$.

Recall that an ideal $I \subseteq R$ in a ring R is called *prime* if $I \neq R$ and for $x, y \in R$, $xy \in I$ implies $x \in I$ or $y \in I$. The ideal I is called *maximal* if $I \neq R$ and for every ideal $J \subseteq R$ with $I \subseteq J$, we have $J = I$ or $J = R$. An ideal $I \subseteq R$ is maximal if and only if the quotient ring R/I (consisting of all cosets $x + I$, $x \in R$) is a field. Similarly, $I \subseteq R$ is prime if and only if R/I is an *integral domain* which means that it satisfies $0 \neq 1$ and has no *zero divisors* (i.e. non-zero elements x, y such that $xy = 0$).

Any ideal $I \neq R$ is contained in a maximal ideal by a principle called *Zorn's lemma*, which states that any partially ordered set P (such as the set of ideals in a ring R ordered with respect to inclusion) contains a maximal element (i.e. an element $m \in P$ such that $a \in P$ and $m \leq a$ implies $m = a$), provided that for any subset L which is *totally ordered* (i.e. $a, b \in L$ implies $a \leq b$ or $b \leq a$) there exists an element $\ell \in P$ greater or equal than all elements of L.

Now it is easy to see that an affine algebraic set X is irreducible (i.e. is an affine variety) if and only if the ideal $I(X)$ is prime. Indeed, if $I(X)$ is not prime, then there exists $f, g \notin I(X)$ such that $fg \in I(X)$, so X is a union of the two closed subsets $Z(f) \cap X$, $Z(g) \cap X$ neither of which is equal to X. On the other hand, if $X = X_1 \cup X_2$ where $X_i \neq X$ are closed, then by definition, there are $f_i \in I(X_i) \smallsetminus I(X)$, while $f_1 f_2 \in I(X)$.

In particular, since polynomials over a field obviously form an integral domain, the 0 ideal is prime, and thus, the affine space $\mathbb{A}_{\mathbb{C}}^n$ is irreducible (and hence, an affine variety).

1.3.2 Projective and Quasi-Projective Varieties

The *n-dimensional projective space* $\mathbb{P}_{\mathbb{C}}^n$ is the set of *ratios*

$$[x_0 : \cdots : x_n]$$

of complex numbers. In a ratio, the numbers $x_0, \ldots, x_n$ are not allowed to all be 0 (although some may be 0), and a ratio is considered the same if we multiply all the numbers by the same non-zero number:

$$[x_0 : \cdots : x_n] = [ax_0 : \cdots : ax_n]$$

with $a \neq 0 \in \mathbb{C}$.

A *projective algebraic set* is a set of zeros in $\mathbb{P}_{\mathbb{C}}^n$ of a set of homogeneous polynomials. (A polynomial is *homogeneous* if all its monomials have the same *degree*, which is defined as the sum of exponents of all its variables.) Projective algebraic sets are, by definition, the closed sets in the *Zariski topology* on $\mathbb{P}_{\mathbb{C}}^n$. Irreducible projective algebraic sets are called *projective varieties*. A Zariski open subset (i.e. complement of a Zariski closed subset) in a projective variety is called a *quasi-projective variety*.

One can, for many practical purposes, define an *algebraic variety* as a quasi-affine or quasi-projective variety. The necessity to always refer to an ambient affine or projective space in such a definition, however, is unsatisfactory, and it is a part of what motivates schemes. However, we must learn about varieties, and some other mathematics, first, before discussing schemes.

1.4   Regular Functions on Different Types of Varieties

Regular functions is a common name for the type of functions on a variety which we study in algebraic geometry. By a *function*, we mean a mapping into $\mathbb{C}$.

1.4.1   Regular Functions on a Quasiaffine and Quasiprojective Variety

Let V be a quasiaffine variety (or, more generally, a Zariski open set in an affine algebraic set). A *regular function on V at a point p* is a function

$$f : U \to \mathbb{C}$$

where U is a Zariski open set in V with $p \in U$ such that

$$f(x) = \frac{g(x)}{h(x)} \tag{1.4.1}$$

where $g(x)$, $h(x)$ are polynomials, and $h(x) \neq 0$ for all $x \in U$ (here we write x for an n-tuple: $x = (x_1, \ldots, x_n)$).

A *regular function on V* is a function

$$f : V \to \mathbb{C}$$

which is regular at every point $p \in V$, i.e. for every $p \in V$, there exists a Zariski open neighborhood U of p such that on U, f is of the form (1.4.1).

A *regular function on a quasiprojective variety* (or at a point of a quasiprojective variety) is defined the same way as a regular function on a quasiaffine variety with the exception that $g(x)$, $h(x)$ are homogeneous polynomials of equal degree (so that $f(x)$ is well defined on ratios). The definition also applies to Zariski open subsets of projective algebraic sets.

Regular functions on an algebraic variety X form a commutative ring (i.e. we can add and multiply them). This ring is denoted by $\mathbb{C}[X]$. We will now compute the ring of regular functions for varieties of certain kinds.

1.4.2 Regular Functions on $\mathbb{A}_{\mathbb{C}}^n$

The ring of regular functions on the affine space is simply the ring of polynomials in n variables:

$$\mathbb{C}[\mathbb{A}_{\mathbb{C}}^n] = \mathbb{C}[x_1, \ldots, x_n].$$

(1.4.2)

To see this, first note that since $\mathbb{A}_{\mathbb{C}}^n$ is irreducible, two polynomials f, g which coincide on a non-empty Zariski open set $U \subseteq \mathbb{A}_{\mathbb{C}}^n$ coincide (since $\mathbb{A}_{\mathbb{C}}^n = (\mathbb{A}_{\mathbb{C}}^n \setminus U) \cup Z(f - g)$). Now since $\mathbb{C}[x_1, \ldots, x_n]$ has unique factorization (see Theorem 4.1.3 below), the same is true for rational functions: Suppose on a non-empty Zariski open set $U \subseteq \mathbb{A}_{\mathbb{C}}^n$,

$$\frac{g_1}{h_1} = \frac{g_2}{h_2}$$

where g_i, h_i have greatest common divisor 1 for $i = 1, 2$, and h_i are non-zero on U. Then

$$g_1 h_2 = g_2 h_1,$$

and hence there exists a $u \in \mathbb{C}^\times$ such that $g_1 = u g_2, h_1 = u h_2$.

Now let f be a regular function on $\mathbb{A}_{\mathbb{C}}^n$. But by what we just observed, in Zariski open neighborhoods of all points, we can write $f = g/h$ with the same polynomials g, h which, moreover, have greatest common divisor 1. However, if $h \notin \mathbb{C}^\times$, by the Nullstellensatz, then, the set of zeros $Z(h)$ of h would be non-empty, so at a point $x \in Z(h)$, we would have a contradiction. Thus, $h \in \mathbb{C}^\times$, and f is a polynomial.

1.4.3 Regular Functions on an Affine Variety

Regular functions on an affine variety (or, more generally, an affine algebraic set) $X \subseteq \mathbb{A}_{\mathbb{C}}^n$ are also polynomials in the sense that they do not have a denominator. More precisely, we have

$$\mathbb{C}[X] = \mathbb{C}[x_1, \ldots, x_n]/I(X)$$

(1.4.3)

where $I(X)$ is the ideal of all polynomials which are 0 at every point of X. Recall that the division symbol in (1.4.3) denotes *cosets*, i.e. the elements of the ring (1.4.3) are cosets, which are sets of the form

$$p + I(X) = \{p + q \mid q \in I(X)\}.$$

Recall that cosets are the algebraic device for setting the elements of $I(X)$ equal to 0 in the ring (1.4.3), which is what we want, since they are constantly zero as functions on X.

To see that (1.4.3) is the correct formula for the ring of regular functions on X, first note that the right hand side of (1.4.3) maps injectively into the left hand side by the definition of the ideal $I(X)$. To show that the map is onto, we need to show that if we cover X with Zariski open sets U_i in $\mathbb{A}^n_{\mathbb{C}}$ and exhibit rational functions g_i/h_i where $h_i \neq 0$ on $U_i, i \in S$, which such that $g_i/h_i = g_j/h_j$ on $U_i \cap U_j \cap X$, then there exists a polynomial ϕ which restricts to each g_i/h_i on $U_i \cap X$. To this end, first note that we can assume that the set S is finite. This is because by the Nullstellensatz,

$$1 \in I(X) + \sum_{i \in S} I(\mathbb{A}^n_{\mathbb{C}} \setminus U_i),$$

and so the indexing set S can be made finite, since only finite sums of elements are allowed. Suppose, then,

$$S = \{1, \ldots, n\}.$$

Now also by the Nullstellensatz, there exist polynomials $a_1, \ldots, a_n, q$ such that $q \in I(X)$ and

$$a_1 h_1 + \cdots + a_n h_n + q = 1.$$

Then one verifies that the polynomial

$$\phi = a_1 g_1 + \cdots + a_n g_n \tag{1.4.4}$$

restricts to g_i/h_i on each of the Zariski open sets $U_i \cap X$ (See Exercise 4.)

1.4.4 Regular Functions on the Complement of a Set of Zeros $Z(f)$

Let X be an affine variety and let $f \in \mathbb{C}[X]$. Then the complement $X \setminus Z(f)$ is a special kind of quasiaffine variety. (We will see later that, in some sense, "it is still affine," although not according to our current definition.) We have

$$\mathbb{C}[X \setminus Z(f)] = \mathbb{C}[X][f^{-1}] = \mathbb{C}[X][\tfrac{1}{f}]. \tag{1.4.5}$$

Note that it is alright to take the reciprocal of f, because $X \setminus Z(f)$ does not contain any zeros of f. This construction is a special case of *localization* $S^{-1}R = R[S^{-1}]$ *of a ring R with respect to a subset $1 \in S \subseteq R$ closed under multiplication.* On the set of "fractions" r/s (i.e., formally, pairs (r, s)) with $r \in R$ and $s \in S$, define an equivalence relation where $r_1/s_1 \sim r_2/s_2$ when there exists some $t \in S$ such that $r_1 s_2 t = r_2 s_1 t$. The set of equivalence classes is $R[S^{-1}]$. In our case, we let S be the set of all $f^n, n \in \mathbb{N}_0$. We can alternately describe

$$R[f^{-1}] = R[t]/(ft - 1).$$

The proof of (1.4.5) is actually essentially the same as in Sect. 1.4.3, with the exception that we have

$$a_1 h_1 + \cdots + a_n h_n = f^N$$

for some natural number N (which is true by the Nullstellensatz). We then put

$$\phi = a_1 g_1 / f^N + \cdots + a_n g_n / f^N.$$

1.4.5 Regular Functions on a Quasiaffine Variety

Let $0 \neq f_1, \ldots, f_m \in \mathbb{C}[X]$ where X is an affine variety. Then

$$\mathbb{C}[X \smallsetminus Z(f_1, \ldots f_m)] = \mathbb{C}[X][f_1^{-1}] \cap \cdots \cap \mathbb{C}[X][f_m^{-1}]. \qquad (1.4.6)$$

Note that the intersection on the right hand side of (1.4.6) is formed in the field of rational functions $K(X)$, which is the field of fractions of the ring $\mathbb{C}[X]$. The *field of fractions* QR of an integral domain R is the localization with respect to the set of all non-zero elements, which is closed under multiplication because R is an integral domain. Also, the canonical map $R \to QR$ is injective, by cancellation. Conversely, a subring of a field (more generally an integral domain) is obviously an integral domain. Thus, integral domains are precisely those rings which are subrings of fields.

The reason $\mathbb{C}[X]$ is an integral domain is that X is irreducible. Now we have

$$X \smallsetminus Z(f_1, \ldots, f_m) = \bigcup_{i=1}^{m} X \smallsetminus Z(f_i).$$

Thus, we can characterize a regular function on $X \smallsetminus Z(f_1, \ldots, f_m)$ by a collection of regular functions on $X \smallsetminus Z(f_i)$ which coincide on intersections, but that is equivalent to coinciding in $K(X)$ since $\mathbb{C}[X]$, and hence also $\mathbb{C}[X][f_i^{-1}]$, are integral domains. Thus, (1.4.6) follows.

1.4.6 Regular Functions on a Projective Variety

Essentially, the above discussion of rings of regular functions has an exact analogue on quasiprojective varieties (or Zariski open subsets of projective algebraic sets), if we restrict attention to homogeneous rational functions of degree 0. In particular, if X is a projective variety (or, more generally, projective algebraic set), then $\mathbb{C}[X]$ consists of homogeneous polynomials of degree 0, but those are always constant! Thus,

$$\mathbb{C}[X] = \mathbb{C},$$

i.e. *all regular functions on a projective variety are constant.*

1.5   Sheaves

The fact that there are "not enough" regular functions on some varieties X (such as projective varieties) forces us to study regular functions on open subsets U of X, instead of just on X. Let us state two basic properties of $\mathbb{C}[U]$ as U varies over open subsets of X which follow immediately from the definition:

1. **Restriction:** For $V \subseteq U$, we have a *restriction map*

$$\mathbb{C}[U] \to \mathbb{C}[V]$$

   (given by restriction of the domain of a regular function). The restriction is transitive (i.e. for $W \subseteq V \subseteq U$, restriction from $\mathbb{C}[U]$ to $\mathbb{C}[V]$ and then to $\mathbb{C}[W]$ is the same thing as restricting to $\mathbb{C}[W]$ directly). Also, the restriction from $\mathbb{C}[U]$ to itself is just the identity.

2. **Gluing:** If we have regular functions $f_i \in \mathbb{C}[U_i]$ where U_i are open sets, such that f_i and f_j restrict to the same regular function in $\mathbb{C}[U_i \cap U_j]$, then there exists a unique regular function $f \in \mathbb{C}[\bigcup U_i]$ which restricts to all the functions f_i.

It is convenient to formalize this property. Suppose that for every open set U of a topological space X, we are given a set $\mathcal{F}(U)$ which has the restriction and gluing properties described above with $\mathbb{C}[U]$ replaced by $\mathcal{F}(U)$. This data is then called a *sheaf* (of sets) $\mathcal{F}$, and the sets $\mathcal{F}(U)$ are called *sets of sections*. We will use sheaves heavily in the next chapter when defining schemes. The sheaf

$$U \mapsto \mathbb{C}[U]$$

on an algebraic variety X (with the Zariski topology) is denoted by $\mathcal{O}_X$, and called the *structure sheaf* of X.

2   Categories, and the Category of Algebraic Varieties

It is, however, important to note that the sets of sections of the structure sheaf are *not* just sets: they are *rings*. To introduce the concept of a sheaf in proper generality, we need to talk about *categories*.

2.1   Categories, Functors and Algebraic Structures

2.1.1   The Definition of a Category, and an Example: The Category of Sets

In a category C, we have a *class of objects* $Obj(C)$ and a *class of morphisms* $Mor(C)$, satisfying certain axioms.

Explaining the need to distinguish between *sets* and *classes* takes us on a brief detour into *set theory*. It comes from the fact that the naive interpretation of the notation

$$\{X \mid \dots\}$$

(2.1.1)

as "the set of all sets X such that . . ." leads to a contradiction in

$$\{X \mid X \notin X\}$$

(2.1.2)

where neither $X \in X$ nor $X \notin X$ are possible; because of that, we distinguish between sets and classes and interpret (2.1.1) as "the class of all sets X such that . . .," and define a set as a class which is an element of another class. Otherwise, it is called a *proper class*. Note that then (2.1.2) is just an example of a proper class; in fact, it is the class of all sets.

The axioms of a category say that $Obj(C)$ and $Mor(C)$ satisfy all the formal properties of the most basic example: the category *Sets* whose objects are sets and morphisms are mappings of sets. Thus, we have two mappings

$$S, T : Mor(C) \to Obj(C)$$

(called *source and target*, which in the category of sets are the domain and codomain of a mapping). A morphism $f \in Mor(C)$ with $S(f) = X$, $T(f) = Y$ (where X, Y are objects) is called a *morphism from X to Y*, and denoted by

$$f : X \to Y$$

or

$$X \xrightarrow{\ f\ } Y,$$

same as for mappings of sets. We have a mapping $Obj(C) \to Mor(C)$ called the *identity morphism*

$$Id_X : X \to X$$

Also like for mappings, the structure of a category specifies, for two morphisms

$$f : X \to Y, \quad g : Y \to Z,$$

the *composition*

$$g \circ f : X \to Z$$

(note the reversal of order of f and g, motivated by mappings: when we apply mappings to an element, we write $g \circ f(x) = g(f(x))$, even though we apply f first).

Morphisms, however, may not always be mappings, (although in the category *Sets*, and many other examples, they are), and so they cannot be, in the context of pure category theory, applied to elements. So instead, we must define a category by *axioms*. These axioms are simple: they say that the source and target of Id_X are equal to X, and that the composition of morphisms is associative

$$(h \circ g) \circ f = h \circ (g \circ f)$$

(when applicable) and unital, i.e. for $f : X \to Y$,

$$Id_Y \circ f = f \circ Id_X = f.$$

Lastly, we require that the class $C(X, Y)$ of all morphisms $f : X \to Y$ be a set. We call the category C *small* if the class $Obj(C)$ is a set. (Then necessarily also $Mor(C)$ is a set.)

To see that morphisms do not always have to be mappings, note that to every category C, there is the *opposite* (sometimes also called *dual*) category C^{Op} which "turns around the arrows": $Obj(C^{Op}) = Obj(C)$, $Mor(C^{Op}) = Mor(C)$ and Id is C and C^{Op} are the same, but S in C^{Op} is T in C and vice versa, and composition of morphisms $\alpha \circ \beta$ in C^{Op} is $\beta \circ \alpha$ in C.

2.1.2 Categories of Algebraic Structures

One purpose of categories is to be able to discuss, and relate, mathematical structures of the same kind. For example, all sets, all groups, all abelian groups, all rings, all topological spaces, all algebraic varieties. (Recall that a *group* has one operation which is associative, unital and has an inverse; an *abelian group* is a group which is also commutative.) So we want a category whose objects are the given structures, i.e. the category of groups, rings, etc. But what should the morphisms be?

Of course, we may be able to define the morphisms in a fairly arbitrary way, as long as they satisfy the axioms of a category, which we learned in Sect. 2.1.1. For example, we could define the only morphisms to be identities, but that would not be very useful for understanding the given mathematical structure. This is why, usually, there is a standard choice of morphisms of mathematical structures of a given kind, which are, vaguely speaking, *mappings which preserve the given structure*. Making this precise requires different techniques in different cases.

The case which is the easiest to handle are categories of *algebraic structures*. An algebraic structure comes with *operations* (example: addition or multiplication). In this case, the default choice of morphisms are *homomorphisms* of the given algebraic structures, which means mappings which preserve the operations.

For example, a homomorphism of groups $f : G \rightarrow H$, written multiplicatively, is required to satisfy

$$f(x \cdot y) = f(x) \cdot f(y).$$

(Philosophically, the unit and inverse are also operations, so we should include $f(1) = 1$ and $f(x^{-1}) = (f(x))^{-1}$, but in the case of groups, it follows from the axioms.) The category of groups and homomorphisms is denoted by Grp, the category of abelian groups and homomorphisms is denoted by Ab.

Analogously, a homomorphism of rings satisfies

$$f(x + y) = f(x) + f(y),$$

and

$$f(xy) = f(x)f(y).$$

A non-zero ring is *not* a group with respect to multiplication (because one cannot divide by 0), so we must also require

$$f(1) = 1,$$

since it does not follow automatically.

One must be careful not to confuse a homomorphism of rings with a homomorphism of R-modules over a fixed ring R. (Recall that a *module* over a commutative ring R is an abelian group M with an operation of taking *multiples* by elements $r \in R$ which satisfies distributivity from both sides, unitality and associativity; an example of an R-module is R itself or more generally an ideal of R, which is the same thing as a submodule of the R-module R.)

Thus, a homomorphism $f : M \rightarrow N$ of R-modules satisfies

$$f(x + y) = f(x) + f(y),$$

$$f(rx) = rf(x) \text{ for } r \in R.$$

Sometimes, the same algebraic object may be used for two different purposes. For example, as already remarked, a ring R is a module over itself. In such cases, we must be careful to specify which category we are working in.

2.1.3 Isomorphisms

The most important morphisms are *isomorphisms*. An isomorphism $f : X \rightarrow Y$ in a category C is a morphism for which there exists an *inverse* with respect to composition, i.e. a morphism $g : Y \to X$ such that $f \circ g = Id_Y, g \circ f = Id_X$. As expected, we then write $f^{-1} = g$. If there exists an isomorphism between two objects of a category, they are called *isomorphic*. To complicate things, there can exist isomorphisms between an object and itself which may not be the identity. Such isomorphisms are called *automorphisms*. For example, in the category of abelian groups, (*not* rings), there is a non-trivial automorphism $f : \mathbb{Z} \to \mathbb{Z}$ given by $f(n) = -n$. The automorphisms of a given object of a given category form a group, called the *automorphism group*.

2.1.4 Functors and Natural Transformations

Let C, D be categories. A *functor* $F : C \to D$ consists of maps $F = Obj(F) : Obj(C) \to Obj(D), F = Mor(F) : Mor(C) \to Mor(D)$ which preserves identity, source, target and composition: For $X \in Obj(C), f, g \in Mor(C)$,

$$F(Id_X) = Id_{F(X)},$$

$$F(S(f)) = S(F(f)),$$

$$F(T(f)) = T(F(f)),$$

$$F(g \circ f) = F(g) \circ F(f).$$

when applicable.

A *natural transformation* $\eta : F \to G$ is a collection of morphisms

$$\eta_X : F(X) \to G(X), \ X \in Obj(C),$$

such that for every morphism $f : X \to Y$ in C, we have a *commutative diagram*:

$$
\begin{array}{ccc}
F(X) & \xrightarrow{\eta_X} & G(X) \\
{\scriptstyle F(f)} \downarrow & & \downarrow {\scriptstyle G(f)} \\
F(Y) & \xrightarrow{\eta_Y} & G(Y).
\end{array}
$$

Commutativity means that the two compositions of arrows (i.e. morphisms) indicated in the diagram are equal.

An *equivalence of categories* C, D is a pair of functors $F : C \to D$ and $G : D \to C$ and natural isomorphisms (i.e. natural transformations which have inverses)

$$F \circ G \cong Id_D,$$

$$G \circ F \cong Id_C.$$

2.2 Categories: Topological Spaces and Algebraic Varieties

Not all mathematical objects are algebraic structures. In greater generality, we may need to consider on a case by case basis what it means to "preserve the structure." In this subsection, we will discuss two examples, topological spaces and algebraic varieties.

2.2.1 The Category of Topological Spaces

Topological spaces are not really algebraic structures. Still, we have distinguished maps $f : X \to Y$ which "preserve topology." They are called *continuous maps*, defined by the property that for each open set $U \subseteq Y$, the set $f^{-1}(U) \subseteq X$ is open. Here

$$f^{-1}(U) = \{x \in X \mid f(x) \in U\}$$

is the *inverse image of U*. The category of topological spaces and continuous maps will be denoted by Top.

Note that we now have learned three different meanings of the symbol f^{-1} which occur in three different contexts: the reciprocal of f (inverse under multiplication), the inverse under composition, and the inverse image. This is a classic example of an imperfection of mathematical notation. Such inconsistencies occur for historical reasons, and cannot be completely avoided.

Also for historical reasons, an isomorphism in the category of topological spaces and continuous maps is called a *homeomorphism* (not to be confused with "homomorphism.").

2.2.2 The Category of Algebraic Varieties

Recall that at this moment, our definition of an algebraic variety includes affine, quasi-affine, projective and quasiprojective varieties, defined in Sects. 1.3.1 and 1.3.2. These are the objects of the category of algebraic varieties. Following our general guiding principle, morphisms of algebraic varieties $f : X \to Y$ are the mappings which preserve the structure.

The structure consists of the Zariski topology, and regular functions. It is correct to say that morphisms of varieties are those mappings which preserve topology and regular functions.

In more detail, then, morphisms of algebraic varieties

$$f : X \to Y$$

are maps which

1. are continuous with respect to the Zariski topology
2. have the property that if $g : U \to \mathbb{C}$ is a regular function, then the composition

$$g \circ f : f^{-1}(U) \to \mathbb{C},$$

or, more precisely,

$$g \circ f|_{f^{-1}(U)} : f^{-1}(U) \to \mathbb{C},$$

is a regular function. Note the inverse image in both formulas. The extra notation in the second formula means the restriction of a function.

Note that in particular, by the second property, a morphism of varieties $f : X \to Y$ specifies (we sometimes say: *induces*) a homomorphism of rings

$$\mathbb{C}[f] : \mathbb{C}[Y] \to \mathbb{C}[X].$$

In fact, both rings contain $\mathbb{C}$ (thought of as constant functions), and the homomorphism $\mathbb{C}[f]$ fixes $\mathbb{C}$, i.e. satisfies

$$\mathbb{C}[f](\lambda) = \lambda \text{ for } \lambda \in \mathbb{C}.$$

Commutative rings R with a homomorphism of rings

$$A \to R$$

for some other commutative ring A are called (commutative) *A-algebras*. Therefore, $\mathbb{C}[f]$ is a *homomorphism of $\mathbb{C}$-algebras*. In general, a homomorphism of A-algebras $R \to R'$ is defined as a homomorphism of rings which commutes with the homomorphisms from A to R, R'.

2.3 The Morphisms into an Affine Variety

An easy but powerful theorem states that a morphism in the category of varieties

$$f : X \to Y \tag{2.3.1}$$

where Y *is affine* is characterized by the induced homomorphism of $\mathbb{C}$-algebras

$$\mathbb{C}[X] \leftarrow \mathbb{C}[Y] \tag{2.3.2}$$

(note the reversal of the arrow, called *contravariance*).

This means, in more detail, that for every homomorphism of rings (2.3.2), there exists a *unique* morphism of varieties (2.3.1) which induces it, provided that Y is affine. In particular:

The category of affine varieties over $\mathbb{C}$ and morphisms of varieties is equivalent to the opposite category of the category of finitely generated $\mathbb{C}$-algebras and homomorphisms of $\mathbb{C}$-algebras.

To see why (2.3.1) and (2.3.2) are equivalent for Y affine, note that the passage from (2.3.1) to (2.3.2) is immediate from the definition. On the other hand, given a homomorphism

$$\mathbb{C}[x_1, \ldots, x_n]/I(Y) \to \mathbb{C}[X],$$

we can take the images of the generators $x_1, \ldots, x_n$ as coordinates of a mapping from X to $\mathbb{A}^n_{\mathbb{C}}$, which lands in Y. It is readily checked that this is a morphism of varieties, and that both passages between (2.3.1) and (2.3.2) are inverse to each other (although note that in our current setting, there are several cases for X to consider!).

Roughly speaking, we think of affine varieties as those which have "enough regular functions." From this point of view, they are the opposite of projective varieties: The only algebraic variety which is both affine and projective is a single point (and in some definitions, the empty set, but our definition of irreducibility excludes the empty set, so we do not count it).

2.3.1 Quasiaffine Varieties which are not Isomorphic to Affine Varieties

The theorem described at the beginning of Sect. 2.3 can be useful in deciding which varieties are isomorphic to affine varieties. For example, note that in $\mathbb{A}^n_{\mathbb{C}}$,

$$\{(0, \ldots, 0)\} = Z(x_1, \ldots, x_n)$$

(Z denotes the set of zeros, see Sect. 1.1.2). By (1.4.6), (1.4.2), for $n \geq 2$, we have

$$\mathbb{C}[\mathbb{A}^n_{\mathbb{C}} \setminus \{(0, \ldots, 0)\}] =$$
$$= \mathbb{C}[x_1, \ldots, x_n][x_1^{-1}] \cap \cdots \cap \mathbb{C}[x_1, \ldots, x_n][x_n^{-1}] =$$
$$= \mathbb{C}[x_1, \ldots, x_n] = \mathbb{C}[\mathbb{A}^n_{\mathbb{C}}]$$

Since we know that $\mathbb{A}^n_{\mathbb{C}}$ is affine, this means that

$$\mathbb{A}^n_{\mathbb{C}} \setminus \{(0, \ldots 0)\}$$

is not affine for $n \geq 2$: its inclusion into $\mathbb{A}^n_{\mathbb{C}}$ induces an isomorphism of rings of regular function, but is not an isomorphism of varieties (it is not onto on points), so if both varieties were isomorphic to affine varieties, it would contradict the theorem at the beginning of Sect. 2.3).

We will be able to generalize this observation later using the concept of *dimension* (see Comment 4 at the end of Sect. 5.5).

2.3.2 Quasiaffine Varieties which are Isomorphic to Affine Varieties

Notice that Sect. 2.3.1 excludes the case $n = 1$, and in fact, $\mathbb{A}^1_{\mathbb{C}} \smallsetminus \{0\}$ *is isomorphic to an affine variety*. In fact, a more general statement is true.

Let X be an affine variety, and let $0 \neq f \in \mathbb{C}[X]$. Then we claim that $X \smallsetminus Z(f)$ is always isomorphic to an affine variety. To see this, write

$$\mathbb{C}[\mathbb{A}^{n+1}_{\mathbb{C}}] = \mathbb{C}[x_1, \ldots, x_n, y],$$

and in these variables, consider the affine variety

$$Y = Z(I(X) \cup \{f(x_1, \ldots, x_n) \cdot y - 1)\})$$

(see Sect. 1.1.2 for $I(X)$). We claim that Y is isomorphic to $X \smallsetminus Z(f)$. Indeed, the morphism of varieties

$$X \smallsetminus Z(f) \to Y$$

given by

$$(x_1, \ldots, x_n) \mapsto (x_1, \ldots, x_n, \frac{1}{f(x_1, \ldots, x_n)})$$

is inverse to the morphism

$$Y \to X \smallsetminus Z(f)$$

given by

$$(x_1, \ldots, x_n, y) \mapsto (x_1, \ldots, x_n).$$

This construction in particular implies for any Zariski open set $U \subseteq X$ in any variety X and a point $P \in U$, there exists a Zariski open subset $V \subseteq U$ which is affine in the broader sense (i.e. isomorphic to an affine variety), such that $P \in V$. We say that *any Zariski open neighborhood of a point P contains an affine Zariski open neighborhood.*

To see this, it suffices to consider the case of X quasiaffine, $U = X$ (see Exercise 13). Then X is the complement of a Zariski closed set in an affine variety Y, i.e.

$$X = Y \smallsetminus Z((f_1, \ldots, f_m)).$$

For a point $P \in X$, not all of the functions $f_1, \ldots, f_m$ can be zero on P. Let $f_i(P) \neq 0$. Then $P \in Y \smallsetminus Z((f_i))$, which is affine in the broader sense by the above argument.

3 Rational Maps, Smooth Maps and Dimension

In this section, we will introduce more geometric concepts which are central in algebraic geometry, and also motivate some of the algebra covered in the remaining section of this chapter. Rational maps are an important alternative choice of morphisms in the category of varieties, which are more general than the morphisms of varieties we introduced so far. Smooth maps, on the other hand, are a more special kind of morphisms of varieties, which capture algebraically the concept of non-singularity. We already hinted at the importance of the algebraic concept of *dimension*, which we will introduce in this section as well.

3.1 Definition of a Rational Map

For two algebraic varieties X, Y, a *rational map from X to Y* is morphism of varieties

$$f : U \to Y$$

where U is a non-empty Zariski open subset of X. The rational map f is considered equal to a rational map

$$g : V \to Y$$

if

$$f(x) = g(x) \text{ for all } x \in U \cap V.$$

Recall that (by irreducibility), a non-empty Zariski open set in a variety X is *dense* which means that its complement does not contain any non-empty open set. This implies that for non-empty Zariski open sets $U, V \subseteq X$, $U \cap V$ is non-empty and Zariski open.

A rational map $f : X \to Y$ is called *dominant* if for $W \subseteq Y$ non-empty Zariski open, the inverse image $f^{-1}(W)$ is non-empty (note that it is by definition open).

3.2 The Field of Rational Functions

For an affine variety X, the field $K(X)$ of rational functions on X can be defined as the field of fractions of the ring of regular functions $\mathbb{C}[X]$.

Note that for a general variety X this definition does not work, since there may not be enough regular functions on X. However, X always contains a non-empty open set U which is affine in the broader sense (see Exercise 13). We may define $K(X)$ as the field of fractions of $\mathbb{C}[U]$. One easily proves that this is (functorially) independent of the choice of the non-empty affine open subset $U \subseteq X$.

3.2.1 The Category of Varieties and Dominant Rational Maps

One can consider the category whose objects are varieties, and morphisms are dominant rational maps. An isomorphism in this category is called a *birational equivalence*. A variety is called *rational* if it is birationally equivalent to an affine (equivalently, projective) space. Note also that any rational map of varieties that has an inverse as a rational map is necessarily dominant. This means that if we consider the larger category of varieties and all rational maps (not necessarily dominant), it has the same isomorphisms as the category of varieties and rational dominant maps.

Additionally, one sees that mapping

$$X \mapsto K(X)$$

where X is a variety and $K(X)$ is its field of rational functions gives rise to an equivalence of categories between the category of varieties and rational dominant maps, and the opposite of the category of fields containing $\mathbb{C}$ which are finitely generated (as fields) over $\mathbb{C}$, and homomorphisms of $\mathbb{C}$-algebras (or, equivalently, homomorphisms of fields which fix $\mathbb{C}$). Such fields are sometimes known as *function fields* over $\mathbb{C}$. Functoriality follows from functoriality of the field of fractions with respect to injective homomorphisms of integral domains.

To go the other way, select, for a function field K over $\mathbb{C}$, elements $x_1, \ldots x_n$ which generate K as a field containing $\mathbb{C}$. This defines a map

$$h : \mathbb{C}[x_1, \ldots, x_n] \to K.$$

Then send K to $Z(I)$ where I is the kernel of h, i.e. the ideal of all polynomials p such that $p(h) = 0$. This is an affine variety since the ideal I is prime (because the quotient by I, which is the image of h, is an integral domain). By definition, homomorphisms of fields give rise to rational maps, and the two constructions are inverse to each other.

In particular, a variety X is rational if and only if

$$K(X) \cong \mathbb{C}(x_1, \ldots, x_n)$$

as $\mathbb{C}$-agebras for some n. (Note: the right hand side means the field of rational functions on $\mathbb{A}^n$, i.e. the field of fractions of the ring of polynomials $\mathbb{C}[x_1, \ldots, x_n]$.) We also say that the field $K(X)$ is rational.

It is generally a hard problem to decide if a variety (or equivalently a field) is rational. For example, if D is the discriminant of the polynomial

$$x^n + a_{n-1}x^{n-1} + \cdots + a_0,$$

then it is not known in general whether the field

$$\mathbb{C}(a_1, \ldots, a_n, \sqrt{D})$$

is rational. It is known to be true for $n \leq 5$.

3.3   Standard Smooth Homomorphisms of Commutative Rings

A homomorphism of commutative rings

$$f : A \to B$$

is called *standard smooth of dimension* $k \geq 0$ if f can be expressed as

$$A \to A[x_1, \ldots, x_n]/(f_1, \ldots, f_m) \cong B$$

where $n = m + k$, the first homomorphism sends $a \in A$ to a, and f_i are polynomials such that the ideal in

$$A[x_1, \ldots, x_n]/(f_1, \ldots, f_m)$$

generated by the determinants of the $m \times m$ submatrices of the Jacobi matrix

$$\begin{pmatrix} \frac{\partial f_1}{\partial x_1} & \cdots & \frac{\partial f_1}{\partial x_n} \\ \cdots & \cdots & \cdots \\ \frac{\partial f_m}{\partial x_1} & \cdots & \frac{\partial f_m}{\partial x_n} \end{pmatrix}$$

is $A[x_1, \ldots, x_n]/(f_1, \ldots, f_m)$ (or, equivalently, contains 1). This is equivalent to saying that the ideal in $A[x_1, \ldots, x_n]$ generated by the determinants and $f_1, \ldots, f_m$ contains 1. If A is a field, this can be tested using Gröbner basis algorithm, which we will learn in the next Section.

3.4   Smooth Morphisms of Varieties

A morphism of varieties $f : X \to Y$ is called *smooth* if for every point $P \in X$ there exists an affine open set $U \subseteq X$ with $P \in U$ and an affine open set $V \subseteq Y$ such that $f(U) \subseteq V$

(and hence, necessarily, $f(P) \in V$) and the induced homomorphism of rings

$$\mathbb{C}[f] : \mathbb{C}[V] \to \mathbb{C}[U]$$

is standard smooth.

3.4.1   Smooth Varieties

A variety X is called *smooth* if the unique morphism of varieties

$$X \to \mathbb{A}_{\mathbb{C}}^0$$

which sends every point of X to the single point of $\mathbb{A}_{\mathbb{C}}^0$ is smooth.

3.5   Regular Rings and Dimension

In commutative algebra, the *Krull dimension $dim(R)$* of a commutative ring R is the maximal number d such that there exist prime ideals

$$p_0 \subsetneq p_1 \subsetneq \cdots \subsetneq p_d \tag{3.5.1}$$

in R. If no such maximum exists, we say that the dimension is infinite: $dim(R) = \infty$. This can actually occur even when R is Noetherian (see Exercise 38). For a given prime ideal p in a Noetherian ring, the maximal d for which $p = p_d$ in a chain of the form (3.5.1) is called the *height* of the prime ideal p.

A commutative ring R is called *local* if it has a unique maximal ideal m (recall that being a maximal ideal means that $m \neq R$ and every ideal which contains m is either m or R; equivalently, this means that R/m is a field). We will see in Sect. 5.3 that R being local Noetherian guarantees that d exists (i.e. is finite).

Consider a commutative ring R and a prime ideal $p \subset R$. Then we may consider R_p, the ring R *localized at p*, which means $S^{-1}R$ with $S = R \smallsetminus p$. Note that R_p is a local ring whose maximal ideal is generated by p.

For a Noetherian local ring R with maximal ideal m, $k = R/m$ is a field called the *residue field*. Then m/m^2 is a vector space over k of finite dimension (m^2 is the ideal generated by $x \cdot y$ with $x, y \in m$). We will show in Sect. 5.3 (Theorem 5.3.5) that we always have

$$dim(R) \leq dim_k(m/m^2).$$

We say that R is *regular* if equality arises, i.e. if

$$dim(R) = dim_{R/m}(m/m^2).$$

For an arbitrary commutative ring R, we say that R is *regular* if it is Noetherian and if R_p is regular for every prime ideal p. We will see later (Exercise 19 of Chap. 5) that it suffices to verify this condition for maximal ideals (since the localization of a regular ring is regular).

3.5.1 Smooth Varieties and Regular Rings

For a variety X, and a point $P \in X$, we consider the ring $\mathbb{C}[X]_P$ of all regular functions at the point P (i.e. defined on some open subset containing P). If X is affine, then $\mathbb{C}[X]_P$ can be thought of as the ring $\mathbb{C}[X]$ localized at the ideal of all functions which are 0 at P. (In more advanced algebraic geometry, as we shall see in the next chapter, this prime ideal is considered to be the same thing as the point P.) If X is not affine, replace it by an open affine neighborhood of P. In any event, $\mathbb{C}[X]_P$ are local rings.

It is worth noting that for every point P in a variety as defined in this chapter, $dim(\mathbb{C}[X]_P)$ is equal to $dim(X)$, which is defined as the maximum d such that there exist closed subvarieties

$$X_0 \subsetneq X_1 \subsetneq \cdots \subsetneq X_d. \tag{3.5.2}$$

(We say that every variety is *equidimensional*; see Exercise 36.)

The main theorem of smooth varieties is that a variety X over $\mathbb{C}$ is smooth if and only if all the rings $\mathbb{C}[X]_P$ are regular. We shall postpone the proof of this theorem to Chap. 3, Sect. 3.

4 Computing with Polynomials

By now, it is clear that we need more algebra to understand fully the geometric concepts introduced so far. In this section, we will learn how to work with rings of polynomials in several variables over a field. We will show that they have unique factorization, and that they are Noetherian. We will also prove the Nullstellensatz. On the computational side, we will discuss Gröbner bases. As applications, we will discuss algorithms for deciding whether one ideal is contained in another ideal (or its radical). We will also give a criterion for deciding whether a multivariable polynomial ideal is prime.

4.1 Divisibility of Polynomials

We begin with some very basic facts about divisibility.

4.1.1 Proposition (Chinese Remainder Theorem) *Let* $I_1, \ldots, I_n$ *be ideals in a commutative ring* R *such that* $1 \in I_i + I_j$ *for all* $i \neq j$. *Then* $I_1 \cap \cdots \cap I_n = I_1 \cdots I_n$ *(the product of ideals is the ideal generated by* $x_1 \cdots x_n$ *with* $x_i \in I_i$) *and the product of* *projections*

$$R/(I_1 \cap \cdots \cap I_n) \to \prod_{i=1}^{n} R/I_i \qquad (4.1.1)$$

is an isomorphism.

Proof It suffices to consider the case $n = 2$ (then we can use induction). For $n = 2$, we always have $I_1 I_2 \subseteq I_1 \cap I_2$. To show the opposite inequality, let $1 = a_1 + a_2$ where $a_i \in I_i$. Then for $x \in I_1 \cap I_2$, $x = xa_1 + xa_2 \in I_1 I_2$. Now (4.1.1) is always injective since an element goes to 0 on the right hand side if and only if it is in every I_i. To show surjectivity for $n = 2$, choosing $x_1, x_2 \in R$, the element $x_1 a_2 + x_2 a_1$ is congruent to x_i modulo I_i for $i = 1, 2$, which proves surjectivity. □

An element $u \in R$ of a commutative ring is called a *unit* if there exists another element $u^{-1} \in R$ such that $uu^{-1} = 1$. Let R be an integral domain. An *irreducible element* is an element $x \in R$ which is not zero or a unit such that $yz = x$ implies that one of the elements y, z is a unit. An integral domain R is called a *unique factorization domain* (or UFD) if every element $x \in R$ which is not 0 or a unit factors uniquely into irreducible elements up to order and multiplication by units, i.e.

$$x = x_1 \ldots x_n$$

where x_i are irreducible, and whenever

$$x = y_1 \ldots y_m$$

where y_i are irreducible, we have $m = n$ and there exists a permutation σ and units u_i such that

$$x_i = u_i y_{\sigma(i)}.$$

In a UFD, any set of elements S has a greatest common divisor (GCD) which is an element x dividing all elements of S such that every other elements dividing all elements of S divides x. The GCD is, of course, uniquely determined up to multiplication by a unit. A particular type of example of a UFD is a *principal ideal domain* (or PID) which means an integral domain whose every ideal is principal (i.e. generated by a single element). In particular, then R is Noetherian, which guarantees that a decomposition into irreducible elements exists. Then the principal ideal property guarantees that an irreducible element a generates a prime ideal: if $xy \in (a)$ and $x \notin (a)$, then $(x, a) = (b)$ for some element b, but b has to be a unit by irreducibility. Thus, $y \in (yx, ya) \subseteq (a)$.

A particular type of example of a PID is a *Euclidean domain* which means an integral domain R on which there exists a function $f : R \smallsetminus \{0\} \to \mathbb{N}_0$ such that for any $a, b \in R \smallsetminus \{0\}$ there are elements $c, r \in R$ such that $a = bc + r$ and $r = 0$ or $f(r) < f(b)$. The element r is called the *remainder*. In any ideal $I \subseteq R$, then, we can produce a sequence of elements $x_1, x_2, \ldots$ where x_{i+1} is a non-zero remainder (if one exists) of dividing some element $y \in I$ by x_i. Since $f(x_1) > f(x_2) > \ldots$, the sequence must terminate eventually, yielding a generator of I. Examples of Euclidean domains include $\mathbb{Z}$ and the ring of polynomials in one variable $k[x]$ where k is a field, where f is the absolute value and degree of a polynomial, respectively.

Let R be a UFD. A non-zero polynomial $f(x) \in R[x]$ is called *primitive* if the GCD of its coefficients is 1.

4.1.2 Proposition (Gauss Lemma)

(1) Let R be a UFD and let $f(x), g(x) \in R[x]$ be primitive polynomials. Then $f(x)g(x)$ is primitive.

(2) Let K be the field of fractions of R, and let $g(x)h(x) \in R[x]$ where $g(x), h(x) \in K[x]$. Then there exists a $u \in K \smallsetminus \{0\}$ such that $ug(x), u^{-1}h(x) \in R[x]$. (In particular, an irreducible polynomial in $R[x]$ of degree ≥ 1 is also irreducible in $K[x]$.)

Proof

(1) Let $f(x) = a_m x^m + \cdots + a_0$, $g(x) = b_n x^n + \cdots + b_0$ with $a_m, b_n \neq 0$, and let $f(x)g(x) = c_{m+n} x^{m+n} + \cdots + c_0$. Let the coefficients $c_0, \ldots, c_{m+n}$ all be divisible by an irreducible element c. Consider the smallest i such that a_i is not divisible by c, and the smallest j such that b_j is not divisible by c. Now the coefficient c_{i+j} is a sum of $a_i b_j$, which is not divisible by c by unique factorization, and terms divisible by c. Thus, c_{i+j} is not divisible by c, which is a contradiction.

(2) Without loss of generality, $g(x)h(x)$ is a primitive polynomial. Using common denominators, we can find $u, v \in K \smallsetminus \{0\}$ such that $ug(x), vh(x) \in R[x]$ are primitive polynomials. Then $uvg(x)h(x)$ is a primitive polynomial, which implies that uv is a unit in R.

$\square$

4.1.3 Theorem If R is a UFD, then $R[x]$ is a UFD. In particular, rings of the form $\mathbb{Z}[x_1, \ldots, x_n]$ and $k[x_1, \ldots, x_n]$ where k is a field are UFD's.

Proof Let R be a UFD. Clearly, units in $R[x]$ are precisely units in R. Suppose $f(x) \in R[x]$ is non-zero, and not a unit, and suppose

$$f(x) = g_1(x) \ldots g_m(x) = h_1(x) \ldots h_n(x)$$

where g_i, h_i are irreducible polynomials in $R(x)$. Without loss of generality, there are numbers $1 \leq \overline{m} \leq m$, $1 \leq \overline{n} \leq n$ such that g_i (resp. h_i) is of degree ≥ 1 if and only if $i \leq \overline{m}$ (resp. $i \leq \overline{n}$). Now g_i, $i \leq \overline{m}$, and h_j, $j \leq \overline{n}$ are primitive and, by Proposition 4.1.2, also irreducible in $K[x]$. Thus, since $K[x]$ is a UFD, $\overline{m} = \overline{n}$, and for some permutation σ on $\{1, \ldots, \overline{m}\}$, $g_i(x) = h_{\sigma(i)}(x)u_i$ for a unit u_i in R. (The last statement follows from the fact that the polynomials are primitive.) Thus, we can divide f by

$$\prod_{i \leq \overline{m}} g_i(x),$$

and therefore without loss of generality assume $\overline{m} = \overline{n} = 0$. But then all the polynomials g_i, h_i are of degree 0, and hence our statement follows from the fact that R is a UFD. $\qquad \square$

4.2 Gröbner Basis

We shall now prove that rings of polynomials over a Noetherian ring are Noetherian. In the special case of multivariable polynomials over a field, we can be a lot more explicit, with computational applications.

4.2.1 Theorem (Hilbert Basis Theorem) *If a ring R is a Noetherian, then so is the ring of polynomials $R[x]$.*

COMMENT In this context, the term "basis" refers to a set of generators of an ideal, no linear independence is implied.

Proof Assume R is Noetherian. Let $I \subseteq R[x]$ be an ideal. Then the top coefficients (i.e. coefficients of the highest power of x) of all the polynomials $f \in I$ form an ideal $J \subseteq R$ (since two nonzero polynomials of unequal degrees can be brought to the same degree by multiplying the polynomial of lesser degree by a power of x). By assumption, then, the ideal J is generated by the top coefficients of some polynomials $f_1, \ldots, f_n \in I$.

Let d be the maximum of the degrees of the polynomials $f_1, \ldots, f_n$. Then by construction, for any polynomial $g \in I$ of degree $\geq d$, there exist $a_1, \ldots a_n \in R$, $m_1 \ldots, m_n \in \mathbb{N}_0$ such the top coefficients of $g(x)$ and $a_1 f_1(x)x^{m_1} + \ldots a_n f_n(x)x^{m_n}$ coincide. By induction, then, there exists an $R[x]$-linear combination g_0 of the polynomials $f_1, \ldots, f_n$ such that $g(x) - g_0(x)$ is either 0 or is of degree $< d$.

Now consider for each fixed $i \in \mathbb{N}_0$ the ideal $J_i \subseteq R$ of all the top coefficients of all polynomials in I of degree i. Then each of these ideals J_i is finitely generated, so by taking finitely many polynomials $h_1, \ldots, h_\ell$ in I of degrees $i = 0, \ldots, d-1$ whose top coefficients are the generators of all the J_i's, $0 \leq i < d$, we see that every polynomial in I of degree $< d$ is an R-linear combination of $h_1, \ldots, h_\ell$. Thus, we are done. $\qquad \square$

Next, we shall discuss the ring $k[x_1, \ldots, x_n]$ of polynomials in n variables over a field k. Even though this ring is not a Euclidean domain for $n > 1$ (because it is not a PID—think, for example, of the ideal $(x_1, \ldots, x_n)$) there is a certain analog of the long division algorithm which allows us decide, for example, whether a polynomial is an element of a given ideal, or whether two ideals are the same.

By a *monomial*, we shall mean an expression of the form $x_1^{m_1} \ldots x_n^{m_n}$, i.e. equivalently, the n-tuple $a = (m_1, \ldots, m_n) \in \mathbb{N}_0^n$, which are sometimes referred to as *multidegrees*. For what follows, we need to fix a *monomial order*. This means a total ordering $\geq$ on n-tuples of non-negative integers (i.e. for any two n-tuples a, b we have $a \geq b$ or $b \geq a$) which satisfies the descending chain condition (or DCC), i.e. any sequence $a_1 \geq a_2 \geq \ldots$ is eventually constant. A totally ordered set satisfying the DCC is also sometimes called *well-ordered*. In addition, we require that for multidegrees a, b, c, if $a \geq b$, then $a + c \geq b + c$.

Note that this implies that the multidegree $(0, \ldots, 0)$ is the smallest (since otherwise, the DCC would be violated). This implies that when $m_i \leq p_i$ for all $i = 1, \ldots n$, then $(m_1, \ldots, m_n) \leq (p_1, \ldots, p_n)$.

An example is the *lexicographical order* where

$$(m_1, \ldots, m_n) \geq (p_1, \ldots, p_n)$$

when either the two n-tuples are equal, or there exists an i such that $m_j = p_j$ for $j < i$ and $m_i > p_i$. We indicate this lexicographical order by the symbol $x_1 > \cdots > x_n$. Obviously, we may order lexicographically by using the variables in a different order. Another example of a monomial ordering orders, say, by total degree and uses a lexicographic order to break ties.

Given a fixed monomial order and a non-zero polynomial $f \in k[x_1, \ldots, x_n]$, the *leading term* $L(f)$ is the monomial of f which is the greatest in the given monomial order. Its multidegree is called the *multidegree of the polynomial f*, and its coefficient is called the *leading coefficient*. For convenience, put also $L(0) = 0$. Now if $I \subseteq k[x_1, \ldots, x_n]$ is an ideal, the *leading term ideal* $L(I)$ is the ideal generated by the leading terms of the elements of I.

4.2.2 Definition A system of generators $\{f_1, \ldots, f_m\}$ of an ideal $I \subseteq k[x_1, \ldots, x_n]$ is called a *Gröbner basis* if $L(f_1), \ldots, L(f_m)$ generate the ideal $L(I)$.

If $\{f_1, \ldots, f_m\}$ is a Gröbner basis of I, then for any polynomial

$$g \in k[x_1, \ldots, x_n],$$

we can find polynomials $h_1, \ldots h_m, r \in k[x_1, \ldots, x_n]$ such that

$$g = h_1 f_1 + \ldots h_m f_m + r \tag{4.2.1}$$

and no monomial summand of r is in $L(I)$. Moreover, the polynomial r is *uniquely determined* (and is called the *remainder* of g with respect to the Gröbner basis $\{f_1, \ldots, f_m\}$). To see this, let g be any polynomial. If no monomial summand of g is in $L(I)$, then we can put $r := g$. Otherwise, let $d(g)$ be the highest multidegree of a monomial summand of g which is in $L(I)$. Then subtracting a certain linear combination of the f_i's (with polynomial coefficients) from g creates a polynomial $\overline{g}$ where $d(\overline{g}) < d(g)$. (In fact, we just need to subtract a multiple of one of the f_i's by a monomial and a coefficient.) Because of the well-ordering property, this process must eventually terminate, yielding the remainder. To prove uniqueness of the remainder, if there were two different possible remainders r, s, then $r - s \in I$. However, by the definition of a remainder, $L(r - s) \notin L(I)$, which is a contradiction.

Note that the definition of the remainder makes sense for any set of generators f_i of an ideal I, not necessarily a Gröbner basis, if we weaken the condition on the remainder to say that none of its monomial terms is divisible by any of the $L(f_i)$'s. The proof of existence is the same, but uniqueness fails.

Our proof of the existence of a remainder with respect to a Gröbner basis is constructive. In other words, it also gives a way of finding the remainder, and therefore, if we have a Gröbner basis of an ideal I in $k[x_1, \ldots, x_n]$, we can determine for a given polynomial g whether $g \in I$ (since this happens if and only if its remainder is 0).

Therefore, the question becomes whether Gröbner bases exists, and if so, how to find one. Suppose $f, g \in k[x_1, \ldots, x_n]$. Let the multidegrees of f resp. g be $(m_1, \ldots, m_n)$ resp. $(p_1, \ldots, p_n)$ and let the leading coefficients of f, g be a, b, respectively. Denoting

$$q_i = max(m_i, p_i),$$

$$f * g = b x_1^{q_1 - m_1} \ldots x_n^{q_n - m_n} f,$$

$$g * f = a x_1^{q_1 - p_1} \ldots x_n^{q_n - p_n} g,$$

put

$$S(f, g) := f * g - g * f.$$

(Basically, we multiply by constants and the smallest monomials to make the leading terms cancel out—the multidegrees of the multiplier monomials are the smallest in the coordinate-wise partial ordering, and hence also in our chosen monomial ordering.)

The *Buchberger algorithm* produces, for an ideal $I \subseteq k[x_1, \ldots, x_n]$ with a given finite set of generators S_0, an increasing (with respect to inclusion) sequence of finite sets

$$S_0, S_1, \ldots, S_m$$

of generators of I where S_m is a Gröbner basis. Given S_i, we take a pair of polynomials $f, g \in S_i$, and find any remainder of $S(f, g)$ by S_i. If we find a 0 remainder for all pairs $f, g \in S_i$, we are done. Otherwise, if we find a non-zero remainder r_i, we let $S_{i+1} :=$ $S_i \cup \{r_i\}$.

The reason this works can be explained as follows: First, let us show that the algorithm terminates. Suppose it does not, i.e. we produce an infinite sequence of remainders r_i. Then for any $i < j \in \mathbb{N}$, $L(r_i)$ cannot divide $L(r_j)$. But we can easily see that this is impossible: In any infinite sequence of natural numbers, we can pick an infinite non-decreasing subsequence. By iterating this selection, we can prove by induction on $m \leq n$ that there is an infinite subsequence of the multidegrees of the $L(r_i)$'s whose first m coordinates form non-decreasing sequences. For $m = n$, this is a contradiction with our definition of a remainder.

Next, we need to prove that last term S_m of our sequence forms a Gröbner basis. Thus, assume that some linear combination

$$f = h_1 r_1 + \cdots + h_m r_m$$

has a leading term which is not in $(L(r_1), \ldots, L(r_m))$. Let α denote the highest of the leading monomials $L(h_1 r_1), \ldots, L(h_m r_m)$. We will proceed by induction on α. Certainly $\alpha \geq L(f)$, but equality is ruled out by the assumption $L(f) \notin (L(r_1), \ldots, L(r_m))$. Now let, without loss of generality, $i = 1, \ldots, p$ be precisely the numbers for which $L(h_i r_i) = \alpha$. We will now proceed by induction on p. Note that $p = 1$ is actually impossible (since then $L(h_1 r_1)$ would have nothing to cancel against). If $p \geq 2$, then, there exists an $i < p$ such that $L(h_p r_p) = L(h_i r_i)$. This implies that $L(h_p r_p)$ is divisible by $L(r_p * r_i)$. Thus, adding an appropriate multiple of $S(r_i, r_p)$ decreases p (or else α). But we assumed that $S(r_i, r_p)$ is a linear combination of $r_1, \ldots, r_m$ (i.e., a 0 remainder can be achieved). Thus, we are done by the induction hypothesis.

4.2.3 Definition A Gröbner basis $\{f_1, \ldots, f_m\}$ of an ideal $I \subseteq k[x_1, \ldots, x_n]$ is called *reduced* if the leading coefficient of each f_i is 1, and no monomial summand of any f_i is divisible by $L(f_j)$ for any $j \neq i$.

Clearly, a reduced Gröbner basis of any ideal I exists, since we can just multiply by constants to make the leading coefficients 1, and then make the basis reduced by taking remainders. However, it is useful to note that up to order, the reduced Gröbner basis is also *unique*. To see this, first note that clearly, the leading monomials of the elements of a reduced Gröbner basis of an ideal I must be exactly the minimal monomials in $L(I)$ with respect to the partial order given by comparing each coordinate of the multi-degree (i.e. $\alpha = (m_1, \ldots, m_n)$ is less or equal than $\beta = (p_1, \ldots, p_n)$ if and only if $m_i \leq p_i$ for each i). Now suppose we have two reduced Gröbner bases S, T and let $f \in S, g \in T$ have the same leading monomial. If $f \neq g$, then $L(f - g)$ is a monomial of f or g. Suppose it is a monomial of f. But then, of course, $f - g \in I$, so there must be some element $h \in S$ such that $L(h)$ divides $L(f - g)$ which is a monomial of f. This contradicts the assumption that the Gröbner basis S is reduced.

Thus, reduced Gröbner bases can be used directly to check whether two ideals in $k[x_1, \ldots, x_n]$ are the same, although note that given Gröbner bases of both ideals (not necessarily reduced), we can also check both inclusions of the ideals by repeatedly finding remainders of one basis with respect to the other.

Given an ideal $I \subseteq k[x_1, \ldots, x_n]$, consider its reduced Gröbner basis $f_1, \ldots, f_m$ with respect to the lexicographic order $x_1 > \cdots > x_n$, arranged in decreasing order of multidegrees. Let m_i be the smallest number p such that f_p contains no variables $x_1, \ldots, x_{i-1}$. Then, using the division algorithm, we see that $f_{m_i}, f_{m_i+1}, \ldots, f_m$ form a reduced Gröbner basis of the ideal $I \cap k[x_i, \ldots, x_n]$ with respect to the lexicographical order $x_i > \cdots > x_n$. This ideal is called the *elimination ideal* of the ideal I with respect to the variables $x_i, \ldots, x_n$. Elimination ideals are useful in solving systems of algebraic equations, although we must be more careful than in the case of systems of linear equations. This is because a projection of an algebraic set in $\mathbb{A}_k^n$ is not necessarily Zariski closed. An elimination ideal only defines a closure of the projection. Thus, for example, if the elimination ideal corresponding to the variable x_n is (0), it means that there is no constraint on the last variable in a solution of the system of equations $f_i = 0$, but it does *not* mean that any value of x_n can be plugged in to get a solution (see Exercise 9).

Using elimination ideals, one can test if an ideal $I \subseteq k[x_1, \ldots, x_n]$ is prime. This is based on the following

4.2.4 Proposition *Let R be an integral domain and let $I \subset R$ be an ideal. Suppose $a \in R \setminus I$, and suppose $\overline{I} = (a^{-1}R) \cdot I \subset a^{-1}R$ is a prime ideal. Then I is prime if and only if $\overline{I} \cap R = I$.*

Proof Since $\overline{I}$ is assumed to be prime, $\overline{I} \cap R$ is prime. To prove the converse, let $p, q \subset R$ be prime ideals not containing a such that $\overline{p} = \overline{q}$ (using the same notation as above). Then for $x \in p$, by definition, $a^m x \in q$ for some m, but then $x \in q$ because q is prime and $a \notin q$. Thus, $p \subseteq q$, and by symmetry, $p = q$. Applying this to $p = I, q = \overline{I} \cap R$ proves the claim. □

Now let R be an integral domain and let $I \subset R[x]$ be an ideal, and assume that the elimination ideal $J = I \cap R$ is prime (for otherwise we know that I is not prime). Then $S = R/J$ is an integral domain. Let F be its field of fractions. Let $\overline{I}$ be the ideal generated in $F[x]$ by I. Then $\overline{I}$ is principal since $F[x]$ is a PID, and by multiplying by a common denominator in S, we can represent the generator of $\overline{I}$ by an element $h \in I$. Note that $\overline{I} \neq (1)$, and thus, the degree of h is > 0. Further, if $h \in F[x]$ is not irreducible, then I is not prime. (Since then a non-zero R-multiple of h is a product of two polynomials $g_1, g_2 \in R[x]$ of lower degree, which are therefore not in I, while their product is.)

Thus, assume $h \in F[x]$ is irreducible. Let a be the coefficient of h at the highest power of x. Then in $a^{-1}R[x]$, $a^{-1}R \cdot I = \overline{I}$ is prime, since it is the pullback of the ideal

$(h) \subset a^{-1} S[x]$, which is prime, since $h \in a^{-1} R[x]$ is irreducible. Thus, I is irreducible if and only if $\overline{I} \cap R[x] = I$.

If $R = k[x_1, \ldots, x_{n-1}]$, then we can check inductively whether the elimination ideal $I \cap R$ is prime and if it is, we can apply the above method by obtaining the reduced Gröbner basis of the ideal J generated by I and $at - 1$ in $k[x_1, \ldots, x_n, t]$ with respect to the lexicographical order $t > x_n > \cdots > x_1$ and check whether the elimination ideal $J \cap k[x_1, \ldots, x_n]$ is equal to I. (For an example, see Exercise 20.)

4.3   Nullstellensatz

The Nullstellensatz used in Sect. 1 can be rephrased (and generalized) as follows:

4.3.1 Theorem *A finitely generated algebra R over a field k satisfies a Nullstellensatz, which means that for every ideal I, its radical $\sqrt{I} = \{x \in R \mid \exists n \in \mathbb{N} \ x^n \in I\}$ is an intersection of maximal ideals.*

This is brought to the familiar context by the following

4.3.2 Proposition *Suppose k is an algebraically closed field. Then every maximal ideal $I \subset k[x_1, \ldots, x_n]$ is of the form*

$$(x_1 - a_1, \ldots, x_n - a_n) \tag{4.3.1}$$

for some $a_1, \ldots, a_n \in k$.

Proof We will show that any ideal $I \subsetneq k[x_1, \ldots, x_n]$ is contained in an ideal of the form (4.3.1). This is proved by induction on n. Suppose the statement is true with n replaced by any lower number. (For $n = 1$, the assumption is vacuous.) Then there are two possibilities:

Case 1: The ideal $J = I \cap k[x_n]$ in $k[x_n]$ is non-zero. Then, since $k[x_n]$ is a PID, $J = (f)$ is a principal ideal, and since k is algebraically closed, f factors into powers of linear factors $(x_n - b_i)^{\ell_i}$, $i = 1, \ldots, m$. By the Chinese Remainder Theorem, $k[x_1, \ldots, x_n]/I$ is isomorphic to the product of the rings $k[x_1, \ldots, x_n]/(I + (x_n - b_i)^{\ell_i})$. Then, for some i, $I + (x_n - b_i)^{\ell_i} \neq (1)$, but this implies $I + (x_n - b_n) \neq (1)$ (since an ideal whose radical is (1) is itself (1)). Therefore, we can pass to the ring $k[x_1, \ldots, x_n]/(x_n - b_i) \cong k[x_1, \ldots, x_{n-1}]$ and use the induction hypothesis.

Case 2: $I \cap k[x_n] = (0)$. Therefore, if we set $R = k(x_n)[x_1, \ldots, x_{n-1}]$ (recall that $k(x)$ denotes the field of rational functions in k in one variable), then $I \cdot R \neq R$. Now we can apply the induction hypothesis to the ring of polynomials $\overline{R} = \overline{k(x_n)}[x_1, \ldots, x_{n-1}]$ where $\overline{k(x_n)}$ denotes the algebraic closure of $k(x_n)$. Thus, the ideal $I \cdot \overline{R}$ is contained in an ideal of the form $(x_1 - b_1, \ldots, x_{n-1} - b_{n-1})$ for $b_i \in \overline{k(x_n)}$. Thus, each b_i is the root of a polynomial with coefficients in $k(x_n)$. Now since k is algebraically closed, all of the coefficient polynomials factor into linear factors, and there are only finitely many values of $x_n \in k$ for which either the denominator or numerator of any of the coefficient polynomials is 0. Since k is algebraically closed, it is infinite, and we can choose an element $a_n \in k$ which is different from any of those values. Plugging in $x_n = a_n$, all the expressions for b_i give meaningful formulas for elements $a_i \in k$. Then, the ideal I is contained in (4.3.1). □

It follows from Theorem 4.3.1 and Proposition 4.3.2 that two affine algebraic sets over an algebraically closed field k defined by ideals I, J coincide if and only if $\sqrt{I} = \sqrt{J}$. To see this, first note that the set of polynomials p satisfying

$$p(x_1, \ldots, x_n) - p(a_1, \ldots, a_n) \in (x_1 - a_1, \ldots, x_n - a_n)$$

is a subalgebra of $k[x_1, \ldots, x_n]$ since

$$p(x_1, \ldots x_n)q(x_1, \ldots, x_n) - p(a_1, \ldots a_n)q(a_1, \ldots, a_n) =$$
$$p(x_1, \ldots x_n)(q(x_1, \ldots, x_n) - q(a_1, \ldots, a_n)) +$$
$$(p(x_1, \ldots x_n) - p(a_1, \ldots, a_n))q(a_1, \ldots, a_n),$$

and thus is equal to $k[x_1, \ldots, x_n]$, since it contains $x_1, \ldots, x_n$. This implies that

$$I \subseteq (x_1 - a_1, \ldots, x_n - a_n)$$

if and only if $(a_1, \ldots, a_n) \in Z(I)$, which in turn implies our statement by Theorem 4.3.1 and Proposition 4.3.2.

Note that using Gröbner bases, for any field k (not necessarily algebraically closed), this can be checked as follows: First, for an ideal $I \subseteq k[x_1, \ldots, x_n]$, $f \in \sqrt{I}$ occurs if and only if

$$(k[x_1, \ldots, x_n]/I)[f^{-1}] = 0 \qquad (4.3.2)$$

(since both conditions are equivalent to $f^{-1}m = k[x_1, \ldots, x_n][f^{-1}]$ for every maximal ideal $I \subseteq m$). Now (4.3.2) can be checked by considering the ideal generated by I and $ft - 1$ in $k[x_1, \ldots, x_n]$, and checking whether its reduced Gröbner basis is $\{1\}$.

For two ideals $I = (f_1, \ldots, f_m)$, $J = (g_1, \ldots, g_m)$, we have $\sqrt{I} = \sqrt{J}$ if and only if all $f_i \in \sqrt{J}$ and $g_j \in \sqrt{I}$, which can be checked using the above method. For an example, see Exercise 3.

The reader may have noticed that we have not proved Theorem 4.3.1 yet. In fact, our approach is to introduce some related facts first, which will be useful later. Theorem 4.3.1 will be proved at the end of this subsection.

In a commutative ring R, the intersection $Jac(R)$ of all maximal ideals is called the *Jacobson radical* and the intersection $Nil(R)$ of all prime ideals is called the *nilradical*. Therefore, by definition, the Jacobson radical contains the nilradical.

4.3.3 Lemma *The nilradical consists of all elements $x \in R$ which are nilpotent (i.e. satisfy $x^n = 0$ for some $n \in \mathbb{N}$). The Jacobson radical consists precisely of those elements $x \in R$ such that for all $y \in R$, $1 + xy$ is a unit.*

Proof Clearly, a prime ideal contains all nilpotent elements. Thus, to prove the statement about the nilradical, we need to show that if $x \in R$ is not nilpotent, then there exists a prime ideal $p \subset R$ not containing x. Clearly, a union of a set of ideals linearly ordered by inclusion and not containing a power of x is an ideal not containing a power of x. Thus, by Zorn's lemma, the set of all ideals not containing x has a maximal element I with respect to inclusion. We claim that I is prime. If I is not prime, then we have some $y, z \in R$ with $yz \in I$ and $y, z \notin I$. But then either the ideal generated by I and y, or the ideal generated by I and z, contains no power of x. For otherwise, $x^m \in ay + I$, $x^n \in bz + I$, and thus $x^{m+n} \in (ay + I)(bz + I) = I$, which is a contradiction. Thus, we have a larger ideal than I not containing a power of x, which is also a contradiction.

To prove the statement about the Jacobson radical, suppose $x \in R$ is contained in every maximal ideal, but for some $y \in R$, $1 + xy$ is not a unit. Then by Zorn's lemma, $1 + xy$ is contained in some maximal ideal, which then cannot contain x (for otherwise it would contain 1)—a contradiction. If $x \notin m$ for a maximal ideal m, then since R/m is a field, there exists a $y \in R$ such that $1 + xy \in m$, and hence is not a unit. $\square$

Let $R \subseteq S$ be an inclusion of commutative rings. An element $x \in S$ is said to be *integral* over R if it is a root of a *monic polynomial* with coefficients in R, i.e. if

$$x^n + a_1 x^{n-1} + \cdots + a_n = 0$$

with $a_i \in R$. Clearly, then the subring R' of S generated by R and x is a finitely generated R-module (with generators $1, x, \ldots, x^{n-1}$). The converse is also true: suppose some linear combinations $r_1, \ldots, r_m$ of powers of x generate R' as an R-module. Let $n \in \mathbb{N}$ be a number greater than all the powers of x involved in any of the r_i's. Then x^n must be an R-linear combination of $r_1, \ldots, r_m$, which gives a monic polynomial with root x.

Now if $y \in S$ is another element integral over R, then the subring $R'' \subseteq S$ is a finitely generated R'-module, and hence a finitely generated R-module. If R is Noetherian, then every submodule of R'' is also finitely generated, and hence, in particular, $x + y$ and xy are integral elements over R. If R is not Noetherian, the same conclusion still holds, since an element integral over a ring is also integral over some finitely generated subring.

Thus, the subset of S of elements integral over R is a subring of S, called the *integral closure* of R in S. When the integral closure is equal to S, we say that S is an *integral extension* of R.

4.3.4 Lemma *Let $R \subseteq S$ be an integral extension where S is an integral domain. Then R is a field if and only if S is a field.*

Proof If R is a field and $x \in S$ is integral, then the subring Rx of S generated by R and x is an integral domain, and a quotient of $R[x]$. But $R[x]$ is a PID, so every non-zero prime ideal is maximal (see Exercise 24). But the ideal of polynomials with coefficients in R which are 0 on x is prime since Rx is an integral domain, and is not 0, since x is integral. Thus Rx is a field. Since a union of an increasing sequence of fields is a field, we are done by Zorn's lemma.

Suppose conversely S is a field. Suppose that $0 \neq x \in R$. Then $x^{-1} \in S$ is integral over R, i.e.

$$x^{-n} + a_1 x^{-n+1} + \cdots + a_n = 0$$

for $a_i \in R$. Thus, $x^{-1} = -(a_1 + \cdots + a_n x^{n-1}) \in R$. $\square$

4.3.5 Theorem (Noether's Normalization Lemma) *Let R be a finitely generated k-algebra where k is a field. Then R is isomorphic to an integral extension of a polynomial algebra $k[x_1, \ldots, x_m]$.*

Proof Induction on the number of generators $y_1, \ldots, y_n$ of R. If $n = 0$, there is nothing to prove. Otherwise, we may assume that there exists a non-zero polynomial f with coefficients in k such that

$$f(y_1, \ldots, y_n) = 0$$

(since otherwise we would also be done). Now let

$$z_i = y_i - y_n^{r_i}, \ i = 1, \ldots, n-1$$

for some $r_1, \ldots, r_{n-1} \in \mathbb{N}$. Then we have

$$f(z_1 + y_n^{r_1}, \ldots, z_{n-1} + y_n^{r_{n-1}}, y_n) = 0.$$

Thus, if d is the degree of the polynomial f, and $r_i > d r_{i-1}$ for $i = 1, \ldots, n-1$ (where we set $r_0 = 1$), then the highest powers of y_n occurring when expanding the $y_1, \ldots, y_n$-monomials of f into sums of monomials in $z_1, \ldots, z_{n-1}, y_n$ are all different. Thus, the highest of those powers survives, and hence, y_n is integral over the subalgebra of R generated by $z_1, \ldots, z_{n-1}$. Since R is generated by $z_1, \ldots, z_{n-1}, y_n$, we can use the induction hypothesis. $\square$

COMMENT Note that our argument actually proves a little bit more, namely that the number m can be chosen to be equal to the cardinality of any maximal finite set of algebraically independent elements of R.

4.3.6 Corollary *Let k be a field. If R is a finitely generated k-algebra which is an integral domain but not a field, and $0 \neq f \in R$, then $f^{-1}R$ is not a field.*

Proof By Noether's normalization lemma, R is an integral extension of a polynomial ring $k[x_1, \ldots, x_n]$. Since R is not a field, $n \geq 1$ by Lemma 4.3.4. Now suppose $f^{-1}R$ is a field. Then f is integral over $k[x_1, \ldots, x_n]$, so it is a root of a monic polynomial over $k[x_1, \ldots, x_n]$. Thus, f divides the constant term $a \in k[x_1, \ldots, x_n]$ of that polynomial, and hence $a^{-1}R$ is a localization of the field $f^{-1}R$, and thus is equal to it. But now also $a^{-1}R$ is an integral extension of $a^{-1}k[x_1, \ldots, x_n]$, which, by Lemma 4.3.4, is then also a field. This, however, is clearly absurd, since there exists a non-zero polynomial $b \in k[x_1, \ldots, x_n]$ which is not a factor of a power of a, and hence is not inverted in $a^{-1}k[x_1, \ldots, x_n]$. (Let $\bar{k}$ be the algebraic closure of k. All algebraically closed fields are infinite. Therefore, there exists an $\alpha \in \bar{k}$ such that $x_1 - \alpha$ is not an irreducible factor of $a \in \bar{k}[x_1, \ldots, x_n]$. Let, for example, $b \in k[x_1]$ be the minimal polynomial of α.) □

Proof of Theorem 4.3.1 Consider a finitely generated k-algebra. By taking the quotient by I, we may assume without loss of generality that $I = 0$. Thus, it suffices to prove that in a finitely generated k-algebra R, we have

$$Jac(R) = Nil(R). \tag{4.3.3}$$

Since $\supseteq$ is automatic, all we need to prove is that if an element $f \in R$ is not nilpotent, then there is a maximal ideal in R which does not contain it. Thus, let $m \subset f^{-1}R$ be a maximal ideal (it can be 0). Now consider the ideal $m \cap R$. We claim that this ideal in R is maximal (and obviously, it does not contain f). We know that the ideal $R \cap m$ in R is prime. Suppose it is not maximal. Then $R/(R \cap m)$ is a finitely generated k-algebra which is an integral domain but not a field, and $f^{-1}(R/(R \cap m)) = (f^{-1}R)/m$ is a field, which contradicts Corollary 4.3.6. □

5   Introduction to Commutative Algebra

Commutative algebra is the study of properties of commutative rings. As we already saw, many of those properties have a geometric meaning. The kinds of rings we encountered so far are mostly finitely generated algebras over a field (and their localizations), which is a rather special type of a ring. We discussed many special properties of those rings in the previous section.

In this section, we will turn to methods valid for more general Noetherian rings, including primary decomposition of ideals. We will use these methods to develop the concept of *dimension*, and exhibit applications to varieties over fields. The greater generality of the methods of commutative algebra, however, also motivates the idea of developing algebraic geometry in the context of general rings. This leads to Grothendieck's concept of a *scheme*, which we will discuss in the next chapter.

5.1   Primary Decomposition

An ideal q in a ring R is called *primary* if $q \neq R$, and whenever $xy \in q$, we have either $x \in q$ or $y^n \in q$ for some $n \in \mathbb{N}$. This definition may seem unnatural at first, because of its asymmetry. In particular, it is not the same thing as a power of a prime ideal (see Exercises 27 and 28). It is, however, obviously true that for a primary ideal q, the radical $p = \sqrt{q}$ is prime. We often call q a *p-primary* ideal.

It turns out that the concept of a primary ideal behaves better than many similar notions. Perhaps it could be motivated by noting that an ideal q is primary if and only if in the ring R/q, every zero divisor x (which, recall, means a non-zero element x for which there is a nonzero element y with $xy = 0$) is nilpotent (i.e. satisfies $x^n = 0$ for some $n \in \mathbb{N}$).

Note that this implies the following

5.1.1 Lemma *If q is an ideal in a ring R such that $m = \sqrt{q}$ is a maximal ideal, then q is m-primary.*

Proof Every element of the image $\overline{m}$ of m in R/q is, by assumption, nilpotent. Therefore $\overline{m} = Nil(R/q) = Jac(R/q)$. Therefore, the ring R/q is local, and every element not in $\overline{m}$ is a unit, and hence cannot be a zero divisor. Thus, every zero divisor in R/q is nilpotent, as we needed to prove.   □

To further demonstrate the utility of primary ideals, consider the concept of *decomposition of ideals*: A *decomposition* of an ideal $I \neq R$ in a ring R is an expression of the form

$$I = J_1 \cap \cdots \cap J_n \tag{5.1.1}$$

where $J_1, \ldots, J_n \neq R$ are ideals. An ideal $I \neq R$ is called *indecomposable* if it cannot be expressed as $I = J \cap K$ for ideals $J, K \supsetneq I$. Recall that since a Noetherian ring satisfies the ascending chain condition (ACC) with respect to ideals, there cannot be an infinite sequence of ideals

$$I_1 \subsetneq I_2 \subsetneq I_3 \cdots \cdots .$$

Thus, in a Noetherian ring, an ideal always has a decomposition into indecomposable ideals. Now we have the following

5.1.2 Lemma *In a Noetherian ring R, an indecomposable ideal I is primary.*

Proof Suppose $xy \in I$. Let, for $n \in \mathbb{N}$, $I_n = \{z \in R \mid zy^n \in I\}$. By the ACC, we must have $I_n = I_{n+1}$ for some n. Now let J be the ideal generated by I and x, and K be the ideal generated by I and y^n. We claim that

$$J \cap K = I,$$

$$(5.1.2)$$

which proves our assertion, since I is indecomposable. To prove (5.1.2), note that $\supseteq$ is obvious. To prove $\subseteq$, let $a \in J \cap K$. Then we have $a + by^n \in I$, $a + cx \in I$ for some $b, c \in R$. Thus, $by^n - cx \in I$, and hence $by^{n+1} \in I$, which, however, by our choice of n, implies $by^n \in I$, and hence $a \in I$. $\qquad\square$

Thus, in particular, in a Noetherian ring R, any ideal $I \neq R$ has a decomposition into primary ideals. Uniqueness is more delicate. For example, even when I is a prime ideal, nothing prevents us from throwing in a larger prime ideal into its primary decomposition. On the other hand, Exercise 28 shows that the radicals of a decomposition of a given ideal into primary ideals may in some cases necessarily contain two prime ideals one of which is contained in another. (In fact, this always happens when the ideal is not primary, but has a prime radical.)

Nevertheless, uniqueness statements can be made. For example, we have the following

5.1.3 Proposition *The set of radicals $\{\sqrt{J_1}, \ldots, \sqrt{J_n}\}$ of a primary decomposition of I which is minimal with respect to inclusion does not depend on the decomposition, and is equal to the set of all prime ideals of the form $\sqrt{I_x}$ where $I_x = \{y \in R \mid xy \in I\}$, for $x \in R$.*

Proof Consider any primary decomposition (5.1.1), and suppose that $\sqrt{I_x}$ is a prime ideal for some $x \in R$. Then $\sqrt{I_x}$ is the intersection of the radicals of all the ideals $\{y \mid yx \in J_i\}$. Those radicals are either R (when $x \in J_i$) or else, by definition, are equal to $\sqrt{J_i}$. Therefore, the (by assumption prime) ideal $J = \sqrt{I_x}$ is equal to an intersection of finitely many prime ideals, which means that it is equal to one of them (since otherwise each of the primes would contain an element not contained in J, which therefore cannot contain their product—a contradiction).

On the other hand, since the decomposition is minimal, there exists an $x \notin J_i$ which is in the intersection of J_j with $j \neq i$. Then $I_x = \{y \in R \mid xy \in J_i\}$, so $\sqrt{I_x} = \sqrt{J_i}$. $\qquad\square$

5.2   Artinian Rings

A commutative ring R is called *Artinian* if its ideals satisfy the descending chain condition (or DCC), i.e. if every sequence of ideals

$$I_1 \supseteq I_2 \supseteq \cdots$$

in R is eventually constant. We will see that this is actually a very restrictive condition (more so than the ACC for ideals).

5.2.1 Lemma *An integral domain R which is Artinian is a field.*

Proof Let $0 \neq x \in R$. Then by the DCC, $(x^n) = (x^{n+1})$ for some $n \in \mathbb{N}$. Therefore, x^n is a multiple of x^{n+1}, and since R is an integral domain, x is a unit.   □

Since a quotient of an Artinian ring is obviously Artinian, every prime ideal in an Artinian ring R is maximal, and hence $dim(R) = 0$. Also, obviously, R satisfies (4.3.3).

5.2.2 Lemma *The nilradical of an Artinian ring R is nilpotent, i.e. there exists a $k \in \mathbb{N}$ such that $Nil(R)^k = 0$.*

Proof By the DCC, there is some $k \in \mathbb{N}$ such that $a = Nil(R)^k = Nil(R)^{k+1}$. We will show that $a = 0$. Assume this is false. Note that then $a \cdot a \neq 0$, i.e. there exists an element $x \in a$ with $x \cdot a \neq 0$. By the DCC, we may further assume that if this is also true with x replaced by xy for some $y \in R$, then $(x) = (xy)$. But now if $x \cdot a \neq 0$, then $x \cdot a \cdot a = x \cdot a \neq 0$, so indeed, there exists a $y \in a$ such that $xy \cdot a \neq 0$. Thus, $(x) = (xy)$, and inductively, $(x) = (xy^n)$ for every $n \in \mathbb{N}$. However, y is by assumption nilpotent, and hence $x = 0$, which is a contradiction.   □

5.2.3 Proposition *An Artinian ring R is a product of finitely many local Artinian rings.*

(Note that since a product of finitely many Artinian rings is obviously Artinian, this is an if and only if condition.)

Proof By Lemma 5.2.1, every prime ideal of R is maximal. Thus, $Nil(R)$ is an intersection of maximal ideals. By the DCC, it is an intersection of finitely many maximal ideals:

$$Nil(R) = m_1 \cap \cdots \cap m_n.$$

By the Chinese Remainder Theorem, $Nil(R) = m_1 \cdots \cdots m_n$. Hence, by Lemma 5.2.2, for some $k \in \mathbb{N}$, $m_1^k \cdots \cdots m_n^k = 0$, and hence, by the Chinese Remainder Theorem again,

$$m_1^k \cap \cdots \cap m_n^k = 0,$$

and

$$R \cong \prod_{i=1}^{n} R/m_i^k.$$

However, a ring of the form R/m^k where m is a maximal ideal is always local, since the image $\overline{m}$ of m is nilpotent in R/m^k and hence, $\overline{m} = Nil(R/m^k) = Jac(R/m^k)$. □

A module $M \neq 0$ over a commutative ring R is called *simple* if every submodule $N \subseteq M$ is either equal to M or 0. A *composition series* of a module M is a sequence

$$M = M_0 \supset M_1 \supset \cdots \supset M_\ell = 0 \tag{5.2.1}$$

where each M_{i-1}/M_i is simple. The number ℓ is called the *length* of the composition series. If a module M has a composition series of length ℓ and $N \subseteq M$ is a simple module, then by considering the smallest i such that $N \subseteq M_i$ in (5.2.1) and taking quotients M_j/N for $j \geq i$, we see that M/N has a composition series of length $< \ell$. Thus, we can find a composition series of M starting with any chosen simple module N of length $\leq \ell$. Thus, by induction, we see that if a module has a composition series, then all composition series have equal length, which is called the *length of the module M* and denoted by $\ell(M)$. We then also call M a module of *finite length*. Also, by the same argument, length is additive, meaning that for any R-modules $N \subseteq M$ of finite length,

$$\ell(M) = \ell(N) + \ell(M/N) \tag{5.2.2}$$

By (5.2.2), a module of finite length satisfies both the ACC and the DCC with respect to submodules. In fact, the converse is also true (see Exercise 31 below).

If R is a local Artinian ring with maximal ideal m, then we already saw that $m^k = 0$ for some k. Now R/m is a field, and m^i/m^{i+1} is an R/m-vector space, which has to be finite-dimensional by the DCC. Thus, R is of finite length, and hence is also Noetherian. By Proposition 5.2.3, every Artinian ring is Noetherian.

5.2.4 Theorem *A (commutative) ring is Artinian if and only if it is Noetherian of dimension 0.*

Proof We have already seen that Artinian rings are Noetherian and have dimension 0. Conversely, let R be a Noetherian ring of dimension 0. Let

$$0 = q_1 \cap \cdots \cap q_n$$

where q_i are primary. Let $m_i = \sqrt{q_i}$. Since $dim(R) = 0$, m_i are maximal ideals. Since R is Noetherian, there exists a $k \in \mathbb{N}$ such that for all i, $m_i^k \subseteq q_i$. By the Chinese Remainder Theorem, then, R is isomorphic to the product of the rings $R_i = R/m_i^k$. Now m_i^j/m_i^{j+1} is an R/m_i-vector space, which is finite-dimensional by the ACC. Thus, each of the rings R_i is an R_i-module of finite length, and hence R_i is Artinian. Hence, R is Artinian. $\qquad\square$

5.3 Dimension

Let A be a Noetherian local ring with maximal ideal m and an m-primary ideal q with s generators. We are interested in studying powers of the ideal q, but for inductive purposes, a more general concept must be introduced. Let M be a finitely generated A-module. A *q-stable filtration on M* is a sequence $\mathcal{M}$ of submodules

$$M = M_0 \supseteq M_1 \supseteq M_2 \supseteq M_3 \supseteq \cdots$$

such that

$$q M_i \subseteq M_{i+1} \tag{5.3.1}$$

for all $i \in \mathbb{N}$, and there exists a k such that equality arises for all $i \geq k$. The key point about q-stable filtrations is the following

5.3.1 Lemma (Artin-Rees Lemma) *Let M be a finitely generated A-module with a q-stable filtration $\mathcal{M}$, and let $N \subseteq M$ be a submodule. Then the submodules $N_i = M_i \cap N$ form a q-stable filtration on N (denoted by $\mathcal{M} \cap N$).*

Proof Consider the ring

$$A^* = \bigoplus_{i \in \mathbb{N}_0} q^i$$

(where we set $q^0 = A$). The ring structure is by the product from q^i and q^j to q^{i+j}. Then the ring A^* is a finitely generated A-algebra, and hence is Noetherian by the Hilbert basis theorem, and

$$M^* = \bigoplus_{i \in \mathbb{N}_0} M_i$$

is a finitely generated module (since the filtration on M is q-stable). Now consider the submodules

$$N_k^* = \bigoplus_{i \leq k} (N \cap M_k) \ \oplus \ \bigoplus_{j \in \mathbb{N}} q^j (N \cap M_k).$$

We have $N_k^* \subseteq N_{k+1}^*$, so by the ACC, equality arises for large enough k, which is what we were trying to prove.

$\square$

We are interested in measuring the growth of the A-modules M/M_k. Since the ring A/q is Artinian, the finitely generated A/q-modules M_i/M_{i+1} have finite length, and hence the A-modules M/M_k have finite length. We put

$$\chi_q^{\mathcal{M}}(k) = \ell(M/M_k).$$

5.3.2 Proposition *There exists a $k_0 \in \mathbb{N}$ and a polynomial $p(k)$ of degree $\leq s$ such that $\chi_q^{\mathcal{M}}(k) = p(k)$ for all $k \geq k_0$. (It is called the* characteristic polynomial.*)*

Proof We form the *Poincaré series*

$$P_{\mathcal{M}}(t) = \sum_{i=0}^{\infty} \ell(M_i/M_{i+1})t^i.$$

Now by induction on the number s of generators of q, one can show that there exists a polynomial $g(t)$ such that

$$P_{\mathcal{M}}(t) = \frac{g(t)}{(1-t)^s}. \tag{5.3.2}$$

Indeed, if $q = 0$, the statement is obvious. Let x be one of the generators of q, and consider the map

$$M \xrightarrow{\ x\ } M. \tag{5.3.3}$$

Then the kernel K and the cokernel C of (5.3.3) are A-modules, but also $A/(x)$-modules. If we let $\bar{q}, \bar{m}$ be the images of the ideals q, m in $R/(x)$, then $\bar{m}$ is maximal, $\bar{q}$ is $\bar{m}$-primary, and has a set of $s - 1$ generators. Additionally, $\mathcal{K} = K \cap \mathcal{M}$ is a $\bar{q}$-stable filtration (by

Lemma 5.3.1). Also, we have a filtration $\mathcal{C}$ given by the images of the modules $M_n/(x)$ in $M/(x)$. By the additivity of length, we have

$$P_{\mathcal{M}}(t) - P_{\mathcal{C}}(t) = t(P_{\mathcal{M}}(t) - P_{\mathcal{K}}(t)),$$

both being equal to the Poincaré series of the q-stable filtration by $x M_{k-1}$ on $x M$. Now the induction hypothesis can be applied to $P_{\mathcal{C}}(t)$ and $P_{\mathcal{K}}(t)$, thus proving the existence of the expression (5.3.2).

Now using the expansion

$$\frac{1}{(1-t)^s} = \sum_{n \geq 0} \binom{-s}{n}(-t)^n = \sum_{n \geq 0} \binom{n+s-1}{s-1} t^n,$$

we see that there exists a polynomial $h(n)$ of degree $\leq s - 1$ such that the coefficient of $P_{\mathcal{M}}(t)$ at t^n is equal to $h(n)$ for n large enough. The polynomial $h(n)$ is known as the *Hilbert polynomial*. The statement of the Proposition follows. □

5.3.3 Lemma *Suppose* $\mathcal{M}, \mathcal{M}'$ *are q-stable filtrations on a finitely generated A-modules* M, *and suppose that* $\mathcal{Q}$ *is an m-stable filtration on* M *(recall that q is an m-primary ideal). Then*

$$\lim_{k \to \infty} \frac{\chi_q^{\mathcal{M}}(k)}{\chi_q^{\mathcal{M}'}(k)} = 1, \tag{5.3.4}$$

$$0 < \lim_{k \to \infty} \frac{\chi_q^{\mathcal{M}}(k)}{\chi_m^{\mathcal{Q}}(k)} < \infty. \tag{5.3.5}$$

Proof Let $k_0 \in \mathbb{N}$ be such that for $k \geq k_0$, we have $M_{k+1} = q M_k$ and $M'_{k+1} = q M'_k$ where M'_k are the modules of the filtration $\mathcal{M}'$. Then by definition, for every $k \in \mathbb{N}$,

$$M_k \supseteq M'_{k+k_0},$$

$$M'_k \supseteq M_{k+k_0}.$$

Thus,

$$\chi_q^{\mathcal{M}}(k) \leq \chi_q^{\mathcal{M}'}(k + k_0),$$

$$\chi_q^{\mathcal{M}'}(k) \leq \chi_q^{\mathcal{M}}(k + k_0).$$

Thus, (5.3.4) follows from the fact that $\chi_q^{\mathcal{M}}$ and $\chi_q^{\mathcal{M}'}$ are polynomials (by Proposition 5.3.2).

Similarly, we have $m^r \subseteq q \subseteq m$ for some $r \in \mathbb{N}$, so if we let simply $M_k = q^k M$, $Q_k = m^k M$, we have

$$\chi_m^Q(rk) \geq \chi_q^{\mathcal{M}}(k) \geq \chi_m^{\mathcal{M}}(k)$$

so again (5.3.5) follows from Proposition 5.3.2.

$\square$

Now let R again be a local Noetherian ring with maximal ideal m. Denote by $\delta(R)$ the minimum number d for which there exists an m-primary ideal q with d generators. Denote by $d(R)$ the degree of the polynomial $\chi_q^{\mathcal{M}}(k)$ for an m-primary ideal q and a stable q-filtration $\mathcal{M}$ (which is independent of the choice of q and $\mathcal{M}$ by Lemma 5.3.3).

5.3.4 Lemma *Let R be a Noetherian local ring with maximal ideal m. Then for any ideal $I \subseteq m$, we have*

$$d(R/I) \leq d(R).$$

Let $x \in m$ be a regular element (i.e. not a divisor of 0). Then

$$d(R/(x)) \leq d(R) - 1.$$

Proof The first statement is clear. For the second statement, let $\mathcal{M} = (M_k)_k$ be an m-stable filtration on the R-module R, and let $\overline{\mathcal{M}} = (M_k/(x))_k$ be the induced m-stable filtration on $R/(x)$ (which is also an $\overline{m}$-stable filtration where $\overline{m} = m/(x)$ is the maximal ideal in $R/(x)$). Then by additivity of length, we have

$$\chi_m^{\mathcal{M}}(k) - \chi_m^{\mathcal{M} \cap xR}(k) = \chi_{\overline{m}}^{\overline{\mathcal{M}}}(k).$$

By Lemma 5.3.3, the two functions on the left hand side are polynomials in k of degree $d(R)$ with the same coefficient at k^d. Thus, their difference is a polynomial of degree $\leq d - 1$. But this degree is, by definition, $d(R/(x))$.

$\square$

5.3.5 Theorem *We have $\delta(R) = d(R) = dim(R)$.*

Proof By Proposition 5.3.2, $\delta(R) \geq d(R)$. Next, we will prove that $dim(R) \leq d(R)$ (and hence, in particular, $dim(R)$ is finite). For this purpose, we use induction on $d(R)$. Consider a chain of prime ideals $p_0 \subset \cdots \subset p_d$ in R. Now if we replace R by R/p_0, clearly, it will not increase $d(R)$, and the projections of p_i still form a chain of length d. Thus, without loss of generality, $p_0 = 0$, and in particular, R is an integral domain. Now if $d = 0$, there is nothing to prove. Otherwise, let $0 \neq x \in p_1$. By Lemma 5.3.4, $d(R/(x)) \leq d(R) - 1$, while the projections of $p_1, \ldots p_d$ form a chain of prime ideals of length $d - 1$ in $R/(x)$. Thus, by the induction hypothesis, $d - 1 \leq d(R) - 1$, and hence $d \leq d(R)$.

Thus, it remains to prove that $\delta(R) \leq dim(R)$ or in other words to construct an m-primary ideal q with $dim(R)$ generators. To this end, we will construct, by induction, for $i \leq dim(R)$, an ideal $(x_1, \ldots, x_i)$ whose every prime (from its minimal primary decomposition) has height $\geq i$. Suppose this is done with i replaced by $i - 1$. Then let U be the union of all primes of $(x_1, \ldots, x_{i-1})$ of height $i - 1$. We claim that

$$U \neq m. \tag{5.3.6}$$

In fact, in general, when an ideal a is contained in a union of finitely many primes $p_1, \ldots, p_k$, it is contained in one of them. To see this, consider a counterexample with smallest k. Then there exist elements $x_1, \ldots, x_k \in a$ where x_j is not contained in x_m for $m \neq j$. But then

$$\sum_{j=1}^{k} x_1 \ldots x_{j-1} x_{j+1} \ldots x_k \in a \smallsetminus (p_1 \cup \cdots \cup p_k),$$

which is a contradiction.

Thus, (5.3.6) is proved. Now let $x_i \in m \smallsetminus U$. Then every prime in a minimal primary decomposition of $(x_1, \ldots, x_i)$ contains a prime of $(x_1, \ldots, x_{i-1})$, and hence cannot have height $i - 1$ since $x_i \notin U$.

Thus, the induction is complete. For $i = dim(R)$, the ideal m must be a prime of $(x_1, \ldots, x_i)$, for otherwise it would have height $> i$.   □

COMMENTS

1. When proving he inequality $d(R) \geq dim(R)$, we used induction, reducing, in each step, to the case when R is an integral domain by factoring out the smallest prime ideal in a chain. Because of this, the proof fails to produce an actual *regular sequence* of length $d = dim(R)$ (meaning a sequence $(x_1, \ldots, x_d)$ where x_i is a non-zero divisor in $R/(x_1, \ldots, x_{i-1})$). A Noetherian local ring R for which there exists a regular sequence of elements of the maximal ideal of length $dim(R)$ is called a *Cohen-Macaulay ring*. See Exercise 32.

2. For R a Noetherian local ring with maximal ideal m, $dim(R) = d$, the generators $x_1, \ldots, x_d$ of an m-primary ideal are called *parameters*.

3. By definition, for a field k, we have

$$d(k[x_1, \ldots, x_n]_{(x_1, \ldots, x_n)}) = n. \tag{5.3.7}$$

5.3.6 Corollary *Let R be a Noetherian local ring and let $x \in R$ be a regular element. Then*

$$dim(R/(x)) = dim(R) - 1.$$

Proof The inequality $\leq$ follows from Lemma 5.3.4. For the opposite inequality, let m be the maximal ideal of R, let $d = dim(R/(x))$, and let $x_1, \ldots, x_d \in R$ modulo (x) generate an $m/(x)$-primary ideal in $R/(x)$. Then the ideal $(x, x_1, \ldots, x_d) \subset R$ is m-primary, thus showing $\geq$.

$\square$

5.3.7 Theorem *Let R be a Noetherian ring. Let p be a prime ideal minimal (with respect to inclusion) among those containing an ideal $(x_1, \ldots, x_r)$. Then p has height $\leq r$.*

Proof In R_p, by minimality, $R_p p$ must be the only radical of an ideal occurring in any primary decomposition of $(x_1, \ldots, x_r)$, and thus must be its radical. Thus, $(x_1, \ldots, x_r)$ is $R_p p$-primary, and hence has height $\leq r$ by Theorem 5.3.5.

$\square$

COMMENT Theorem 5.3.7 is known as *Krull's Height Theorem.* Note that if $(x_1, \ldots, x_r) \neq R$, a minimal prime containing it exists by Zorn's lemma. If $r = 1$, this is equivalent to $x = x_1$ not being a unit. For $r = 1$, if, additionally, x is not a zero divisor, then we know that the height of p is exactly 1, since a prime of height 0 cannot contain a non-zero divisor by Corollary 5.3.6 (in fact, only the inequality of Lemma 5.3.4 is needed). With these assumptions on x for $r = 1$, the theorem is known as *Krull's Hauptidealsatz*, which in German means "principal ideal theorem."

5.3.8 Proposition *Let k be a field and let R be a finitely generated k-algebra with a maximal ideal m. Let $x_1, \ldots, x_d$ be parameters of the localization R_m. Then $x_1, \ldots, x_d$ are algebraically independent over k (i.e. we do not have $f(x_1, \ldots, x_d) = 0$ for any non-zero polynomial with coefficients in k).*

Proof Let $n = (x_1, \ldots, x_d)$. By assumption, n is m-primary. Thus, R/n is a local Artinian ring. Let its length (as an R/n-module, or, equivalently, R_m-module), be ℓ. Now let A be the image of the polynomial ring $k[x_1, \ldots, x_d]$ in R_m under the inclusion of the generators. Denote by $\mathcal{N}$ the filtration of R_m by powers of n. Further, denote by q the maximal ideal of $A_{(x_1, \ldots, x_d)}$, and by $\mathcal{Q}$ the filtration of $A_{(x_1, \ldots, x_d)}$ by powers of q. Then since homogeneous polynomials in the x_i's of degree r generate n^r/n^{r+1} as an R/n-module, we have

$$\chi_n^{\mathcal{N}}(k) \leq \ell \cdot \chi_q^{\mathcal{Q}}(k).$$

However, if $f(x_1, \ldots, x_d) = 0$ for a non-zero polynomial f over k, then the degree of the right hand polynomial is less or equal to the dimension of

$$k[x_1, \ldots, x_d]_{(x_1,\ldots,x_d)}/(f),$$

$\square$

which is $\leq d - 1$ by Lemma 5.3.4.

5.4   Regular Local Rings

We will be using the following fact:

5.4.1 Lemma (Nakayama Lemma) *Let I be an ideal in a commutative ring R, and let M be a finitely generated R-module such that $IM = M$. Then there exists an element $r \in R$ such that r projects to 1 in R/I and $rM = 0$.*

Proof Let $x_1, \ldots, x_n$ be a set of generators of M as an R-module. By assumption, there exists an $n \times n$ matrix A with entries in I which acts by the identity on the column vector

$$\begin{pmatrix} x_1 \\ \vdots \\ x_n \end{pmatrix}.$$

$\square$

Then $r = det(Id - A)$ works (by the Cramer rule).

It implies the following

5.4.2 Proposition (Krull) *Let R be a local Noetherian ring with maximal ideal m. Then*

$$\bigcap_{i \in \mathbb{N}} m^i = 0. \tag{5.4.1}$$

Proof Denote the ideal (5.4.1) by I. It is finitely generated since R is Noetherian. We have $mI = I$ ($\subseteq$ is trivial, and $\supseteq$ follows from the Artin-Rees Lemma applied to $I \subset R$), so there exists a unit modulo m (hence a unit) annihilating I. Thus, $I = 0$. $\square$

Let R be a local ring with maximal ideal m and residue field k. Then we can form a k-algebra

$$E_0 R = \bigoplus_{i \in \mathbb{N}_0} m^i / m^{i+1}$$

where for $x \in m^i/m^{i=1}$, $y \in m^j/m^{j+1}$, $xy \in m^{i+j}/m^{i+j+1}$ is given by the multiplication in R.

5.4.3 Theorem *Let R be a Noetherian local ring with maximal ideal m and residue field k. Let $x_1, \ldots, x_d$ be a basis of the k-vector space m/m^2. Then R is regular if and only if the canonical mapping*

$$\phi : k[x_1, \ldots, x_d] \to E_0 R$$

(sending $x_i \mapsto x_i$) is an isomorphism.

Proof If ϕ is an isomorphism, then, by definition, the number $d(R) = d(E_0 R)$ defined in the last section is equal to d, and thus, R is regular. Since, also by definition, ϕ is onto, if R is regular, all we need to prove is that ϕ is injective. But if $0 \neq f \in Ker(\phi)$, then, since $k[x_1, \ldots, x_d]$ is an integral domain,

$$d(R) = d(E_0 R) \leq d(k[x_1, \ldots, x_d]/(f)) = d - 1$$

by Corollary 5.3.6 and Theorem 5.3.5. Thus, R cannot be regular. $\qquad \square$

5.4.4 Corollary *A regular local ring R is an integral domain.*

Proof Let R be a regular local ring with maximal ideal m and residue field k. Let $0 \neq x, y \in R$ such that $xy = 0$. Let $s, t \in \mathbb{N}_0$ be the smallest numbers such that $x \in m^s$, $y \in m^t$. (Such numbers exist by Proposition 5.4.2.) Then by Theorem 5.4.3, xy has a non-zero image in m^{s+t}/m^{s+t+1}, which is a contradiction. $\qquad \square$

5.5 Dimension of Affine Varieties

For an inclusion $k \subseteq K$ of fields, the *transcendence degree* $td(K/k)$ is the maximal possible cardinality of a set of algebraically independent elements of K over k.

5.5.1 Theorem *Let k be a field and let R be a finitely generated k-algebra which is an integral domain. Then for any maximal ideal m of R, $dim(R_m)$ is equal to the transcendence degree d of the field of fractions $K = QR$ over k.*

Proof By Proposition 5.3.8, $d = td(K/k) \geq dim(R_m)$, since parameters are algebraically independent over k. To prove the converse, by clearing denominators, we can assume that the elements $x_1, \ldots, x_d$ algebraically independent over k are in R. By Noether's normalization lemma, we may then assume that R is an integral extension of $A = k[x_1, \ldots, x_d]$.

Let $n = m \cap A$. We claim that n is a maximal ideal of A. To this end, note that the field R/m is an integral extension of A/n. Since R/m is a field, A/n is a field by Lemma 4.3.4 and hence n is a maximal ideal in A. We will show that

$$dim(A_n) = dim(R_m). \tag{5.5.1}$$

Suppose (5.5.1) is proved. Then if $n = (x_1 - a_1, \ldots x_d - a_d)$ for some $a_1, \ldots a_d \in k$, we are done by Comment 3 after Theorem 5.3.5. In general,

$$n \subseteq m = (x_1 - a_1, \ldots, x_d - a_d) \subset E[x_1, \ldots, x_d]$$

for some finite extension E of k. Then we are done by using (5.5.1) again with R replaced by $E[x_1, \ldots, x_d]$.

To prove (5.5.1), note that the ring $R/n^k R$ is integral over A/n^k, and hence is Artinian. Since an Artinian ring is a product of its localizations at maximal ideals, the map

$$R/n^k R \to R_m/n^k R_m \tag{5.5.2}$$

is onto, and $n^k R_m$ is $m R_m$-primary. Also, the ring $R_n = (A \smallsetminus n)^{-1} R$ is integral over A_n, but it is also finitely generated as an A_n-algebra, and hence is finitely generated as an A_n-module (say, on g generators).

Let M be the image of the finitely generated A_n-module R_n in R_m, and consider the $n A_n$-stable filtration $\mathcal{M} = (n^k M)$ on M. Also consider the $n R_m$-stable filtration $\mathcal{P} = (n^k R_m) = ((n R_m)^k)$ on R_m. Since (5.5.2) is onto, we have

$$\chi_{n R_m}^{\mathcal{P}}(k) = \chi_{n A_n}^{\mathcal{M}}(k). \tag{5.5.3}$$

Now since $A_n \subseteq M$ is injective (since localization is injective in an integral domain), we have

$$\chi_{n A_n}^{\mathcal{M} \cap A_n}(k) \leq \chi_{n A_n}^{\mathcal{M}}(k),$$

which, together with (5.5.2), implies $dim(A_n) \leq dim(R_m)$.

On the other hand,

$$\chi_{n A_n}^{\mathcal{M}}(k) \leq g \chi_{n A_n}^{\mathcal{M} \cap R_n}(k)$$

and thus $dim(R_m) \leq dim(A_n)$.   □

COMMENT

1. In (5.5.1), R_m is often *not* an integral extension of A_n. See Exercises 33 and 34.

2. A Noetherian integral extension does not have to be finitely generated (for example, the algebraic closure is an integral extension of a field). Also, an integral extension of a Noetherian ring does not have to be Noetherian (consider, for example, the integral closure of $\mathbb{Z}$ in $\mathbb{C}$). Nevertheless, more general results on dimension of Noetherian integral extensions of Noetherian rings, without assuming finiteness, hold (see the last chapter of [2]).

3. Note that, in particular, we have shown that the number d does not depend on the maximal set of algebraically independent elements chosen, i.e. that in a finitely generated k-algebra $R = k[x_1, \ldots, x_n]/I$ which is an integral domain, all maximal sets of elements algebraically independent over k have the same cardinality, equal to the dimension. Thus, we can use elimination ideals to compute dimension: Suppose I_i is the elimination ideal of I with respect to the variables $x_i, \ldots, x_n$. Put $I_{n+1} = 0$. Then d is the cardinality of the set S of all $i = 1, \ldots, n$ such that $I_i = I_{i+1}$. This is because by definition, no non-zero polynomial in the variables x_i, $i \in S$ with coefficients in k is in I, so they are algebraically independent over k in R, while our proof of Noether's normalization lemma lets us construct a set of $|S|$ variables over which all the other generators (and hence all elements) of R are integral.

4. If X is an affine algebraic variety, then every affine subvariety not equal to X has dimension $\leq dim(X) - 1$. Additionally, if f is a regular function on X, then every irreducible component (see Exercise 2) of the zero set of f has, by Corollary 5.3.6, dimension $dim(X) - 1$. (Localize at a closed point contained in one irreducible component, but not the others.) For a quasiaffine variety of the form $X \smallsetminus S$ where S is a Zariski closed subset, let S' be the union of all irreducible components of S of dimension $dim(X) - 1$ (see Exercise 2). Then we have $\mathbb{C}[X \smallsetminus S] = \mathbb{C}[X \smallsetminus S']$. In particular, if $S' \neq S$, $X \smallsetminus S$ cannot be affine.

6 Exercises

1. The *connected components* of a topological space are maximal subsets which are connected with respect to the induced topology. Prove that every topological space is, as a set, a disjoint union of its connected components.

2. Prove that an affine algebraic set S can be uniquely expressed as a union of a finite set of irreducible affine algebraic sets none of which is included in another. The elements of this set are called *irreducible components of S*. [The key point is that there is no infinite sequence of Zariski closed subsets of S strictly decreasing with respect to inclusion.] How about projective algebraic sets?

3. Are the ideals

$$(x^2 - y^3, z^2 - t^3, xt), \quad (x^2 - y^3, z^2 - t^3, yz) \subseteq \mathbb{C}[x, y, z, t]$$

equal? Do they define the same affine algebraic set?

4. Prove formula (1.4.4) in Sect. 1.4.3. [Multiply by h_i.]

5. Let S be a projective algebraic set in $\mathbb{P}_{\mathbb{C}}^n$ and let $I \subseteq \mathbb{C}[x_0, \ldots, x_n]$ be the ideal generated by all homogeneous polynomials which are 0 on S. The affine algebraic set of zeros in $\mathbb{A}_{\mathbb{C}}^{n+1}$ defined by the same set of polynomials is called the *affine cone* CS on S. Prove that S is irreducible if and only if CS is irreducible (with respect to the Zariski topology).

6. The *Veronese embedding* of degree d is a morphism of projective varieties $\phi : \mathbb{P}_{\mathbb{C}}^n \to \mathbb{P}_{\mathbb{C}}^N$ where the projective coordinates in the target are all the different monomials of degree d in the projective coordinates of $\mathbb{P}_{\mathbb{C}}^n$.

    (a) Calculate the number N.

    (b) Prove that the image of the Veronese embedding is a projective variety, and that ϕ is an isomorphism of projective varieties.

7. Consider the algebraic set X in $\mathbb{A}_{\mathbb{C}}^3$ which is the set of zeros of the polynomial $x^2 - (y^2 + z^2)$. Prove that X is an affine variety. Now consider the quasiaffine variety $Y = X \smallsetminus \{(0, 0, 0)\}$. Compute the ring of regular functions $\mathbb{C}[Y]$. Is Y affine?

8. Prove that $\mathbb{A}_{\mathbb{C}}^1$ and its open affine subvariety $\mathbb{A}_{\mathbb{C}}^1 \smallsetminus \{0\}$ are birationally equivalent, but not isomorphic as algebraic varieties.

9. Give an example of an ideal $(f_1, \ldots, f_m)$ in $\mathbb{C}[x, y]$ whose elimination ideal with respect to the variable y is 0, but there exists a value $b \in \mathbb{C}$ such that there is no solution to the system of equations $f_i = 0$ with $y = b$.

10. Consider the affine algebraic sets S, T in 2-dimensional affine space over $\mathbb{C}$ given by the polynomials $x^2 - y^3$, $x^2 - y^3 - y^2$, respectively. Prove that they are affine algebraic varieties. Prove that they are both rational, but not isomorphic to an affine space.

11. Let X be the projective algebraic set in $\mathbb{P}_{\mathbb{C}}^n$ defined by the homogeneous polynomial $x_0^d + x_1^d + \cdots + x_n^d$, where $d \geq 1$. Prove that X is a smooth algebraic variety. What is its dimension?

12. Let P be any point of the projective variety E discussed in Exercise 11, and let Y be the quasi-projective variety given as the complement of P in E. Prove that Y is affine (i.e. isomorphic to an affine variety).

13. Prove that the map $\phi : \mathbb{A}_{\mathbb{C}}^n \to \mathbb{P}_{\mathbb{C}}^n$ which sends an n-tuple $(x_1, \ldots, x_n)$ to the ratio $[1 : x_1 : \cdots : x_n]$ is a morphism of algebraic varieties whose image is open. Conclude that every quasiaffine variety is isomorphic to a quasiprojective variety, and that in every quasiprojective variety, every point has an affine open neighborhood.

14. Prove that if $Z \subseteq \mathbb{P}_{\mathbb{C}}^n$ is a projective variety and ϕ is the map of Exercise 13, then $\phi^{-1}(Z)$ is an affine variety and the restriction $\phi^{-1}(Z) \to Z$ of ϕ is a morphism of algebraic varieties.

15. Prove that if $Z \subseteq \mathbb{A}_{\mathbb{C}}^n$ is an affine variety, then its Zariski closure in $\mathbb{P}_{\mathbb{C}}^n$ is a projective variety.

16. Prove that there exists a morphism of algebraic varieties

$$\mathbb{A}_{\mathbb{C}}^2 \smallsetminus \{(0, 0)\} \to \mathbb{P}_{\mathbb{C}}^1$$

which does not extend to a morphism of algebraic varieties

$$\mathbb{A}_\mathbb{C}^2 \to \mathbb{P}_\mathbb{C}^1.$$

In particular, there exists a rational function on $\mathbb{A}_\mathbb{C}^2$ which is not equal to an affine coordinate on $\mathbb{P}_\mathbb{C}^1$ composed with a morphism of algebraic varieties $\mathbb{A}_\mathbb{C}^2 \to \mathbb{P}_\mathbb{C}^1$.

17. Prove that a morphism of algebraic varieties over $\mathbb{C}$ is continuous in the analytic topology.

18. An *affine quadric* is the set of zeros of an irreducible degree 2 polynomial f in $\mathbb{A}_\mathbb{C}^n$. A *projective quadric* is the set of zeros of an irreducible homogeneous degree 2 polynomial f in $\mathbb{P}_\mathbb{C}^n$. A projective quadric is called *non-degenerate* when the symmetric matrix formed by the coefficients of f (dividing mixed coefficients by 2) is non-degenerate. (Note that the 2-multiple of this matrix is the matrix of second partial derivatives of f at 0). For an affine quadric, introduce a new variable z and make all monomials of f quadratic by multiplying them by the appropriate powers of z. The affine quadric is called *non-degenerate* if the resulting matrix of coefficients, including the new variable z, (with mixed coefficients again divided by 2) is non-degenerate. For non-degenerate quadrics, applying linear changes of variables in the projective case we can assume $f = x_0^2 + \cdots + x_n^2$, and applying an affine change of variables (linear transformations and shifts) in the affine case we can assume $f = x_1^2 + \cdots + x_k^2 + x_{k+1} + \cdots + x_n + 1, n - 1 \leq k \leq n$. This is proved by forming a symmetric matrix of coefficients of the quadratic part (dividing mixed coefficients by 2), and then performing symmetric row and column operations; in the affine case, we also form perfect squares, shift, and scale as needed.

(a) Prove that all non-degenerate affine and projective quadrics are smooth varieties. Is this true for all affine and projective quadrics?

(b) Prove that an intersection of a projective quadric in $\mathbb{P}_\mathbb{C}^n$ with $\mathbb{A}_\mathbb{C}^n$ is an affine quadric. Similarly for non-degenerate quadrics.

(c) Prove that the closure in $\mathbb{P}_\mathbb{C}^n$ of an affine quadric in $\mathbb{A}_\mathbb{C}^n$ is a projective quadric. Similarly for non-degenerate quadrics.

(d) Prove that all (affine or projective) quadrics (over $\mathbb{C}$) are rational.

(e) Prove that the non-degenerate affine quadric for $k = n - 1$ is isomorphic to the affine space.

(f) Prove that the non-degenerate affine quadric for $k = n$ is homeomorphic to an affine space with the origin removed in the analytic topology, and conclude that it is not isomorphic to the affine space as an algebraic variety.

19. (a) Let R be an integral domain and let $p \subset R$ be a prime ideal. A monic polynomial

$$p(x) = x^n + a_{n-1}x^{n-1} + \cdots + a_0 \in R[x]$$

is called *Eisenstein* if $a_i \in p$ for all $i < n$, and $a_0 \notin p^2$. Prove that an Eisenstein polynomial is irreducible in $R[x]$.

(b) Prove that if R is a UFD and Q is its field of fractions, then an Eisenstein polynomial is irreducible in $Q[x]$.

20. Is the polynomial $x^2y^3 + yx^3 + xz \in \mathbb{C}[x, y, z]$ irreducible? [Use the Gröbner basis criterion.]

21. Prove that if R is a UFD, so is $a^{-1}R$ for any non-zero element $a \in R$.

22. Prove that any quotient R/I and localization $S^{-1}R$ of a Noetherian ring R is Noetherian.

23. Prove that a finitely generated module over a Noetherian ring satisfies ACC with respect to submodules. Conclude that a submodule of a finitely generated module over a Noetherian ring is finitely generated.

24. Prove that in a PID, every nonzero prime ideal is maximal.

25. Is $x + y + z \in \sqrt{I}$ where

$$I = (x^2 - y^2 - z^2, xz^2 + z^3, xy + xz + yz + y^2 + z^2, y^2z^2)?$$

26. Give an example of a Noetherian ring R with $Jac(R) \neq Nil(R)$.

27. Show that in the ring $R = \mathbb{C}[x, y]$, for $n > 1$, the ideal (x, y^n) is primary, but is not a prime power.

28. Show that in the ring $R = \mathbb{C}[x, y]/(xy - z^2)$, the ideal (x^2, z^2, xz) is a prime power, but is not primary.

29. Prove that an intersection of finitely many p-primary ideals is p-primary.

30. Prove that in $\mathbb{C}[x, y]$, the ideal (x^2, xy) has at least two different minimal primary decompositions.

31. Prove that a module which satisfies both the ACC and DCC with respect to submodules is of finite length.

32. Consider the Noetherian local ring $R = (\mathbb{C}[x, y]/(x^2, xy))_{(x,y)}$ with maximal ideal $m = (x, y)$.
    (i) Calculate $dim(R)$.
    (ii) Prove that R is not Cohen-Macaulay.

33. Consider the integral extension $\mathbb{Z} \subset \mathbb{Z}[i] = \mathbb{Z}[x]/(x^2 + 1)$.
    (i) We have $5 = (2+i)(2-i) \in \mathbb{Z}[i]$. Prove that $(2+i)$ and $(2-i)$ are prime ideals in $\mathbb{Z}[i]$.
    (ii) Prove that $(2 + i) \cap \mathbb{Z} = (5)$.
    (iii) Prove that $\mathbb{Z}[i]_{(2+i)}$ is not an integral extension of $\mathbb{Z}_{(5)}$.

34. Consider the rings $R = \mathbb{C}[x] \subset S = \mathbb{C}[x, y]/(y^2 + y + x)$. Show that if $m = (y) \subset S$, $n = (x) \subset R$ then m is a maximal ideal, and $m \cap R = n$. Prove that S_m is not an integral extension of R_n.

35. (i) Prove that the ideal $I = (xz - y^2, yt - z^2, xt - yz) \subset \mathbb{C}[x, y, z, t]$ is prime.
    (ii) Find the dimension of the affine variety of zeros of the ideal I.

36. Prove that the dimension of a local ring $\mathbb{C}[X]_P$ at any point P of a variety over $\mathbb{C}$ (in the sense of this chapter) is equal to its dimension in the sense of the maximum number d for which closed subvarieties (3.5.2) exist. [To prove $\geq$, consider a point in

X_0. To prove $\leq$, consider a point $P \in X$, and its open affine neighborhood U. Intersect the prime ideals guaranteed by the Krull dimension of $\mathbb{C}[X]_P$ with $\mathbb{C}[U]$, and take the closures of their zero sets in X.]

37. Let R be a regular local ring with maximal ideal m, and let $I \subset R$ be an ideal such that R/I is a regular (local) ring. Prove that there exists a minimal (with respect to inclusion) set of generators $x_1, \ldots, x_d$ of m such that for some $0 \leq c \leq d$, $I = (x_1, \ldots, x_c)$. [First of all, $I \subseteq m$. Next, let $dim(R/I) = e$. Then $e = dim_{R/m}(m/(I + m^2))$. Thus, we can choose generators $x_1, \ldots, x_c$ of I, $c = d - e$, which form a subset of a basis of m/m^2 over R/m. We obtain an onto homomorphism of regular local rings $R/(x_1, \ldots, x_c) \to R/I$ of the same dimension. This has to be an isomorphism, since a regular local ring is an integral domain.]

38. (The Nagata Example) Consider the ring $R = k[x_1, x_2, \ldots]$ of polynomials in countably many variables with coefficients in a field k. Let $1 = n_0 < n_1 < \ldots$ be natural numbers. Consider the prime ideals $p_i = (x_{n_i}, \ldots, x_{n_{i+1}-1})$. Let S be the multiplicative set which is the complement of the union of the p_i's, $i \geq 0$.
    (a) Prove that an ideal in R contained in the union of the p_i's is contained in one fixed p_i.
    (b) Prove that the ring $S^{-1}R$ is Noetherian.
    (c) Prove that if

$$\lim_{i \to \infty} (n_i - n_{i-1}) = \infty,$$

    then for every $n \in \mathbb{N}$, there exists a prime ideal in $S^{-1}R$ of height $> n$.

39. Let R be a commutative ring. The *total ring of fraction* QR is the ring $S^{-1}R$ where S is the set of all non-zero divisors of R. Observe that the localization $R \to S^{-1}R$ is injective. Prove that if R is Noetherian and *reduced* (meaning that it has no non-zero nilpotent elements), then the canonical homomorphism

$$QR \to \prod_i Q(R/p_i)$$

is an isomorphism where p_i are the minimal prime ideals of R. (Note that on the right hand side, we have ordinary fields of fractions.) [Show that in any commutative ring, every prime ideal contains a minimal prime ideal by Zorn's lemma. Prove that for a reduced Noetherian ring, the minimal primes form a primary decomposition of 0. In particular, S is the complement of the union of minimal primes. Then observe that QR is an Artinian ring.]

40. Prove that if R is an integral domain, then R is equal to the intersection of all the subrings R_m of the field of fractions QR, where m are maximal ideals in R. [Let a be an element of the intersection. Consider the ideal $\{x \in R \mid ax \in R\}$.]

41. A *normal ring* is a commutative ring R which is reduced and integrally closed in its total ring of fractions. Prove that a reduced Noetherian ring is normal if and only if

the canonical homomorphism

$$R \to \prod_i R/p_i$$

is an isomorphism and each of the integral domains R/p_i is normal (where again p_i are the minimal primes). [For necessity, note that each of the elements $(0, \ldots, 0, 1, 0, \ldots, 0) \in QR$ are idempotent, i.e. are roots of the monic polynomial $e^2 - e$.]

42. Prove that a reduced Noetherian normal local ring is an integral domain.

43. Let R be a reduced Noetherian ring. Prove that R is normal if and only if for every maximal ideal m in R, the localization R_m is normal. [Use Exercises 40 and 41.]

Schemes

2

We have now seen the basic idea of what algebraic geometry aims to investigate, and also some of the commutative algebra needed to prove its basic facts. However, it is clear that the concept of a variety, as we introduced it in Chap. 1, is not satisfactory: It is based on two examples, the affine and projective space, and their subobjects. This would be like defining a topological space as a subset of $\mathbb{R}^n$. For proper foundations, a general concept, based on abstract axioms, is needed.

In algebraic geometry, this is why Grothendieck introduced the concept of a *scheme*. The main purpose of this chapter is to introduce the definition of a scheme, and its basic categorical properties. This will also give us some very basic tools for constructing examples of schemes. We will conclude the chapter with by describing "finiteness" properties of schemes and their morphisms, which will be necessary pieces of the language we will use later.

1 Sheaves and Schemes

1.1 Sheaves Revisited

Recall that a *sheaf of sets* $\mathcal{F}$ can be defined on a topological space X. It assigns to every open set $U \subseteq X$ the set of *sections* $\mathcal{F}(U)$. The following properties are required:

1. **Restriction:** For $V \subseteq U$, we have a *restriction map*

$$\mathcal{F}(U) \to \mathcal{F}(V).$$

© The Author(s), under exclusive license to Springer Nature Switzerland AG 2021
I. Kriz, S. Kriz, *Introduction to Algebraic Geometry*,
https://doi.org/10.1007/978-3-030-62644-0_2

The restriction is required to be transitive (i.e. for $W \subseteq V \subseteq U$, restriction from $\mathcal{F}(U)$ to $\mathcal{F}(V)$ and then to $\mathcal{F}(W)$ is the same thing as restricting to $\mathcal{F}(W)$ directly). Also, the restriction from $\mathcal{F}(U)$ to itself is just the identity.

2. **Gluing:** If we have sections $s_i \in \mathcal{F}(U_i)$ where U_i are open sets, such that s_i and s_j restrict to the same section in $\mathcal{F}(U_i \cap U_j)$, then there exists a unique section $s \in \mathcal{F}(\bigcup U_i)$ which restricts to all the functions s_i.

The *stalk* $\mathcal{F}_x$ of a sheaf $\mathcal{F}$ at a point $x \in X$ is the set of equivalence classes of sections in $\mathcal{F}(U)$ with $x \in U$ where U is any open set containing x, where two sections $s \in \mathcal{F}(U)$, $t \in \mathcal{F}(V)$, are equivalent if they restrict to the same section in $\mathcal{F}(U \cap V)$.

We can also have sheaves of algebraic structures such as groups, abelian groups or rings defined analogously except that $\mathcal{F}(U)$ are groups, abelian groups or rings, and restrictions are homomorphisms.

A morphism of sheaves $\phi : \mathcal{F} \to \mathcal{G}$ gives for an open set U a map (resp. homomorphism of whatever algebraic structures we are considering)

$$\phi(U) : \mathcal{F}(U) \to \mathcal{G}(U)$$

such that ϕ of a restriction of a section s is the restriction of $\phi(s)$.

A morphism of sheaves $\phi : \mathcal{F} \to \mathcal{G}$ induces, for every $x \in X$, a map (or homomorphism of whatever algebraic structures we have) $\phi_x : \mathcal{F}_x \to \mathcal{G}_x$.

If $f : X \to Y$ is a continuous map and $\mathcal{F}$ is a sheaf on X, we have a sheaf $f_* \mathcal{F}$ (sometimes called the *pushforward*) on Y where

$$f_* \mathcal{F}(U) = \mathcal{F}(f^{-1}(U))$$

for every open set $U \subseteq Y$.

1.2   Ringed Spaces and Locally Ringed Spaces

Recall that, unless otherwise specified, by a ring, we mean a commutative ring. A *ringed space* is a topological space X with a sheaf of rings $\mathcal{O}_X$ (called the *structure sheaf*). A *morphism of ringed spaces* $f : X \to Y$ is a continuous map together with a morphism of sheaves of rings

$$\phi : \mathcal{O}_Y \to f_* \mathcal{O}_X. \tag{1.2.1}$$

A *locally ringed space* is a ringed space where every stalk $\mathcal{O}_{X,x} = (\mathcal{O}_X)_x$ is a local ring. (Recall that a local ring is a ring which has a unique maximal ideal; a maximal ideal of a ring R is an ideal $m \neq R$ such that there exists no ideal I with $m \subsetneq I \subsetneq R$. Equivalently, an ideal m is maximal if and only if R/m is a field.)

A *morphism of locally ringed spaces* $f : X \to Y$ is a morphism of ringed spaces such that for every point $x \in X$,

$$\mathcal{O}_{Y, f(x)} \xrightarrow{\phi_{f(x)}} (f_* \mathcal{O}_X)_{f(x)} \longrightarrow \mathcal{O}_{X,x}$$

is a morphism of local rings (where the second map is defined in the obvious way). Here by a *morphism of local rings* $\phi : R \to S$ where the maximal ideal of R is m and the maximal ideal of S is n, we mean a homomorphism of rings such that $\phi^{-1}(n) = m$ or, equivalently, $\phi(m) \subseteq n$ (see Exercise 2); an example of a homomorphism between local rings which is *not* a morphism of local rings is the inclusion $\mathbb{Z}_{(p)} \subset \mathbb{Q}$, where $\mathbb{Z}_{(p)}$ is $\mathbb{Z}$ localized at the prime ideal (p) for p prime, or, in other words, the set of rational numbers whose denominators are not divisible by p.)

Note that if X is a locally ringed space and $U \subseteq X$ is an open set, then U with $\mathcal{O}_U$ equal to the restriction $\mathcal{O}_X|_U$ of the sheaf $\mathcal{O}_X$ to U (given by $\mathcal{O}_X|_U(V) = \mathcal{O}_X(V)$ for $V \subseteq U$ open) is a locally ringed space. Let us call it the *restriction* of the locally ringed space X to U.

1.3   Schemes

An *affine scheme* is a locally ringed space of the form

$$Spec(R) = \{p \mid p \text{ is a prime ideal in } R\}.$$

The topology is the *Zariski topology* where closed sets are of the form

$$Z_I = \{p \in Spec(R) \mid I \subseteq p\}$$

for an ideal I.

As in Chap. 1, we have $Z_{I \cdot J} = Z_I \cup Z_J$ and

$$Z_{\sum I_i} = \bigcap_i Z_{I_i},$$

thus showing that we have indeed defined a topology. Denote by

$$U_I = Spec(R) \smallsetminus Z_I$$

the complementary open set.

A *distinguished open set* is a set of the form $U_{(r)}$ for $r \in R$ (i.e. U_I where I is a principal ideal). Every open set is a union of distinguished open sets. The structure sheaf $\mathcal{O}_{Spec(R)}$

is uniquely determined by its sections on distinguished open sets by gluing. We have

$$\mathcal{O}_{Spec(R)}(U_{(r)}) = r^{-1}R \tag{1.3.1}$$

(recall that $r^{-1}R$ is the set of equivalence classes of fractions s/r^n, $s \in R$ by the equivalence relation $s/r^n \sim t/r^m$ when $r^{n+k}t = r^{m+k}s$ for some $k = 0, 1, 2, \dots$). It is possible to use (1.3.1) as a definition. Some consistency checks are needed. We prefer, however, a definition using actual functions; using our definition, we will prove (1.3.1) in Sect. 2.2 below (see Lemma 2.2.2).

More concretely, recall that for a commutative ring R and a prime ideal p, the *localization* R_p of R at p is the set of equivalence classes

$$\{r/s \mid r, s \in R, s \notin p\}/\sim$$

where

$$\frac{r_1}{s_1} \sim \frac{r_2}{s_2}$$

when

$$r_1 s_2 u = r_2 s_1 u \ \text{ for some } u \notin p.$$

Then for any Zariski open set $U \subseteq Spec(R)$, $\mathcal{O}_{Spec(R)}(U)$ is defined to be to the ring of all functions

$$f : U \to \coprod_{p \in Spec(R)} R_p \tag{1.3.2}$$

such for every $p \in U$, $f(p) \in R_p$, there exists an open subset $V \subseteq U$ such that $p \in V$, and there exist $g, h \in R$ where for every $q \in V$, we have $h \notin q$ and

$$f(q) = \frac{g}{h} \in R_q. \tag{1.3.3}$$

(Note: in this notation, we use the same symbol for an element of R and its image in R_q.) Using this definition, it is obvious that $\mathcal{O}_{Spec(R)}$ is a sheaf.

A *scheme* is a locally ringed space X where for every $x \in X$, there exists an open set $U \ni x$ such that the restriction of the locally ringed space X to U is isomorphic to an affine scheme as a locally ringed space. Morphisms of schemes are defined to be morphisms of locally ringed spaces $f : X \to Y$ where X and Y are schemes. If we have any category C and a class of objects $S \subseteq C$, the category with objects S and all C-morphisms between them is called the *full subcategory of* C *on* S. The category of schemes is a full subcategory of the category of locally ringed spaces. The category of affine schemes is, in turn, a full subcategory of the category of schemes on affine schemes. It turns out (Theorem 2.2.3 below) that the category of affine schemes is equivalent to the opposite category of the category of commutative rings and homomorphisms. The equivalence is given by the functors $Spec(R)$ and $\mathcal{O}_X(X)$. (Note: for a sheaf $\mathcal{F}$ on X, elements of $\mathcal{F}(X)$ are also referred to as *global sections*.)

Recall from Chap. 1 the general concept of *localization*. In the above discussion, we encountered two cases of localization of a commutative ring R: $r^{-1}R$ and R_p for a prime ideal p. The most general localization of a ring R is at a *multiplicative set*, which means a subset $S \subset R$ which is closed under multiplication ($x, y \in S \Rightarrow x \cdot y \in S$), contains 1 and does not contain 0. The *localization $S^{-1}R$ of the ring R at the set S* is the set of equivalence classes of the set

$$\{\frac{r}{s} \mid r \in R,\ s \in S\}$$

with respect to the equivalence relation $\sim$ where

$$\frac{r_1}{s_1} \sim \frac{r_2}{s_2}$$

when

$$r_1 s_2 u = r_2 s_1 u$$

for some $u \in S$.

Localization has the following universal property: We have a canonical homomorphism of rings $i : R \to S^{-1}R$ which sends every element of S to a unit. For a homomorphism of rings $h : R \to R'$, there exists a homomorphism of rings $\overline{h} : S^{-1}R \to R'$ with $h = \overline{h} \circ i$ if and only if h sends every element of S to a unit. Moreover, if $\overline{h}$ exists, it is unique.

Note that $r^{-1}R$ is the localization of R at the multiplicative set $S = \{r^n \mid n = 0, 1, 2, \ldots\}$, and, for a prime ideal p, R_p is localization at the multiplicative set $S = R \setminus p$ (which is multiplicative because the ideal p is prime). More generally, for any set T of prime ideals in a ring R, the complement

$$R \setminus \bigcup_{p \in T} p$$

is a multiplicative set.

1.4 Category Theory Revisited: Adjoints, Limits and Colimits, Universality

First note that for every category C, we have a functor

$$C(?, ?) : C^{Op} \times C \to Sets. \tag{1.4.1}$$

Recall that *Sets* is the category of sets and mappings. In (1.4.1), ?, as often, denotes a variable which we do not want to name. For $x, y \in Obj(C)$, $C(x, y)$ is the set of morphisms $\alpha : x \to y$ in C. On morphisms, for $f : x' \to x$ and $g : y \to y'$ in C, considering f as a morphism from x to x' in C^{Op}, we let

$$C(f, g)(\alpha) = g \circ \alpha \circ f.$$

Notice the contravariance in the first variable. We encounter this first in high school algebra: to transform a graph of a function by transforming the x coordinate, we need to apply the inverse transformation to the independent variable x. A functor $C^{Op} \to D$ to another category D is called a *contravariant functor* from C to D. A functor $C \to D$ is then sometimes called a *covariant functor* from C to D.

The functor (1.4.1), sometimes also referred to as the Hom-functor, is important because it is used in the definition of adjoint functors. Let $F : C \to D$, $G : D \to C$ be functors. We say that F is *left adjoint* to G (equivalently, G is *right adjoint* to F) if there exists a natural bijection (i.e. isomorphism in *Sets*)

$$D(Fx, y) \cong C(x, Gy). \tag{1.4.2}$$

Any two left adjoint functors to the same functor are naturally isomorphic. Similarly for right adjoints. Adjoints can be characterized by a *universal property*: A functor $G : C \to D$ has a left adjoint if and only if for every object x of C, there exists an object $y_x \in Obj(D)$ and a morphism

$$\alpha_x : x \to G(y_x) \tag{1.4.3}$$

(called the *unit of adjunction*) such that for every other morphism $\beta : x \to G(z)$, there exists a unique morphism $\widetilde{\beta} : y_x \to z$ in D such that the following diagram commutes:

$$
\begin{array}{ccc}
x & \xrightarrow{\ \alpha_x\ } & G(y_x) \\
& \beta \searrow & \Big\downarrow G(\widetilde{\beta}) \\
& & G(z)
\end{array}
\tag{1.4.4}
$$

It is easy to see that we can then define $F : C \to D$ by putting $F(x) = y_x$, and the universal property lets us uniquely define F on morphisms. The unit of adjunction then becomes a natural transformation. (And, of course, turning around the arrows, we can characterize a functor which has a right adjoint by a *co-unit of adjunction*.)

From this point of view, for example, the free group $F(S)$ on a set S is the left adjoint to the *forgetful functor* U from the category *Grp* of groups and homomorphisms to *Sets*:

the unit of adjunction is the inclusion of S as the set of free generators of $F(S)$, and (1.4.4) is the familiar universal property in this setting. Similarly, the free abelian group is the left adjoint to the forgetful functor from the category Ab of abelian groups to $Sets$.

Similarly as for groups or abelian groups, this actually works in any category of *universal algebras and homomorphisms*. A universal algebra has certain prescribed operations (such as $+, \cdot, (?)^{-1}, 1$, etc.) and relations among those operations which are expressed as equations valid for all choices of input elements.

Examples of universal algebras include rings and abelian groups, but not fields, local rings or integral domains: those structures include axioms which cannot be expressed purely as equations on elements. For example, a field is a non-zero commutative ring where every non-zero element has a multiplicative inverse: the condition of being non-zero cannot be expressed by an equation.

For any category $\mathcal{A}$ of universal algebras and homomorphisms (i.e. mappings preserving the operations), the forgetful functor

$$U : \mathcal{A} \rightarrow Sets$$

always has a left adjoint, the *free $\mathcal{A}$-algebra on a set*. However, again, this does not apply to fields, since they are not universal algebras: there is no such thing as a free field on a set.

There are many other interesting examples of adjoint functors. For example, the inclusion of the category of affine schemes into the category of schemes is right adjoint to the functor $Spec(\mathcal{O}_X(X))$ (where X runs through all schemes). We will prove this in Theorem 2.2.3 below.

One important case of adjoint functors is the definition of limits and colimits. It is good to formalize diagrams in a category C as functors $F : I \rightarrow C$ where I is a small category (this means that $Obj(I)$ and $Mor(I)$ are sets, not just classes; recall that sets are those classes which are elements of other classes). For example, a diagram in a category C in the shape

$$
\begin{array}{ccc}
 & x & \\
 & \downarrow v & \\
y & \longrightarrow & z
\end{array}
$$

$$(1.4.5)$$

is formally a functor $F : I \rightarrow C$ where I is the category

$$(1.4.6)$$

and F takes the dots to x, y, z and the arrows to the arrows of (1.4.5). The loops in (1.4.6) are identities, and go to identities in C by F (every category has an identity morphism on every object). The *diagram category* C^I has objects which are diagrams $F : I \to C$; the morphisms are natural transformations.

Now for an object $x \in Obj(C)$, there is a *constant diagram*

$$Const_x : I \to C$$

which sends every object of I to x and every morphism of I to Id_x. Obviously,

$$Const : C \to C^I \qquad (1.4.7)$$

is a functor where a morphism $f : x \to y$ in C is sent to the natural transformation

$$Const_f : Const_x \to Const_y$$

which is f on every object of I.

The right adjoint of (1.4.7), if one exists, is called the (categorical) *limit*, (also called the *inverse limit* or *projective limit*) and denoted by either one of the symbols

$$\lim, \ \varprojlim .$$

The unit of adjunction is sometimes called the *diagonal*. The left adjoint of (1.4.7), if one exists, is called the (categorical) *colimit*, (also called *direct limit* or *inductive limit*) and denoted by either one of the symbols

$$\text{colim}, \ \varinjlim .$$

The counit of this adjunction is sometimes called the *codiagonal*.

So for example the limit u of the diagram (1.4.5) is commutative diagram

$$
\begin{array}{ccc}
u & \longrightarrow & x \\
\downarrow & & \downarrow \\
y & \longrightarrow & z
\end{array}
\qquad (1.4.8)
$$

which is *universal* in the sense that for any commutative diagram

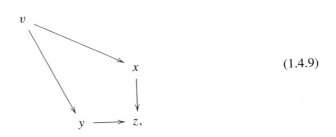

(1.4.9)

a unique dotted arrow combines (1.4.8) and (1.4.9) into a commutative diagram:

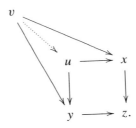

This particular limit is called a *pullback*. A limit of a diagram with no arrows (except the identities) is called the *product* and denoted by $\prod$. A limit of a diagram of the form

$$x \rightrightarrows y \qquad\qquad (1.4.10)$$

is called an *equalizer*.

Dually, the colimit of a diagram with no arrows except the identities is called a *coproduct*, denoted by $\coprod$. A colimit of the diagram (1.4.10) is called the *coequalizer*. The colimit of the diagram (1.4.5) is actually just z, but the colimit of the dual diagram (with arrows turned around) is called the *pushout*.

On a related note, a *monomorphism* in a category is a morphism $f : X \to Y$ such that for any two morphisms $g, h : Z \to X$, $f \circ g = f \circ h$ implies $g = h$. Symmetrically, f is an *epimorphism* if for any $g, h : Y \to Z$, $g \circ f = h \circ f$ implies $g = h$. Monomorphisms and epimorphisms are designed to model injective and surjective maps in a category, but the analogy is not always perfect (for example $\mathbb{Z} \subset \mathbb{Q}$ is an epimorphism of commutative rings—see Exercise 3).

It is possible to characterize monomorphisms by means of equalizers (when equalizers exist): it is easy to see that $f : x \to y$ is a monomorphism if and only if for any two morphisms $g, h : z \to x$, the equalizer of g and h is the same as the equalizer of $f \circ g$ and $f \circ h$. There is a symmetrical characterization of epimorphisms in terms of coequalizers. Thus, functors which preserve equalizers preserve monomorphisms and functors which preserve coequalizers preserve epimorphisms.

We should mention that a colimit or limit of an individual diagram can be defined even if the category in question does not have all colimits (resp. limits), using the universal property (1.4.3) or its dual. For example, the category of schemes does not have all limits (or all colimits), but there are some important cases of limits and colimits of schemes. Categories of universal algebras and homomorphisms always have limits and colimits. The product in *Sets* is the Cartesian product (i.e., say, the product of a set X and a set Y is the set of all pairs (x, y) where $x \in X$ and $y \in Y$), the coproduct in *Sets* is the disjoint union.

There is an important connection between limits, colimits and adjoints: Left adjoints preserve colimits and right adjoints preserve limits. This means that, for example, if $G : C \to D$ is a right adjoint and in the diagram (1.4.8), u is the limit (pullback) of (1.4.5), then $G(u)$ is the limit of G applied to (1.4.5), i.e. of

$$
\begin{array}{ccc}
 & & G(x) \\
 & & \downarrow \\
G(y) & \longrightarrow & G(z)
\end{array}
\qquad (1.4.11)
$$

(One can easily see this using the adjunction property and the universal property of the limit.)

As previously mentioned, the forgetful functor

$$U : \mathcal{A} \to Sets$$

where $\mathcal{A}$ is a category of universal algebras and homomorphisms always is a right adjoint, and hence preserves limits, in particular products. This is why products of any kind of universal algebras (such as groups, abelian groups or rings) are just Cartesian products on the underlying sets. This is not true for coproducts (as, for example, a disjoint union of groups would not be a group in an obvious way).

Since the category of affine schemes is equivalent to the category opposite to the category of commutative rings (as we will show in Theorem 2.2.3), which is a category of universal algebras, it has all limits and colimits. Since, further, the inclusion functor from the category of affine schemes to the category of schemes is a right adjoint, it preserves limits. This means that a limit of any diagram of affine schemes always exists in the category of schemes, and is an affine scheme.

It is also worth considering the limit (if one exists) of the empty diagram (i.e. the product of an empty set of factors). This is called the *terminal object*. Explicitly, there is a unique morphism from any object to the terminal object. Dually, the *initial object*, if one exists, is the colimit of the empty diagram, or the coproduct over an empty set. There is a unique morphism from the initial object to any given object.

Universality is even more general than adjunction. For a category C and a functor

$$F : C \to Sets,$$

a *universal element* of F consists of an object $u \in Obj(C)$ and an element $a \in F(u)$ such that for every object $v \in Obj(C)$ and every element $x \in F(v)$, there exists a unique morphism $f : u \to v$ in C such that

$$F(f)(a) = x.$$

This can be used, for example, to characterize the *tensor product*. Let R be a commutative ring, let M, N be R-modules. Let, for an R-module P, $F(P)$ consist of all bilinear maps

$$f : M \times N \to P, \tag{1.4.12}$$

which means that for every $m \in M$,

$$f_m : N \to P$$

given by $f_m(n) = f(m, n)$ is a homomorphism of R-modules, and for every $n \in N$,

$$_n f : M \to P$$

given by

$$_n f(m) = f(m, n)$$

is a homomorphism of R-modules. Then F becomes a functor from R-modules to Sets by noticing that composing a homomorphism of modules with a bilinear map produces a bilinear map. The universal element of this functor F consist of the *tensor product* $M \otimes_R N$ and the universal bilinear map

$$M \times N \to M \otimes_R N$$

which, by definition, sends

$$(m, n) \mapsto m \otimes n.$$

1.5   Abelian Categories: Abelian Sheaves

As we already observed, there is a unique morphism from the initial to the terminal object (if they exist). A category is called *based*, or alternately a *category with zero*, if the initial and terminal object exist, and the unique morphism o from the initial to the terminal object is an isomorphism. The initial object (which is then also the terminal object) is then often denoted by 0. In a based category, a morphism $f : X \to Y$ is called *zero* if it factors through 0:

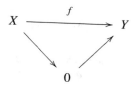

It is easy to see that there is precisely one zero morphism between any two objects of a based category. In a based category, the equalizer of a morphism f with the zero morphism is called the *kernel of* f and denoted by $Ker(f)$, and the coequalizer of a morphism f with the zero morphism is called the *cokernel of* f and denoted by $Coker(f)$.

For example, in the category of abelian groups (or more generally R-modules over, say, a commutative ring R), we have for $f : A \to B$,

$$Ker(f) = \{x \in A \mid f(x) = 0\},$$

$$Coker(f) = B/Im(f)$$

where

$$Im(f) = \{f(x) \mid x \in A\}$$

is the *image* of f.

In fact, more generally, a category is called *abelian* if it is a based category, has finite limits and colimits (i.e. limits and colimits of diagrams on finitely many objects and morphisms) and if every monomorphism is a kernel and every epimorphism is a cokernel. The last condition is equivalent to requiring that if a morphism $f : x \to y$ satisfies $Ker(f) = 0$, then f is the kernel of the cokernel morphism $g : y \to Coker(f)$, and dually, if a morphism $g : y \to z$ satisfies $Coker(g) = 0$, then g is the cokernel of the kernel morphism $f : Ker(g) \to y$. (The advantage of rephrasing the condition this way is that it refers only to finite limits and colimits.)

A category C is called *preadditive* if the set of C-morphisms $X \to Y$ is given the structure of an abelian group, and composition is bilinear (i.e. becomes a homomorphism of abelian groups upon fixing either coordinate). For a pre-additive category C, we denote the abelian group of morphisms $X \to Y$ by $Hom_C(X, Y)$ (to distinguish it from the *set* $C(X, Y)$). For pre-additive categories $\mathcal{A}, \mathcal{B}$, an *additive functor* $F : \mathcal{A} \to \mathcal{B}$ is a functor which on morphisms gives a homomorphism of abelian groups

$$Hom_{\mathcal{A}}(X, Y) \to Hom_{\mathcal{B}}(F(X), F(Y)).$$

A pre-additive category is called *additive* if it has finite products and coproducts and the canonical morphism from the coproduct to the product

$$A \amalg B \to A \sqcap B$$

is an isomorphism. For this reason, the product (or coproduct) of two objects is sometimes called the *biproduct* and denoted by $\oplus$. The biproduct is characterized up to canonical isomorphism as an object C together with morphisms

$$A \xrightarrow{\ i\ } C \xleftarrow{\ j\ } B,$$

$$A \xleftarrow{\ p\ } C \xrightarrow{\ q\ } B$$

such that $p \circ i = Id_A$, $q \circ j = Id_B$, $i \circ p + j \circ q = Id_C$. This is also true in a pre-additive category provided that the object C exists (i.e. when C exists, it is then automatically a biproduct—see Exercise 4). For this reason, an additive functor between additive categories automatically preserves biproducts.

An abelian category is automatically additive. The proof of this fact is a part of *homological algebra*. We will discuss it in the context of cohomology in Sect. 2.2 of Chap. 5. However, we use additivity sooner, as abelian categories are an important tool in the theory of schemes. For the time being, the reader can treat it as an additional axiom.

An *abelian subcategory* of an abelian category is a full subcategory with finite limits and colimits whose inclusion preserves finite limits and colimits. (It is then automatically an abelian category.) Note: a functor which preserves finite limits and colimits is also called *exact*. A functor which preserves finite limits is called *left exact* and a functor which preserves finite colimits is called *right exact*.

The category Ab of abelian groups (or, more generally, the category $R\text{-}Mod$ of R-modules for a fixed ring R), is an abelian category. Thus, again, a product and coproduct of finitely many objects $A_1, \ldots, A_n$ are both the Cartesian product, and n-tuples are added one coordinate at a time:

$$(a_1, \ldots, a_n) + (b_1, \ldots, b_n) = (a_1 + b_1, \ldots, a_n + b_n).$$

Because of this, the product/coproduct of finitely many abelian groups (or R-modules) is denoted by

$$A_1 \oplus \cdots \oplus A_n$$

and is again called the *biproduct*.

While the product of infinitely many abelian groups (or R-modules) A_i, $i \in I$, is still the Cartesian product, the coproduct, however, is given by the subset of systems $(a_i)_{i \in I}$ where for all but at most finitely many i's, $a_i = 0$. Because of that, the general product is denoted by

$$\prod_{i \in I} A_i$$

and called the *direct product* while the general coproduct is denoted by

$$\sum_{i \in I} A_i \text{ or } \bigoplus_{i \in I} A_i.$$

and called the *direct sum*.

Abelian sheaves (i.e. sheaves of abelian groups) on a topological space X also form an abelian category. In fact, the stalk

$$\mathcal{F} \mapsto \mathcal{F}_x$$

at a point x is left adjoint to pushforward via the inclusion

$$\{x\} \to X.$$

The stalk is also an exact functor. A monomorphism of abelian sheaves is a morphism which is a monomorphism on sections, which is equivalent to inducing monomorphisms on stalks. An epimorphism of sheaves is a morphism which induces an epimorphism on stalks (see Exercise 1).

We shall further elaborate on these points in Sect. 1 of Chap. 4 below.

2   Beginning Properties and Examples of Schemes

2.1   Connected, Irreducible, Reduced and Integral Schemes

Recall that a topological space is *connected* if it is not a union of two non-empty disjoint open subsets. A scheme is connected if it is connected as a topological space.

Recall also that a topological space is *irreducible* if it is not a union of finitely many closed subsets other than itself. A scheme is irreducible if it is irreducible as a topological space.

A scheme X (or more generally a ringed space) is *reduced* if all the rings $\mathcal{O}_X(U)$ for $U \subseteq X$ open are reduced: A ring is *reduced* if it has no nilpotent element. An element x of a ring is *nilpotent* if $x \neq 0$ and $x^n = 0$ for some $n \in \{2, 3, \dots\}$.

A scheme is called *integral* if it is reduced and irreducible.

2.2   Properties of Affine Schemes

It is now time to prove (1.3.1). Note that this statement is, in a way, analogous to the statements of Sect. 1.3.2 of Chap. 1. There, the main tool was the Nullstellensatz. But we do not have a Nullstellensatz for general rings. This is, in fact, the main reason why in schemes, *points* are defined as prime ideals and not just maximal ideals! This is explained by the following result:

2.2.1 Lemma *Let I be an ideal in a commutative ring R. Then the radical $\sqrt{I}$ is equal to the intersection of all prime ideals containing I.*

Proof Apply Lemma 4.3.3 of Chap. 1 to the ring R/I.   □

2.2.2 Lemma *In an affine scheme $Spec(R)$, the equality (1.3.1) holds.*

Proof By definition of the structure sheaf of $Spec(R)$ (i.e. the sheaf of regular functions), we have a homomorphism of rings

$$r^{-1}R \to \mathcal{O}_{Spec(R)}(U_{(r)}). \tag{2.2.1}$$

We first prove that (2.2.1) is injective. This says that for $0 \neq f \in r^{-1}R$, there exists a prime ideal $p \in U_{(r)}$ such that $f \neq 0 \in R_p$. To see this, let A be the annihilator of f in $r^{-1}R$, i.e. the ideal of all $s \in R$ such that $sf = 0$. Then $A \neq r^{-1}R$, and hence there exists a maximal ideal m in $r^{-1}R$ with $f \notin m$. However, then $m \cap R$ is a prime ideal in R, proving what we need.

Now we prove that (2.2.1) is onto. To this end, note that without loss of generality, $r = 1$ (since we may replace R by $r^{-1}R$). Therefore, we must show that if open sets U_i, $i \in I$, cover $Spec(R)$, and there are elements $g_i, h_i \in R$ with $h_i \neq 0$ on U_i (meaning in R_p for all $p \in U_i$) such that

$$g_i/h_i = g_j/h_j \text{ on } U_i \cap U_j \tag{2.2.2}$$

then there is an $f \in R$ such that f is equal to g_i/h_i on every U_i. First note that the sets U_i may as well be distinguished, and $U_i = Spec(R) \setminus Z(h_i)$. Next, we note that by Lemma 2.2.1, there are some elements $\gamma_i \in R$ whose powers are divisible by h_i and whose sum is $1 \in R$. Thus, 1 is in fact a sum of finitely many such elements, and we may assume I is finite, i.e., say, $I = \{1, \dots, n\}$.

Next, we note that $g_i h_j - g_j h_i$ is annihilated by some power of $h_i h_j$ by the injectivity part of our statement, and hence, by replacing each h_i by its power, we may assume

$$g_i h_j = g_j h_i \text{ for all } i, j = 1, \dots, n. \tag{2.2.3}$$

Next, by taking powers, we may assume $\gamma_i = h_i$, i.e. there exist $a_i \in R$ such that

$$a_1 h_1 + \cdots + a_n h_n = 1.$$

Then we claim that the function

$$f = a_1 g_1 + \cdots + a_n g_n$$

restricts to g_i / h_i on U_i. In effect, without loss of generality, $i = 1$. Then compute by (2.2.3),

$$f h_1 = a_1 g_1 h_1 + a_2 g_2 h_1 + \cdots + a_n g_n h_1 = a_1 g_1 h_1 + a_2 g_1 h_2 + \cdots + a_n g_1 h_n = g_1.$$

□

2.2.3 Theorem　*The mapping*

$$R \mapsto Spec(R) \tag{2.2.4}$$

gives rise to a functor which defines an equivalence between the opposite of the category of commutative rings and the category of affine schemes. Moreover, for a scheme X, the functor

$$X \mapsto Spec(\mathcal{O}_X(X)) \tag{2.2.5}$$

is a left adjoint to the inclusion functor from the category of affine schemes to the category of schemes.

Proof　By Lemma 2.2.2, an inverse functor to (2.2.4) is defined by

$$X \mapsto \mathcal{O}_X(X) \tag{2.2.6}$$

for an affine scheme X. The two compositions of the functors thus defined are isomorphic to the identity in an obvious way. However, we must prove that the isomorphisms are natural. This follows on the side of commutative rings, but on the side of schemes, we must prove that every morphism of schemes

$$f : Spec(R) \to Spec(S)$$

is induced by the homomorphism of rings $h : S \to R$ it specifies. The key point is to show that for $p \in Spec(R)$,

$$f(p) = h^{-1}(p). \tag{2.2.7}$$

To this end, we must remember that schemes form a full subcategory of locally ringed spaces. Therefore, from the sheaf data, we obtain a diagram of the form

$$
\begin{array}{ccc}
S & \xrightarrow{\ h\ } & R \\
\downarrow & & \downarrow \\
S_{f(p)} & \dashrightarrow & R_p
\end{array}
$$

where the vertical arrows are localizations. By universality of localization, the existence of such a diagram implies $f(p) \supseteq h^{-1}(p)$. The requirement that the dotted arrow be a morphism of local rings requires that this be an equality, thus proving (2.2.7). After this, the fact that f coincides with the morphism induced by h on sheaf data follows from Lemma 2.2.2.

To prove the adjunction, we must prove that for a general scheme X, specifying a morphism of schemes

$$X \mapsto Spec(R) \tag{2.2.8}$$

is naturally the same thing as specifying a homomorphism of rings

$$R \to \mathcal{O}_X(X). \tag{2.2.9}$$

It is obvious by definition that (2.2.8) specifies (2.2.9). In the other direction, suppose we are given (2.2.9) and an affine open set $U \subseteq X$. Then we obtain by composition a homomorphism of rings

$$R \to \mathcal{O}_X(U),$$

and hence, by what we already proved, a morphism of schemes

$$Spec(U) \to Spec(R).$$

It is easy to see that these morphisms glue to give a morphism (2.2.8), and that these constructions are inverse to each other. (In fact, we shall study this principle in more detail in Sect. 2.5.) □

2.3   Open and Closed Subschemes

If $\mathcal{F}$ is a sheaf on a topological space X and $j : U \subseteq X$ is the inclusion of an open set, then the *restriction* $\mathcal{F}|_U$ of the sheaf $\mathcal{F}$ to U is the sheaf on U defined by taking only the values of $\mathcal{F}$ on open subsets of U. It is worth noting that j_* is the right adjoint to restriction to an open subset, but the restriction also has a left adjoint $j_!$. If $\mathcal{S}$ is a sheaf of rings (or abelian groups) on U, and $V \subseteq X$ is an open subset, then

$$j_!(\mathcal{S})(V) = \begin{cases} \mathcal{S}(V) & \text{if } V \subseteq U \\ 0 & \text{otherwise.} \end{cases}$$

The functor $j_!$ is called the *extension by* 0.

If X is a scheme and $j : U \subseteq X$ is open, then U with the restriction of the structure sheaf is called an *open subscheme* of X.

A *closed immersion* is a morphism of schemes $f : Z \to X$ which gives a homeomorphism (see Sect. 2.2.1 of Chap. 1) from Z onto a closed subset of X with the induced topology (this means that the open sets in Z are of the form $U \cap Z$ where U is an open set in X), such that the map of sheafs

$$\phi : \mathcal{O}_X \to f_*\mathcal{O}_Z$$

is a onto (which means it is onto on stalks).

Two closed immersions $f : Z \to X$, $f' : Z' \to X$ are called equivalent if there exists an isomorphism of schemes which makes the following diagram commute:

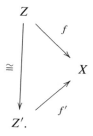

An equivalence class of closed immersions $f : Z \to X$ is called a *closed subscheme of* X.

2.3.1 Theorem *A closed subscheme Z of an affine scheme $Spec(R)$ is always of the form $Spec(R/I)$ for some ideal I in R.*

Before proving the Theorem, let us state the following

2.3.2 Lemma *Suppose an affine scheme Spec(S) is a closed subscheme of an affine scheme Spec(R). Then the corresponding homomorphism of rings $R \to S$ is onto.*

Proof A morphism ϕ of sheaves of abelian groups (i.e. abelian sheaves) is by definition onto if and only if it is onto on stalks (see Exercise 1). But this implies that for every section s there is an open cover U_i such that the restriction of s to each U_i is in the image of ϕ. This implies that under our assumptions, for every $s \in S$, there are finitely many $f_1, \ldots f_n \in R$ such that the ideal in R generated by $f_1, \ldots, f_n$ is R, i.e. there exist elements $a_i \in R$ with

$$a_1 f_1 + \cdots + a_n f_n = 1,$$

and if we denote by $\psi : R \to S$ the homomorphism of rings realizing the closed immersion $Spec(S) \subseteq Spec(R)$, then $s \in f_i^{-1}s = \psi(r_i)$ for some $r_i \in f_i^{-1}R$. By replacing f_i by its power, if necessary, we may assume without loss of generality that then

$$f_i s = \psi(r_i)$$

for some $r_i \in R$. Then

$$s = \psi(a_1 r_1 + \cdots + a_n r_n).$$

$\square$

Proof of Theorem 2.3.1. Z is covered by open affine subschemes $V_i = Spec(S_i)$. This means that $V_i = U_i \cap Z$ where V_i are open subsets of $Spec(R)$. Every V_i is covered by distinguished open subsets $U_j = Spec(f_j^{-1}R)$, of which we may choose finitely many to cover V_i, as in the proof of Lemma 2.2.2. Take all the U_j's together with an open cover of $Spec(R) \setminus Z$ by distinguished open sets. Again, we can choose finitely many, as in the proof of Lemma 2.2.2. Therefore, without loss of generality, we have a finite cover $U_1, \ldots, U_n$ of $Spec(R)$ such that every intersection $U_j \cap Z$ is an affine open set in Z. (Note that if $U_j \cap Z \subseteq V_i$, then $U_j \cap Z = Spec(f_j^{-1}S_i)$.)

Now this means that there exist finitely many elements $f_j \in \mathcal{O}_Z(Z)$ such that the sets $U_j \cap Z = Z \setminus Z(f_j)$ are affine and the ideal generated by the f_j's in $\mathcal{O}_Z(Z)$ is $\mathcal{O}_Z(Z)$. This implies that Z is affine (one shows that the unit of the adjunction morphism $Z \to Spec(\mathcal{O}_Z(Z))$ is an isomorphism). By Lemma 2.3.2, then, $\mathcal{O}_Z(Z)$ is a factor ring of R.

$\square$

2.4   Limits of Schemes

2.4.1 Theorem *A diagram of schemes*

$$
\begin{array}{ccc}
 & & X \\
 & & \downarrow {\scriptstyle f} \\
Y & \xrightarrow{\ g\ } & Z
\end{array}
\qquad\qquad (2.4.1)
$$

has a limit denoted by $X \times_Z Y$. (Recall that this limit is called a pullback.)

We will return to this statement at the end of the next subsection, after discussing some preliminary facts.

Recall that the limit of a diagram

$$
X \qquad\qquad Y
$$

with no morphisms is called the *product*. In the category of sets, the product is the Cartesian product

$$
X \times Y = \{(x, y) \mid x \in X, y \in Y\}.
$$

The pullback of sets, i.e. the limit of a diagram of the form (2.4.1), is

$$
X \times_Z Y = \{(x, y) \mid x \in X, y \in Y, f(x) = g(y)\}.
$$

We have a "forgetful" functor

$$
U : Schemes \to Sets
$$

But the forgetful functor from schemes to sets does not preserve the pullback (or even the product)!

Recall that the category of affine schemes is equivalent to the opposite category of the category of commutative rings. Furthermore, the inclusion functor from affine schemes to schemes is right adjoint to the functor

$$
X \mapsto Spec(\mathcal{O}_X(X)).
$$

So the inclusion functor from affine schemes to schemes preserves limits, and limits of affine schemes are *Spec* of colimits of the corresponding commutative rings.

The coproduct of commutative rings R, S is $R \otimes S = R \otimes_{\mathbb{Z}} S$. (To see this, recall that the operations in a ring R can be characterized as homomorphisms of abelian groups $prod : R \otimes_{\mathbb{Z}} R \to R$, $unit : \mathbb{Z} \to R$ and that the associativity, commutativity and unit axioms can be characterized as commutative diagrams: Associativity is

$$
\begin{array}{ccc}
R \otimes R \otimes R & \xrightarrow{\ prod \otimes Id\ } & R \otimes R \\
{\scriptstyle Id \otimes prod}\Big\downarrow & & \Big\downarrow{\scriptstyle prod} \\
R \otimes R & \xrightarrow[\ prod\ ]{} & R,
\end{array}
$$

commutativity is

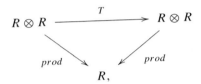

$$
\begin{array}{ccc}
R \otimes R & \xrightarrow{\ T\ } & R \otimes R \\
 & {\scriptstyle prod}\searrow \quad \swarrow {\scriptstyle prod} & \\
 & R, &
\end{array}
$$

where T switches the factors of the tensor product, the left unit is

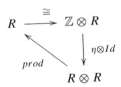

$$
\begin{array}{ccc}
R & \xrightarrow{\ \cong\ } & \mathbb{Z} \otimes R \\
 & {\scriptstyle prod}\nwarrow \quad \Big\downarrow {\scriptstyle \eta \otimes Id} & \\
 & R \otimes R &
\end{array}
$$

and the right unit is analogous. The unit on $R \otimes S$ is $1 \otimes 1$, or, in the above notation, $\eta \otimes \eta$, and the product is

$$
\begin{array}{c}
(R \otimes S) \otimes (R \otimes S) \\
\Big\downarrow {\scriptstyle Id \otimes T \otimes Id} \\
R \otimes R \otimes S \otimes S \\
\Big\downarrow {\scriptstyle prod \otimes prod} \\
R \otimes S.
\end{array}
$$

So for homomorphisms of commutative rings $R \to Q$, $S \to Q$, the bilinearity test gives a unique homomorphism of abelian groups $R \otimes S \to Q$, and one checks that it is a homomorphism of commutative rings.)

So in the category of affine schemes (and hence schemes)

$$Spec(R) \times Spec(S) = Spec(R \otimes S). \tag{2.4.2}$$

More generally, the pushout of a diagram of commutative rings of the form

$$
\begin{array}{ccc}
S & \longrightarrow & R_1 \\
\downarrow & & \\
R_2 & &
\end{array}
$$

is $R_1 \otimes_S R_2$. Therefore, in the category of schemes,

$$Spec(R_1) \times_{Spec(S)} Spec(R_2) = Spec(R_1 \otimes_S R_2).$$

Now contemplate the fact that this is completely different from the underlying sets.

For example, in (2.4.2), let $R = \mathbb{Z}[x]$, $S = \mathbb{Z}[y]$, so $R \otimes S = \mathbb{Z}[x, y]$. A prime ideal in $\mathbb{Z}[x, y]$ can be for example a principal ideal generated by an irreducible polynomial (because $\mathbb{Z}[x, y]$ is a unique factorization domain), for example, $(x + y)$. This has nothing to do with a pair of prime ideals in $\mathbb{Z}[x]$ and $\mathbb{Z}[y]$.

More strangely still, recall that the *characteristic* of a field is the smallest positive number p such that

$$\underbrace{1 + \cdots + 1}_{p \text{ times}} = 0;$$

it is always a prime number. If such a p does not exist, we say the characteristic is 0. If E, F are two fields of different positive characteristics, then $E \otimes F = 0$ because we are forcing $p \cdot 1 = q \cdot 1 = 0$ for two different primes p, q. This means that in the category of schemes,

$$Spec(E) \times Spec(F) = \emptyset.$$

(This is also true when the two fields have different characteristics, one of which is 0.)

In schemes, the terminal object is $Spec(\mathbb{Z})$. This is because this is true in affine schemes, since in commutative rings, the initial object is $\mathbb{Z}$; the unique morphisms of the open affine subschemes of a scheme into $Spec(\mathbb{Z})$ are compatible. Recall that as a set, $Spec(\mathbb{Z}) = \{(0), (2), (3), (5), (7), (11), \dots\}$.

2.5   Gluing of Schemes: Colimits

Let X be a scheme. Recall that an *open subscheme* of X is just an open subset $U \subseteq X$ where the structure sheaf on U is the restriction of the structure sheaf on X (i.e. the structure sheaf of X only applied to open subsets of U). This is always a scheme because an open subset of an affine scheme $Spec(R)$ is always a union of affine open subsets of the form $Spec(f^{-1}R)$ for some choices of $f \in R$.

When U and V are schemes and W is an open subscheme of U and W' is an open subscheme of V and we have an isomorphism of schemes $\phi : W \cong W'$ then we have a pushout X of the diagram

where the horizontal arrow is the inclusion and the vertical arrow is the inclusion composed with ϕ. As a set, X is the pushout of the sets U and V (disjoint union with the subsets W, W' identified). A subset of X is open if its intersection with U is open in U and its intersection with V is open in V. The ring of sections of the structure sheaf of X on subsets of U or V are the rings of sections of the corresponding structure sheaves, and we can use gluing property of the sheaf to figure out the ring of sections on any open set of X by intersecting it with $U, V, U \cap V$.

More generally, if U_i are schemes which have open subschemes U_{ij}, $i \in I$, and we have isomorphisms of schemes $\phi_{ij} : U_{ij} \to U_{ji}$ such that $\phi_{ii} = Id_{U_i}$ and $\phi_{ij} = \phi_{ji}^{-1}$ and furthermore there is an isomorphism of schemes

$$\widetilde{\phi_{ijk}} : U_{ij} \cap U_{ik} \to U_{ji} \cap U_{jk}$$

which is a restriction of ϕ_{ij} such that the diagrams

$$
\begin{array}{ccc}
U_{ij} \cap U_{ik} & \xrightarrow{\widetilde{\phi_{ijk}}} & U_{ji} \cap U_{jk} \\
{\scriptstyle \widetilde{\phi_{ikj}}} \downarrow & \swarrow {\scriptstyle \widetilde{\phi_{jki}}} & \\
U_{ki} \cap U_{kj} & &
\end{array}
$$

then there is a colimit X of the diagram of schemes

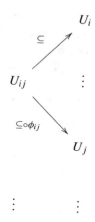

whose underlying topological space is the corresponding colimit of topological spaces etc. We say that X is *glued* from U_i, $i \in I$.

In particular, we have a disjoint union $\coprod X_i$ of any set of schemes X_i whose underlying topological space is the disjoint union of the topological spaces X_i.

One uses this construction to prove Theorem 2.4.1. One glues $X \times_Z Y$ from $Spec(R_1) \times_{Spec(S)} Spec(R_2)$ where $Spec(R_1)$, $Spec(R_2)$, $Spec(S)$ are open affine subsets of X, Y, Z where $Spec(R_i)$ map to $Spec(S)$ under the given maps. (See Exercise 21.)

One can show similarly that any finite diagram of schemes has a limit. (Actually, this is formal. Any category which has pullbacks and a terminal object has finite limits—see Exercise 22.) However, infinite diagrams of schemes may not have limits!

2.6 A Diagram of Schemes Which Does Not Have a Limit

Recall the one-dimensional projective space $\mathbb{P}^1_{\mathbb{C}}$. We will soon describe how to define projective schemes in general. For now, we can describe $\mathbb{P}^1_{\mathbb{C}}$ as a scheme as the colimit (pushout) of the diagram

$$
\begin{array}{ccc}
Spec(\mathbb{C}[x, x^{-1}]) & \xrightarrow{\;f\;} & Spec(\mathbb{C}[x]) \\
\downarrow g & & \\
Spec(\mathbb{C}[y]) & &
\end{array}
\qquad (2.6.1)
$$

where f is given by $x \mapsto x$, and g is given by $y \mapsto x^{-1}$. (Think of $\mathbb{P}^1_{\mathbb{C}}$ as a complex line with a point at ∞.)

2.6.1 Theorem *Consider the diagram D:*

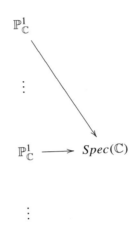

with countably many copies of $\mathbb{P}^1_{\mathbb{C}}$. *Then there does not exist*

$$\lim_{\leftarrow} D \tag{2.6.2}$$

in the category of schemes.

Proof Let $A = Spec(\mathbb{C}[x])$, $B = Spec(\mathbb{C}[y])$. Let D_1 be the diagram

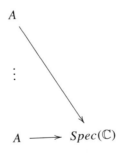

and let D_2 be the diagram

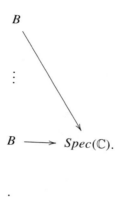

Then D_1 and D_2 are diagrams of affine schemes, so

$$\varprojlim D_1 = Spec(\mathbb{C}[x_1, x_2, \dots]), \quad \varprojlim D_2 = Spec(\mathbb{C}[y_1, y_2, \dots]). \qquad (2.6.3)$$

But recall that any non-empty Zariski open subset of $Spec(R)$ contains $Spec(f^{-1}R)$ for some $f \in R$ (not nilpotent), and any polynomial f can only contain finitely many of the variables x_i or y_i. This means that open subsets of the limits (2.6.3) contain open sets of the form

$$\widetilde{U} = U \times_{Spec(\mathbb{C})} A \times_{Spec(\mathbb{C})} \times A \times_{Spec(\mathbb{C})} \dots,$$

$$\widetilde{V} = V \times_{Spec(\mathbb{C})} B \times_{Spec(\mathbb{C})} \times B \times_{Spec(\mathbb{C})} \dots.$$

Now there are natural transformations from the diagrams D_i, $i = 1, 2$ to D given by the inclusions $A \subset \mathbb{P}^1_{\mathbb{C}}$, $B \subset \mathbb{P}^1_{\mathbb{C}}$, which induce morphisms

$$\phi : \varprojlim D_1 \to \varprojlim D,$$

$$\psi : \varprojlim D_2 \to \varprojlim D.$$

This implies that $\phi^{-1}(W)$ contains a set of the form $\widetilde{U}$ and $\psi^{-1}(W)$ contains a set of the form $\widetilde{V}$ as above. This implies that W must contain a set of the form

$$\{P_1\} \times \dots \times \{P_{n-1}\} \times \mathbb{P}^1_{\mathbb{C}} \times \{P_{n+1}\} \times \dots$$

for some closed points $P_i \in \mathbb{P}^1_{\mathbb{C}}$. But this is impossible because there is no non-trivial morphism of schemes from $\mathbb{P}^1_{\mathbb{C}}$ to an affine scheme over $Spec(\mathbb{C})$. (See, for example, Exercise 27.) □

2.7 *Proj* **Schemes**

In the beginning of the last example, we noted that we should define a scheme analogue of projective varieties. This can be done very generally. In algebraic geometry, a *graded ring* is a commutative ring R where, as abelian groups,

$$R = \bigoplus_{n \in \{0,1,2,\dots\}} R_n.$$

(Here recall that $\bigoplus$ is the coproduct of abelian groups, i.e. the colimit of a diagram with no arrows. It is like the Cartesian product $\prod$, except we only allow elements which have at most finitely many non-zero coordinates.) The elements of the abelian groups R_n are called *homogeneous elements of R of degree n*. It is required that if $z \in R_m, t \in R_n$, then $zt \in R_{m+n}$.

A typical example of a graded ring is the ring of polynomials

$$\mathbb{Z}[x_1, \dots, x_k].$$

Then homogeneous elements of degree n are simply homogeneous polynomials in the variables $x_1, \dots, x_k$ of degree n. A *homogeneous ideal I* of a graded ring is an ideal generated by homogeneous elements. Then R/I is also a graded ring where a coset $x + I$ for a homogeneous element $x \in R_n$ is homogeneous of degree n.

A graded ring always has the ideal

$$R_+ = \bigoplus_{n>0} R_n.$$

This is called the *augmentation ideal*. Now for a graded ring R, we define the scheme $Proj(R)$. As a set, $Proj(R)$ is the set of all homogeneous prime ideals in R which do not contain R_+. The *Zariski topology* on $Proj(R)$ has closed sets

$$Z(I) = \{p \in Proj(R) \mid I \subseteq p\}$$

for homogeneous ideals I. Let $U_I = Proj(R) \smallsetminus Z(I)$.

We have distinguished open sets

$$U_{(f)} = Proj(R) \smallsetminus Z(f)$$

for a homogeneous element $f \in R$ (i.e. $Proj(R) \setminus Z(I)$ where I is a principal ideal).
Now one can define a structure sheaf on $Proj(R)$ by requiring that

$$U_{(f)} = Spec((R_{(f)})_0). \tag{2.7.1}$$

where the subscript 0 on the right hand side means the subring of elements of degree 0.
Since the sets $U_{(f)}$ form a basis of topology, one can define sections on any other open set
by gluing. (See Exercise 24.)

More concretely, for any Zariski open set $U \subseteq Proj(R)$, $\mathcal{O}_{Spec(R)}(U)$ is canonically
isomorphic to the ring of all functions

$$f : U \rightarrow \coprod_{p \in Proj(R)} (R_p)_0 \tag{2.7.2}$$

such that $f(p) \in (R_p)_0$ and every point $p \in U$ is contained in an open subset $V \subseteq U$ and
there exists an n and elements $g, h \in R_n$ where for every $q \in V$, we have $h \notin q$ and

$$f_q = \frac{g}{h} \in (R_q)_0. \tag{2.7.3}$$

(Note: Again, in this notation, we use the same symbol for an element of R and its image
in R_q. In (2.7.3), f_q means the q'th coordinate of f in the product (2.7.2).)

Note that (2.7.1) also implies that this is a locally ringed space and that every point is
contained in an open subset isomorphic to an affine scheme. Thus, $Proj(R)$ is a scheme.
In particular, for $f \in R$ homogeneous, note that $U_{(f)} \subseteq Proj(R)$ is isomorphic to an
affine scheme.

2.8 The Affine and Projective Space Over a Scheme: Projective Schemes

When discussing varieties in Chap. 1, we used the affine space over $\mathbb{C}$. It can be realized
in schemes as

$$\mathbb{A}^n_{\mathbb{C}} = Spec(\mathbb{C}[x_1, \ldots, x_n]).$$

(Of course, considered as a scheme, $\mathbb{A}^n_{\mathbb{C}}$ has more points; in Chap. 1 we only considered
closed points.)

However, in schemes, the most universal affine space is over $\mathbb{Z}$:

$$\mathbb{A}^n_{\mathbb{Z}} = Spec(\mathbb{Z}[x_1, \ldots, x_n]).$$

We have

$$\mathbb{C}[x_1, \ldots, x_n] = \mathbb{Z}[x_1, \ldots, x_n] \otimes \mathbb{C},$$

so we have

$$\mathbb{A}_\mathbb{C}^n = \mathbb{A}_\mathbb{Z}^n \times Spec(\mathbb{C}).$$

For a general scheme S, we then put

$$\mathbb{A}_S^n = \mathbb{A}_\mathbb{Z}^n \times S.$$

This is important because the product comes with a projection

$$\mathbb{A}_S^n = \mathbb{A}_\mathbb{Z}^n \times S \to S.$$

(Of course, this is true of the categorical product in any category.) More generally, a scheme X with a morphism

$$X \to S$$

is called a *scheme over S*. So, $\mathbb{A}_S^n$ is an example of a scheme over S. Later, many concepts for schemes will actually be defined for morphisms of schemes, i.e. for schemes over another scheme. This is called the *relative* approach to concepts in algebraic geometry.

Another example of a scheme over S is the *projective space over S*. Again, we start with the projective space over $\mathbb{Z}$:

$$\mathbb{P}_\mathbb{Z}^n = Proj\left(\mathbb{Z}[x_0, \ldots, x_n]\right)$$

where $\mathbb{Z}[x_0, \ldots, x_n]$ is considered to be a graded ring where $x_0, \ldots, x_n$ have degree 1. Recall that the extra variable also came up when we treated the projective space over $\mathbb{C}$ as a variety. In fact, if we see that

$$U_{(x_0)} = (x_0^{-1}\mathbb{Z}[x_0, \ldots, x_n])_0$$

(the subscript means elements of degree 0). But note that

$$(x_0^{-1}\mathbb{Z}[x_0, \ldots, x_n])_0 \cong \mathbb{Z}[x_1, \ldots x_n]$$

where the isomorphism is

$$x_0 \mapsto 1, \ x_i \mapsto x_i \text{ for } i = 1, \ldots, n.$$

Thus, we see that in $\mathbb{P}^n_{\mathbb{Z}}$,

$$U_{(x_0)} \cong \mathbb{A}^n_{\mathbb{Z}}.$$

Now we can define the n-dimensional projective space over a scheme S by

$$\mathbb{P}^n_S = \mathbb{P}^n_{\mathbb{Z}} \times S.$$

It is customary to use the term *projective scheme over S* for a closed subscheme of $\mathbb{P}^n_S$. Note that from this point of view, the construction in Sect. 2.7 is simultaneously more, and also less, general: For an arbitrary scheme S, $\mathbb{P}^n_S$ certainly may not be isomorphic to $Proj(R)$ for a graded ring R. On the other hand, $Proj(R)$ may not, in general, be isomorphic to a closed subset of $\mathbb{P}^n_{Spec(R_0)}$.

It is true however that *if a graded ring R is generated, as a ring, by finitely many elements $r_0, \ldots, r_n \in R_1$, then $Proj(R)$ is a closed subscheme of*

$$\mathbb{P}^n_{Spec(R_0)} = Proj(R_0[x_0, \ldots, x_n]),$$

by sending

$$x_i \mapsto r_i, \; i = 0, \ldots, n.$$

Thus, in this case, $Proj(R)$ is a projective scheme over $Spec(R_0)$.

It is not difficult to see that all the theory of varieties as discussed in Chap. 1 can be completely described within the theory of schemes. It follows directly from Theorem 2.2.3 that the category of affine varieties as described in Chap. 1 is equivalent to the full subcategory of the category of schemes over $Spec(\mathbb{C})$ on closed integral subschemes of the affine spaces $\mathbb{A}^n_{Spec(\mathbb{C})}$. As already noted, the affine scheme actually has more points than the variety as considered in Chap. 1, because the points correspond to prime and not just maximal ideals. In fact, maximal ideals correspond exactly to *closed* points. However, in this case, all the information about the prime ideals can be recovered from the maximal ideals by the Nullstellensatz. Similar observations also extend to the projective, quasi-affine and quasi-projective varieties considered in Chap. 1. (See Exercise 27.)

3 Finiteness Properties of Schemes and Morphisms of Schemes

3.1 Quasicompactness

At this point, we must remind ourselves of the fact that in analysis, we usually do not study the Zariski topology. A typical example of the topology we study in analysis is the *analytic topology* on $\mathbb{R}^n$ (here by $\mathbb{R}^n$, we mean simply the set of all n-tuples of real numbers). On

$\mathbb{R}^n$, we have a notion of *distance*

$$\rho((x_1, \ldots, x_n), (y_1, \ldots, y_n)) = \sqrt{(x_1 - y_1)^2 + \cdots + (x_n - y_n)^2}.$$

A set U is open in the analytic topology when for every $z \in U$, there exists a $\delta > 0$ where for every $t \in \mathbb{R}^n$ with $\rho(z, t) < \delta$, we have $t \in U$. (This means, with every point z, the set U also contains all points near z.)

There is a notion of *compactness* of topological spaces, which is very important in the analytic topology:

3.1.1 Definition A topological space X is *compact* if whenever

$$X = \bigcup_{i \in I} U_i$$

where U_i are open subsets of X, there exists a finite subset $F \subseteq I$ such that

$$X = \bigcup_{i \in F} U_i.$$

In other words, we say that a topological space X is compact if every open cover of X has a finite sub-cover.

Compactness captures a very specific and important property of certain spaces. For example, in the analytic topology on $\mathbb{R}$, one can show that a closed interval $[a, b]$ is compact, but an open interval (a, b) or a half-open interval $[a, b)$ is not for $a < b$. Note that in the analytic topology, $\mathbb{R}$ itself is homeomorphic to $(0, 1)$, and hence is not compact, even though it is, of course, embedded as a closed subset of, say, $\mathbb{R}^2$ with the induced topology.

But now let us remember that the topology of schemes is different in that the Zariski topology has "far fewer open sets" (the quotation marks signify that this statement is imprecise, since even the sets of points are different in schemes from the points considered in classical analysis). The notion of compactness behaves differently in spaces with "very few open sets." Because of this, some literature sources call the property of Definition 3.1.1 *quasi-compactness*, reserving the term *compact* for spaces which have "enough open sets" in the sense that any two distinct points are separated by disjoint open sets. (This is called the *Hausdorff property*; it will be studied in more detail in Sect. 1.1 of Chap. 3.) The term *quasi-compactness* for the property of Definition 3.1.1 became established in algebraic geometry, where the topology of most schemes is not Hausdorff.

Another point is that the category of schemes also has "far fewer morphisms" than the category of spaces, so quasi-compactness of schemes is, in a sense, an even weaker property than quasi-compactness in the category of all (not necessarily Hausdorff) spaces. In schemes, the appropriate analog of quasi-compactness in the category of spaces is the concept of being *universally closed*, and the appropriate analog of compact Hausdorff spaces are *proper schemes*. For a discussion of these concepts, see Sect. 2 of Chap. 3. Quasi-compactness in the category of schemes is important, but has a different significance, as illustrated by the following fact.

3.1.2 Theorem *Every affine scheme X is quasi-compact.*

Proof Roughly, this is due to the fact that a linear combination of a set of generators is always a linear combination of a finite subset, which we already used in our proof of Lemma 2.2.2 and also Theorem 2.3.1. Let us recast the argument in terms of the quasi-compactness property we now introduced:

We have $X = Spec(R)$ for some commutative ring R. Suppose

$$Spec(R) = \bigcup_{i \in I} U_i \tag{3.1.1}$$

where U_i are Zariski open. But recall that every Zariski open set is a union of distinguished open sets, which are sets of the form

$$U_{(f)} = Spec(R) \smallsetminus Z(f)$$

where

$$Z(f) = \{p \in Spec(R) \mid f \in p\}.$$

Thus, we may assume that the sets U_i are distinguished,

$$U_i = U_{(f_i)}, \quad f_i \in R.$$

Then (3.1.1) is equivalent to

$$\emptyset = \bigcap_{i \in I} Z(f_i) = Z(f_i \mid i \in I).$$

This means that the ideal $(f_i \mid i \in I)$ is not contained in any prime ideal. But every ideal, other than R itself, is contained in a maximal ideal, which is prime. This means that

$$(f_i \mid i \in I) = R,$$

which is equivalent to saying that 1 is a linear combination of the f_i's. But only finitely many f_i's, say, for $i \in F \subseteq I$ finite, are involved in this linear combination. This means that

$$\emptyset = \bigcap_{i \in F} Z(f_i) = Z(f_i \mid i \in F)$$

and hence

$$Spec(R) = \bigcup_{i \in F} U_i,$$

$\square$

which is what we were trying to prove.

An example of a scheme which is not quasi-compact is

$$\coprod_{i \in I} Spec(\mathbb{Z})$$

for any I infinite.

3.2   Noetherian Schemes

Let us recall that a (commutative) ring R is called *Noetherian* if every ideal $I \subseteq R$ is finitely generated. This means that there exist elements $r_1, \ldots, r_k \in R$ such that

$$I = (r_1, \ldots, r_k).$$

(Recall that the right hand side consists of all *linear combinations* of the elements $r_1, \ldots, r_k$ with coefficients in R.) Hilbert's basis theorem tells us that if R is Noetherian, then the ring $R[x]$ of polynomials in one variable with coefficients in R is Noetherian (Theorem 4.2.1 of Chap. 1). Therefore, the polynomial rings

$$\mathbb{Z}[x_1, \ldots, x_m]$$

or

$$F[x_1, \ldots, x_m]$$

where F is a field are Noetherian. A polynomial ring over $\mathbb{Z}$ with infinitely many variables

$$\mathbb{Z}[x_1, x_2, x_3, \ldots]$$

is not Noetherian.

The point of the notion of Noetherian rings is that it is a finiteness property which is weaker than being finitely generated as a ring. Recall that a ring R is generated by elements $r_1, \ldots, r_k$ if R is equal to the set of all polynomials with coefficients in $\mathbb{Z}$ in the generators $r_1, \ldots r_k$. This does *not* mean that $R = \mathbb{Z}[r_1, \ldots, r_k]$, but it means that $R = \mathbb{Z}[r_1, \ldots, r_k]/I$ for some ideal I. Since the property of being Noetherian passes from a ring to its factor rings, a finitely generated ring is always Noetherian, but not vice versa: an example of a Noetherian ring which is not finitely generated is $\mathbb{Q}$, or $\mathbb{Z}_{(p)}$ (the set of all rational numbers whose denominators are not divisible by p).

3.2.1 Definition A scheme X is Noetherian if X is a union of finitely many affine open sets $V_1, \ldots, V_n$, $V_i = Spec(R_i)$, where all the rings $R_1, \ldots, R_n$ are Noetherian.

3.2.2 Lemma *Every Noetherian scheme X is quasi-compact.*

Proof Every affine scheme is quasicompact by Theorem 3.1.2. Let $X = V_1 \cup \cdots \cup V_n$ where $V_1, \ldots, V_n$ are affine. Now let

$$X = \bigcup_{i \in I} U_i$$

for U_i open. Then $U_i \cap V_j$ is open in V_j and

$$V_1 = \bigcup_{i \in I} (U_i \cap V_1),$$
$$\vdots$$
$$V_n = \bigcup_{i \in I} (U_i \cap V_n).$$

Since V_j is quasicompact, there exists a finite subset $F_j \subseteq I$ such that

$$V_j = \bigcup_{i \in F_j} (U_i \cap V_j).$$

Then, since $U_i \supseteq U_i \cap V_j$,

$$X = \bigcup_{i \in F_1 \cup \cdots \cup F_n} U_i.$$

$\square$

In fact, in the same way, one can prove the following

3.2.3 Corollary *A scheme X is quasi-compact if and only if it is a union of finitely many open affine subsets.*

□

Proof Exercise 28.

Maybe surprisingly, the following is harder:

3.2.4 Lemma *An affine scheme $X = Spec(R)$ is Noetherian if and only if the ring R is Noetherian.*

Proof If the ring R is Noetherian, then, by definition, the scheme X is Noetherian. Assume conversely that X is a Noetherian scheme. Then $X = V_1 \cup \cdots \cup V_n$ where V_i are open subsets which are isomorphic to affine schemes. We are *not* a priori guaranteed that V_i are distinguished open subset, i.e. that

$$V_i = Spec(f_i^{-1}R), \text{ for some } f_i \in R. \tag{3.2.1}$$

However, every open set is a union of distinguished open sets, and the open subsets V_j are affine therefore quasi-compact, so we can make each of those unions finite. Note that if $V_i = Spec(R_i)$ then

$$Spec(f^{-1}R) \cap V_i = Spec(f^{-1}R_i)$$

is also distinguished in V_i and hence $f^{-1}R_i$ is Noetherian. Therefore, (3.2.1) can be assumed without loss of generality, after all. Now the fact that

$$Spec(R) = \bigcup_{i=1}^{n} V_i$$

means, on the level of rings, that the ideal $(f_1, \ldots, f_n)$ in R is R itself, which means that

$$1 = a_1 f_1 + \cdots + a_n f_n \text{ for some } a_i \in R. \tag{3.2.2}$$

Now let I be an ideal in R. Since $f_i^{-1}R$ is Noetherian, the ideal $f_i^{-1}I$ in $f_i^{-1}R$ is finitely generated, say, by elements $r_{i,1}, \ldots, r_{i,k_i}$. Since f_i is invertible (has a reciprocal) in $f_i^{-1}R$, we may assume without loss of generality that $r_{i,j} \in R$.

We claim that the ideal I is generated by the finitely many elements $r_{i,j}$. Indeed, let $x \in I$. Then, in $f_i^{-1}R$, x is a linear combination of the elements $r_{i,1}, \ldots, r_{i,k_i}$. Note that the coefficients may have denominators which are powers of f_i. Therefore, after clearing denominators, this is telling us that $f_i^{N_i}x$ is a linear combination of $r_{i,1}, \ldots, r_{i,k_i}$ in R. But $(f_i^{N_i})^{-1}R = f_i^{-1}R$, so we may replace f_i by $f_i^{N_i}$. Thus, without loss of generality, $N_i = 1$, i.e. in R we have

$$f_i x = b_{i,1} r_{i,1} + \cdots + b_{i,k_i} r_{i,k_i}, \quad b_{i,j} \in R.$$

Then using (3.2.2),

$$x = a_1 f_1 x + \cdots + a_n f_n x = \sum_{i=1}^{n} a_i (b_{i,1} r_{i,1} + \cdots + b_{i,k_i} r_{i,k_i}),$$

which is a linear combination of the $r_{i,j}$'s with coefficients in R, as claimed. $\qquad\square$

Recall that a commutative ring R is Noetherian if and only if it satisfies the ascending chain condition (ACC) with respect to ideals. Now by contravariance, an affine scheme X is Noetherian if and only if it satisfies the descending chain condition (DCC) for closed subschemes. By this we mean that there do not exist closed immersions

$$Z_1 \leftarrow Z_2 \leftarrow \cdots \leftarrow Z_n \leftarrow \cdots \qquad\qquad (3.2.3)$$

of closed subschemes of X none of which are isomorphisms.

3.2.5 Lemma *If a scheme X is Noetherian, then it satisfies the DCC for closed subschemes.*

Proof Suppose X is Noetherian. Then it is covered by finitely many open affine $U_1, \ldots U_k$ which are Noetherian, and hence satisfy the DCC with respect to closed subschemes. If we had a chain of closed subschemes (3.2.3) in X, then at least for one U_i, $Z_n \cap U_i$ (interpreted as a pullback) would form an infinite chain of decreasing closed subschemes in U_i, which is a contradiction. (Note carefully that some of the closed immersions $Z_n \cap U_i \leftarrow Z_{n+1} \cap U_i$ can be isomorphisms, but for some $i = 1, \ldots, k$, there will be infinitely many non-isomorphisms among them.)

$\qquad\square$

3.2.6 Lemma *If X is a Noetherian scheme, then every open subscheme $U \subseteq X$ and every closed subscheme $Z \subseteq X$ are Noetherian.*

Proof Let X be covered by finitely many affine open Noetherian subschemes U_i. For a closed subscheme Z, we can just cover Z by $U_i \cap Z$ (interpreted as a pullback), and use Theorem 2.3.1.

For an open subscheme $U \subseteq X$, the subschemes $U \cap U_i$ are then covered by distinguished open subsets U_{ij} of U_i, which are Noetherian. However, U_i must then be covered by finitely many U_{ij}'s, since otherwise ACC on open subsets of X would be violated. This would imply a violation of DCC on closed subsets of X, and hence closed subschemes of X (since with every closed subset we can associate a unique reduced closed subscheme—Exercise 15). This would contradict Lemma 3.2.5. Thus, $U \cap U_i$ are Noetherian, and hence so is U.

$\qquad\square$

3.3   Morphisms Locally of Finite Type and Morphisms of Finite Type

A morphism of schemes $f : X \to Y$ is called *locally of finite type* if Y is covered by open affine subsets $U_i = Spec(A_i)$, $i \in I$, such that for every $i \in I$, $f^{-1}(U_i)$ is covered by open affine subsets $V_{i,j} = Spec(B_{i,j})$, $j \in J_i$ where the corresponding homomorphism of rings $A_i \to B_{i,j}$ makes $B_{i,j}$ a finitely generated A_i-algebra (this means that every element of $B_{i,j}$ can be expressed as a polynomial, with coefficients in A_i, of finitely many given elements in B_{ij}). The morphism f is called *of finite type* if the sets J_i, $i \in I$, can be chosen to be finite.

As already mentioned, when we have a morphism of schemes $f : X \to Y$, we also sometimes refer to X as a *scheme over* Y. If the morphism f is of finite type, we say that the scheme X is of finite type over Y. A scheme is *of finite type* if it is of finite type over $Spec(\mathbb{Z})$. Similarly for locally finite type.

For example, the projection $\mathbb{A}^n_S \to S$ for any scheme S is of finite type: without loss of generality, S is affine, say, $S = Spec(R)$. Then $\mathbb{A}^n_S = Spec(R[x_1, \ldots, x_n])$ which is generated by the finitely many elements $x_1, \ldots, x_n$.

Similarly, the projective space $\mathbb{P}^n_S \to S$ is of finite type over S. Again, without loss of generality, $S = Spec(R)$ is affine. Then $\mathbb{P}^n_S$ is covered by the finitely many sets $Spec((x_i^{-1}R[x_0, \ldots, x_n])_0)$. (Recall from Sect. 2.8 that the subscript 0 means the subring of elements of degree 0.)

A closed subscheme of a scheme S (see Sect. 2.3) is always of finite type over S. Again, without loss of generality, $S = Spec(R)$ is affine. Then closed subschemes are of the form $Spec(R/I)$ for an ideal I. R/I is generated by the empty set as an algebra over R.

An open subscheme of a scheme S is always locally of finite type over S. Again, without loss of generality, $S = Spec(R)$ is affine. Then every open subset is a union of distinguished open subsets $Spec(f_i^{-1}R)$. The ring $f_i^{-1}R$ is generated by the single element

$$\frac{1}{f_i} = f_i^{-1}$$

as an R-algebra.

An open affine subscheme U of an affine scheme S is of finite type over S. This is because an affine scheme is quasi-compact, so when U is affine, we can find a finite subcover of the cover of U by distinguished open sets in S.

The morphism from a disjoint union of infinitely many copies of a scheme $Spec(\mathbb{Z})$ to $Spec(\mathbb{Z})$ is locally of finite type but not of finite type.

A composition of two morphisms locally of finite type is locally of finite type, and a composition of two morphisms of finite type is of finite type. Therefore, for example, a closed subscheme of an affine space over S is always of finite type over S.

The morphism of schemes

$$Spec(\mathbb{Q}) \to Spec(\mathbb{Z})$$

(given by the inclusion $\mathbb{Z} \subseteq \mathbb{Q}$) is not locally of finite type. This is because $\mathbb{Z}$ is a principal ideal domain, which means that every ideal is principal, so every open set in $Spec(\mathbb{Z})$ is distinguished, of the form $Spec(n^{-1}\mathbb{Z})$. But $\mathbb{Q}$ is not generated by finitely many elements as an algebra over $n^{-1}\mathbb{Z}$. This is because $\mathbb{Q}$ can have any nonzero integer as a denominator, and there are infinitely many primes.

3.4   Finite Morphisms

A morphism of schemes $f : X \to Y$ is called *finite* if for every $Spec(R) \cong U \subseteq Y$ affine open, $Spec(S) \cong f^{-1}(U)$ is affine (it is automatically open) and the corresponding homomorphism of rings $R \to S$ makes S into a *finitely generated R- module.*

One can show that if $X = Spec(S)$, $Y = Spec(R)$, this is equivalent to the homomorphism of rings $R \to S$ making S a finitely generated R-module. (See Exercise 37.)

Note: this is a much stronger property than finite type. Essentially, we should think of X "covering Y with finitely many sheets," but that is a bit oversimplified. Let us give some examples:

Consider the morphism

$$f : X = Spec(\mathbb{C}[x]) \to Spec(\mathbb{C}[y]) = Y \qquad (3.4.1)$$

given by $y = x^2$. Note that both X and Y are isomorphic to $\mathbb{A}^1_{\mathbb{C}}$. This is a morphism of affine schemes. The corresponding homomorphism of rings is $\mathbb{C}[y] \to \mathbb{C}[x]$ which sends y to x^2. This makes $\mathbb{C}[x]$ a finitely generated module over $\mathbb{C}[y]$ where the generators are 1 and x: any polynomial over x can be grouped into a sum of monomials with even powers of x, which is in $\mathbb{C}[x^2]$, and odd powers of x, which is in $x \cdot \mathbb{C}[x^2]$. So (3.4.1) is a finite morphism of schemes.

Note that if we imagine $\mathbb{A}^1_{\mathbb{C}}$ as a complex number line, then every element has two square roots except 0. This means that the morphism f has "two sheets," but they merge to one sheet at 0. So this singularity does not spoil finiteness.

Now consider in $\mathbb{A}^1_{\mathbb{C}}$ the distinguished set

$$\mathbb{A}^1_{\mathbb{C}} \setminus \{0\} = Spec(\mathbb{C}[x]) \setminus Z(x).$$

The inclusion

$$Spec(x^{-1}\mathbb{C}[x]) \to Spec(\mathbb{C}[x]) \qquad (3.4.2)$$

can be written as *Spec* of the inclusion of rings

$$\mathbb{C}[x] \subset x^{-1}\mathbb{C}[x].$$

Note that the ring $x^{-1}\mathbb{C}[x]$ is the ring of "finite Laurent series," which means they are linear combinations, with coefficients in $\mathbb{C}$, of powers x^n where $n \in \mathbb{Z}$ (i.e. negative powers of x are allowed). Since there are infinitely many negative powers, this is not finitely generated as a $\mathbb{C}[x]$-module. So the morphism of schemes (3.4.2) is not finite.

Now consider the projection

$$\mathbb{A}_\mathbb{C}^2 \to \mathbb{A}_\mathbb{C}^1 \qquad (3.4.3)$$

to the first coordinate. This can be written as *Spec* of the inclusion of rings

$$\mathbb{C}[x] \to \mathbb{C}[x, y].$$

If we put $R = \mathbb{C}[x]$, we can write $\mathbb{C}[x, y] = R[y]$, i.e. polynomials in the variable y with coefficients in $\mathbb{C}[x]$. Again, there are infinitely many powers of y, so this is not finitely generated as a $\mathbb{C}[x]$-module.

Finally, consider the morphism

$$\mathbb{P}_\mathbb{C}^1 \to Spec(\mathbb{C}). \qquad (3.4.4)$$

This is not finite either, since $Spec(\mathbb{C})$ only has one non-empty affine subset (itself), and its inverse image is not finite!

Note that all the examples of morphisms of schemes given in this subsection, including the ones which were not finite, are of finite type.

4   Exercises

1. (a) Prove that for two morphisms $\phi, \psi : \mathcal{F} \to \mathcal{G}$ of sheaves of sets on a topological space X, $\phi = \psi$ if and only if ϕ and ψ induce the same maps on stalks.
   (b) Deduce that a morphism of sheaves of sets or abelian sheaves on X is an epimorphism if and only if it is onto on stalks.
2. Let R be a local ring with maximal ideal m, and S be a local ring with maximal ideal n. Let $f : R \to S$ be a homomorphism of rings. Prove that $f^{-1}(n) = m$ if and only if $f(m) \subseteq n$. (Then we say that f is a *morphism of local rings*.)
3. Prove that a localization of commutative rings $R \to S^{-1}R$ where S is a multiplicative set is an epimorphism of rings.
4. Prove that in a pre-additive category, for objects A, B, C and morphisms

$$A \xrightarrow{\ i\ } C \xleftarrow{\ j\ } B,$$

$$A \xleftarrow{\ p\ } C \xrightarrow{\ q\ } B$$

such that $p \circ i = Id_A, q \circ j = Id_B, i \circ p + j \circ q = Id_C$, the object C is a biproduct of the objects A and B.

5. Let R be the ring of all complex polynomials f in one variable such that $f(0) = f(1)$. Prove that R is a finitely generated $\mathbb{C}$-algebra, and thus, $Spec(R)$ is a scheme of finite type over $Spec(\mathbb{C})$. Give a presentation of the $\mathbb{C}$-algebra R in terms of (finitely many) generators and defining relations. Describe $Spec(R)$ explicitly as a topological space. [Look for an infinite, but good, basis of R as a $\mathbb{C}$-vector space first.]

6. In the setup of Exercise 5, consider the morphism $h : \mathbb{A}^1_{Spec(\mathbb{C})} \to Spec(R)$ given by the embedding of $\mathbb{C}$-algebras $R \subset \mathbb{C}[x]$. Prove that this morphism is a coequalizer of the embeddings $Spec(\mathbb{C}) \to \mathbb{A}^1_{Spec(\mathbb{C})}$ given by the maps of rings $x \mapsto 0$ and $x \mapsto 1$ in the category of affine schemes.

7. Is the morphism h considered in Exercise 6 of finite type? Is it finite?

8. Let $P \in Spec(R)$ be the image of the point $Q_1 = (x)$ (or, equivalently, $Q_2 = (x - 1)$) in $\mathbb{A}^1_{Spec(\mathbb{C})}$. Prove that the homomorphism h restricts to an isomorphism of open affine subschemes

$$\widetilde{h} : \mathbb{A}^1_{Spec(\mathbb{C})} \smallsetminus \{Q_1, Q_2\} \to Spec(R) \smallsetminus \{P\}.$$

9. Let S be the ring of rational functions of the form $f(x) = p(x)/(x - 2)^n$ where p is a polynomial, such that $f(0) = f(1)$. Consider the morphism $g : Spec(S) \to Spec(R)$ given by the inclusion of rings $R \subset S$. Prove that the image U of g is an open subscheme of $Spec(R)$, but that $g : Spec(S) \to U$ is not an isomorphism of schemes. Is U affine?

10. Let $Q_3 = (x - 2) \subset \mathbb{C}[x]$ and let

$$k : \mathbb{A}^1_{Spec(\mathbb{C})} \smallsetminus \{Q_3\} \to Spec(S)$$

be the morphism of affine schemes given by inclusion of rings. Let $\overline{P} \in Spec(S) = k(P_1) = k(P_2)$ (see Exercise 8). Prove that k restricts to an isomorphism of schemes

$$\widetilde{k} : \mathbb{A}^1_{Spec(\mathbb{C})} \smallsetminus \{Q_1, Q_2, Q_3\} \to Spec(S) \smallsetminus \{\overline{P}\}.$$

Now let X be the pushout of the diagram

$$
\begin{array}{ccc}
\mathbb{A}^1_{Spec(\mathbb{C})} \smallsetminus \{Q_1, Q_2, Q_3\} & \xrightarrow{\subset} & \mathbb{A}^1_{Spec(\mathbb{C})} \smallsetminus \{Q_1, Q_2\} \\
{\scriptstyle Co\widetilde{k}} \downarrow & & \\
Spec(S) & &
\end{array}
$$

in the category of schemes (see Sect. 2.5). Then we have a pushout morphism ϕ : $X \to Spec(R)$. Is ϕ an isomorphism of schemes? Is X affine?

11. Let R be the ring of pairs of polynomials (f, g) in one variable with coefficients in $\mathbb{C}$ satisfying $f(0) = g(0)$. Prove that $R \cong \mathbb{C}[x, y]/(xy)$. Is $Spec(R)$ a pushout of two copies of the inclusion $Spec(\mathbb{C}) \to \mathbb{A}^1_{Spec(\mathbb{C})}$ given by $x \mapsto 0$?

12. Consider the setup of Exercise 11. Consider the morphism

$$h : \mathbb{A}^1_{Spec(\mathbb{C})} \to Spec(R)$$

given by the embedding of $\mathbb{C}$-algebras $R \subset \mathbb{C}[x]$. Is this morphism a coequalizer of the embeddings $Spec(\mathbb{C}) \to \mathbb{A}^1_{Spec(\mathbb{C})}$ given by the maps of rings $x \mapsto 0$ and $x \mapsto 1$ in the category of

   (i) affine schemes

   (ii) schemes?

   [Investigate, again, what happens to the construction when you remove finitely many closed points from the affine lines to be glued.]

13. Prove that for a non-empty scheme X, the following are equivalent:

   (i) X is integral

   (ii) $\mathcal{O}_X(U)$ is an integral domain for all $U \subseteq X$ non-empty open affine

   (iii) $\mathcal{O}_X(U)$ is an integral domain for all $U \subseteq X$ open.

14. Let Y be a scheme. The category of *schemes over Y* has morphisms of schemes $X \to Y$ as objects, and commutative diagrams

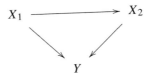

as morphisms. Prove that the category of schemes over Y has finite limits.

15. Give an example of two non-isomorphic closed subschemes of the same scheme X which have the same points. Prove that for a given closed subset S of a scheme X, there exists a unique reduced closed subscheme of X whose set of points is S.

16. Consider a countable infinite coproduct X of copies of $Spec(\mathbb{Z})$ in the category of schemes.

   (a) Is X (isomorphic to) an affine scheme?

   (b) Is X of finite type over $Spec(\mathbb{Z})$?

17. (a) Prove that a scheme X is reduced if and only if the ring $\mathcal{O}_X(U)$ is reduced for every U open affine.

   (b) Prove that a scheme X is reduced if and only if the stalks of the structure sheaf $\mathcal{O}_X$ are reduced rings.

18. Is every reduced scheme which has only one point isomorphic to $Spec(k)$ for some field k?

19. Is the forgetful functor from the category of schemes to the category of topological spaces (a) a right adjoint? (b) a left adjoint?

20. Do there exist non-empty schemes X, Y such that there is no morphism of schemes $X \to Y$ or $Y \to X$ and the underlying space of the product of X and Y is the product of the underlying spaces of X and Y?

21. Using gluing, finish the proof that the category of schemes has pullbacks.

22. Prove that any category which has pullbacks and a terminal object has finite limits.

23. Prove that if $f : X \to Y$ is a closed immersion of schemes and $g : Z \to Y$ is any morphism of schemes, then $f \times_Y Z : X \times_Y Z \to Z$ is a closed immersion.

24. Prove that for a graded ring R and homogeneous elements $f \in R_m, g \in R_n$,

$$((fg)^{-1}R)_0 \cong (g^m/f^n)^{-1}((f^{-1}R)_0)_0.$$

Use this to fill in the details in the definition of $Proj(R)$.

25. Let R be a commutative ring. The scheme

$$\mathbb{P}_R(a_0, \ldots, a_n)$$

defined as $Proj(R[x_0, \ldots, x_n])$ where x_i is homogeneous of degree $a_i \in \mathbb{N}$ is called a *weighted projective space*.

(a) Prove that the scheme $\mathbb{P}_R(da_0, da_1, \ldots, da_n)$ is isomorphic to the scheme $\mathbb{P}_R(a_0, a_1, \ldots, a_n)$ for any natural number d.

(b) Prove that $\mathbb{P}_{\mathbb{C}}(2, 3, 5)$ is not isomorphic to a projective space over $\mathbb{C}$.

26. Prove that an integral closed subscheme of $\mathbb{A}^1_{\mathbb{Z}}$ is either an integral closed subscheme of the closed subscheme $\mathbb{A}^1_{Spec(\mathbb{F}_p)} \subset \mathbb{A}^1_{\mathbb{Z}}$ (where $\mathbb{F}_p$ is the finite field with p elements) or the closure of an integral closed subscheme of the subscheme $\mathbb{A}^1_{Spec(\mathbb{Q})} \subset \mathbb{A}^1_{\mathbb{Z}}$. [Recall Sect. 4.2 of Chap. 1.] Is an analogous statement true with $\mathbb{A}^1$ replaced by $\mathbb{P}^1$?

27. Prove that the category of quasiprojective varieties over $\mathbb{C}$ considered in Chap. 1 is equivalent to the full subcategory of the category of schemes $X \to Spec(\mathbb{C})$ over $Spec(\mathbb{C})$ where X is an open subscheme of an integral closed subscheme of $\mathbb{P}^n_{Spec(\mathbb{C})}$ for some $n \in \mathbb{N}_0$.

28. Prove Corollary 3.2.3.

29. (The Segre embedding) Let $\sigma : \mathbb{P}^m_{\mathbb{Z}} \times \mathbb{P}^n_{\mathbb{Z}} \to \mathbb{P}^N_{\mathbb{Z}}, N = (m+1)(n+1) - 1$, be the morphism of schemes where the projective coordinates of the target are $x_i y_j$ where $x_0, \ldots, x_m$ are the projective coordinates of $\mathbb{P}^m_{\mathbb{Z}}$ and $y_0, \ldots, y_n$ are the projective coordinates of $\mathbb{P}^n_{\mathbb{Z}}$. Prove that the image is a closed subscheme $Z \subset \mathbb{P}^N_{\mathbb{Z}}$ and that σ restricts to an isomorphism of the source to Z. Conclude that a product over a scheme S of two projective schemes over S is projective over S.

30. Let $f : Y \to X$ be a morphism of schemes of finite type and let $U \subseteq X$ be open. Prove that the restriction $f^{-1}(U) \to U$ is of finite type.

31. Let $f : Y \to X$ be a morphism of schemes of finite type and let $Spec(R) \cong U \subseteq X$ be open affine. Prove that there exists a cover of $f^{-1}U$ by finitely many open affine subsets $V_i = Spec(R_i)$ such that R_i is a finitely generated R-algebra.

32. Find an example of a scheme X and an affine open subscheme $U \subset X$ such that U is not of finite type over X. [First, find an example of a closed subset Z of an affine scheme Y such that $\phi : Y \smallsetminus Z \subset Y$ is not of finite type. Then, consider a pushout of two copies of the morphism ϕ.]

33. Is a scheme of finite type over a Noetherian scheme always Noetherian?

34. Give an example of an affine scheme which satisfies the DCC with respect to closed subsets but is not Noetherian.

35. Prove that if $f : Y \to X$ is a morphism of schemes and X has a cover by open subschemes U_i such that each restriction $f^{-1}(U_i) \to U_i$ is finite, then f is finite. [Let R be a commutative ring, and let the unit ideal be $(f_1, \ldots, f_n)$. Let M be an R-module such that $f_i^{-1}M$ is a finitely generated $f_i^{-1}R$-module. Prove that then M is a finitely generated R-module.]

36. Prove that every closed immersion is finite.

37. Prove that for a homomorphism of rings $f : R \to S$, $Spec(f) : Spec(S) \to Spec(R)$ is finite if and only if f makes S a finitely generated R-module.

38. The *dimension* $dim(X)$ of a scheme X is defined as the (possibly infinite) supremum of n such that there exist irreducible closed subschemes

$$X_0 \subsetneqq X_1 \subsetneqq \cdots \subsetneqq X_n \subseteq X$$

(where $\subsetneqq$ means a closed subscheme where the sets of points are not equal). If $dim(X) < \infty$ and $Z \subseteq X$ is a closed irreducible subscheme, then *codimension* $cod_X(Z)$ of Z in X is defined as the supremum of numbers n such that there exist irreducible closed subschemes

$$Z = Z_0 \subsetneqq Z_1 \subsetneqq \cdots \subsetneqq Z_n \subseteq X$$

(using the same interpretation of $\subsetneqq$ as above). Let $R = \mathbb{Z}_{(p)}[x]$ where $p \in \mathbb{N}$ is a prime number.

(a) Prove $dim(Spec(R)) = 2$.

(b) Find $f \in R$ such that $(f) \subset R$ is a maximal ideal and

$$dim(Spec(R/(f))) = 0.$$

(c) Prove that for any scheme X and an irreducible closed subscheme Z,

$$dim(Z) + cod_X(Z) \leq dim(X).$$

(d) Does equality always arise in (c)?
   (Compare with the comments at the end of Sect. 5.5 of Chap. 1.)

Properties of Schemes

3

It is immediately apparent from the definition, and the basic examples we studied, that the concept of a scheme is far more general than the concept of a variety as introduced in Chap. 1, just as a topological space is much more general than a subset of $\mathbb{R}^n$. What are the properties of schemes we should study?

As it turns out, topology, and analysis, provide good starting points for answering this question, and many of the scheme-theoretical properties we study are analogues of concepts from topology. For example, in general topology, important kinds of spaces include spaces which are *Hausdorff* and *compact*. We will review these properties here, and introduce analogous concepts for schemes, i.e. *separated*, *universally closed*, and *proper* schemes. In analysis, we have a concept of *smoothness*. We will also study analogous concepts in algebraic geometry.

Developing scheme-theoretical analogues of topological concepts is not always technically straightforward. This is due to the fact that topologies on schemes have "far fewer" open and closed sets than we are used to in analysis, and in addition, the category of schemes has "far fewer" morphisms than the category of topological spaces. Because of that, the definitions of concepts expressing the same geometrical idea, and the proofs, are often quite different, and additional commutative algebra will be needed.

We will also revisit varieties, developing an abstract definition of a variety not relying on having a familiar ambient space. We will give a better (more general) proof of the Nullstellensatz, and a classification of 1-dimensional varieties (i.e. curves) over an algebraically closed field up to birational equivalence.

Finally, we will begin exploring some basic features of doing algebraic geometry over non-algebraically closed fields, and the role of group actions, specifically the action of the Galois group. (A review of Galois theory is contained in our narrative and exercises.) Continuing with the topological analogues, we will also mention the topological concept of a fundamental group, and its algebraic analogue, the *étale fundamental group*. One of

© The Author(s), under exclusive license to Springer Nature Switzerland AG 2021
I. Kriz, S. Kriz, *Introduction to Algebraic Geometry*,
https://doi.org/10.1007/978-3-030-62644-0_3

the most striking examples of the use of Galois actions are the Weil conjectures proved by P. Deligne based on the work of A. Grothendieck and others, which will be briefly mentioned at the end of the chapter.

1 Separated Schemes and Morphisms

The concept of a separated scheme models in schemes a certain concept in topology. The reader may wonder what this means: a scheme is, after all, a topological space. But in mathematics (not just applied), concepts are defined to model phenomena we see. A topologist will typically look at examples close to the *analytic topology*, i.e. $\mathbb{R}^n$ where the basis of topology are open balls $B_\epsilon(x)$, the set of all points y of distance $< \epsilon$ from x. (A *basis of topology* is a set S of open sets such that every open set is a union of elements of S—for example, the set of distinguished open set in an affine scheme is a basis of the Zariski topology.) A topologist may also typically think of a subset X of $\mathbb{R}^n$ with the *induced topology* which means that the sets considered open in X are precisely sets of the form $U \cap X$ where U is open in $\mathbb{R}^n$. (Note that we have also already considered the induced topology when looking at open and closed subschemes, and even subvarieties.)

The point is, however, that a topologist will typically not (at least not primarily) think about the Zariski topology, or the topology of a scheme. We have already noticed that the notion of a compact space, while it makes sense in schemes (giving rise to the meaningful notion of a quasicompact scheme), does not carry the same geometrical meaning as topologists would expect from their examples. (We are yet to define the scheme-theoretical concept which correctly models the topologist's intuition of compactness.)

In this section, we will introduce another concept from topology which needs to be modeled differently in schemes, and we will show how it is done. Let us define the topological concept first.

1.1 Hausdorff and T_1 Topological Spaces

A topological space X is called *Hausdorff* when for every two points $x \neq y \in X$, there exists an open set $x \in U$ and an open set $y \in V$ such that $U \cap V = \emptyset$. The Hausdorff property is also sometimes called T_2. The space $\mathbb{R}^n$ with the analytic topology, and all its subsets with the induced topology, are clearly Hausdorff: For two points x, y of distance $d > 0$, let $U = B_{d/2}(x)$, $V = B_{d/2}(y)$. More generally, in a Hausdorff space, a subset with the induced topology is Hausdorff.

It is also useful to say that a topological space X is T_1 if every point $x \in X$ is closed. (This means, more precisely, that the one element set $\{x\}$ is closed.) It is easy to see that every Hausdorff space is T_1: Let $x \in X$, and let for every $y \in X$ with $x \neq y$, $U_y \ni y$ be an open set disjoint with some open set containing x, hence, in particular, not containing x. Then

$$\bigcup_{y \neq x} U_y = X \setminus \{x\}$$

is an open set, since it is a union of open sets.

Example Consider the pushout $\widetilde{\mathbb{R}}$ of the diagram

$$\begin{array}{ccc} \mathbb{R} \setminus \{0\} & \longrightarrow & \mathbb{R} \\ \downarrow & & \\ \mathbb{R} & & \end{array}$$

(1.1.1)

where both of the arrows are inclusions. The space $\widetilde{\mathbb{R}}$ can be thought of as a real line where 0 is replaced by two of its copies 0_1 and 0_2. A subset $U \subseteq \widetilde{\mathbb{R}}$ is open if and only if U is open in both copies of $\mathbb{R}$ sitting in $\widetilde{\mathbb{R}}$: this means that the set U_1 obtained from U by omitting 0_2 (if present) and replacing 0_1 with 0, and also the set U_2 obtained from U by omitting 0_1 (if present) and replacing 0_2 with 0, are both open in $\mathbb{R}$. We see than that the space $\widetilde{\mathbb{R}}$ is T_1, but is not Hausdorff: there are no disjoint open sets containing the points $0_1, 0_2$, respectively.

The following example may seem more troubling:

1.1.1 Lemma *The topology of a Noetherian scheme X is never T_1 unless it is discrete, i.e. such that every subset is open.*

Proof Without loss of generality, we may assume that $X = Spec(R)$ is affine. Additionally, we may assume that R is reduced, which means it has no nonzero nilpotent elements, since otherwise we may factor it by the nilradical (i.e. ideal generated by all the nilpotent elements) without altering $Spec$. (Recall that the nilradical is the intersection of all prime ideals.)

Now if all points of $Spec(R)$ are closed, it means that all prime ideals of R are maximal, and thus R is an Artinian ring. Now an Artinian ring is a product of finitely many local Artinian rings by Proposition 5.2.3 of Chap. 1, and a reduced local Artinian ring is by definition a field. Thus, $Spec(R)$ is a finite discrete set.   □

1.2   The Product of Topological Spaces: Reformulating the Hausdorff Property

As often, to find an appropriate analogue of our concept in another context, we want to phrase it as much as possible in the language of category theory. For this purpose, let us examine a little more closely the categorical product of topological spaces. If X, Y are topological spaces, recall that their categorical product is the limit of the diagram consisting of X and Y and no arrows. One easily sees that as a set, this is the Cartesian product $X \times Y = \{(x, y) \mid x \in X, y \in Y\}$ and that a basis of its topology is the set of sets

$$\{U \times V \mid U \subseteq X \text{ open}, V \subseteq Y \text{ open}\}.$$

1.2.1 Lemma *A topological space is Hausdorff if and only if the diagonal*

$$\Delta_X = \{(x, x) \mid x \in X\} \subseteq X \times X$$

is a closed set.

Proof The diagonal being a closed set is equivalent to its complement

$$X \times X \smallsetminus \Delta_X \tag{1.2.1}$$

being open. This is equivalent to any (x, y) in (1.2.1) being contained in an open subset of (1.2.1) which belongs to a basis of topology, i.e. to a set $U \times V$ where U is open in X and V is open in Y and

$$U \times V \subseteq X \times X \smallsetminus \Delta_X. \tag{1.2.2}$$

But (1.2.2) is equivalent to $U \cap V = \emptyset$.

$\square$

1.3   Separated Schemes and Morphisms

Now we are ready for the scheme-theoretical concept. Note that in any category with products there is a diagonal morphism

$$\Delta : X \to X \times X \tag{1.3.1}$$

where the right hand side is the categorical product. This is because the product is, after all, a limit of a diagram with no arrows, so (1.3.1) is gotten from the two maps $Id : X \to X$, $Id : X \to X$. Similarly, for a morphism $f : X \to Y$ in any category with pullbacks, we have a diagonal

$$\Delta_Y : X \to X \times_Y X \qquad (1.3.2)$$

1.3.1 Definition A scheme X is called *separated* if the image of the diagonal morphism (1.3.1) in the category of schemes is a closed set. More generally, a morphism of schemes $f : X \to Y$ is called *separated* if the image of the diagonal (1.3.2) is a closed set.

Of course, a scheme X is separated if and only if the unique morphism $f : X \to Spec(\mathbb{Z})$ is separated. Note that this definition makes sense in spite of Lemma 1.1.1 because the topological space underlying a product of schemes is typically *not* the categorical product of the underlying spaces. In fact, recall that it is not even true for the underlying sets. For example, if E, F are fields of different characteristic, then $Spec(E) \times Spec(F) = \emptyset$. But perhaps more to the point, while, say, in $\mathbb{A}^1_\mathbb{C}$ (considered as a scheme), all Zariski open sets consist precisely of complements of finite sets of closed points together with the point (0), in $\mathbb{A}^2_\mathbb{C} = \mathbb{A}^1_\mathbb{C} \times \mathbb{A}^1_\mathbb{C}$, we have a lot more Zariski open sets (e.g., complements of curves).

1.3.2 Lemma *Every affine scheme is separated.*

Proof For an affine scheme $Spec(R)$, the diagonal is $Spec$ of the codiagonal in commutative rings, which is the multiplication homomorphism

$$R \otimes R \to R.$$

Since this homomorphism is always onto, the diagonal is a closed subscheme. □

We see from this proof that when a scheme or more generally a morphism of schemes is separated, the inclusion (1.3.1) (resp. (1.3.2)) is in fact a closed immersion.

1.3.3 Lemma *An open or closed subscheme of a separated scheme is separated.*

Proof Let $f : Y \to X$ be a closed immersion. If X is separated, then the composition

$$Y \xrightarrow{\ f\ } X \xrightarrow{\ \Delta\ } X \times X$$

is a closed immersion. If we can prove that

$$f \times f : Y \times Y \to X \times X$$

is injective on points, then the image of $\Delta : Y \to Y \times Y$ is the inverse image of $(f \times f)^{-1} Im(\Delta \circ f)$, which is a closed set in Y by continuity. To show injectivity on points, it suffices to consider the case where $X = Spec(R)$ is affine. Then, by Theorem 2.3.1 of Chap. 2, $Y = Spec(R/I)$ for some ideal I, so

$$Y \times Y = Spec(R/I \otimes R/I) = Spec((R \otimes R)/(I \otimes I)),$$

which injects into $X \times X = Spec(R \otimes R)$. Thus, we are done.

For an open subscheme $Y \subseteq X$, one first observes that by construction, $Y \times Y$ is an open subscheme of $X \times X$. Thus, it suffices to prove that the image of the diagonal $\Delta : X \to X \times X$, intersected with $Y \times Y$, is contained in the image of the diagonal $\Delta : Y \to Y \times Y$ (since it is then automatically equal). For this purpose, it suffices to consider the case when $X = Spec(R)$ is affine. Then we have

$$r^{-1}R \otimes s^{-1}R = (R \otimes R)[(r \otimes 1)^{-1}, (1 \otimes s)^{-1}],$$

so if the image of a point $p \in Spec(R)$ in $Spec(R \otimes R)$, which is the prime ideal $p \otimes R + R \otimes p$, is in

$$Spec(r^{-1}R \otimes s^{-1}R) = Spec(r^{-1}R) \times Spec(s^{-1}R)$$

where $Spec(r^{-1}R), Spec(s^{-1}R) \subseteq Y$, then $r, s \notin p$, and thus

$$p \in Spec(R[r^{-1}, s^{-1}]) \subseteq Y,$$

which is what we are trying to prove.

$\square$

1.3.4 Proposition *If a scheme X is separated then an intersection of two affine open sets in X is affine. (Recall that an intersection of two open sets is always open.)*

Proof Let $U, V \subseteq X$ be affine open. Then

$$U \cap V \cong (U \times V) \cap \Delta_X \subseteq X \times X$$

where Δ_X is the diagonal. The open subscheme $U \times V \subset X \times X$ is affine, and if X is separated, by definition the set $(U \times V) \cap \Delta_X$ is closed in it. Now a closed subset $Z(I)$ in an affine scheme $Spec(R)$ is affine: it is of the form $Spec(R/I)$. (Although we don't need its full strength here, it may be worthwhile recalling Theorem 2.3.1 of Sect. 2.3 of Chap. 2.)

$\square$

From the Proposition, we see that the pushout of the diagram of schemes

$$\mathbb{A}_\mathbb{C}^2 \setminus Z(x, y) \longrightarrow \mathbb{A}_\mathbb{C}^2 = V$$

$$\downarrow$$

$$U = \mathbb{A}_\mathbb{C}^2$$

$$(1.3.3)$$

where the arrows are inclusions is not separated. Obviously, $Z(x, y)$ is the one point set containing the point $(0, 0)$ in $\mathbb{A}_\mathbb{C}^2$. The sets U, V are affine open, but their intersection, $\mathbb{A}_\mathbb{C}^2 \setminus Z(x, y)$ is not affine. We saw this in the language of varieties in Sect. 2.3.1 of Chap. 1, but this is sufficient by Exercise 27 of Chap. 2. (In fact, we will discuss varieties in a greater generality in Sect. 4 below; see also Exercise 23.)

Of course, the example (1.3.3) produces a non-separated scheme when generalized to any n: The pushout X_n of

$$\mathbb{A}_\mathbb{C}^n \setminus Z(x_1, \ldots, x_n) \longrightarrow \mathbb{A}_\mathbb{C}^n = V$$

$$\downarrow$$

$$U = \mathbb{A}^n$$

is not separated for the same reason, even though for $n = 1$, the set $\mathbb{A}^1 \setminus Z(x)$ is distinguished open, and therefore affine.

One can see that X_n is not separated directly as follows: X_n looks like $\mathbb{A}_\mathbb{C}^n$ with the origin $(0, \ldots, 0)$ removed and replaced with two copies of itself. The product $X_n \times X_n$ then has "two times two, i.e. four, copies of the origin," but only two will be on the diagonal. The remaining two "copies of the origin" will not be contained in a Zariski open set disjoint with the diagonal, and therefore the diagonal is not closed.

We see that the concept of a separated scheme mimics the Hausdorff property well in the context of schemes, even though it had to be defined differently because, again, the examples we are interested in are so different from the usual examples in topology and morphisms of schemes are not the same as morphisms of spaces.

2   Universally Closed Schemes and Morphisms

We already learned about compact topological spaces and Hausdorff topological spaces. So far we have a scheme-theoretical analog of Hausdorff (which is separated), but not of compact. The notion of a universally closed scheme, which we will study in this section, is the scheme-theoretical analog of a compact space.

2.1   More Facts About Compact and Hausdorff Spaces

To motivate what the definition of a universally closed scheme says, let us learn a few more properties of compact and Hausdorff spaces. Recall that the *induced topology* on a subset Z of a topological space X has open sets of the form $U \cap Z$ where U is an open set in X.

2.1.1 Theorem *A subset Z of a Hausdorff space X which is compact with the induced topology is closed.*

Proof Let $Z \subseteq X$ be compact (with the induced topology). Let $x \in X \smallsetminus Z$. It suffices to show that there exists $U_x \ni x$ open with $U_x \cap Z = \emptyset$, for then

$$\bigcup_{x \notin Z} U_x = X \smallsetminus Z.$$

To this end, let $z \in Z$. Then, since X is Hausdorff, there exists $U_{x,z} \ni x$ open and $V_{x,z} \ni z$ open such that

$$U_{x,z} \cap V_{x,z} = \emptyset.$$

The sets $V_{x,z}$, $z \in Z$ form an open cover of Z. (Note: we should, technically, say that the sets $V_{x,z} \cap Z$ form an open cover of Z, but this slight inaccuracy is common in topological proofs.)

In any case, since Z is compact, this cover has a finite subcover,

$$V_{x,z_1} \cup \cdots \cup V_{x,z_n} \supseteq Z.$$

But then

$$x \in U_x := U_{x,z_1} \cap \cdots \cap U_{x,z_n}$$

is open, and we have

$$U_x \cap Z \subseteq U_x \cap (V_{x,z_1} \cup \cdots \cup V_{x,z_n}) = \emptyset,$$

so we are done.

$\square$

Let us list one important consequence. A map of topological spaces $f : X \to Y$ (not necessarily continuous) is called *closed* if for every closed set $Z \subseteq X$, the direct image $f(Z) \subseteq Y$ is closed.

2.1.2 Theorem *Every continuous map of spaces $f : X \to Y$ where X is compact and Y is Hausdorff is closed.*

Proof First let us realize that every closed subset Z of a compact space X is compact (with the induced topology). Indeed, if U_i, $i \in I$ is a cover of Z by open sets in X, then U_i, $i \in I$ together with the set $X \setminus Z$ form a cover of X. Since X is compact, that cover has a finite subcover. If this subcover contains the set $X \setminus Z$, which is disjoint from Z, delete it, and you have a finite subcover of the original cover U_i, $i \in I$ of Z.

The next step is to realize that if $f : X \to Y$ is a continuous map, and $Z \subseteq X$ is compact (with the induced topology), then the direct image $f(Z)$ is compact. Indeed, let U_i, $i \in I$ be an open cover of $f(Z)$. This is *equivalent* to $f^{-1}(U_i)$, $i \in I$ being an open cover of Z. But since Z is compact, the latter cover has a finite subcover: $f^{-1}(U_1), \ldots f^{-1}(U_n)$. For the same reason, then, $U_1, \ldots U_n$ is a cover of $f(Z)$.

Finally, let us prove that if $f : X \to Y$ is continuous where X is compact and Y is Hausdorff, then f is closed: Let $Z \subseteq X$ be closed. Then, as we saw, Z is compact with the induced topology. Therefore, $f(Z) \subseteq Y$ is compact. By Theorem 2.1.1, then, it is closed since Y is Hausdorff. □

2.1.3 Theorem *A topological space X (not necessarily Hausdorff) is compact if and only if for every topological space Y, the projection $p : X \times Y \to Y$ is closed.*

Proof Suppose X is compact, and let $Z \subseteq X \times Y$ be closed. Let $y \in Y \setminus p(Z)$. Then for every $x \in X$, there exist open sets $x \in U_x \subseteq X$, $y \in V_x \subseteq Y$ such that

$$(U_x \times V_x) \cap Z = \emptyset.$$

Since X is compact, there exists a finite subcover $U_{x_1}, \ldots, U_{x_n}$ of the open cover U_x, $x \in X$. Then

$$\left(X \times \bigcap_{i=1}^{n} V_{x_i} \right) \cap Z = \emptyset,$$

so

$$\bigcap_{i=1}^{n} V_{x_i} \cap p(Z) = \emptyset.$$

To prove the converse, it is useful to recall the notion of an *ordinal*, which is a set α totally ordered with respect to the relation $\in$. This definition is due to von Neumann. Note that transitivity of $\in$ can be rephrased to say that every $\beta \in \alpha$ also satisfies $\beta \subseteq \alpha$. Ordinals are well-ordered by axioms of set theory, which also imply that every well-ordered set is isomorphic, as a totally ordered set, to an ordinal. Zorn's lemma can be used to prove that

every set is bijective to an ordinal. The first ordinals are $0 = \emptyset$, $1 = \{0\}$, $2 = \{0, 1\}$, ...
For every ordinal α, we have the ordinal $\beta = \alpha + 1$, which is the set of all ordinals $\leq \alpha$.
We also write $\alpha = \beta - 1$. Ordinals which cannot be expressed as $\alpha + 1$ for another ordinal
α are called *limit ordinals*.

Now if X is not compact, then there exists a limit ordinal α and open sets $U_\beta \subsetneq X$,
$\beta < \alpha$ such that $U_\beta \subseteq U_\gamma$ for $\beta < \gamma < \alpha$, and

$$\bigcup_{\beta < \alpha} U_\beta = X.$$

(To see this, let U_δ, $\delta < \gamma$ be an open cover which does not have a finite subcover. We
shall produce a decreasing sequence of ordinals $\gamma = \alpha_0 > \alpha_1 > \ldots$ where U_δ, $\delta < \alpha_n$ is
a cover of $X_n = X \smallsetminus \bigcup_{i=1}^n U_{\alpha_i}$. Suppose this is done for a given $n \geq 0$. Let $\gamma < \alpha_n$ be
the smallest ordinal such that U_δ, $\delta < \gamma$ cover X_n. If $\gamma = 0$, we have a contradiction, if
$\gamma > 0$ is a limit ordinal, we are done. Otherwise, put $\gamma_{n+1} = \gamma - 1$.)

Now on a totally ordered set T, we have the *order topology* where open sets are unions
of *open intervals* $(a, b) = \{c \in T \mid a < c < b\}$ with $a, b \in T \sqcup \{-\infty, \infty\}$, where
$-\infty$ (resp. ∞) is a formally defined element smaller (resp. greater) than all elements of T.
(Note: these elements are introduced to avoid special cases in the definition when T has a
smallest resp. greatest element.)

Then let $Y = \alpha + 1$, and let $Z \subset X \times Y$ be the complement of the open set

$$\bigcup_{\beta < \alpha} U_\beta \times (\beta, \alpha].$$

We then see that $p(Z) = \alpha$ (considered as a set) which is not closed in $\alpha + 1$. □

2.2 Universally Closed Schemes and Morphisms

The concept of a universally closed scheme is analogous, in the category of schemes, to
the concept of a compact space. Theorem 2.1.3 then motivates the following

2.2.1 Definition A morphism of schemes $f : X \to Y$ of schemes is called *universally
closed* if for every morphism of schemes $g : Y' \to Y$, the morphism f' in the pullback
diagram of schemes

$$
\begin{array}{ccc}
X \times_Y Y' & \xrightarrow{\ f'\ } & Y' \\
\downarrow & & \downarrow{\scriptstyle g} \\
X & \xrightarrow[\ f\ ]{} & Y
\end{array}
\qquad (2.2.1)
$$

is closed (as a map of topological spaces). A scheme X is called *universally closed* if the unique morphism of schemes $X \to Spec(\mathbb{Z})$ is universally closed. A scheme or morphism of schemes is called *proper* if it is of finite type and both separated and universally closed.

Example The morphism $f : \mathbb{A}^1_{\mathbb{C}} \to Spec(\mathbb{C})$ is not universally closed, since if we pull it back via $g = f$, $f' : \mathbb{A}^2_{\mathbb{C}} \to \mathbb{A}^1_{\mathbb{C}}$ is the projection to one of the coordinates (i.e. *Spec* of the inclusion of rings $\mathbb{C}[x] \subset \mathbb{C}[x, y]$). But in this projection, say, the image of the "hyperbola" $Z(xy - 1)$ is $\mathbb{A}^1_{\mathbb{C}} \setminus \{0\}$, which is not Zariski closed in $\mathbb{A}^1_{\mathbb{C}}$.

2.3   Specialization

We saw that it is easy to prove that a morphism of schemes is not universally closed (if that is the case), since all we need is to find one example. But it is not so easy to prove directly that a morphism of schemes is universally closed, since that involves checking all morphisms of schemes $g : Y' \to Y$. To make the concept useful, we therefore have to develop other tools for checking that a morphism is universally closed.

An important tool is *specialization*. For an affine scheme $Spec(R)$, we say that a point $q \in Spec(R)$ is a *specialization* of a point $p \in Spec(R)$ (we write $p \rightsquigarrow q$) when these prime ideals satisfy $p \subseteq q$. Note that this, of course, is equivalent to q belonging to the Zariski closure of p, which is $Z(p)$. Thus, we can define specialization in any scheme X by writing $p \rightsquigarrow q$ for $p, q \in X$ when q is in the closure of p. Points in an affine scheme are, of course, closed if and only if they are maximal ideals.

Generic Points As already noted, when considering affine varieties as schemes, we include more points in the Zariski topology than just the closed points. We remarked in Sect. 2.2 that the additional points are necessary in general schemes to have a usable analog of the Nullstellensatz for arbitrary rings. But these non-closed points are also interesting geometrically. For example, the affine space $\mathbb{A}^2_{\mathbb{C}}$, considered as a scheme, contains the point $(x + y)$ (this is a prime ideal, since in a unique factorization domain, the principal ideal generated by an irreducible element is prime). Of course $(x + y)$ is not a closed point, since it is contained, for example, in the maximal ideal (x, y) (which, in the language of affine varieties, represents the origin), as well as, of course, in all the points $(x + a, x - a)$ for $a \in \mathbb{C}$ (which, in the language of affine varieties, represents the point $(-a, a)$). The way we think of the Zariski point $(x + y)$ then is as a *generic* point of the affine variety $Z(x + y)$. More generally in an affine scheme $Spec(R)$, a prime ideal $p \in Spec(R)$ is called the *generic point* of the closed set $Z(p)$. The concepts of generic points and specialization have no analog in the analytic topology.

2.3.1 Lemma *Let* $f : Spec(A) \to Spec(B)$ *be a morphism of affine schemes and let a point* $p \in Spec(B)$ *be in the Zariski closure of* $f(Spec(A))$. *Then there exists a point* $p' \in f(Spec(A))$ *which specializes to* p.

Proof Without loss of generality, $f(Spec(A))$ is dense in $Spec(B)$ (recall that a subset $S \subset X$ of a topological space X is *dense* when for all $U \subseteq X$ nonempty, $U \cap S$ is nonempty). This is because we may, if necessary, replace $Spec(B)$ with the Zariski closure of $f(Spec(A))$ in $Spec(B)$, which is also affine. Without loss of generality, also, A and B are reduced, since factoring out the nilradical does not change $Spec$.

Now recall that the category of affine schemes is equivalent to the opposite category of the category of commutative rings, so $f = Spec(h)$ for some homomorphism of rings

$$h : B \to A,$$

which is injective because $f(Spec(A))$ is dense in $Spec(B)$ (otherwise, $f(Spec(A))$ would be contained in $Z(Ker(h))$).

Now assume $f(Spec(A))$ is closed under specialization. Let $p \in Spec(B)$. Let $p' \subseteq p$ be any minimal prime ideal. To see that minimal prime ideals exist, for a totally ordered set I, ($i \neq j \in I$ implies $i < j$ or $j < i$), we have inclusions of prime ideals $p_i \subseteq p_j$ for all $i < j \in I$, then $\bigcap_{i \in I} p_i$ is also prime. By Zorn's lemma, this lets us conclude that there is a minimal prime ideal (i.e. one that no other prime ideal is contained in it).

Now review localization of rings at ideals (Sect. 1.3 of Chap. 2). The localization $B_{p'}$ of a reduced ring B at the minimal prime ideal p' is a field. Actually, in general, for a prime ideal q in a ring R, the prime ideals in R_q correspond bijectively to those prime ideals in R which are contained in q. Thus, the ring $B_{p'}$ has only one prime ideal, which is therefore the nilradical. Since the ring $B_{p'}$ is reduced, the nilradical is (0), and hence $B_{p'}$ is a field.

Now localization of rings preserves injective morphisms. Thus,

$$h \otimes_B B_{p'} : B_{p'} \to A \otimes_B B_{p'}$$

is injective. Let q_0' be any prime ideal in $A \otimes_B B_{p'}$ (which exists, since it is not the 0 ring). We have

$$q_0' \cap B_{p'} = (0)$$

since $B_{p'}$ is a field (and q_o' does not contain 1). Now let

$$\phi : A \to A \otimes_B B_{p'}$$

be the localization morphism, and let $q' = \phi^{-1}(q'_0)$. Then taking $Spec$ of the diagram of rings

$$
\begin{array}{ccc}
B & \longrightarrow & B_{p'} \\
\downarrow & & \downarrow \\
A & \xrightarrow{\quad\phi\quad} & A \otimes_B B_{p'},
\end{array}
$$

we get a correspondence of prime ideals

$$
\begin{array}{ccc}
p' & \longleftarrow\!\!\!| & (0) \\
\uparrow & & \uparrow \\
q' & \longleftarrow\!\!\!| & q'_0.
\end{array}
$$

$\square$

Thus, $p' \in f(Spec(A))$, as required.

Example It is not true in general that every subset of $Spec(B)$ which is closed under specialization is closed, even though note that every subset is an image of a morphism of schemes $X \to Spec(B)$ (as we may take X to be the disjoint union of all points of $Spec(B)$). X may not, however, be an affine scheme. To give an example, consider

$$
B = \prod_{n=1,2,\dots} \mathbb{F}_2,
$$

(where $\mathbb{F}_2$ is the field with two elements) and let

$$
\phi : \coprod_{n=1,2,\dots} Spec(\mathbb{F}_2) \to Spec(B) \tag{2.3.1}
$$

be the morphism of schemes which sends the n'th point $Spec(\mathbb{F}_2)$ to the maximal ideal p_i of B consisting of all sequences $(a_1, a_2, \dots)$, $a_i \in \mathbb{Z}/2$, such that $a_n = 0$. (That morphism of schemes is $Spec$ of the n'th projection from the product.) Then the image of the morphism of schemes (2.3.1) is certainly closed under specialization, as every point of its image is closed. The image of ϕ is also dense, since its closure is

$$
Z(\bigcap_{i=1,2,\dots} p_i) = Z(0) = Spec(B).
$$

However, there are closed points in $Spec(B)$ which are not in the image of ϕ: For example, let I be the ideal in B consisting of all sequences $(a_1, a_2, \ldots)$ where for all but finitely many n, $a_n = 0$. Let p be any maximal ideal containing I (which exists by Zorn's lemma). Then, clearly, for any given n, p contains elements which are not in p_n, and hence is not in the image of ϕ.

2.4 Valuation Rings

Let us now model the situation

$$p \rightsquigarrow q \in Spec(B) \tag{2.4.1}$$

by morphisms of schemes. We know that $p \subseteq q$ are prime ideals of B. Then B/p is an integral domain (i.e. has no zero divisors), so it has a field of fractions

$$Q(B/p) = (Q/p)_{(0)}$$

(i.e. its localization at the 0 ideal, which is prime). On the other hand, by Noether's theorem, ideals of B/p correspond bijectively to ideals of B which contain p, so we can (using q also to denote the corresponding ideal in B/p) also consider the local ring

$$(B/p)_q,$$

and clearly, its fraction field is $(B/p)_{(0)} = Q(B/p)$. Thus, the situation (2.4.1) can be modeled by a diagram of morphisms of affine schemes

$$\begin{array}{ccc} Spec(K) & \longrightarrow & Spec(B) \\ \downarrow{\scriptstyle \subseteq} & \nearrow & \\ Spec(R) & & \end{array} \tag{2.4.2}$$

where

> K is a field of fractions of an integral domain R, which is a local ring not equal to K. $\hspace{2em}$ (2.4.3)

But it turns out that the situation (2.4.3) is still too general, since there are too many local rings which are integral domains (for example, all stalks of structure sheaves of integral schemes). However, for the purposes of (2.4.1), we may make the situation (2.4.3) much more manageable by the following trick.

For a fixed field K, and rings R, $R' \subseteq K$ local subrings, we say that R' *dominates* R if $R \subseteq R'$, and the inclusion is a morphism of local rings (in other words, denoting the maximal ideals of R, R' by m_R, $m_{R'}$, if $m_{R'} \cap R = m_R$). We will see that maximal local rings with respect to the relation of domination can be characterized nicely.

2.4.1 Definition A *partially ordered abelian group* is an abelian group Γ together with a partial ordering $\leq$ such that for all $x, y, z \in \Gamma$,

$$x \leq y \Rightarrow x + z \leq y + z.$$

We say that Γ is a *totally ordered abelian group* if the ordering $\leq$ is a total ordering (i.e. $x \leq y$ or $y \leq x$ for any two elements x, y). We say that a ring R as in (2.4.3) is a *valuation ring* if there exists a totally ordered abelian group Γ, and a *valuation*

$$v : K^{\times} = K \smallsetminus \{0\} \to \Gamma,$$

which means a function satisfying the following properties for all $x, y, z \in K^{\times}$:

1. $v(x) \geq 0$ if and only if $x \in R$.
2. $v(x \cdot y) = v(x) + v(y)$.
3. $v(x + y) \geq \min(v(x), v(y))$ (assuming $x + y \neq 0$).

Example We say that a ring in the situation (2.4.3) is a *discrete valuation ring* if it satisfies Definition 2.4.1 for $\Gamma = \mathbb{Z}$. A typical example of a discrete valuation ring in $\mathbb{Z}_{(p)}$. The valuation

$$v : \mathbb{Q}^{\times} \to \mathbb{Z}$$

is defined as follows: assuming $a, b \in \mathbb{Z}$, $gcd(a, b) = 1$, $b \neq 0$,

$$v(\frac{a}{b}) = \begin{cases} 0 \text{ if } p \nmid a \text{ and } p \nmid b \\ n \text{ if } p^n \mid a, \ p^{n+1} \nmid a, n > 0 \\ -n \text{ if } p^n \mid b, \ p^{n+1} \nmid b, n > 0. \end{cases}$$

In fact, more generally, a Noetherian ring R is called a *Dedekind domain* if it is an integral domain, has Krull dimension 1, and is normal. (This means that R is integrally closed in its field of fractions, i.e. that every root $x \in QR$ of a monic polynomial with coefficients in R satisfies $x \in R$.) Every unique factorization domain is normal (see Exercise 6), although it, of course, does not have to be of dimension ≤ 1 (think about $\mathbb{C}[x, y]$). On the other hand, Dedekind domains do not have to be unique factorization domains. They have, however, unique factorization of *ideals* (with respect to product), which is a basic starting point of algebraic number theory. Unique factorization of ideals in Dedekind domains is fairly easy to prove directly. We will give a geometric proof in Sect. 2.3 of Chap. 4 below using coherent sheaves of ideals. However, it is also possible to give a direct proof using commutative algebra. (See Exercises 7, 8.) As a key step in proving unique factorization of ideals for Dedekind domains, one proves the following result. For any ideal I in an integral domain R with field of fractions K, define $I^{-1} \subseteq K$ by

$$I^{-1} = \{a \in K \mid aI \subseteq R\}. \tag{2.4.4}$$

2.4.2 Lemma

1. If p is the maximal ideal of a local integral domain A and $p \cdot p^{-1} = A$, then p is principal (and hence A is a discrete valuation ring).
2. For a prime ideal p in a Dedekind domain R, the localization R_p is a discrete valuation ring.

Proof To prove 1, we use here Krull's theorem which asserts that in a Noetherian integral domain A, and $I \subsetneq A$ an ideal not equal to A,

$$\bigcap_{n=1}^{\infty} I^n = \{0\}$$

(Proposition 5.4.2 of Chap. 1). Applying this to p, we see that $p^2 \neq p$. Let $a \in p \smallsetminus p^2$. Then $ap^{-1} \subseteq A$. We shall see that equality arises. If not, then $ap^{-1} \subseteq p$, which would imply

$$aA = ap^{-1}p \subseteq p^2,$$

which is a contradiction with the choice of a. Thus, $ap^{-1} = A$, so

$$aA = ap^{-1}p = p,$$

and so $p = (a)$.

To prove 2, without loss of generality, $p \neq 0$. Let $A = R_p$. By 1, it suffices to show that

$$pp^{-1} = A. \tag{2.4.5}$$

Of course, again, by definition we have $pp^{-1} \subseteq A$, so if (2.4.5) is false, then $pp^{-1} = p$, which implies

$$p(p^{-1})^n = p \tag{2.4.6}$$

for all $n \in \mathbb{N}$. Since A is Noetherian, p is finitely generated as an ideal, so using these generators, (2.4.6) can be used to write polynomial equations for all elements of p^{-1} with highest coefficient 1, and hence all elements of p^{-1} are integral over A, and hence $A = p^{-1}$ since A is normal. This is a contradiction, since $a^{-1} \in p^{-1} \smallsetminus A$. □

At the moment, however, our interest in valuation rings comes from the following result. Note that by Zorn's lemma, every ring R in the situation (2.4.3) is dominated by a maximal one.

2.4.3 Theorem *Maximal elements of the set of local subrings of a field K with respect to domination are precisely valuation rings.*

Proof First, note that a subring $R \subseteq K$ is a valuation ring if

$$\text{for all } x \in K, x \in R \text{ or } x^{-1} \in R. \tag{2.4.7}$$

The reason is that we then may let

$$\Gamma = K^{\times}/R^{\times}$$

for a ring R, $R^{\times}$ denotes the group of units, i.e. invertible elements, of R, and let

$$v : K^{\times} \to \Gamma$$

be the projection (although to conform with Definition 2.4.1, we would need to write Γ additively). The ordering on Γ is

$$a \leq b \text{ if there exist } c \in R \text{ such that } ac = b.$$

Now let $R \subseteq K$ be a local ring with maximal ideal m, and let $x \in K^{\times}$, $x \notin R$. Denote by $R[x] \subseteq K$ the subring of K generated by R and x. Also denote by $m[x]$ the ideal in $R[x]$ generated by m. We claim that

$$m[x] \neq R[x] \text{ or } m[x^{-1}] \neq R[x^{-1}]. \tag{2.4.8}$$

Suppose the contrary, i.e. $m[x] = R[x]$ and $m[x^{-1}] = R[x^{-1}]$. Then there exist $u_i, v_j \in m$ such that in K we have

$$u_0 + u_1 x + \cdots + u_m x^m = 1, \tag{2.4.9}$$

$$v_0 + v_1 x^{-1} + \cdots + v_n x^{-n} = 1. \tag{2.4.10}$$

Now assume $m + n$ to be as small as possible. Without loss of generality, $m \geq n$. Then from (2.4.10),

$$(v_0 - 1)x^n + v_1 x^{n-1} + \cdots + v_n = 0. \tag{2.4.11}$$

Further, since R is a local ring, $v_0 - 1$ is a unit in R. So multiplying (2.4.11) by

$$\frac{u_m x^{m-n}}{v_0 - 1}$$

and subtracting from (2.4.9), we can decrease m, which contradicts our assumption that $m + n$ was as small as possible.

Now suppose a local ring $R \subseteq K$ is maximal with respect to domination. We claim that (2.4.7) occurs. Otherwise, let $x \in K$ be such that $x, x^{-1} \notin R$. Then we have (2.4.8). Assume, without loss of generality, $m[x] \neq R[x]$. But then $R[x]$ dominates R, contradicting maximality. Thus, f satisfies (2.4.7), and hence is a valuation ring. $\square$

2.5   Valuation Criteria for Universally Closed and Separated Morphisms

Our main interest in valuation rings is a general principle where to test a property for (2.4.1), we only need to test, in terms of morphisms of affine schemes, (2.4.2) in the case when R is a valuation ring with field of fractions K. We are specifically interested in the following

2.5.1 Theorem *Let X be a scheme of finite type. Then X is universally closed if and only if every diagram*

$$
\begin{array}{ccc}
Spec(K) & \longrightarrow & X \\
\Big\downarrow{\scriptstyle \subseteq} & \nearrow & \\
Spec(R) & &
\end{array}
\tag{2.5.1}
$$

where R is a valuation ring with field of fractions K can be completed with the dotted arrow (existence). On the other hand, X is separated if every diagram (2.5.1) where R is a valuation ring with field of fractions K can be completed with the dotted arrow in at most one way (uniqueness).

Remark An analogous statement holds for morphisms of schemes $f : X \to Y$. The testing diagram (2.5.1) then becomes

$$
\begin{array}{ccc}
Spec(K) & \longrightarrow & X \\
\downarrow \subseteq & \nearrow & \downarrow f \\
Spec(R) & \longrightarrow & Y
\end{array}
\qquad (2.5.2)
$$

(when Y is the terminal object $Spec(\mathbb{Z})$, we do not have to write it). When X is Noetherian, uniqueness of the dotted arrow for every testing diagram is equivalent to the morphism f being separated, and when f is of finite type and X is quasicompact, the existence for every testing diagram is equivalent to f being universally closed. While the exact assumptions needed are subtle, for our purposes, they matter less than one may think. This is because the main application we need is to prove that projective schemes are proper, and it suffices to verify that for $\mathbb{P}^n_{\mathbb{Z}}$.

Proof of the First Part of Theorem 2.5.1 We will prove that X satisfies the criterion of being able to complete every diagram (2.5.1) if and only if it is universally closed.

For necessity, suppose we have a diagram (2.5.1) for a valuation ring R with field of fractions K. Consider the product map

$$
f : Spec(K) \to X \times Spec(R).
$$

Let Z be the closure of the image q' of $(0) \subset K$ under f. Consider the projection

$$
\pi : X \times Spec(R) \to Spec(R).
$$

Since X is universally closed, we must have $p \in \pi(Z)$ where $p \in Spec(R)$ is the maximal ideal in R. Thus, there must exist a point $p' \in Z$ such that $\pi(p') = p$. Now we must have $p' \in Spec(B \otimes R)$ where $Spec(B) \subseteq X$ is an open affine subscheme. In particular, we must also have $q' \in Spec(B)$ (since $Spec(B)$ is an open subset and p' is in the closure of q'). Thus, we obtain a homomorphism

$$
h : B \otimes R \to K
$$

and a prime ideal $p' \subset B \otimes R$ such that the pullback of p' to R is p. We claim that the composition

$$
\overline{h} : B \to B \otimes R \to K
$$

where the first homomorphism is the coproduct injection and the second is h lands in R. Then we are done, since $Spec(\overline{h})$ is a restriction of the testing map. Otherwise, there would be an element $x \in B$ with $v(\overline{h}(x)) < 0$ (where v is the valuation on K corresponding to R). Then

$$h((x \otimes (\overline{h}(x))^{-1}) - 1) = 0,$$

and thus,

$$(x \otimes (\overline{h}(x))^{-1}) - 1 \in p'.$$

On the other hand, by assumption, $1 \otimes (\overline{h}(x))^{-1} \in p'$ and thus

$$x \otimes (\overline{h}(x))^{-1} \in p'.$$

Thus, $1 \in p'$, which is a contradiction.

To prove sufficiency, we need to prove that for all schemes Y, the projection

$$\pi : X \times Y \to Y$$

is closed. Without loss of generality, $Y = Spec(B)$ is affine. As spaces, we have

$$X = Spec(A_1) \cup \ldots Spec(A_n)$$

because X is quasicompact, so the image of π is the image of the corresponding morphism of schemes

$$Spec(A_1 \times \cdots \times A_n) = Spec(A_1) \amalg \cdots \amalg Spec(A_n) \to Spec(B).$$

The source is an affine scheme, so by Lemma 2.3.1, it suffices to show that the image of π is closed under specialization. Let, then, $p \in \pi(X \times Spec(B))$, $p \rightsquigarrow q$. This means that p is in the image of $Spec(A_i) \times Spec(B) = Spec(A_i \otimes B)$ under π, and $q \supseteq p$. This means that there exists a prime ideal

$$p' \subset A_i \otimes B$$

where $\iota^{-1}(p') = p$ where

$$\iota : B \to A_i \otimes B$$

is $1 \otimes ?$. Now denote by K the field of fractions of B/p, and by K' the field of fractions of $(A_i \otimes B)/p'$. Let

$$R = B_q.$$

Then we have a diagram

$$
\begin{array}{ccc}
 & & K' \\
 & & \uparrow{\scriptstyle \subseteq} \\
R & \xrightarrow{\subseteq} & K.
\end{array}
$$

Let R' be a maximal local subring of K' dominating R. By the criterion of the diagram (2.5.1), we can complete the diagram

$$
\begin{array}{ccc}
Spec(K') & \longrightarrow & X \\
\downarrow{\scriptstyle \subseteq} & \nearrow & \\
Spec(R'). & {\scriptstyle \phi} &
\end{array}
$$

where the top row has

$$(0) \mapsto p'.$$

On the other hand, the homomorphism ι induces a morphism of schemes

$$Spec(R') \to Spec(R).$$

Denote by ψ its composition with the canonical morphism

$$Spec(R) = Spec(B_q) \to Spec(B).$$

Then we have a product morphism

$$(\phi, \psi) : Spec(R') \to X \times Y.$$

By definition, if q' is the maximal ideal in R', then

$$(\phi, \psi)(q') = q,$$

so q is in the image of π, which we were trying to prove.

□

Example We will prove that the projective space $\mathbb{P}_{\mathbb{Z}}^n$ is proper. (It then follows that the projection $\mathbb{P}_S^n \to S$ is proper for any scheme S.) We will use Theorem 2.5.1. Let us recall some of the concepts. We have

$$\mathbb{P}_{\mathbb{Z}}^n = Proj(\mathbb{Z}[x_0, \ldots, x_n]).$$

Recall that $\mathbb{Z}[x_0, \ldots, x_n]$ is a graded ring (which is why we may apply *Proj*), where a homogeneous element of degree k is a homogeneous polynomial of degree k. For example, for $n = 2$, a homogeneous polynomial of degree 2 is

$$3x_0^2 + 5x_0x_1. \tag{2.5.3}$$

The distinguished affine open sets of $\mathbb{P}_{\mathbb{Z}}^n$ are then

$$Spec((z^{-1}\mathbb{Z}[x_0, \ldots, x_n])_0)$$

where z is a homogeneous polynomial. For example,

$$Spec((x_i^{-1}\mathbb{Z}[x_0, \ldots, x_n])_0), \quad i = 0, \ldots, n,$$

are distinguished open affine sets. Recall that the subscript means homogeneous elements of degree 0. For example,

$$Spec((x_1^{-1}\mathbb{Z}[x_0, x_1, x_2])_0 \ni \frac{3x_0^2 + 5x_0x_1}{x_1^2} = 3\left(\frac{x_0}{x_1}\right)^2 + 5\left(\frac{x_0}{x_1}\right). \tag{2.5.4}$$

Now we will prove that $\mathbb{P}_{\mathbb{Z}}^n$ is proper. Let us first prove that $\mathbb{P}_{\mathbb{Z}}^n$ is universally closed. This means that for every scheme Y', the projection

$$p : \mathbb{P}_{\mathbb{Z}}^n \times Y' \to Y' \tag{2.5.5}$$

is closed, which in turn means that for every $Z \subseteq \mathbb{P}_{\mathbb{Z}}^n \times Y'$ closed, $p(Z)$ is closed.

This is a powerful statement. For example, recall that a closed subset of a scheme again has the structure of a scheme, so we know that every $p(Z)$ as above has the structure of a scheme!

Now our strategy is not to prove that (2.5.5) is closed directly. Instead, we shall use Theorem 2.5.1. This means that we need to prove that for a valuation ring $R \subseteq K$ where

K is the field of fractions of R, every diagram of the form

$$\begin{array}{ccc} Spec(K) & \xrightarrow{\;f\;} & \mathbb{P}^n_{\mathbb{Z}} \\ {\scriptstyle \subseteq}\downarrow & \nearrow & \\ Spec(R) & & \end{array}$$

$$(2.5.6)$$

can be completed. Denote the valuation of R by

$$v : K^\times \to \Gamma$$

where Γ is a totally ordered abelian group. Note that $Spec(K)$ has only one point. Assume by induction that we have already proved that $\mathbb{P}^{n-1}_{\mathbb{Z}}$ is universally closed. ($\mathbb{P}^0_{\mathbb{Z}} = Spec(\mathbb{Z})$ is obviously universally closed, since for $n = 0$ the projection (2.5.5) is the identity). We may then assume that $f(Spec(K))$ is not contained in the image of any of the maps

$$\iota_i : \mathbb{P}^{n-1}_{\mathbb{Z}} \to \mathbb{P}^n_{\mathbb{Z}}, \; i = 0, \dots, n$$

where $\iota_i = Proj(h_i)$ where

$$h_i : \mathbb{Z}[x_0, \dots, x_n] \to \mathbb{Z}[x_0, \dots, \widehat{x_i}, \dots, x_n]$$

is the homomorphism of graded rings which sends $x_i \mapsto 0$ (recall that the notation on the right hand side means that the variable x_i is omitted).

The complement of the images of all the ι_i's is a distinguished affine open set in $\mathbb{P}^n_{\mathbb{Z}}$ of the form

$$Spec(((x_0 \dots x_n)^{-1}\mathbb{Z}[x_0, \dots, x_n])_0),$$

so we have a morphism of schemes

$$Spec(K) \to Spec(((x_0 \dots x_n)^{-1}\mathbb{Z}[x_0, \dots, x_n])_0),$$

which is the $Spec$ of a homomorphism of rings

$$\phi : ((x_0 \dots x_n)^{-1}\mathbb{Z}[x_0, \dots, x_n])_0 \to K.$$

$$(2.5.7)$$

Now the domain of (2.5.7) is generated as a ring by elements of the form

$$\frac{x_i}{x_j}$$

(recall (2.5.4) as an example). Now choose i in such a way that the valuation

$$v(\phi(\frac{x_i}{x_0}))$$

is the smallest among

$$v(\phi(\frac{x_0}{x_0})), \ldots, v(\phi(\frac{x_n}{x_0})).$$

This means that for all $j \in \{0, \ldots, n\}$,

$$v(\phi(\frac{x_j}{x_i})) = v(\phi(\frac{\frac{x_j}{x_0}}{\frac{x_i}{x_0}})) = v(\phi(\frac{x_j}{x_0})) - v(\phi(\frac{x_j}{x_0})) \geq 0,$$

which means

$$\phi(\frac{x_j}{x_i}) \in R \subseteq K,$$

so

$$\phi((x_i^{-1}\mathbb{Z}[x_0, \ldots, x_n])_0) \subseteq R \subseteq K.$$

Taking $Spec$, we managed to complete the diagram

$$
\begin{array}{ccc}
Spec(K) & \xrightarrow{\ f\ } & Spec((x_i^{-1}\mathbb{Z}[x_0, \ldots, x_n])_0) \\
\downarrow{\subseteq} & \nearrow & \\
Spec(R) & &
\end{array}
$$

which means we completed Diagram (2.5.6) also. Thus, we verified the assumptions of Theorem 2.5.1, and hence we know that $\mathbb{P}^n_{\mathbb{Z}}$ is universally closed.

We shall now prove that $\mathbb{P}^n_{\mathbb{Z}}$ is separated. It then follows that for any scheme S, $\mathbb{P}^n_S$ is separated over S, by Exercise 23 of Chap. 2. Thus, $\mathbb{P}^n_S$ is proper over S.

To see that $\mathbb{P}^n_{\mathbb{Z}}$ is separated, we need to show that the diagonal in $\mathbb{P}^n_{\mathbb{Z}} \times \mathbb{P}^n_{\mathbb{Z}}$ is closed. Cover $\mathbb{P}^n_{\mathbb{Z}} = Proj[x_0, \ldots, x_n]$ by the affine open subsets

$$U_i = Spec((x_i^{-1}\mathbb{Z}[x_0, \ldots, x_n])_0)$$

$$= Spec(\mathbb{Z}[\frac{x_0}{x_i}, \ldots, \widehat{\frac{x_i}{x_i}}, \ldots, \frac{x_n}{x_i}]), \ i = 0, \ldots, n.$$

This notation is slightly imprecise, but it is true that

$$U_i \cong \mathbb{A}_{\mathbb{Z}}^n = Spec(\mathbb{Z}[y_{0,i}, \ldots, \widehat{y_{i,i}}, \ldots, y_{n,i}])$$

where $y_{j,i}$ corresponds to $\frac{x_j}{x_i}$. Similarly, we may then identify

$$U_i \times U_j \cong$$

$$Spec(\mathbb{Z}[y_{0,i}, \ldots, \widehat{y_{i,i}}, \ldots, y_{n,i}, z_{0,j}, \ldots, \widehat{z_{j,j}}, \ldots, z_{n,j}]).$$

We also know that the intersection of $U_i \times U_j$ with the diagonal is

$$U_i \cap U_j = Spec(((x_i x_j)^{-1} \mathbb{Z}[x_0, \ldots, x_n])_0).$$

In particular, this is also an affine scheme, and the inclusion

$$U_i \cap U_j \to U_i \times U_j$$

is the $Spec$ of the homomorphism of rings which sends

$$y_{k,i} \mapsto \frac{x_k}{x_i}, \quad z_{\ell,j} \mapsto \frac{x_\ell}{x_j}.$$

The images generate the ring

$$((x_i x_j)^{-1} \mathbb{Z}[x_0, \ldots, x_n])_0,$$

which is why $U_i \cap U_j$ is closed in $U_i \times U_j$, and $\mathbb{P}_{\mathbb{Z}}^n$ is separated.

2.5.2 Lemma *A closed subscheme Z of a universally closed scheme X is universally closed. (Similarly for schemes over a fixed scheme S.)*

Proof Consider the diagram

$$\begin{array}{ccc} Z \times Y' & \longrightarrow & Y' \\ \downarrow & & \downarrow {\scriptstyle Id} \\ X \times Y' & \longrightarrow & Y'. \end{array}$$

Clearly, all we need to prove is that the left vertical arrow is closed. To this end, it suffices to consider the case when X and Y' are affine (although note that this may break the assumption of X being universally closed, but we no longer need that assumption for what we are proving now). Thus, let $X = Spec(R)$, $Y' = Spec(R')$. We have $Z = Spec(R/I)$ (by Theorem 2.3.1 of Chap. 2), so $X \times Y' = Spec(R \otimes R')$, $Z \times Y' = Spec(R/I \otimes R')$. Now

$$(R/I) \otimes R' \cong (R \otimes R')/(I),$$

so $Z \times Y'$ is a closed subset of $X \times Y'$, as claimed. The case over a scheme S is analogous.

$\square$

We see therefore that a projective scheme over a scheme S is proper over S.

We saw that the part of Theorem 2.5.1 about separatedness is less important, since it is easier to prove directly that a scheme is separated, but for completeness, we should include an argument. For the necessity, see Exercise 13.

Proof of Sufficiency of the Separatedness Criterion of Theorem 2.5.1. Suppose X is a Noetherian scheme and for every valuation ring R with field of fractions K and a morphism of schemes $f : Spec(K) \to X$, the diagram

$$
\begin{array}{ccc}
Spec(K) & \xrightarrow{f} & X \\
\Big\downarrow{\subseteq} & \nearrow & \\
Spec(R) & &
\end{array}
$$

can be completed at most in one way.

Since X is Noetherian, so it is covered by finitely many Noetherian open affine subschemes $U_1, \ldots, U_k$. By Lemma 3.2.6 of Chap. 2, the open subschemes $U_i \cap U_j$ are also Noetherian and hence quasicompact. Now the diagonal morphism $\Delta : X \to X \times X$ satisfies $\Delta^{-1}(U_i \times U_j) = U_i \cap U_j$ where $U_i \times U_j$ cover $X \times X$. Let $p \in U_i \times U_j$ be in the closure of $\Delta(X)$. Then p is in the closure of $\Delta(U_i \cap U_j)$. However, since $U_i \cap U_j$ is quasicompact, it is a union of finitely many affine open subsets, and thus there exists an affine open subset $V \subseteq U_i \cap U_j$ such that p is in the closure of $f(V)$. Therefore, by Lemma 2.3.1, p is a specialization of a point in $f(V)$.

Therefore, it suffices to show that the diagonal is closed under specialization. We may test specialization by a diagram of the form

$$
\begin{array}{ccc}
Spec(K) & \xrightarrow{f} & X \\
\Big\downarrow & & \Big\downarrow{\Delta} \\
Spec(R) & \xrightarrow{g} & X \times X
\end{array}
\tag{2.5.8}
$$

for a valuation ring R with field of fractions K. In a general category, the morphism g is always given as $g = (g_1, g_2)$ where $g_i : Spec(R) \to X$. By the uniqueness, however,

$$g_1 = g_2$$

and therefore Diagram (2.5.8) completes to

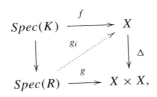

thus showing that the diagonal is closed under specialization.   □

2.6   More Observations on Universally Closed and Separated Morphisms

Note that in the case of the projective space, we in fact applied another criterion of being separated, which is:

2.6.1 Proposition *Let X be a scheme which is a union of open affine sets $U_i = Spec(R_i)$, $i \in I$. Then X is separated if and only if for all $i, j \in I$, $U_i \cap U_j = Spec(R_{i,j})$ is affine, and the homomorphism of rings*

$$R_i \otimes R_j \to R_{i,j}$$

corresponding to the diagonal inclusion $U_i \cap U_j \subseteq U_i \times U_j$ is onto. (Similarly for schemes over a scheme S, if we replace $\times$ by $\times_S$.)

□

Proof See Exercise 14.

The following theorem is a scheme-theoretic analog of Theorem 2.1.2:

2.6.2 Theorem *If $f : X \to Y$ is a morphism of schemes where X is universally closed and Y is separated, then f is closed. (Similarly for schemes over a scheme S.)*

Proof Consider the pullback diagram

$$
\begin{array}{ccc}
\Gamma & \xrightarrow{\ \ g\ \ } & Y \\
{\scriptstyle h}\downarrow & & \downarrow{\scriptstyle \Delta} \\
X \times Y & \xrightarrow[\ f\times Id\ ]{} & Y \times Y.
\end{array}
\tag{2.6.1}
$$

The scheme Γ is called the *graph* of the morphism f. (It is tempting to imagine that it consists of pairs $(x, y) \in X \times Y$ where $f(x) = y$, but note that this is of course incorrect in the category of schemes.) The morphism h is closed since we are assuming that Δ is closed (note that Y is separated). Thus, the composition of h with the projection

$$
p_2 : X \times Y \to Y
\tag{2.6.2}
$$

to the second factor is closed, since (2.6.2) is closed since X is universally closed. Now we claim that we actually have a diagram

$$
\begin{array}{ccc}
X & \xrightarrow{\ f\ } & Y \\
{\scriptstyle \cong}\downarrow{\scriptstyle j} & \nearrow{\scriptstyle g} & \\
\Gamma & &
\end{array}
\tag{2.6.3}
$$

To see (2.6.3), consider also the pullback diagram

$$
\begin{array}{ccc}
X \times Y & \xrightarrow{\ f\times Id\ } & Y \times Y \\
{\scriptstyle p_1}\downarrow & & \downarrow{\scriptstyle p_1} \\
X & \xrightarrow[\ f\ ]{} & Y.
\end{array}
\tag{2.6.4}
$$

Composition of pullback diagrams is a pullback diagram, and composing (2.6.1) with (2.6.4) is therefore a pullback. Note, however, that this is the diagram

$$
\begin{array}{ccc}
\Gamma & \xrightarrow{\ \ g\ \ } & Y \\
{\scriptstyle p_1\circ h}\downarrow & & \downarrow{\scriptstyle Id} \\
X & \xrightarrow[\ f\ ]{} & Y.
\end{array}
$$

which implies (2.6.3) by the categorical uniqueness of pullback. It also shows that

$$p_2 \circ h \circ j = f,$$

since (2.6.1) commutes, which implies $g = p_2 \circ h$. However, we just showed that the left hand side is closed, so f is closed, as claimed. $\quad\square$

2.6.3 Proposition *A finite morphism of schemes $f : X \to Y$ is proper.*

Proof We see right away that f is separated (since it is *affine*, i.e. the inverse image of every affine open subset is affine).

We want to use the valuation criterion (Theorem 2.5.1) to see that f is universally closed. However, a subtlety is that we cannot apply the result directly, since we did not assume that X is quasicompact. This can be remedied as follows: first, one proves that a pullback of a finite morphism is finite (Exercise 35 of Chap. 2). Therefore, it suffices to show that a finite morphism is always closed. For this purpose, without loss of generality, $Y = Spec(A)$, and thus also $X = Spec(B)$ are affine, and f is *Spec* of a homomorphism of rings $h : A \to B$ which makes B into a finitely generated A-module. Now we can use the valuation criterion.

Let R be a valuation ring with field of fractions K and valuation v. The testing diagram is the *Spec* of

$$
\begin{array}{ccc}
A & \longrightarrow & R \\
{\scriptstyle h}\downarrow & & \downarrow{\scriptstyle \subseteq} \\
B & \underset{g}{\longrightarrow} & K.
\end{array}
\qquad (2.6.5)
$$

Since the ring B is finitely generated as an A-module, all its elements are integral over A and hence their images in K are integral over R. However, a valuation ring is integrally closed in its field of fractions by maximality with respect to domination (see Exercise 10). Thus, $g(B) \subseteq R$. $\quad\square$

---

3 Regular Schemes and Smooth Morphisms

Algebraic geometry knows two basic concepts of 'non-singularity': regularity and smoothness. There are some relationships between them, but they are not equivalent.

3.1 Regular Schemes

A scheme X is *regular* when for every point $x \in X$, the stalk $\mathcal{O}_{X,x}$ is a regular local ring and x has an open affine neighborhood U such that $\mathcal{O}_X(U)$ is Noetherian. (We express this last condition by saying that X is *locally Noetherian*. Compare with Sect. 3.5 of Chap. 1.) It is sufficient to verify the regularity condition for closed points $x \in X$, since if x is a closed point which is a specialization of a point y (i.e. x is in the closure of y), then the ring $\mathcal{O}_{X,y}$ is a localization of the ring $\mathcal{O}_{X,x}$ at the ideal corresponding to y, and a localization of a regular local ring is regular. (This follows from the cohomological characterization of regularity which we will prove in Sect. 2.6 of Chap. 5. See also Exercise 19 of Chap. 5.)

3.2   Smooth Morphisms

A separated morphism of schemes $f : X \to Y$ is called *smooth* when for every $y \in Y$ there exists an affine open set $U \subseteq Y$, $y \in U \cong Spec(A)$, such that $f^{-1}(U)$ is a union of open affine subsets $V_i \cong Spec(B_i)$ where the homomorphism of rings $A \to B_i$ associated with f is standard smooth. (See Sect. 3.3 of Chap. 1.) If all of the homomorphisms $A \to B_i$ are standard smooth of a constant dimension k, we say that that the morphism f is *smooth of dimension k*. A smooth morphism of dimension 0 is also called an *étale morphism*.

Example

1. For any scheme X, the projection $\mathbb{A}^n_X \to X$ is obviously smooth of dimension n, and hence so is the projection $\mathbb{P}^n_X \to X$.
2. Now consider the scheme

$$X = Spec(R), \quad R = \mathbb{C}[x, y]/(x^2 - y^3 - y). \tag{3.2.1}$$

We will show that the projection $X \to Spec(\mathbb{C})$ associated with the inclusion homomorphism of rings $\mathbb{C} \to R$ is smooth. Indeed, the partial derivatives of $x^2 - y^3 - y$ by x and y are

$$2x, \quad 3y^2 - 1. \tag{3.2.2}$$

Thus, we need to show that the ideal in R generated by (3.2.2) contains 1. This is equivalent to showing that the ideal I in $\mathbb{C}[x, y]$ generated by

$$2x, 3y^2 - 1, x^2 - y^3 - y$$

contains 1. For this, we have an algorithm (Gröbner basis). Specifically, since 2 is a unit, I contains $y^3 + y$, but it also contains $3y^3 - 3y$, so it contains $6y$, and hence y, and hence 1.

On this example, we see why smoothness is intrinsically a property of morphisms: smooth schemes over $Spec(\mathbb{Z})$ are, in fact, quite rare (although there are some, such as the affine and projective space). See also Exercise 22.

3. Let us consider the embedding of the scheme X from the last example into $\mathbb{A}^2_{\mathbb{C}} = Spec(\mathbb{C}[x, y])$. Let us compose this with the embedding $\mathbb{A}^2_{\mathbb{C}} \to \mathbb{P}^2_{\mathbb{C}}$, and let us consider the closure E of the image of X in $\mathbb{P}^2_{\mathbb{C}} = Proj(\mathbb{C}[x, y, z])$. We will show that the projection $p : E \to Spec(\mathbb{C})$ is also smooth. Since smoothness is a local property, it suffices to consider the restriction of p to the affine sets $U_{xy} = Spec(\mathbb{C}[\frac{x}{z}, \frac{y}{z}])$, $U_{xz} = Spec(\mathbb{C}[\frac{x}{y}, \frac{z}{y}])$ and $U_{yz} = Spec(\mathbb{C}[\frac{y}{x}, \frac{z}{x}])$ of $\mathbb{P}^2_{\mathbb{C}}$. We already dealt with the set U_{xy} by identifying it with $\mathbb{A}^2_{\mathbb{C}}$ via $z = 1$. Thus, the ideal generator (3.2.1) becomes

$$\frac{x^2}{z^2} - \frac{y^3}{z^3} - \frac{y}{z},$$

which over $U_{xy} \cap U_{xz}$ can be written as

$$\frac{x^2}{y^2} - \frac{y}{z} - \frac{z}{y}.$$

We see that the intersection of the ideal generated by this element in $\mathbb{C}[\frac{x}{y}, \frac{z}{y}, \frac{y}{z}]$ with $\mathbb{C}[\frac{x}{y}, \frac{z}{y}]$ is generated by

$$\frac{x^2 z}{y^3} - \frac{z^2}{y^2} - 1,$$

which, using the isomorphism with $\mathbb{C}[x, z]$ by setting $y = 1$, becomes

$$x^2 z - z^2 - 1.$$

The partial derivatives are

$$2xz, x^2 - z^2.$$

These three elements generate the unit ideal in $\mathbb{C}[x, z]$, since they generate $z^2 + 1$, and also $z^3 - z^2 - 1$. Similarly, over $\mathbb{C}[y, z] \cong \mathbb{C}[\frac{y}{x}, \frac{z}{x}]$, we obtain the ideal generated by

$$z - y^3 - yz^2,$$

with partial derivatives

$$1 - 2yz, 3y^2 - z^2,$$

which again generate 1, since they generate $2y^3 - z, z - 4y^3$, and hence z.

The projective scheme E is an example of an *elliptic curve*. An elliptic curve over a field k is a closed subscheme of $E \subset \mathbb{P}^2_{Spec(k)}$ which is smooth over $Spec(k)$, has a point with residue field k and such that $E \cap \mathbb{A}^2_{Spec(k)}$ is isomorphic, over $Spec(k)$, to $Spec(k[x, y]/(p(x, y)))$ where $p(x, y)$ is a cubic polynomial.

3.3 Smooth Morphisms Over Regular Schemes

3.3.1 Theorem *Let $f : X \to Y$ be a smooth morphism of schemes where Y is a regular scheme. Then X is a regular scheme.*

Proof Clearly, it suffices to consider the case when both X and Y are affine, i.e. a smooth homomorphism of rings

$$h : A \to B$$

where A is regular. Let us assume that

$$B = A[x_1, \ldots, x_{m+n}]/(p_1, \ldots, p_m)$$

where the determinants of the $m \times m$ submatrices of the Jacobi matrix

$$\left(\frac{\partial p_i}{\partial x_j} \right)_{i,j}$$

generate the ideal B in B. We will proceed by induction on m. Assume the statement is true with m replaced by $m - 1 \geq 0$.

First, note that the tuples $p_1, \ldots, p_\ell$ for $1 \leq \ell \leq m$ satisfy the same condition, since the $m \times m$ determinants obviously are in the ideal generated by the $\ell \times \ell$ determinants in the first ℓ rows. Next, we claim that $p_1, \ldots, p_m$ is a regular sequence in $A[x_1, \ldots, x_{m+n}]$. (Recall that a sequence of elements $r_1, \ldots, r_m$ in a commutative ring R is a *regular sequence* when r_ℓ is a non-zero divisor in $R/(r_1, \ldots, r_{\ell-1})$.)

To show that $p_1, \ldots, p_m$ form a regular sequence, we may assume by the induction hypothesis that it is true with m replaced by $m - 1$. Assume, therefore, that

$$p_m \cdot r = a_1 p_1 + \cdots + a_{m-1} p_{m-1} \in A[x_1, \ldots, x_{m+n}], \qquad (3.3.1)$$

$$r \notin (p_1, \ldots, p_{m-1}), a_i \in A[x_1, \ldots, x_{m+n}].$$

Since we know by the induction hypothesis that the ring

$$A[x_1, \ldots, x_{m+n}]/(p_1, \ldots, p_{m-1})$$

is regular, it cannot have nilpotent elements, so the image of r in that ring is not nilpotent, and hence there exists a prime ideal

$$q \supseteq (p_1, \ldots, p_{m-1})$$

in $A[x_1, \ldots, x_{m+n}]$ such that $r \notin q$. Computing from (3.3.1), we get

$$\frac{\partial p_m}{\partial x_j} \cdot r = a_1 \frac{\partial p_1}{\partial x_j} + \cdots + a_{m-1} \frac{\partial p_{m-1}}{\partial x_j}$$

$$\in A[x_1, \ldots, x_{m+n}]/(p_1, \ldots, p_m),$$

so localizing at q (so r becomes a unit), we see that the determinants of the $m \times m$ submatrices of the Jacobi matrix are 0, and hence the ideal generated by them does not contain 1, which is a contradiction. Thus, $p_1, \ldots, p_m$ are a regular sequence in $A[x_1, \ldots, x_{m+n}]$. Assume also without generality that $1 \notin (p_1, \ldots, p_m)$.

Next, since A is regular and hence Noetherian, the ring

$$A[x_1, \ldots, x_{m+n}]$$

is regular. (This is actually a non-trivial statement. We will prove it using cohomological methods in Chap. 5, Corollary 2.6.8.) Let q be its maximal ideal containing $(p_1, \ldots, p_m)$. Let

$$K = A[x_1, \ldots, x_{m+n}]/q$$

be the residue field. Consider the homomorphism of K-modules

$$\phi : q/q^2 \to K^{m+n}$$

given by

$$f \mapsto (\frac{\partial f}{\partial x_1}, \ldots, \frac{\partial f}{\partial x_{m+n}}).$$

Then the Jacobi matrix condition implies that the images of $p_1, \ldots, p_m$ under ϕ are K-linearly independent, and hence $p_1, \ldots, p_m \in q/q^2$ are K-linearly independent. This means that if we denote by $\bar{q}$ the ideal $q/(p_1, \ldots, p_m)$ in

$$A[x_1, \ldots, x_{m+n}]/(p_1, \ldots, p_m),$$

we have

$$dim_K(\overline{q}/\overline{q}^2) = dim_K(q/(q^2 + (p_1, \ldots, p_m)))$$
$$\leq dim_K(q/q^2) - m = (dim A[x_1, \ldots, x_{m+n}]_q) - m =$$
$$dim(A[x_1, \ldots, x_{m+n}]/(p_1, \ldots, p_m)_{\overline{q}}.$$

The last equality is because $p_1, \ldots, p_m$ is a regular sequence which, by assumption, generates an ideal not containing 1.

$\square$

Example Consider the morphism

$$f : \mathbb{A}^1_{\mathbb{C}} \to \mathbb{A}^1_{\mathbb{C}}$$

given by the homomorphism of rings

$$h : \mathbb{C}[x] \to \mathbb{C}[y], \ x \mapsto y^2.$$

We will show that this morphism is not smooth. Indeed, smooth morphisms are obviously preserved by pullback of schemes, but pulling back to the closed subscheme $\{0\} \cong Spec(\mathbb{C})$, we get the morphism of affine schemes which is $Spec$ of

$$\mathbb{C} \to \mathbb{C}[y]/(y^2),$$

which cannot be smooth, since the target is not a regular ring, while the source is.
On the other hand, if we restrict f to the distinguished open subset $\mathbb{A}^1_{\mathbb{C}} \setminus \{0\}$, it becomes, in fact, étale: It is then $Spec$ of the homomorphism of rings

$$k : \mathbb{C}[x] \to \mathbb{C}[y, y^{-1}], \ x \mapsto y^2.$$

We can represent the target as

$$\mathbb{C}[x, y, z]/(x - y^2, yz - 1).$$

The Jacobi matrix is

$$\begin{pmatrix} -2y & 0 \\ z & y \end{pmatrix}$$

so its determinant is $-2y^2$, which is a unit.

3.3.2 Theorem *A regular scheme X of finite type and separated over $Spec(k)$ for an algebraically closed field k is smooth over $Spec(k)$.*

Proof Without loss of generality, $X = Spec(R)$ is affine of dimension n, $R = k[x_1, \ldots, x_{n+m}]/I$. (Note: by Theorem 5.5.1 of Chap. 1, we have $dim(R) = dim(R_p)$ for all maximal ideals $p \subseteq R$.) Let

$$p = (x - a_1, \ldots, x - a_{m+n})$$

be a closed point of $\mathbb{A}_k^{m+n}$ which is in X. (It is of this form because k is algebraically closed.) Again, consider the homomorphism of k-modules

$$\theta : p/p^2 \to k^{m+n}$$

given by

$$\theta(f) = \left(\frac{\partial f}{\partial x_1}, \ldots, \frac{\partial f}{\partial x_{m+n}} \right)_{(a_1, \ldots, a_{m+n})}.$$

Then the rank of the Jacobi matrix modulo p is just $dim_k(Im(\theta))$. However, θ is an isomorphism since the vectors $\theta(x_i - a_i)$ clearly form the standard basis of k^{m+n}. Since X is regular, we have

$$n = dim_k(p/(I + p^2)),$$

so the rank of the Jacobi matrix modulo p must be m. This implies that if we denote by q the projection of the ideal p in R, then the determinants of the $m \times m$ submatrices of the Jacobi matrix generate an ideal in R_q containing $1 \in R_q$ (since this can be detected modulo q). However, this will then hold in some open neighborhood of q. □

3.4 Étale Schemes Over a Field

Let k be a field. An *algebraic extension* of k is a field $K \supseteq k$ such that every element of k is a root of a non-zero polynomial with coefficients in k. The greatest common divisor of all such polynomials for a given x is called the *minimal polynomial* of x over k. Clearly, the minimal polynomial is irreducible in $k[x]$. If a field $K \supseteq k$ is finite-dimensional as a vector space over k, it is called a *finite extension*. A finite extension $K \supseteq k$ is always algebraic. To see this, note that for $x \in K$, the powers $1, x, x^2, \ldots$ cannot be linearly independent over k. An extension $K \supseteq k$ is called *separable* if the minimal polynomial of every $x \in K$ over k has only simple roots, i.e. roots of multiplicity 1 (i.e. no roots of multiplicity > 1). Recall that this is equivalent to requiring that the greatest common divisor of the polynomial with its derivative is 1. A field k is called *perfect* if all its algebraic extensions are separable. A *separable (resp. algebraic)* closure of a field k is a maximal separable (resp. algebraic) extension of k. It is tempting to say that algebraic (hence separable) closures exist by Zorn's lemma, but this is a little delicate, since algebraic extensions of a given field form a proper class. This can be circumvented by observing that elements algebraic over an algebraic extension of a field k are algebraic over k. (Actually, we already observed this in the greater generality of integral extensions in Chap. 1.) Thus, using the axiom of choice, one may create a transfinite sequence of algebraic extensions of k, attaching, at each step, a root of a polynomial over k which is not already in the extension (and taking unions at limit ordinals). This procedure eventually terminates due to the fact that there is only a set of polynomials with coefficients in k.

3.4.1 Theorem *All fields of characteristic 0 and all finite fields are perfect.*

In characteristic 0, this follows from the fact that the derivative of a non-constant polynomial cannot be 0, so if the polynomial had multiple roots, then the gcd with its derivative (which has lower degree) could not be a unit, which contradicts irreducibility. For finite fields, one can see this pretty much because one knows all their algebraic extensions; we will return to this topic in Sect. 5.7 below.

Let us list one more fact about separable extensions:

3.4.2 Theorem (The Primitive Element Theorem) *Let $F \subseteq E$ be a finite separable field extension. Then E is generated, as a field extension of F or equivalently as an F-algebra, by a single element (called a primitive element).*

Proof If F is a finite field, then so is E, and hence its group of units $E^\times = E \smallsetminus \{0\}$ is cyclic, since otherwise, it would have zero divisors, which would be a contradiction. Thus, we can just take the generator of $E^\times$ as the primitive element.

Thus, assume F has infinitely many elements. By induction, it suffices to treat the case when E is generated by two elements α, β. We claim that for all but finitely many elements λ, the element $\gamma = \alpha + \lambda\beta$ is a primitive element. Let f be a minimal polynomial of α over F, let $h(x) = f(\gamma - \lambda x)$, and let g be the minimal polynomial of β over F. The polynomials g and h have the common root β. Suppose they have no other common root. Then, since E is a separable extension of F, their gcd is of degree 1, and thus β (and hence also α) is in the subfield of E generated by γ and hence, γ is a primitive element.

So suppose g, h have another common root β' in an algebraic closure $\overline{F} \supset E$ of F. But this means that for some root α' of f, we have

$$\gamma = \lambda\beta' + \alpha' = \lambda\beta + \alpha$$

and thus,

$$\lambda = \frac{\alpha' - \alpha}{\beta' - \beta}.$$

But this can only happen for finitely many values $\lambda \in F$. $\square$

To give an example of a field which is not perfect, let $k = \mathbb{F}_p(t)$ be the field of rational functions in one variable t over the field $\mathbb{F}_p = \mathbb{Z}/(p)$ with p elements. Now the ring $\mathbb{F}_p[t]$ is a unique factorization domain, t is irreducible, and the polynomial $x^p - t$ is Eisenstein (it is monic, and all coefficients except the top one are in the prime ideal (t), while the x^0 coefficient is not in (t^2)—see Exercise 19 of Chap. 1), so it is irreducible in $\mathbb{F}_p[t, x]$, and hence it is irreducible over $k[x]$ by the Gauss lemma. Therefore,

$$K = k[\theta]/(\theta^p - t)$$

is a field. But then in $K[x]$, we have

$$x^p - t = (x - \theta)^p \tag{3.4.1}$$

by the binomial theorem (since coefficients which are multiples of p are 0). We see that K is not a separable extension of k and hence k is not perfect.

One can show that a separated morphism $f : X \to Spec(k)$ where k is a perfect field is smooth if and only if X is regular and f is of locally finite type. We prove the following related statement:

3.4.3 Proposition *A morphism $f : X \to Spec(k)$ where k is a field is étale if and only if X is a disjoint union of schemes of the form $Spec(E)$ where E is a finite separable extension of k.*

Proof By the Primitive Element Theorem, a finite separable extension of a field k can always be written in the form

$$E = k[x]/(p(x))$$

where $p(x)$ is an irreducible polynomial such that

$$gcd(p(x), p'(x)) = 1.$$

Thus, by definition, $Spec(E)$ is étale over $Spec(k)$, an a disjoint union of such schemes is étale over $Spec(k)$, since being étale is a local property.

Conversely, it suffices to consider the case of X affine (since then we will show that the topology of X is discrete), and to characterize étale k-algebras

$$B = k[x_1, \ldots, x_n]/(p_1, \ldots, p_n)$$

(3.4.2)

where the determinant of the Jacobi matrix is a unit. First, note that

$$dim(B) = 0,$$

since

$$dim(k[x_1, \ldots, x_n]) = n$$

and $p_1, \ldots, p_n$ is a regular sequence (as already remarked). However, every Noetherian ring of dimension 0 is Artinian, and has no nilpotent elements. By Proposition 5.2.3 of Chap. 1, it is a product of finitely many fields.

Thus, it suffices to prove that a field extension E which is étale over k is finite and separable. Finiteness is obvious, since a field extension which is not finite is not finitely generated as a k-algebra, so by definition it cannot be smooth. Suppose E is not separable over k. This means that there exists an element of E whose minimal polynomial p over k has at least one multiple root $\alpha \in E$. If E were smooth, then, denoting by $\iota : K \to E$ the inclusion,

$$\iota \otimes_k E : E \to E \otimes_k E$$

would have to be smooth and hence, by Theorem 3.3.1, $E \otimes_k E$ would have to be regular. But we see that this ring has nilpotent elements, since there exists a polynomial $f \in E[x]$ and a natural number $n > 1$ where f is not divisible by p, but f^n is divisible by p (since p has a root of multiplicity > 1 in E).

$\square$

By comparison, it turns out that there are no non-trivial finite étale schemes over $Spec(\mathbb{Z})$. While this is a topic of number theory more than algebraic geometry, it is too good to pass up. Additionally a minor variation of the method also gives a fact relevant in the next chapter (finiteness of class groups of number fields). Because of this, a proof is given in Exercises 17—22 below.

4   Abstract Varieties

4.1   The Definition of an Algebraic Variety

It is a good time now to give a general definition of an algebraic variety.

4.1.1 Definition An *abstract variety* over a field k is an integral separated scheme of finite type over $Spec(k)$. An abstract variety is called *complete* if it is also proper over $Spec(k)$. An abstract variety is called *affine*, resp. *projective* if it is isomorphic to a closed subscheme of $\mathbb{A}^n_{Spec(k)}$, $\mathbb{P}^n_{Spec(k)}$ for some n. An abstract variety is called *quasiaffine*, resp. *quasiprojective* if it is isomorphic to an open subscheme of an affine resp. projective variety over k.

From this point on, we will just say "variety," or "algebraic variety," instead of "abstract variety." Morphisms of varieties over k are morphisms of the underlying schemes over $Spec(k)$. We can define *rational maps* again as morphisms of varieties defined on a non-empty open subset (again, two morphisms which coincide on the intersection of their domains are considered to be the same rational map). A *birational map* is an isomorphism in the category of varieties over k and rational maps. Two varieties between which there exists a birational map are called *birationally equivalent*. Again, this is easily seen to be equivalent to their fields of rational functions (i.e. fields of fractions of the rings of sections of the structure sheaves over non-empty open affine subsets) being isomorphic as k-algebras.

The *dimension* of a variety can be defined as the dimension of any open affine subscheme. Therefore, the facts about dimension proved in Sect. 5.5 of Chap. 1 remain valid in the present context. Varieties of dimensions 1, 2, 3 are called *curves*, *surfaces* and *threefolds*.

An analog of the statement of Exercise 27 of Chap. 2 about recovering all information from closed points only also holds for abstract varieties (see Exercise 23). For a variety X over a field k, the subset of X consisting of all maximal ideals with the induced topology is sometimes denoted by X_m, and the pullback of the structure sheaf $\mathcal{O}_X$ is denoted by $\mathcal{O}_{X_m}$.

When speaking of varieties over $\mathbb{C}$, we also have an *analytic topology* on X_m obtained as a pushout of the analytic topologies on affine open subsets. The set X_m with its analytic topology is sometimes denoted by X_{an}. To make a distinction, the scheme-theoretical topology on X or X_m is sometime referred to as the *Zariski topology*, even though strictly speaking, the term "Zariski topology" should only apply to affine schemes.

4.2 A Classification of Smooth Curves

In this subsection, let k be an algebraically closed field. We will classify smooth curves over the field k. Let K be a field extension of k of transcendence degree 1 finitely generated as a field containing k (recall that the transcendence degree of K over k is the largest number n—possibly infinity—such that there is an embedding $k(x_1, \ldots, x_n) \subseteq K$). We shall construct a canonical smooth projective curve C_K over k with function field K. We start with

4.2.1 Lemma *Let $0 \neq h \in K$. Then there are at most finitely many discrete valuation rings $R \subset K$ with field of fractions K such that $k \subset R$ and $h \notin R$.*

Proof Since k is algebraically closed, $h \notin k$. Since k is algebraically closed, h is transcendental (which means not algebraic) over k, so $k[h^{-1}] \subset K$. Let B be the integral closure of $k[h^{-1}]$ in K. Then B is a Dedekind domain, since K is a finite field extension of $k(h)$ and the integral closure of a Dedekind domain in a finite extension of its field of fractions is again a Dedekind domain. (It is normal by the comments on integral extensions in Sect. 4.3 of Chap. 1, and its dimension is 1 by Theorem 5.5.1 of Chap. 1.) Moreover, K is the field of fractions of B. Now let R be a discrete valuation ring with field of fractions K such that $h \notin R$. Let m be the maximal ideal of R. Then $n = m \cap B$ is a maximal ideal of B and $h^{-1} \in n$. Additionally, R dominates B_n in the sense of Sect. 2.4. But B_n is also a discrete valuation ring, so $B_n = R$ by Theorem 2.4.3, so in particular R is determined by n. On the other hand, since $h^{-1} \neq 0$, $Z(h^{-1}) \subset Spec(B)$ is of dimension 0, and hence can contain only finitely many points n. $\square$

Now construct the scheme C_K as follows: The closed points are all the discrete valuation rings with field of fractions K. There is one additional point, namely the generic point. Non-empty open sets are defined to be complements of finite sets of closed points. The structure sheaf $\mathcal{O}_{C_K}$ is defined by

$$\mathcal{O}_{C_K}(U) = \bigcap_{R \in U} R.$$

If B is a Dedekind domain with field of fractions K then

$$B = \bigcap_{p \in Spec B} B_p. \tag{4.2.1}$$

On the other hand, if $R \in C_K$, and we choose an element $x \in R$ of valuation 1, then again the integral closure B of $k[x]$ in K is a Dedekind domain, and the subset $U \subseteq C_K$ of all R with $B \subseteq R$ is open and by (4.2.1), isomorphic to $Spec(B)$, so we see that C_K is a scheme. In fact, C_K is clearly integral and is locally a smooth curve of dimension 1 over k.

To show that C_K is in fact a smooth curve, it remains to prove that C_K is separated. To prove this, we use the valuation criterion (Theorem 2.5.1). If $k \subsetneq R \subseteq L$ is a valuation ring, then a morphism $f : Spec(L) \to C_K$ over k can extend to $Spec(R)$ for a valuation ring with field of fractions L in at most one way: first of all, f factors through $Spec(K)$, thus giving an inclusion $K \subseteq L$, and if f extends to $Spec(R)$, then the closed point maps to a closed point $p \in C_K$, which means that $R \cap K$ is the discrete valuation ring corresponding to the point p. By definition, the point p is unique.

It is also worth noting that *all* valuation rings R contained in K and containing k are, in fact, discrete valuation rings. To this end, (assuming $R \neq k$), we choose, again, a non-zero element $b \in R \smallsetminus k$. Then let B be the integral closure of $k[b]$ in K. We have $B \subset R$, and also B is a Dedekind domain, since it is the integral closure of the Dedekind domain $k[b]$ in a finite extension of its field of fractions. So (assuming $R \neq K$), we have a maximal ideal $m \subset B$ such that the maximal ideal n of R satisfies $n \cap B = m$. But then R dominates the ring B_m, which itself is a discrete valuation ring, and hence $R = B_m$.

To show that C_K is projective, we first prove

4.2.2 Lemma *Let C be a smooth curve over k and let $U \subseteq C$ be a non-empty open set. Then any morphism*

$$f : U \to X$$

to a projective scheme X over $Spec(k)$ extends uniquely to C.

Proof The set $C \smallsetminus U$ consists of finitely many points. Pick one such point p. Let K be the field of rational functions on C. Then

$$
\begin{array}{ccc}
Spec(K) & \xrightarrow{\ f\ } & X \\
\downarrow & \overset{g}{\nearrow} & \downarrow \\
Spec(\mathcal{O}_{C,p}) & \longrightarrow & Spec(k)
\end{array}
$$

is a testing diagram for the valuation criterion of X being proper. Now the maximal ideal of $\mathcal{O}_{C,p}$ maps to some open affine subset $V \subseteq X$, and we have $V \cong Spec(B)$ where B is a finitely generated k-algebra. Therefore, g is $Spec(h)$ for some homomorphism of k-algebras

$$h : B \to \mathcal{O}_{C,p}.$$

Thus, all h-images of generators of B are in some $\mathcal{O}_C(W)$, $p \in W \subseteq C_K$, and g extends to W. Also, the extension is unique because $\mathcal{O}_C(W) \to K$ is an inclusion. Patching the extensions for all points $p \in C \smallsetminus U$ gives the desired extension to C. □

4.2.3 Proposition *For every field extension K of k of transcendental degree 1 finitely generated as a field containing k, C_K is a smooth projective curve.*

Proof By definition, C_K is quasicompact, so it is covered by finitely many open affine sets $U_1, \ldots, U_n$. Since U_i are affine of finite type over $Spec(k)$, we have morphisms from U_i

to an affine space over k, and hence to a projective space over k. By Lemma 4.2.2, these maps extend to morphisms of schemes

$$f_i : C_K \to \mathbb{P}^{n_i}_k.$$

Take the product

$$f = \prod f_i : C_K \to \prod \mathbb{P}^{n_i}_k. \tag{4.2.2}$$

Since a product of finitely many projective spaces is a projective variety by the Segre embedding of a product of projective spaces into a projective space (given by all possible products of projective coordinates, one of each of the projective spaces, see Exercise 29 of Chap. 2), it suffices to show that (4.2.2) is a closed immersion. The fact that the homomorphisms on stalks are onto follows from the fact that it is true at each point p after projecting to the factor corresponding to an open affine neighborhood of p.

Let Y be the closure of the image of f. Since the isomorphic f-image of U_i is dense in Y, Y is a curve (although a priori, it may not be regular). However, every local ring $\mathcal{O}_{Y,q}$ is dominated by a discrete valuation ring R with field of fractions K (take a localization of its integral closure in K at a maximal ideal), so

$$R = \mathcal{O}_{C_K,R} = \mathcal{O}_{Y,f(R)}$$

and hence $q = f(R)$ (this is because any two points of a projective variety are elements of the same open affine set $Spec(A)$). If one of the local rings is contained in the other, one maximal ideal must be contained in another, so they must be equal. Thus, f is onto. But then it is also injective because different points of C_K have different local rings. $\quad\square$

4.2.4 Theorem *Let k be an algebraically closed field. Then there are canonical equivalences between the following categories:*

1. *Smooth projective curves over k and dominant morphisms (inverse image of a non-empty open set is non-empty).*
2. *Smooth curves over k and dominant rational maps.*
3. *The opposite of the category of field extensions of transcendence degree 1 of k finitely generated as fields containing k and homomorphisms of k-algebras.*

Proof The functor from 1 to 2 is forgetful, from 2 to 3 is by taking the function field, and from 3 to 1 by sending K to C_K. The above discussion implies that compositions of these functors are naturally isomorphic to identities where applicable. $\quad\square$

4.2.5 Proposition *Every dominant morphism $f : X \to Y$ of complete smooth curves over an algebraically closed field k is finite.*

Proof By Theorem 4.2.4, we have an inclusion $h : K \subseteq L$ of finitely generated fields of transcendence degree 1 over k such that $X = C_L$, $Y = C_K$, and $Spec(h)$ is the restriction of f to the generic point. Let $U \subset Y$ be an affine open set, $U \cong Spec(A)$, $A \subset K$. Then $f^{-1}U \cong Spec(B)$ where B is the integral closure of A in L. By Proposition 4.2.3, f is proper, and hence is of finite type. Hence, B is a finitely generated A-algebra. But every finitely generated A-algebra which is integral over A is a finitely generated A-module. □

COMMENTS

1. The statement of Proposition 4.2.5 is also true for complete curves which are not smooth, but the proof is harder. See Chap. 6, Sect. 4.2.3 and Exercise 19 of Chap. 6.
2. Regarding the last step of our proof, it is true more generally that for any normal integral domain A with field of fractions K and a finite field extension L of K, the integral closure of A in L is a finitely generated A-module. However, it is surprisingly tricky to prove: the standard proof treats the case of L separable over K first and then reduces to that case.

Now let again $f : X \to Y$ be a dominant morphism of complete smooth curves over k algebraically closed. By Theorem 4.2.4 again, we can assume that $X = C_K$, $Y = C_L$, $K \subseteq L$ is a finite field extension. The *degree* of the morphism f is defined to be the degree of the field extension:

$$deg(f) = [L : K].$$

4.2.6 Theorem *Let $f : X \to Y$ be a dominant morphism of complete smooth curves over an algebraically closed field k, and let p be a closed point of Y. Then*

$$deg(f) = \sum_{q \in f^{-1}(p)} v_q(t) \tag{4.2.3}$$

where $t \in \mathcal{O}_{Y,p}$ is an element of valuation 1. (Note: the choice of an element of valuation 1 does not matter, since any two choices differ by multiplying by a unit. Since the morphism f is finite by Proposition 4.2.5, its pullback to $Z(p)$ is finite, and hence the sum has finitely many terms.)

Proof Use the same notation as in the proof of Proposition 4.2.5, and assume $p \in U$. Consider the multiplicative set $S = A \setminus p$. Then $B' = S^{-1}B$ is a torsion free finite extension of A_p of rank $[L : K] = deg(f)$. The elements $q \in f^{-1}(p)$ are in bijective correspondence with the maximal ideals m_q of B'. By the Chinese remainder theorem (Proposition 4.1.1 of Chap. 1),

$$B'/tB' = \prod_{q \in f^{-1}(p)} B'/(tB'_q \cap B')$$

(since $\bigcap_{q \in f^{-1}(p)} (t B'_q \cap B') = t B'$) so we have

$$deg(f) = dim_k (B'/t B') =$$

$$\sum_{q \in f^{-1}(p)} dim_k (B'/(t B'_q \cap B')) = \sum_{q \in f^{-1}(p)} v_q(t).$$

(Note that $B'/t B'$ is a k-vector space because $k \subset A$.)

$\square$

Example Let k be an algebraically closed field of characteristic 0. In Sect. 3.2 we proved that the closure E in $\mathbb{P}^2_k$ of

$$Spec(R), \quad R = k[x, y]/(x^2 - y^3 - y) \tag{4.2.4}$$

is smooth over k. All of our arguments would work if we replace R with

$$k[x, y]/(x^2 - (y - a)(y - b)(y - c)) \tag{4.2.5}$$

where a, b, c are different elements of k. Such curves E are called *elliptic curves* over k. (Over a field which is not algebraically closed, the same definition of elliptic curve holds if we allow, in place of $(y - a)(y - b)(y - c)$, any cubic polynomial with different roots in the algebraic closure. The definition actually works over fields of characteristic $\neq 2, 3$. The study of elliptic curves is a very important subject of mathematics. For more information, we refer the reader to [23].) We will show that E is not a rational curve (i.e. is not isomorphic to $\mathbb{P}^1$ in the category of curves and rational maps). In fact, if E were rational, by Theorem 4.2.4, it would be actually isomorphic to $\mathbb{P}^1_k$ as a scheme. Since $Spec(R) = E \setminus Z(p)$ where p is the point $[0 : 0 : 1]$, $Spec(R)$ would have to be isomorphic to $\mathbb{A}^1$, so R would have to be isomorphic to $k[t]$. But we will show that in fact, R is not a unique factorization domain. To this end, note that R is integral over $k[y]$. In fact, as a $k[y]$-module, it is free with basis 1 and x. Further, there is obviously an automorphism of rings $\sigma : R \to R$ which sends 1 to 1 and x to $-x$. Put $N(t) = t \cdot \sigma(t)$. Then N preserves multiplication. If $u \in R^\times$, then $N(u) \in k[y]^\times$, so $N(u) \in k^\times$. But if $u = p(y) + x q(y)$ for polynomials $p, q \in k[y]$, then

$$N(u) = p(y)^2 - x^2 q(y)^2 = p(y)^2 - (y^3 + y) q(y)^2. \tag{4.2.6}$$

We see that unless $p(y) \in k, q(y) = 0$, the top terms of the two summands on the right hand side are in even resp. odd degrees, so they cannot match. Thus, $R^\times = k^\times$. Next, we will show that y is irreducible. Note that $N(y) = y^2$, so since $N(u) \in k^\times$ only for $u \in k^\times$, if $y = uv \in R$, we would have to have $N(u) = ay$, $N(v) = y/a$ for some $a \in k^\times$. But by (4.2.6) again, this is only possible when $u \in k^\times$. Similarly, we will show that $x \in R$ is irreducible: We have $N(x) = x^2 = y^3 + y$, so if $x = uv$ then $N(x) = N(u)N(v)$ so either u or v must be in $k[y]$, (otherwise the second terms of (4.2.6) would combine to too high a degree). Say, $u \in k[y]$. Then $v \in x \cdot k[y]$, and so $u \in k^\times$, $v \in x \cdot k^\times$.

Thus, x, $y \in R$ are both different irreducible elements of R, and hence $x^2 = y^3 + y$ contradicts uniqueness of factorization. Thus, E is not rational. Again, this proof applies to any elliptic curve over a field of characteristic $\neq 2, 3$ (Exercise 26).

4.3 The Role of Closed Points

As already remarked, one feature of algebraic varieties is that all their scheme-theoretical information can actually be derived from maximal ideals. This follows from the Nullstellensatz (Corollary 4.3.1 of Chap. 4, see Exercise 23). One corollary is worth stating explicitly.

By an *open subvariety* of an (abstract) variety X over a field k, we mean any open subscheme. By a *closed subvariety* of a variety K over a field k, we mean a closed integral subscheme (it is then automatically a variety).

4.3.1 Corollary *Two subvarieties Y, Z, closed or open, of a variety X over a field k are equal if and only if they contain the same closed points of X.*

Proof First, we shall prove that Y, Z contain the same points as schemes. To this end, without loss of generality, $X = Spec(R)$, where R is a finitely generated algebra over k. But then by the Nullstellensatz, a prime ideal (which is always a radical) is an intersection of maximal ideals, so the points of Y and Z, considered as schemes, are equal. Now by the definition of topology, this means that they are both open or closed subsets, and hence, by our assumptions, open or closed subvarieties. If they are both open, then they are determined by their set of points, so we are done. If they are both closed, we will prove that for every affine open set $U \subseteq X$, $U = Spec(R)$ where R is a finitely generated algebra over k, there is a unique isomorphism of schemes over X

$$\phi_U : Y \cap U \to Z \cap U. \tag{4.3.1}$$

By uniqueness, the isomorphisms (4.3.1) are then compatible with restriction, and thus, their colimit in the category of schemes defines an isomorphism of schemes $Y \cong Z$.

To see (4.3.1), recall from Theorem 2.3.1 of Chap. 2 that

$$Y \cap U = Spec(R/I), \quad Z \cap U = Spec(R/J)$$

for some ideals I, $J \subseteq R$, which have to be radical, since Y, Z are reduced. Then, however, I, J are intersections of maximal ideals, so by our assumption, they are equal. □

We see that the same argument also implies that for a variety X, two closed reduced subschemes of X which contain the same closed points are equal. We shall see that if k is not algebraically closed, closed points are not necessarily what we may think, however.

In the remainder of this section, we present a proof of a generalization of the Nullstellensatz, which includes, for example, schemes of finite type. A ring R is called a *Jacobson ring* if every prime ideal of R is an intersection of maximal ideals. Note that this, in particular, implies that

$$Nil(R) = Jac(R). \tag{4.3.2}$$

An integral domain R is called a *Goldman ring* if there exists a $u \in R$ such that $K = u^{-1}R$ is the field of fractions of R. Every field is obviously a Goldman ring. A discrete valuation ring is also a Goldman ring (we can take u to be any element of valuation 1). Note that a local ring which is an integral domain is Jacobson if and only if it is a field. In fact, we have the following

4.3.2 Lemma *A Goldman ring R is Jacobson if and only if R is a field.*

Proof If R is a field, it is obviously Jacobson. Assume that R is not a field and it is Goldman, i.e. $u^{-1}R = K$ where K is its field of fractions. Then every maximal ideal of R contains u (thus showing that R is not Jacobson, since it is an integral domain, so (0) is a prime ideal, and $u \neq 0$). Indeed, suppose that $m \subsetneq R$ is a maximal ideal, and $u \notin m$. Then u is a unit in the localization R_m, so $R_m = u^{-1}R_m = K$, contradicting our assumption that R is not a field.

$\square$

4.3.3 Lemma *Let R be a commutative ring, $u \in R$, $u \notin Nil(R)$. Then there exists a prime ideal $p \subset R$, $u \notin p$, such that R/p is Goldman.*

Proof Let m be a maximal ideal of $u^{-1}R$. Put $p = m \cap R$. Then p is prime. We claim that R/p is Goldman. Indeed,

$$u^{-1}(R/p) = (u^{-1}R)/m,$$

which is, by assumption, a field. (Note that for every $x \in m$, there exists $n \in \mathbb{N}$ such that $u^n x \in p$.)

$\square$

4.3.4 Corollary *The following are equivalent for a commutative ring R:*

1. *R is Jacobson.*
2. *All quotient Goldman rings of R are fields.*
3. *For every ideal $I \subseteq R$, the ring R/I is Jacobson.*

4. *For every ideal $I \subseteq R$, the radical $\sqrt{I}$ is equal to the intersection of all maximal ideals containing I.*

Proof By Lemma 4.3.3 applied to R/p for $p \subset R$ prime, if $x \notin p$ then (since the reduction $\overline{x}$ of x modulo p is not nilpotent), there exists a prime ideal $q \supseteq p$ such that $x \notin q$ and R/q is Goldman. Thus, (2) implies (1). On the other hand, if R/p is Goldman but not a field, then by Lemma 4.3.2, p is not an intersection of maximal ideals. Thus, (1) and (2) are equivalent. Now since (2) obviously passes on to quotient rings, it is also equivalent to (3). Since we already observed that for a Jacobson ring R, we have (4.3.2), (3) implies (4). If (4) holds, then, in particular, a prime ideal (which is its own radical) is an intersection of maximal ideals, so we have (1).   □

4.3.5 Lemma *If a finitely generated (commutative) algebra S over a commutative ring R is a Goldman ring and the homomorphism*

$$R \to S$$

is injective, then R is a Goldman ring.

Proof Suppose S is Goldman generated by $u_1, \ldots, u_n$ as a ring over R, and $L = u^{-1}S$ is its field of fractions. Note that by assumption, R is an integral domain. Let $K \subseteq L$ be the field of fractions of R. Then L is finitely generated as a ring over K (by $u_1, \ldots, u_n, u^{-1}$), so it is a finite extension of K, which means that every element of L is algebraic (i.e. roots of nonzero polynomials) over K. Without loss of generality (by clearing denominators), these polynomials have coefficients in R. Let $a_1, \ldots, a_n, a$ be the top coefficients of these polynomials. Consider the subring $R_1 \subseteq L$ generated by R and the elements $a_1^{-1}, \ldots, a_n^{-1}, a^{-1}$. Note that

$$R_1 = (a_1 \cdot \ldots \cdot a_n a)^{-1} R. \tag{4.3.3}$$

Then the field L is an integral extension of the ring R_1 (which means that every element of L is integral, i.e. the root of a monic polynomial, i.e. polynomial with top coefficient 1), over R_1. But a ring R_1 whose integral extension is a field L is itself a field by Lemma 4.3.4 of Chap. 1. Here is another proof: Let m be a maximal ideal of R_1; then $(R_1)_m$ is dominated by a valuation ring V, but valuation rings are *integrally closed* (meaning normal, or integrally closed in their fields of fraction, i.e. that any element of the field of fractions integral over the ring is in the ring—see Exercise 10). Thus, $V = L$, which implies $m = 0$, so R_1 is a field. (Note: not necessarily L.) In any case, (4.3.3) now shows that R is Goldman.   □

4.3.6 Theorem *If R is a Jacobson ring, then $R[x]$ is a Jacobson ring.*

Proof Let $p \subset R[x]$ be a prime ideal such that $R[x]/p$ is Goldman. Let

$$q = R \cap p.$$

Then $R[x]/p$ is finitely generated over R/q (by x) as an algebra, and the homomorphism

$$R/q \to R[x]/p$$

is injective by the homomorphism theorem for commutative rings. By Lemma 4.3.5, R/q is Goldman. Since R is a Jacobson ring, R/q is a field. Now the integral domain $R[x]/p$ is a quotient of the principal ideal domain $R/q[x]$ by a prime ideal. Since $R/q[x]$ itself is certainly not Goldman, $R[x]/p$ is its quotient by a nonzero prime ideal, but such an ideal is maximal. Thus, $R[x]/p$ is a field.

$\square$

Proof of Theorem 4.3.1. Use Theorem 4.3.6 and Corollary 4.3.4.

$\square$

---

5 The Galois Group and the Fundamental Group

In this section, we will explore an important role of group actions in algebraic geometry. We will see how closed points of varieties over perfect fields can be understood as orbits of the Galois group, which is another important philosophical point of algebraic geometry. We will also discuss the analogy between the absolute Galois group of a perfect field and the fundamental group of a topological space, and discuss a combination of both concepts, namely the étale fundamental group. We will conclude the section by mentioning a remarkable result of P. Deligne, A. Grothendieck and others on counting points of varieties over finite fields.

5.1 Varieties Over Perfect Fields and G-Sets

Assume that k is a perfect but not necessarily algebraically closed field. Consider a closed point m of the affine space $\mathbb{A}_k^n$. Then m is a maximal ideal of $k[x_1, \ldots, x_n]$, so

$$K = k[x_1, \ldots, x_n]/m \tag{5.1.1}$$

is a field. It is called the *residue field* of m. Further, this field is a finite extension of k: this is because it is contained in the splitting field E of the product $p(x)$ of the minimal polynomials of all its coordinates over the algebraic closure of k (see Proposition 4.3.2 of Chap. 1). Thus, over the field E, the polynomial $p(x)$ is a product of linear factors. Recall that a splitting field of a polynomial over k is the same thing as a *Galois extension*. So we have

$$K \subseteq E,$$

and E is a Galois field over k, since we assumed that k is perfect and hence E is a separable extension. Now consider the projection

$$\pi : \mathbb{A}_E^n \to \mathbb{A}_k^n. \tag{5.1.2}$$

We can say that π is Spec of the inclusion

$$k[x_1, \ldots, x_n] \subseteq E[x_1, \ldots, x_n].$$

But now note that the Galois group $Gal(E/k)$ (the group of automorphisms of E which leave every element of k fixed) permutes the closed points of $\mathbb{A}_E^n$ (it permutes all the points, but let us restrict attention to closed points for now).

5.2   Some Details on G-Sets

Instead of *permutes*, we will say *acts on*. We say in general that a group G *acts on a set S* (or that there is a G-action on S) when for every element $g \in G$, we are given a permutation (i.e. a bijection) $\sigma_g : S \to S$ such that for all $g, h \in G$,

$$\sigma_{gh} = \sigma_g \circ \sigma_h. \tag{5.2.1}$$

There are several other notations we can use to express this. We may say that there is a homomorphism of groups

$$\sigma : G \to Perm(S)$$

where $Perm(S)$ is the group of permutations on S (i.e. automorphisms of S in the category of sets and mappings). Alternately, we may also write

$$g \cdot x = \sigma_g(x),$$

and then require that for all $g, h \in G$,

$$g \cdot (h \cdot x) = (g \cdot h) \cdot x. \tag{5.2.2}$$

From this point of view, the set S is a universal algebra with an operation $g \cdot ?$ for every $g \in G$. Thus, sets with a G-action are also called *G-sets*. The category *G-Sets* of all G-sets is then defined by letting morphisms of G-sets be maps

$$f : S \to T$$

which satisfy for all $g \in G, x \in S$,

$$f(g \cdot x) = g \cdot f(x).$$

The study of G-sets is called *G-equivariant* mathematics.

Thus, the study of varieties over perfect fields k involves $Gal(E/k)$-equivariant mathematics for every Galois extension $E \supseteq k$. Let us return to that context, and see what else we need to learn about G-sets. By definition, the map (5.1.2) is $Gal(E/k)$-equivariant on closed points, but of course, $Gal(E/k)$ acts trivially on the target (i.e. leaves everything fixed). This means that if two maximal points $x, y \in \mathbb{A}_E^n$ satisfy

$$g \cdot x = y$$

for some $g \in Gal(E/k)$, then we must have

$$\pi(x) = \pi(y).$$

This calls for a definition. For a G-set S and $x \in S$, the *G-orbit* (or simply *orbit*) of x is the subset

$$Orb_G(x) = \{g \cdot x \mid g \in G\}.$$

The orbit is also a G-set, and the inclusion $Orb_G(S) \subseteq S$ is a morphism of G-sets. Thus, we can get closed points of $\mathbb{A}_k^n$ as $Gal(E/k)$-orbits of $\mathbb{A}_E^n$ where E is any Galois extension of k. We will still need to refine this somewhat, but for now, let us *classify orbits up to isomorphisms in the category of G-sets.*

The key concept here is the concept of an *isotropy group* (also known as *stabilizer*) of an element $x \in S$ of a G-set S, defined as

$$Iso_G(x) = \{g \in G \mid g \cdot x = x\}.$$

It is also convenient to denote, for a subgroup $H \subseteq G$,

$$S^H = \{x \in S \mid H \subseteq Iso_G(x)\}.$$

Thus, S^H is the set of all elements $x \in S$ such that for all $h \in H, h \cdot x = x$. We call S^H the set of *fixed points* of S with respect to the subgroup H.

Recall that for a subgroup $H \subseteq G$, we may define *left cosets*

$$gH = \{g \cdot h \mid h \in H\}$$

and *right cosets*

$$Hg = \{h \cdot g \mid h \in J\}.$$

When the left cosets are the same as the right cosets, i.e. for all $g \in G$, $Hg = gH$, then the subgroup H is called *normal*, we write $H \triangleleft G$, and we can form the *factor group* G/H which consists of all the cosets. We then define multiplication of cosets by

$$g_1 H \cdot g_2 H = (g_1 \cdot g_2) H$$

(in proving consistency of this definition, we use the fact that $Hg_2 = g_2 H$).

If H is not a normal subgroup of G, we still use the symbol

$$G/H = \{gH \mid g \in G\}$$

to denote the set of left cosets, which still form a G-set, using the formula

$$g_1 \cdot g_2 H = (g_1 \cdot g_2) H$$

for $g_1, g_2 \in G$.

The isotropy group of an element of a G-set does not have to be a normal subgroup of G.

5.2.1 Proposition *For every element x of a G-set S, the orbit $Orb_G(x)$ is isomorphic to $G/Iso_G(x)$ in the category of G-sets.*

Proof Define a morphism of G-sets

$$\phi : G/Iso_G(x) \to Orb_G(x)$$

by

$$g \cdot Iso_G(x) \mapsto g \cdot x \text{ for } g \in G. \tag{5.2.3}$$

To see that this is well defined, we need to prove that if

$$g_1 \cdot Iso_G(x) = g_2 \cdot Iso_G(x) \tag{5.2.4}$$

then

$$g_1 \cdot x = g_2 \cdot x. \tag{5.2.5}$$

We will prove that (5.2.4) is, in fact, equivalent to (5.2.5).

In general, for a subgroup $H \subseteq G$, $g_1 \cdot H = g_2 \cdot H$ occurs if and only if $g_1^{-1} \cdot g_2 \in H$. Thus, (5.2.4) occurs if and only if

$$g_1^{-1} \cdot g_2 \in Iso_G(x),$$

which means

$$g_1^{-1} \cdot g_2 \cdot x = x,$$

which is equivalent to (5.2.5). Thus, ϕ is well defined by (5.2.3) as a map of sets, but then it clearly preserves the G-action, so ϕ is a morphism of G-sets. It is also clearly onto by the definition of an orbit. The fact that (5.2.5) implies (5.2.4) proves that ϕ is injective. Thus, it is an isomorphism of G-sets.

$\square$

We say that the action of G on a set S is *transitive* or *homogeneous* when S is isomorphic to a G-set of the form G/H. Clearly, this is the same thing as requiring that for all $x, y \in S$ there exists a $g \in G$ such that $g \cdot x = y$.

As any category of universal algebras, the category of G-sets has all limits and colimits. The coproduct in the category of G-sets is the disjoint union (just as in the category of sets), and we immediately see that every G-set is a disjoint union of orbits. In some sense, then, we completely understand the category of G-sets if we know the morphisms between orbits.

5.2.2 Theorem *Let H, K be subgroups of G. Then the set*

$$G\text{-Sets}(G/H, G/K)$$

of morphisms from G/H to G/K in the category of G-sets is bijective with the set G/K^H of fixed points of G/K with respect to the subgroup H, which is equal to the set of all cosets gK such that

$$g^{-1} H g \subseteq K. \tag{5.2.6}$$

(Here $g^{-1} H g = \{g^{-1} x g \mid x \in H\}$.)

COMMENTS Because of (5.2.6), G-set morphisms from G/H to G/K are sometimes called *subconjugacies* from H to K. The full subcategory of the category of G-sets on orbits is called the *orbit category*, and denoted by Orb_G. At this point we are most interested in observing that the isotropy groups of different elements of G/H may be different. In fact, more precisely, the isotropy group of $gH \in G/H$ is the subgroup $gHg^{-1} \subseteq G$, which is known as a subgroup of G *conjugate to H*. The isotropy groups of all elements of the orbit are the same if and only if all conjugate subgroups of H are equal to H, which happens if and only if $H \lhd G$ is a normal subgroup.

Proof First notice that the condition (5.2.6) does not depend on the representative g of a coset gK (since replacing g by kg turns (5.2.6) into an equivalent condition). Next, for all $h \in H$,

$$g^{-1}hg \in K$$

is equivalent to

$$hg \in gK, \tag{5.2.7}$$

and thus to

$$hgK = gK, \tag{5.2.8}$$

which is the same as $gK \in G/K^H$. Now let us study morphisms of G-sets

$$\phi : G/H \to G/K. \tag{5.2.9}$$

Since G/H is an orbit, the morphism ϕ is clearly determined by

$$g = \phi(1). \tag{5.2.10}$$

Which choices of g are allowed in (5.2.10)? The requirement is that the isotropy group of 1 (which is H) must be contained in the isotropy group of $\phi(1)$. This is precisely (5.2.8), which we already proved is equivalent to (5.2.6).

Finally, note that replacing g by gk for $k \in K$ in (5.2.10) produces the same morphism ϕ and vice versa, thus defining the required bijection. $\qquad\square$

5.3 Closed Points as Galois Group Orbits

Now let us return to our closed point m of $\mathbb{A}_k^n$ with (5.1.1), and let us look at the set $\pi^{-1}(m)$ of all points of $\mathbb{A}_E^n$ which map to m by (5.1.2). As already observed, $\pi^{-1}(m)$ is a $Gal(E/k)$-set. To be more specific, note that by definition, $\pi^{-1}(m)$ is the set of closed points of the pullback

$$
\begin{array}{c}
Spec(K) \\
\downarrow \\
\mathbb{A}_E^n \longrightarrow \mathbb{A}_k^n,
\end{array}
$$

which is

$$
Spec(K \otimes_{k[x_1,\dots,x_n]} E[x_1,\dots,x_n]),
$$

which, in turn, is isomorphic to $Spec$ of the ring

$$
K \otimes_k E. \tag{5.3.1}
$$

Now since K is a separable extension of k, it is, generated, as a k-algebra, by a single element a.

Thus, by definition,

$$
K \cong k[x]/p(x)
$$

where p is the minimal polynomial of a over k. Now over E, by assumption, p splits into linear factors

$$
p(x) = (x - a_1)\dots(x - a_m)
$$

where $a_1, \dots, a_m \in E$ are all different. Then

$$
K \otimes_k E \cong E[x]/((x - a_1)\dots(x - a_m)). \tag{5.3.2}
$$

But recall the *Chinese remainder theorem* which states that if any ideals $I_1, \dots, I_m$ of a ring have the property that $I_i + I_j = R$ for any $i \neq j$, then $I_1 \cap \dots \cap I_m = I_1 \cdots I_m$ and

$$
R/(I_1 \cap \dots \cap I_m) = R/I_1 \times \dots \times R/I_m.
$$

Thus, the right hand side of (5.3.2) is isomorphic to

$$
\prod_{i=1}^m E[x]/(x - a_i) \cong \prod_{i=1}^m E.
$$

The spectrum of this ring (which, recall, is $\pi^{-1}(m)$), is

$$\coprod_{i=1}^{m} Spec(E),$$

which consists of m points. But how does $Gal(E/k)$ act on those points? We know that $Gal(E/k)$ acts transitively on $\{a_1, \ldots, a_m\}$, so it is isomorphic, as a $Gal(E/k)$-set, to an orbit by Proposition 5.2.1.

What is the isotropy group? By definition, the embedding $K \subseteq E$ sends the generator $a \in K$ to one of the elements $a_1, \ldots, a_m$. Assume, without loss of generality, that it is a_1. Then the isotropy group is the subgroup of all elements $g \in Gal(E/k)$ which satisfy $g(a_1) = a_1$, which, by Galois theory (see Exercise 34), is $Gal(E/K)$. Thus, we have proved the following

5.3.1 Proposition *If X is a variety over a perfect field k, then there is a canonical bijection from the set of all closed points m of X whose residue field is contained in a given Galois extension E of k to the set of orbits of $Gal(E/k)$ acting on the set of closed points of*

$$X \times_{Spec(k)} Spec(E) \tag{5.3.3}$$

with residue field E. Furthermore, if the residue field of m is $K \subseteq E$, then its isotropy group in (5.3.3) with respect to the $Gal(E/k)$-action is $Gal(E/K)$.

$\square$

The word *canonical* is a philosophical, not a precise mathematical term. A mathematical object (not in the categorical sense) is called canonical if there is an obvious choice for it. If a mapping is canonical, it is usually a natural transformation between functors, but not conversely: there are important examples of natural transformations which are not canonical in any reasonable sense (see, for example, the proof of Lemma 3.3.5 of Chap. 5). On the other hand, a *unique* object of a given kind is always canonical, but a canonical object does not have to be unique, as there may be more than one obvious choice (differing, for example, by a sign). While not strictly speaking rigorous, the term canonical is often useful in suggesting that a construction exists from a general principle, without having to make unnatural choices on a case by case basis. There may be no reasonable way of expressing this without presenting the explicit construction as a part of the statement of the theorem, which may be awkward or even impossible for space reasons. The word canonical often simply suggests that a reasonable uniform construction exists.

The statement of the proposition is awkward, and it would be good to simplify it. One observation is that it is actually possible to omit the assumptions on the residue field. This gives a more elegant but somewhat less precise statement (note that the hypothesis of k

being perfect is omitted):

5.3.2 Proposition *If X is a variety over a field k and E is a Galois extension of k, then there is a canonical bijection from the set of closed points of X to the set of orbits of the set of closed points of the scheme (5.3.3) with respect to the $Gal(E/k)$-action.*

Proof Let E be the splitting field of a polynomial $p(x) \in k[x]$. Let K be a finite extension of k. Then a closed point P of X with residue field K is represented by a morphism of schemes

$$Spec(K) \to X$$

which induces an isomorphism of the residue field of X at P to K. Thus, it suffices to study the closed points of

$$Spec(K) \times_{Spec(k)} Spec(E).$$

Denote by L the splitting field of $p(x) \in K[x]$. Then

$$Gal(L/K) \subseteq Gal(E/k).$$

Then the definition of a splitting field gives a diagram of K-algebras

$$
\begin{array}{ccc}
L & \longrightarrow & K \otimes_k E \\
 & {}_{Id}\searrow & \downarrow \\
 & & L
\end{array}
\qquad (5.3.4)
$$

(by sending the roots of $p(x)$ to the same roots). The groups $Gal(E/k)$ acts on the upper right corner, the subgroup $Gal(L/K)$ acts on L. If we denote

$$M = \prod_{Gal(E/k)/Gal(L/K)} L, \qquad (5.3.5)$$

M is a commutative L-algebra (a product in the category of L-algebras), but as an L-module, M is of course isomorphic to

$$\widetilde{M} = \bigoplus_{Gal(E/k)/Gal(L/K)} L. \qquad (5.3.6)$$

The L-algebra (5.3.5) and the L-module (5.3.6) have a canonical $Gal(E/k)$-action which extends the $Gal(L/K)$-action on L. Using this action, we obtain a diagram of L-modules with $Gal(E/k)$-action

$$
\begin{array}{ccc}
\tilde{M} & \longrightarrow & K \otimes_k E \\
 & \searrow^{\phi} & \downarrow \\
 & & M
\end{array}
\tag{5.3.7}
$$

where the vertical map is a homomorphism of L-algebras. In fact, note that by Galois theory (Exercise 34), as k-modules, both (5.3.5) and (5.3.6) become canonically identified with

$$
\bigoplus_{Gal(E/k)} k,
$$

(by considering the images of 1), and the homomorphism of k-modules ϕ becomes identified with Id again.

Thus, the vertical morphism (5.3.7) is onto. On the other hand, note that the dimensions of its source and target as vector spaces over k are both

$$
[K:k] \cdot [E:k] = \frac{[E:k]}{[L:K]}[L:k]
$$

(where $[K:k]$ denotes the degree of a finite field extension, i.e. the dimension of K as a vector space over k). Thus, the vertical morphism (5.3.4) is an isomorphism of rings

$$
E \otimes_k K \cong \prod_{Gal(E/k)/Gal(L/K)} L. \tag{5.3.8}
$$

Note that the map (5.3.8) is not equivariant, as the Galois action twists by the $Gal(L/K)$-action on L. Nevertheless, we see that on the points of

$$
Spec(E \otimes_k K) \cong Spec(\prod_{Gal(E/k)/Gal(L/K)} L) \cong
$$

$$
\coprod_{Gal(E/k)/Gal(L/K)} Spec(L),
$$

$Gal(E/k)$ acts transitively, its orbit representing the point of $Spec(K)$. $\quad\square$

COMMENT The constructions M and $\tilde{M}$ are examples of a more general construction. In any category C, for a group G, we have the category G-C of G-equivariant objects in C. The objects consists of an object x of C and a homomorphism of groups

$$
\phi : G \to Aut(x)
$$

where $Aut(x)$ denotes the group of automorphisms of x (i.e. isomorphisms $x \to x$). Morphisms in G-C are morphisms $f : x \to y$ in C which respect the G-action, which means for every $g \in G$, there is a commutative diagram

$$
\begin{array}{ccc}
x & \xrightarrow{\ \phi(g)\ } & x \\
f\downarrow & & \downarrow f \\
y & \xrightarrow[\ \phi(g)\ ]{} & y.
\end{array}
$$

Then for an inclusion of groups $i : H \subseteq G$, we have a "forgetful" functor

$$ i^* : G\text{-}C \to H\text{-}C. \tag{5.3.9} $$

Not always, but for many categories C, the functor i^* has a left adjoint $i_\sharp$ and a right adjoint i_* called the *left and right Kan extension*. If C is a category of universal algebras, the left Kan extension $i_\sharp X$ of an H-equivariant C-object X is the colimit (coequalizer) in C of the diagram

$$ \coprod_{G \times H} X \rightrightarrows \coprod_{G} X $$

where $\coprod$ denotes the categorical coproduct, and the two arrows send x in the (g, h)'th copy of X to $h(x)$ in the g'th copy of X, and to x in the gh'th copy of X, respectively. The G-action is defined by letting $g \in G$ send x in the k'th copy of X ($k \in G$) into x in the gk'th copy of X. (Note however that while expressing things in this way, the coproduct is generally not the disjoint union.)

Dually, in a category C of universal algebras, the right Kan extension of an H-equivariant C-object X is the limit (equalizer) of the diagram

$$ \prod_{G} X \rightrightarrows \prod_{G \times H} X $$

where the two arrows send a $|G|$-tuple

$$ \prod_{G} x_g $$

to

$$ \prod_{G \times H} y_{(g,h)} $$

where $y_{(g,h)} = x_{gh}$ and $y_{(g,h)} = h(x)_g$, respectively.

In the proof of Proposition 5.3.2, M was the left Kan extension with respect to the inclusion

$$i : Gal(L/K) \subseteq Gal(E/k)$$

in the category of K-algebras, while $\tilde{M}$ was the left Kan extension in the category of K-modules. As already remarked, writing L for the copies on the right hand side of (5.3.8) is somewhat imprecise if we want to think $Gal(E/k)$-equivariantly: for the g'th copy, we should write the subfield gL which is the image of L in $E \otimes_k K$ under g instead. Of course, this field is by definition isomorphic to L, but not canonically, unless we know the representative g of the corresponding left coset in $Gal(E/k)/Gal(L/K)$.

5.4   The Absolute Galois Group: Profinite Groups

Finally, it seems natural to find a statement where E would be replaced by $\bar{k}$, the algebraic closure of k, (k is perfect), which is a minimal algebraically closed field containing k. (An algebraic closure is unique up to noncanonical isomorphism.) If $\bar{k}$ is a finite extension of k, we may use it as E, and the previous propositions apply. This arises for example when $k = \mathbb{R}$, $E = \bar{\mathbb{R}} = \mathbb{C}$ (which is algebraically closed by the *fundamental theorem of algebra*). Thus, for a variety X over $\mathbb{R}$, the closed points are orbits of the set of closed points of $X \times_{Spec(\mathbb{R})} Spec(\mathbb{C})$ with respect to complex conjugation

$$a + ib \mapsto a - ib, \ a, b \in \mathbb{R},$$

which is the generator of $Gal(\mathbb{C}/\mathbb{R})$. The orbits are of the form $\mathbb{Z}/2 = (\mathbb{Z}/2)/\{e\}$ (an orbit $G/\{e\}$ of any group is called the *torsor*), and $(\mathbb{Z}/2)/(\mathbb{Z}/2)$. Obviously, the torsor of $Gal(\mathbb{C}/\mathbb{R})$ has two elements, corresponding to two complex-conjugate points, while the other type of orbit has a single element, corresponding to a point with residue field $\mathbb{R}$. Studying such points in detail is the subject of *real algebraic geometry*.

Example Consider the affine variety

$$Spec(\mathbb{R}[x, y]/(x^2 + y^2 - 1)) \tag{5.4.1}$$

over $\mathbb{R}$ (the unit circle). We have, of course, the usual real points, for example $(\sqrt{2}/2, -\sqrt{2}/2)$. Written as a maximal ideal, this is the ideal of

$$\mathbb{R}[x, y]/(x^2 + y^2 - 1) \tag{5.4.2}$$

generated by $x - \sqrt{2}/2$ and $y - \sqrt{2}/2$, i.e.

$$(x - \sqrt{2}/2, y + \sqrt{2}/2) \subset \mathbb{R}[x, y]/(x^2 + y^2 - 1).$$

This is a maximal ideal (a closed point). Of course, there is a maximal ideal with the same generators in

$$\mathbb{C}[x, y]/(x^2 + y^2 - 1). \tag{5.4.3}$$

It is a fixed point under the action of $Gal(\mathbb{C}/\mathbb{R})$ (i.e. a 1 element orbit). On the other hand, there are of course closed points of

$$Spec(\mathbb{C}[x, y]/(x^2 + y^2 - 1) \tag{5.4.4}$$

which are not real: for example the point written in affine variety notation as $(i, \sqrt{2})$, i.e. the maximal ideal

$$(x - i, y - \sqrt{2}) \subset \mathbb{C}[x, y]/(x^2 + y^2 - 1).$$

The orbit of this ideal under the action of $Gal(\mathbb{C}/\mathbb{R})$ is

$$\{(x - i, y - \sqrt{2}), (x + i, y - \sqrt{2})\}. \tag{5.4.5}$$

By Proposition 5.3.2, this corresponds to a closed point of (5.4.1). Indeed, this closed point is, in maximal ideal notation,

$$(x^2 + 1, y - \sqrt{2}). \tag{5.4.6}$$

(We find the ideal by taking any element of the orbit (5.4.5) and intersecting it with $\mathbb{R}[x, y]/(x^2 + y^2 - 1)$.) Note that the point (5.4.6) does not have a "naive" notation as a pair of elements of $\mathbb{R}$. We can, however, still write it as a pair of points in $\mathbb{A}^1_{\mathbb{R}}$.

Even that is not possible in general, however. Consider the point of (5.4.4) which is given, in the affine variety notation, as the pair $(3/4 + i/4, 3/4 - i/4)$. (Check that this works.) This corresponds to the maximal ideal

$$(4x - 3 - i, 4y - 3 + i) \subset \mathbb{C}[x, y]/(x^2 + y^2 - 1). \tag{5.4.7}$$

In general, we observe that finding the intersection of an ideal with (5.4.2) is not trivial (it can be done by finding the Gröbner basis of the ideal

$$(4x - 3 - i, 4x - 3 + i, x^2 + y^2 - 1, i^2 + 1) \subseteq \mathbb{R}[x, y, i]$$

with respect to the order $i > x > y$). In the case of the ideal (5.4.7), however, it is still elementary, since we can change the generators of $\mathbb{R}[x, y]$ to $x - y$, $x + y$, so the ideal is

$$(2x + 2y - 3, 2x - 2y - i).$$

Thus, we see that the corresponding point in (5.4.1) is

$$(2x + 2y - 3, 4(x - y)^2 + 1).$$

Not a pair of points in $\mathbb{A}^1_{\mathbb{R}}$! This shows that for varieties over k which is not algebraically closed, $? \times_{Spec(k)} ?$ does not in general give the Cartesian product on sets of closed points. (It is true however when k is algebraically closed, by considering the affine case and the Nullstellensatz; see Exercise 36.)

Speaking of the Nullstellensatz, let us give an even more direct example of why "non-naive" closed points come up. Consider the following affine variety over $\mathbb{R}$:

$$Spec(\mathbb{R}[x, y]/(x^2 + y^2 + 1)). \tag{5.4.8}$$

Then there are *no points* with residue field $\mathbb{R}$, since $x^2 + y^2 + 1 \neq 0$ for real numbers x, y. By the Nullstellensatz, if there were no other closed points, (5.4.8) would be $Spec(0)$, which is clearly not the case, since the ideal $(x^2 + y^2 + 1)$ does not contain 1. Following the above method, we can produce, for example, the closed point

$$(x^2 + 1, y)$$

with residue field $\mathbb{C}$.

Note: in our definitions, $Spec(0)$ is not a variety, since an irreducible space is required to be non-empty. In any case, it brings up the necessity to show that the polynomials $x^2 + y^2 \pm 1$ are irreducible in $\mathbb{R}[x, y]$. Note that they are irreducible in $\mathbb{R}(y)[x]$, (or $\mathbb{C}(y)[x]$), since they have no root in $\mathbb{R}(y)$. Note also that of course an irreducible polynomial can become reducible when extending scalars from k to E, thus showing that (5.3.3) may not always be a variety over E, although it is one in our case.

However, fields k for which the algebraic closure $\bar{k}$ is a finite extension of k are, unfortunately, an exception to the rule, and are, in some sense, all analogous to the case $k = \mathbb{R}$ (for example, in all such cases, $Gal(\bar{k}/k) \cong \mathbb{Z}/2$). We must, therefore, develop some technique for generalizing Galois theory to infinite (separable) algebraic extensions. Note that by contravariance in Galois theory, if

$$k \subseteq K \subseteq L$$

are Galois extensions, we have an onto homomorphism

$$Gal(L/k) \longrightarrow Gal(K/k).$$

Thus, the groups $Gal(?/k)$ form an *inverse system* of groups, which means a functor

$$Gal(?/k) : I^{Op} \to Groups \tag{5.4.9}$$

where I is a directed partially ordered set, which means that for any $a, b \in I$ there exists an element $c \in I$ with $a \leq c$ and $b \leq c$. (Recall that a partially ordered set I can be considered as a category where the morphism set $I(a, b)$ is a one element set when $a \leq b$ and is empty otherwise.) In the present case, the set I is the set of all finite extensions of the field k ordered by inclusion.

We could define $Gal(\overline{k}/k)$ as the limit (meaning inverse limit) of the diagram (5.4.9), but if we do this in the category of groups, it turns out to lose too much information. For example, there can be finite orbits which do not factor through orbits of any $Gal(L/k)$ for a finite Galois extension L of k.

There are several solutions of this problem. For example, we could take a limit of the diagram (5.4.9) in the category of *topological groups* which means a group object in the category of topological spaces, or equivalently, a group with a topology where the operations of multiplication and inverse are continuous, and we could restrict attention to continuous group action (a set on which a topology is not specified is considered to have the *discrete topology* which means that every subset is open).

This would actually produce the right answer in our present situation, but it is useful to learn another formalism. A *pro-finite group* is defined to be an inverse system, i.e. a functor

$$G : I^{Op} \to sf\,Groups \tag{5.4.10}$$

where I is a directed set (which may vary) and $sf\,Groups$ is the category of finite groups and surjective homomorphisms. In particular, then, a finite group is a pro-finite group (we can take, for example, I to be a one element set). Two pro-finite groups from (5.4.10) and

$$H : J^{Op} \to sf\,Groups$$

specify a functor

$$Hom(G, H) : I \times J^{Op} \to Sets. \tag{5.4.11}$$

The set of morphisms $\phi : G \to H$ is defined to be the colimit of (5.4.11) in the I coordinate, followed by the limit in the J^{Op}-coordinate.

This means that for every $j \in J$ there exists an $i(j) \in I$ such that for all $i \geq i(j)$, we are given a homomorphism of groups

$$\phi_{i,j} : G_i \to H_j$$

such that the diagram

$$
\begin{array}{ccc}
G_{i'} & \xrightarrow{\phi_{i',j'}} & H_{j'} \\
\downarrow & & \downarrow \\
G_i & \xrightarrow{\phi_{i,j}} & H_j
\end{array}
$$

commutes whenever it is defined. (Note: there are several variants of the definition of profinite groups, but they give equivalent categories.)

Now by an action of a pro-finite group (5.4.10) on a finite set S (i.e. a finite G-set), we simply mean an action of some $G_i, i \in I$, on S. For $i \leq j$, such an action is considered equivalent to an action of G_j on S, $j \leq i$, which factors through the surjection

$$G_j \longrightarrow\!\!\!\!\!\longrightarrow G_i.$$

The concept of an orbit does not depend on the choice of $i \in I$, and there is an obvious concept of an isotropy group, which is a pro-finite subgroup of G. We may also speak of a finite G-set for the pro-finite group G. Any finite G-set is obviously a disjoint union of orbits. We may also consider general (possibly infinite) G-sets for a profinite group G, by which we mean possibly infinite disjoint unions of (finite) orbits. Note that an infinite G-set may not be a G_i-set for a fixed $i \in I$, since i may vary over different orbits.

In this language, for any perfect field k, $Gal(\overline{k}/k)$ is a profinite group, defined as the inverse system (5.4.9). It is sometimes denoted by $Gal(k)$ and called the *absolute Galois group of the perfect field k*. Then our previous discussion implies the following

5.4.1 Theorem *Let X be a variety over a perfect field k. Then there is a canonical bijective map from the set of closed points of X to the set of orbits of the set of closed points of $X \times_{Spec(k)} Spec(\overline{k})$ with respect to the action of the pro-finite group $Gal(\overline{k}/k)$.* $\quad\square$

5.5   The Fundamental Group of a Topological Space

There is a situation in topology which to some degree is analogous to Galois theory, namely the *fundamental group*. Giving all the details here would take us too far afield, so the reader is referred to [13] for further reading. However, the main point is important, as it is another example of our continuing story of connections between topology and algebraic geometry.

Let X be a topological space and let $x \in X$ be a point. A *path* is a continuous map $\omega : [0, 1] \to X$ where on the interval $[0, 1]$, we consider the analytic topology. The points $\omega(0)$ resp. $\omega(1)$ are called the *beginning point* resp. the *end point* of the path ω. We also say that ω is *a path from* $\omega(0)$ *to* $\omega(1)$. A *based homotopy* between two paths ω_1, ω_2 is a continuous map $h : [0, 1]^2 \to X$ (where the domain has the analytic topology) such that for all $s \in [0, 1]$, $h(s, 0) = \omega_1(s)$, $h(s, 1) = \omega_2(s)$, and for all $t \in [0, 1]$, $h(0, t) = h(0, 0)$ and $h(1, t) = h(1, 0)$. Two paths are called *based-homotopic* if there exists a based homotopy between them. This is an equivalence relation. (Thus, in particular, based-homotopic paths have the same beginning points and the same end points.) The equivalence class of a path ω with respect to based homotopy is often denoted by $[\omega]$ (which is a generic notation for the equivalence classes of any equivalence relation).

Now for a topological space X, there is a groupoid (i.e. category whose every morphism is an isomorphism) $\pi_1(X)$ called the *fundamental groupoid* whose objects are the points of X, and morphisms $x \to y$ are based homotopy classes of paths with beginning point x and end point y. For a path ω from x to y and a path η from y to z, we define the $\pi_1(X)$-composition $\omega * \eta$ (we use $*$ here to avoid confusion with composition of maps) by

$$\omega * \eta(s) = \begin{cases} \omega(2s) & \text{if } 0 \leq s \leq 1/2 \\ \eta(2s - 1) & \text{if } 1/2 \leq s \leq 1. \end{cases}$$

Clearly, the number $1/2$ can be replaced by any number between 0 and 1 without altering the definition of composition on based homotopy classes, which is the reason composition in $\pi_1(X)$ is associative. The identity is the constant path from x to x and the inverse to a class $[\omega]$ is the class $[\overline{\omega}]$ where $\overline{\omega}(s) = \omega(1 - s)$. (See Exercise 37.)

More generally, a *homotopy* between continuous maps of topological spaces $f, g : X \to Y$ is a is a map $h : X \times [0, 1] \to Y$ where for all $x \in X$, $h(x, 0) = f(x)$ and $h(x, 1) = g(x)$. We write $h : f \simeq g$, and also call f and g *homotopic*, and write $f \simeq g$. Homotopy of maps is then also an equivalence relation. A *homotopy equivalence* between topological spaces X, Y is a pair of continuous maps $f : X \to Y, g : Y \to X$ such that $g \circ f \simeq Id_X, f \circ g \simeq Id_Y$. We sometimes also say that f or g is a homotopy equivalence, without mentioning the other map, which is called its *homotopy inverse*. Of course, there is also a category whose objects are topological spaces and morphisms are homotopy classes of maps. It is called the *homotopy category*. For the effect of continuous maps, homotopies and homotopy equivalences on the fundamental groupoid, see Exercise 38.

We may be more comfortable with groups than groupoids, so for $x \in X$, we consider the group $\pi_1(X, x)$ of automorphisms of x in $\pi_1(X)$, and call it the *fundamental group* of the space X at $x \in X$. In other words, $\pi_1(X, x)$ is the group of based homotopy classes of paths with beginning point and end point x (which are also called *loops at x*). However, the fundamental group has a somewhat weaker functoriality due to the role of the base point: For $x \in X$ and a continuous map $f : X \to Y$, we get a canonical map $\pi_1(f) : \pi_1(X, x) \to \pi_1(Y, f(x))$.

Our main interest in the fundamental group is its close connection with *covering spaces*. A map $f : Y \to X$ is called a *covering* if for every $x \in X$, there exists an open neighborhood U of x (called the *fundamental neighborhood*) such that we have a commutative diagram

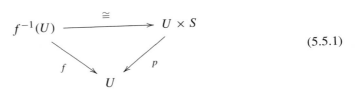

$$(5.5.1)$$

where S is a set with the discrete topology and p is the projection (thus, $U \times S$ is homeomorphic to a disjoint union of copies of U). We also sometimes say (somewhat imprecisely) that *Y is a covering space of* X. We have

5.5.1 Lemma *Let $f : Y \to X$ be a covering and let $f(y) = x$. Then for every path ω in X with beginning point x (or based homotopy h between paths with beginning point x), there exists a unique path $\tilde{\omega}$ with beginning point y (resp. based homotopy $\tilde{h}$ between paths with beginning points y) such that $f \circ \tilde{\omega} = \omega$ (resp. $f \circ \tilde{h} = h$).*

Proof Because $[0, 1]$ is compact, we can break it up into intervals (resp. break $[0, 1] \times [0, 1]$ into squares) which are sent by ω (resp. h) to fundamental neighborhoods. Then existence and uniqueness both follow from Diagram (5.5.1), using the connectedness of the unit interval. $\qquad \square$

Note that Lemma 5.5.1 immediately implies that if f is a covering and $f(y) = x$, then the map $f : \pi_1(Y, y) \to \pi_1(X, x)$ is injective. We can thus ask which subgroups of the fundamental group have covering spaces corresponding to them. To this end, let us assume that X is *path-connected*, which means that for any $x, y \in X$, there is a path from x to y (for otherwise, the fundamental group cannot be expected to recover information about all of X). To formulate our question more precisely, consider the category $Cov_x(X)$ whose objects are pairs (f, y) where $f : Y \to X$ is a covering with Y connected, and $f(y) = x$, and morphisms $(f : Y \to X, y) \to (g : Y' \to X, y')$ are maps $\phi : Y \to Y'$ such that $g \circ \phi = f$ and $\phi(y) = y'$. (Note: morphisms of coverings of X are necessarily also coverings.) It turns out that our question has a good answer when X satisfies two additional hypotheses: X is called *locally path-connected* if every point of X has an open neighborhood which is path-connected; X is called *semilocally simply connected* (SLSC) if every $x \in X$ has an open neighborhood U such that the inclusion induces a 0 map $\pi_1(U, x) \to \pi_1(X, x)$. The fundamental result is the following

5.5.2 Theorem *Let a topological space X be path-connected, locally path-connected and SLSC, and let $x \in X$. Then the correspondence*

$$(f : Y \to X, y) \mapsto \pi_1(Y, y)$$

defines an equivalence of categories between $Cov_x(X)$ and the category of subgroups of $\pi_1(X, x)$ and inclusions.

Proof Essentially, the covering spaces and their maps are constructed by lifting paths in X with beginning point x, using Lemma 5.5.1. However, details are subtle (see [13]). □

There is also a version of this theorem without assuming Y is connected, and without referring to $y \in Y$. Let $Cov(X)$ denote the category of all coverings $f : Y \to X$ and morphisms over X.

5.5.3 Theorem *Let a topological space X be path-connected, locally path-connected and SLSC, and let $x \in X$. Then the correspondence*

$$(f : Y \to X) \mapsto f^{-1}(x)$$

is an equivalence of categories from $Cov(X)$ to $\pi_1(X, x)$-Sets (i.e. sets with $\pi_1(X, x)$-action).

The proof is a formal consequence of Theorems 5.5.2 and 5.2.2. One is led to define $[\omega]y$, for $[\omega] \in \pi_1(X, x)$ and $y \in f^{-1}(x)$, as the endpoint of the lift of the path ω to Y with beginning point y, but because of the way we defined composition of paths, this defines a right action. We can get a left action by replacing ω with $\bar{\omega}$.

5.6 From Coverings to the Fundamental Group: The Étale Fundamental Group

For a path-connected locally path-connected topological space X, we may now ask how much of the fundamental group $\pi_1(X, x)$ can be recovered from the category $Cov(X)$. If X is SLSC, by Theorem 5.5.2, there is a connected covering

$$f : \tilde{X} \to X, \quad f(y) = x$$

$$(5.6.1)$$

where $\pi_1(\widetilde{X}, y) = 0$, and we may simply identify $\pi_1(X, x)$ as the automorphism group of $\widetilde{X}$ in the category $Cov(X)$. The covering (5.6.1) is called a *universal covering of* X.

5.6.1   Regular Coverings

If X is path-connected and locally path-connected but not SLSC, the universal covering does not exist (the condition is if and only if), but we may still have some non-trivial coverings. A covering $f : Y \to X$ is called *regular* if the automorphism group of f in $Cov(X)$ acts transitively on $f^{-1}(x)$. When the universal covering exists, we know from Theorems 5.5.2 and 5.5.3 that a connected covering f is regular if and only if for any (equivalently, all) $y \in f^{-1}(x)$, $\pi_1(Y, y)$ is a normal subgroup of $\pi_1(X, x)$.

In general, however, for a connected covering $f : Y \to X$, *of finite degree* (meaning that $f^{-1}(x)$ is finite), it is possible to construct a diagram

$$
\begin{array}{ccc}
\widetilde{Y} & \longrightarrow & Y \\
 & \scriptstyle{\widetilde{f}} \searrow & \downarrow {\scriptstyle f} \\
 & & X
\end{array}
\tag{5.6.2}
$$

where $\widetilde{Y}$ is a regular covering of finite degree. The construction goes as follows: The category of coverings of X obviously has finite limits (which are the same as in the category of topological spaces over X). Take the product $F : Y \times_X Y \times_X \cdots \times_X Y \to X$ of n copies of $f : Y \to X$, where $n = |f^{-1}(x)|$ is the degree. Then the covering F has an action of the symmetric group Σ_n by switching factors. Now clearly the restriction $\Phi : Z \to X$ of F where $Z \subseteq Y \times_X Y \times_X \cdots \times_X Y$ is the subspace of n-tuples $(y_1, \ldots, y_n)$ where the y_i are all different is also a covering (since its complement, consisting of all n-tuples in which at least two coordinates coincide, is a covering, while both complements are closed and open). Furthermore, by construction, Φ is regular (since the required action is just the action of the symmetric group by switching factors), although it may not be connected. Let, then, $\widetilde{f}$ be the restriction of Φ to one connected component.

Next, we may observe that a morphism h from $(g : Z \to X, z)$ to $(f : Y \to X, y)$ in $Cov_x(X)$ for regular coverings specifies, functorially, a homomorphism of groups

$$
Aut_{Cov(X)}(g) \to Aut_{Cov(X)}(f).
\tag{5.6.3}
$$

In effect, suppose that $h(\alpha(z)) = h(z)$ for some $\alpha \in Aut_{Cov(X)}(g)$. Then $h \circ \alpha = h$ (since the set of all points t for which $h(\alpha(t)) = h(t)$ is an open and closed set in Z). Thus, $W = \{\alpha \in Aut_{Cov(X)}(g) \mid h(\alpha(z)) = h(z)\}$ is a subgroup of $Aut_{Cov(X)}(g)$, and

$$
f^{-1}(x) \cong Aut_{Cov(X)}(g)/W.
\tag{5.6.4}
$$

To prove that the subgroup W is normal and that (5.6.3) is an onto homomorphism of groups, consider an automorphism α of g and an automorphism β of f such that

$$h \circ \alpha(z) = \beta \circ h(z).$$

Then the set of all t such that $h \circ \alpha(t) = \beta \circ h(t)$ is open and closed in Z, and thus is equal to Z. This proves in particular that all automorphisms of Y lift to automorphisms of Z. Since Y is regular, this is equivalent to W being a normal subgroup and (5.6.3) is an onto homomorphism of finite groups, where $Aut(f) \cong f^{-1}(x)$ via the identification which sends an automorphism β to $\beta(y)$.

Indexing over the full subcategory $Reg_x(X)$ of $Cov_x(X)$ on based regular coverings of (X, x) of finite degree, we thereby obtain a pro-finite group which we denote by $\pi_1^{Cov}(X, x)$. We have a "homomorphism"

$$\pi_1(X, x) \to \pi_1^{Cov}(X, x) \tag{5.6.5}$$

by which we mean a compatible system of homomorphisms

$$\pi_1(X, x) \to Aut_{Cov(X)}(Y) \tag{5.6.6}$$

over based regular coverings of (X, x) of finite degree. Now (5.6.5) is a pro-finite completion (i.e. isomorphism on the pro-finite group given by all finite group quotients) when X is SLSC, but not in general. However, by the construction (5.6.4), for a covering $f : Y \to X$ of finite index, we have

$$f^{-1}(x) \cong \operatorname*{colim}_{Z \in Reg_x(X)} Cov(X)(Z, Y), \tag{5.6.7}$$

(where the map from the right to the left hand side is given by the image of the base point). From this, we can deduce that the category of finite degree coverings of X is equivalent to the category of finite $\pi_1^{Cov}(X, x)$-sets.

5.6.2   The Étale Case

Now the method of Sect. 5.6.1 can be used to define the étale fundamental group of a Noetherian scheme X. Just as earlier in this chapter, however, the analogy with topology is not the most straightforward. We do not have any reasonable definition of a path in the algebraic context. The role of finite coverings is played by finite étale morphisms, but it is important to note that such maps are not actually coverings in the topological sense when we use the Zariski topology (see Exercise 43). They play, nevertheless, the same role, and also, over $\mathbb{C}$, define coverings in the analytic topology (see Exercise 44). Note here that the reason we must insist on assuming finiteness is that otherwise, even this analogy would be broken, as the inclusion of an open subscheme is étale, but the inclusion of an open set in a topological space is not in general a covering.

The appropriate analog of the concept of a point is a *geometric point* of X, which consists of a point $x \in X$ together with a choice of a separable closure $\overline{k(x)}$ of the residue field $k(x)$ of x, i.e. a morphism

$$Spec(\overline{k}_x) \to Spec(k_x). \tag{5.6.8}$$

The reason for this definition is to incorporate the Galois group of a field (Exercise 51).

Now for a Noetherian connected scheme X with a geometric point P, we can define the category $Cov_P^{et}(X)$ of *based étale covers of* X to have objects

$$\begin{array}{ccc}
 & & Y \\
 & \nearrow & \downarrow f \\
Spec(K) & \xrightarrow{\;P\;} & X
\end{array} \tag{5.6.9}$$

where f is a finite étale morphism, Y is connected and morphisms given by morphisms of schemes $g : Y \to Z$ which commute with all the arrows in sight. (The morphism g is then automatically finite étale (see Exercise 47).)

Two variants of this construction are the category $Cov^{et}(X)$ of finite étale schemes over X, and the full subcategory $Reg_P^{et}(X)$ on *Galois coverings*, which means based connected finite étale morphisms $f : Y \to X$ where $Aut_{Cov^{et}(X)}(f)$ acts transitively on the set of lifts (5.6.9) of the geometric point P.

Completely analogously as in Sect. 5.6.1, one proves that the finite groups $Aut_{Cov^{et}(X)}(f)$ where f is in $Reg_P^{et}(X)$ form a pro-finite group, which is called the *étale fundamental group of* X *at* P and denoted by $\pi_1^{et}(X, P)$. Also analogously to (5.6.7), the functor from $Cov^{et}(X)$ to finite sets given by the set of lifts of P to the given étale cover is naturally isomorphic to

$$\operatorname*{colim}_{Z \in Reg_P^{et}(X)} Cov^{et}(X)(Z, Y)$$

and consequently the category $Cov^{et}(X)$ is equivalent to the category of finite $\pi_1^{et}(X, P)$-sets.

The foundational details on étale morphisms needed to carry out this analogy rigorously are given in Exercises 45–48.

5.7   Finite Fields and the Weil Conjectures

Let us recall some facts about finite fields. A finite field has characteristic $p > 0$ where p is a prime number, so it contains $\mathbb{F}_p = \mathbb{Z}/p$. This means that it is in particular a vector space over $\mathbb{F}_p$, and as such is isomorphic to $(\mathbb{F}_p)^n$, which means that it has p^n elements. But any two finite fields with $q = p^n$ elements are isomorphic, since any such field $\mathbb{F}_q$ is the splitting field of the polynomial $x^q - x$, the roots being just all the elements of $\mathbb{F}_q$. Note that in characteristic p, the derivative of this polynomial is 1, so it has no multiple roots. Thus, $\mathbb{F}_q$ is a Galois extension of $\mathbb{F}_p$ and furthermore, we have

$$Gal(\mathbb{F}_q/\mathbb{F}_p) \cong \mathbb{Z}/n$$

where the generator is the Frobenius

$$x \mapsto x^p.$$

(Note that by the binomial theorem,

$$(x + y)^p = x^p + \binom{p}{1}x^{p-1}y + \cdots + \binom{p}{p-1}xy^{p-1} + y^p,$$

which is $x^p + y^p$ in characteristic p, since all the other terms are divisible by p.) Thus, we can conclude that the profinite group

$$Gal(\mathbb{F}_q) = Gal(\overline{\mathbb{F}_q}/\mathbb{F}_q)$$

is isomorphic to the profinite group

$$G : I^{Op} \to sf\,Groups$$

where I is the set of natural numbers with ordering $\preceq$ given by

$$m \preceq n \text{ when } n \text{ is divisible by } m,$$

and we have

$$G_n = \mathbb{Z}/n$$

where when $m \preceq n$, the homomorphism $G_n \to G_m$ is the surjective homomorphism sending $k + n\mathbb{Z} \mapsto k + m\mathbb{Z}$. This profinite group is denoted by $\widehat{\mathbb{Z}}$.

It is worth also making a brief note about the actual limit of $\mathbb{Z}/n$ over I^{Op} (which is also sometimes denoted by $\widehat{\mathbb{Z}}$). The ring

$$\mathbb{Z}_p = \lim_{\leftarrow} \mathbb{Z}/p^n$$

(with respect to the unique surjective homomorphisms) is called the *ring of p-adic numbers*. It is a discrete valuation ring, and its field of fractions $\mathbb{Q}_p$ has many properties analogous to the real numbers (although it is not an ordered field). Analysis can be done over the field $\mathbb{Q}_p$, and this field is known as *p-adic analysis*. It is not difficult to compute that in the category of groups (or abelian groups),

$$\widehat{\mathbb{Z}} = \lim_{I^{op}} \mathbb{Z}/n = \prod_{p \text{ prime}} \mathbb{Z}_p.$$

This is a somewhat complicated abelian group, which further makes the case that $\widehat{\mathbb{Z}}$ should be considered in the category of pro-finite (abelian) groups.

Let us now mention the famous Weil conjectures, settled by P. Deligne in 1974 [5, 6], based on previous work by Grothendieck and others. Let X be an n-dimensional smooth projective variety over a finite field $\mathbb{F}_q$ ($q = p^k$). Let

$$\widetilde{N}_m = \frac{N_m}{m}$$

where N_m is the number of "naive" points over $\mathbb{F}_{q^m}$, i.e. ratios in a projective space over $\mathbb{F}_{q^m}$ solving the equations defining X. Now there are deep relations between the numbers $\widetilde{N}_m$. Most concisely, they can be expressed by forming the *zeta function of X*

$$\zeta(X, s) = \exp(\sum_{n=1}^{\infty} \widetilde{N}_m T^m), \quad T = q^{-s} \tag{5.7.1}$$

where $\exp(x) = e^x$. The function (5.7.1) is defined for a complex number s whose real part is sufficiently large (since then the numbers T^m will go to 0 much faster than the numbers $\widetilde{N}_m$ go to ∞. Then the main part of the Weil conjectures states that in fact, $\zeta(X, s)$ is a rational function of s (this was first proved by Dwork), and more specifically,

$$\zeta(X, s) = \frac{P_1(T) P_3(T) \dots P_{2n-1}(T)}{P_0(T) P_2(T) \dots P_{2n}(T)}$$

where $P_0(T), \dots, P_{2n}(T)$ are polynomials in T with integer coefficients such that for some natural numbers β_i and complex numbers α_{ij}, we have

$$P_0(T) = 1 - T,$$

$$P_{2n}(T) = 1 - q^n T,$$

$$P_i(T) = \prod_{j=1}^{\beta_i} (1 - \alpha_{ij} T), \quad |\alpha_{ij}| = |q|^{i/2}.$$

Giving a proof of the Weil conjectures is beyond the scope of this book. However, the starting point of the proof (which was Grothendieck's program, recorded in the SGA volumes) is to push further the analogy between the étale and analytic topology, extending it to cohomology (thus defining *étale cohomology*). We shall mention the construction and some basic facts about étale cohomology in Sect. 4 of Chap. 6.

Much of modern algebraic number theory (including some of the most important unsolved problems of mathematics) are phrased in terms of various variants of zeta functions (also known as *L*-functions). Smooth projective varieties over finite fields are more manageable than, say, over $\mathbb{Q}$ because the Galois group of finite fields is just $\widehat{\mathbb{Z}}$, which is relatively easy. The Galois group of $\mathbb{Q}$ is unknown.

We saw that much additional discussion arises when we do algebraic geometry over fields which are not algebraically closed, and that much of it is related to number theory. Because of this, algebraic geometry over non-algebraically closed fields is also referred to as *arithmetic geometry*.

6   Exercises

1. Is a finite limit of separated (resp. universally closed) schemes always separated (resp. universally closed)?

2. Is the pushout of two copies of the standard inclusion $\mathbb{A}^n_{\mathbb{C}} \to \mathbb{P}^n_{\mathbb{C}}$ in the category of schemes universally closed?

3. A morphism of schemes $f : X \to Y$ is called *quasi-compact* if if Y is covered by open affine subschemes U_i such that each $f^{-1}(U_i)$ is quasi-compact. Prove that if f is a quasi-compact morphism, then $f(X)$ is closed in Y if and only if it is closed under specialization.

4. Is $Spec(k)$ for a field k always proper?

5. Let $f : X \to Y$, $g : Y \to Z$ be morphisms of schemes. Prove that if $g \circ f$ is universally closed (resp. proper) then f is universally closed (resp. proper).

6. Prove that every unique factorization domain is normal. [Suppose r/s is the root of a monic polynomial where $gcd(r, s) = 1$. Clearing denominators, we get that s is a factor of r^n for some n.]

7. Prove that in a Dedekind domain, any non-zero primary ideal I is a (finite) power of a maximal ideal. [First note that primary ideals in a localization $S^{-1}R$ correspond bijectively with primary ideals in R disjoint from S via the inverse maps given by pushforward and intersection. Then localize at $\sqrt{I}$ and use Lemma 2.4.2, 2.]

8. Prove that in a Dedekind domain, any non-zero ideal factors, uniquely up to order of factors, as a product of maximal ideals. [Use primary ideal decomposition, and Exercise 7. To replace intersection by product, use the Chinese Remainder Theorem.]

9. Prove that a valuation ring is Noetherian if and only if it is a discrete valuation ring.

10. Prove that every valuation ring is integrally closed (meaning normal, i.e. integrally closed in its field of fractions).

11. Prove that a discrete valuation ring is the same thing as a local Dedekind domain.

12. Let $f : X \to Y$ be a morphism of Noetherian schemes and suppose Y is covered by open subsets U_i such that the restriction of f to $f^{-1}(U_i) \to U_i$ is universally closed. Prove that then f is universally closed.

13. Prove that if X is a separated scheme and R is a valuation ring with field of fractions K, then a diagram of the form (2.5.1) can be completed in at most one way. [Completing the diagram in two different ways would give a morphism into $X \times X$ which is in the closure of the image of the diagonal.]

14. Prove Proposition 2.6.1. [Use the distributivity of finite products of schemes under gluing of open affine subschemes.]

15. Prove that if X is a regular scheme and $U \subseteq X$ is an open affine set, then $\mathcal{O}_X(U)$ is a Noetherian (and hence regular) ring.

16. (a) Prove that $Spec(\mathbb{C}[x, y, z]/(x^2+y^2+z^2-1, xyz-1))$ is a smooth affine scheme over $Spec(\mathbb{C})$.

    (b) What happens if we replace $\mathbb{C}$ with $\mathbb{F}_2$ (the field with 2 elements)? [You may want to substitute $x = 1/yz$, but then justify it rigorously.]

17. Prove *Minkowski's Theorem* from the *geometry of numbers*: Let $U \subset \mathbb{R}^n$ be a convex set of volume $> 2^n$ (using the standard volume, i.e. the unit cube has volume 1). Assume further that for $x \in U$, we also have $-x \in U$. Then $U \cap (\mathbb{Z}^n \setminus \{0\}) \neq \emptyset$. [Prove first that there exist points $x, y \in U$ such that $x - y \in (2\mathbb{Z})^n$; otherwise, consider $U + z$ for different $z \in (2\mathbb{Z})^n$ are disjoint, which contradicts our assumption on volume.]

18. Let $K \supset \mathbb{Q}$ be a finite extension of degree n (i.e. K is n-dimensional as a vector space over $\mathbb{Q}$). Then K is called a *number field*. Prove that the ring $K_{\mathbb{R}} = K \otimes_{\mathbb{Q}} \mathbb{R}$ is isomorphic to a product of r copies of $\mathbb{R}$ and s copies of $\mathbb{C}$ for some numbers r, s such that $r + 2s = n$. [r is the number of different embeddings of K into $\mathbb{R}$, and s is the number of different embeddings of K into $\mathbb{C}$ whose image is not in $\mathbb{R}$. Observe that every one of the s embeddings of the second kind can be composed with complex conjugation, resulting in a different embedding.]

19. Recalling from Exercise 18 the decomposition of $K_{\mathbb{R}}$ into a product of r copies of $\mathbb{R}$ and s copies of $\mathbb{C}$, let S_t be the set of all points $x \in K_{\mathbb{R}}$ such that the absolute values of the $r + s$ coordinates of x add up to $\leq t$. Using the inequality between the geometric and arithmetic average, observe that the product of the absolute values of coordinates of an element $x \in S_t$ (called the *norm* of x and denoted by $N(x)$) is

$$\leq \left(\frac{t}{n}\right)^n.$$

20. Prove that the volume of S_t (considering the standard volume in $\mathbb{R}^r \times \mathbb{C}^s \cong \mathbb{R}^n$) is

$$2^r \pi^s \frac{t^n}{n!}.$$

[It is possible to consider the volume of the unit ball in the postman's metric in $\mathbb{R}^n$, and then "replace a product of s squares with a product of disks of an equal diameter."]

21. Using the result of Exercise 17 and referring to the terminology of Exercise 18, prove that if K is a number field (i.e. a finite extension of $\mathbb{Q}$) and $\mathcal{O}_K \subset K$ is the set of elements integral over $\mathbb{Z}$, then

$$vol(K_{\mathbb{R}}/\mathcal{O}_K) \geq \frac{n^n}{n!} \left(\frac{\pi}{4}\right)^s.$$

[By Exercise 17, if the volume of S_t is greater than $2^n vol(K_{\mathbb{R}}/\mathcal{O}_K)$, S_t contains a non-zero point of $\mathcal{O}_K$. However, by Exercise 19, we must then have $(t/n)^n \geq 1$ (since a non-zero integral point certainly cannot have a coordinate of absolute value < 1). Putting these observations together gives the estimate.]

22. Noting that $s \leq n/2$, observe that $vol(K_{\mathbb{R}}/\mathcal{O}_K) > 1$ if $[K : \mathbb{Q}] > 1$. Additionally, if p divides the integer $vol(K_{\mathbb{R}}/\mathcal{O}_K)$, observe that $\mathcal{O}_K/(p)$ has nilpotent elements. Conclude that there is no finite étale morphism of schemes $X \to Spec(\mathbb{Z})$ where X is not a finite disjoint union of copies of $Spec(\mathbb{Z})$.

23. State and prove an appropriate analog of the statement of Exercise 27 of Chap. 2 for abstract varieties as defined in this chapter.

24. Prove that every algebraic variety over a field k is birationally equivalent to a hypersurface, i.e. a variety of the form

$$Spec(k[x_1, \ldots, x_n]/f)$$

where f is an irreducible polynomial. [Use Noether's normalization lemma and the Primitive Element Theorem.]

25. Let X be a smooth variety over an algebraically closed field k, and let $Y \subseteq X$ be a closed subscheme which is also a smooth variety over k. Prove that every closed point $P \in Y$, has an open neighborhood U in X together with a smooth morphism $f : U \to \mathbb{A}_k^c$ for some $c \in \mathbb{N}_0$ such that $U \cap Y = f^{-1}(0)$. [Use Exercise 37 of Chap. 1.]

26. Prove that no elliptic curve over a field k of characteristics $\neq 2, 3$ is rational. [Without loss of generality, k is algebraically closed.]

27. Prove that if C is a smooth curve over an algebraically closed field k, then every rational function on C is equal to a composition of an affine coordinate on $\mathbb{P}_k^1$ with a morphism of schemes $C \to \mathbb{P}_k^1$. Is this true when C is not smooth? (Compare with Exercise 16 of Chap. 1.) [In an affine open neighborhood of a closed point P on C, a rational function is given as a ratio f/g where f, g are regular functions. Now we can assume that f, g are non-zero on P, since $\mathcal{O}_{C,P}$ is a unique factorization domain (why?), and factoring out a greatest common factor is valid in an open neighborhood of P.]

28. Let $f : K \subseteq E$ be an inclusion of function fields of curves over an algebraically closed field k. Prove that there exists a finite subset $Z \subset C_K$ such that f restricted to $C_E \smallsetminus f^{-1}(Z)$ is étale.

29. Prove the *Lüroth Theorem* for algebraically closed fields: Let k be an algebraically closed field and let E be a field with $k \subsetneq E \subset k(x)$. (Recall that $k(x)$ is the field of rational functions over k in one variable.) Then E is isomorphic to $k(x)$ as a k-algebra. [First prove that E is finitely generated as a field over k (and thus a function field of a curve). To this end, note that it contains $k(t)$ for some rational function t, and consider a minimal polynomial of x over $k(t)$. Next, consider the corresponding morphism of varieties $f : \mathbb{P}^1_k \to C_E$. Now select a non-empty open subset $U \subset C_E$ on which f is étale, and select two closed points $P, Q \in f^{-1}(U)$ such that $f(P) \neq f(Q)$. Without loss of generality, $P = 0$, $Q = \infty$. Let x be the corresponding affine coordinate on $\mathbb{P}^1$ and let a be the constant term of the monic minimal polynomial of x over E. Prove that $E = k(a)$ by showing that the morphism $C_E \to \mathbb{P}^1_k$ corresponding to a has degree 1.]

30. Describe the orbit category of the symmetric group Σ_3 on 3 elements.

31. Describe the points of the affine scheme $Spec(\mathbb{R}[x, y]/(x^2 + y^2))$.

32. Let $R = S^{-1}\mathbb{C}[x_1, \ldots, x_n]$ where S consists of all non-zero symmetric polynomials in $x_1, \ldots, x_n$. Describe the points of the scheme $Spec(R)$.

33. Prove that any affine conic over $\mathbb{R}$, i.e. scheme of the form $Spec(\mathbb{R}[x, y]/(f))$ where f is an irreducible polynomial of degree 2 with coefficients in $\mathbb{R}$, is isomorphic to $\mathbb{A}^1_{\mathbb{R}}$, $\mathbb{A}^1_{\mathbb{R}} \smallsetminus \{0\}$ or $\mathbb{P}^1_{\mathbb{R}} \smallsetminus \{P\}$ where P is any point with residue field $\mathbb{C}$. Is there a morphism of schemes $\mathbb{A}^1_{\mathbb{R}} \to Spec(\mathbb{R}[x, y]/(x^2 + y^2 - 1))$ given by the formula

$$t \mapsto (\frac{2t}{t^2 + 1}, \frac{t^2 - 1}{t^2 + 1})?$$

34. (Galois Theory) Let E be a finite separable extension of a field F. The extension E is called *Galois* if there exists a finite subgroup G of the group of automorphisms of the field E such that $F = E^G$ (the superscript means fixed points). Recall that the *degree* $[E : F]$ of the extension is the dimension of E as a vector space over F. By the Primitive Element Theorem, E is generated as an F-algebra by a single element α. Let f be its monic minimal polynomial. Then the degree of f is $[E : F]$. Whether E is a Galois extension of F or not, the *Galois group* $Gal(E/F)$ of E over F is the group of automorphisms of E which leave every element of F fixed.

(a) Prove that for any finite separable extension,

$$|Gal(E/F)| \leq [E : F].$$

[The group $Gal(E/F)$ permutes the roots of f in the separable closure of F, and no element other than the unit can leave any root fixed.]

(b) Prove that the extension is Galois if and only if f is a product of linear factors in $E[x]$. (Then we call E the *splitting field* of f.)

(c) Prove that for any finite subgroup H of the group of automorphisms of E, $[E : E^H] \leq |H|$. Thus, by (a), equality always arises, and a finite separable extension is Galois if and only if $|Gal(E/F)| = [E : F]$. [Let $v_1, \ldots, v_m$ be the basis of E as a vector space over $K = E^H$. Let $h_1, \ldots, h_n$ be the different elements of H. Assume for contradiction that $m > n$. Consider the system of linear equations in unknowns $x_1, \ldots, x_m \in E$:

$$x_1 h_1(v_1) + \cdots + x_m h_1(v_m) = 0$$

$$\cdots$$

$$x_1 h_n(v_1) + \cdots + x_m h_n(v_m) = 0.$$

Then there is a non-zero solution $(x_1, \ldots, x_m)$. Since one of the elements h_i is the identity, not all the elements x_j can be in K. Consider a solution with the fewest non-zero elements, i.e. with the largest k such that $x_1 = \cdots = x_k = 0$, and the remaining elements x_j's are non-zero. Dividing by x_{k+1}, we may assume $x_{k+1} = 1$. Now let $x_q \in E \smallsetminus K$. (Thus, necessarily, $q > k + 1$.) Then there is an $h \in H$ such that $h(x_q) \neq x_q$. Further, $(h(x_1), \ldots, h(x_m))$ is also a solution. Of course, $h(x_1) = \cdots = h(x_k) = 0$, $h(x_{k+1}) = 1$. Subtracting both solutions, we obtain a solution with fewer non-zero terms, which is a contradiction.]

(d) Prove that if E is a Galois extension of F, then $H \mapsto E^H$ is an isomorphism of the partially ordered set of fields L with $F \subseteq L \subseteq E$ and the opposite of the partially ordered set of subgroups of $Gal(E/F)$ (with both orderings by inclusion).

35. Prove the following version of *Hilbert's Theorem 90*: Let a finite group G act on a ring $L[x_1, \ldots, x_n]$ where L is a field so that the action restricts to a faithful action of G on L (i.e. no element of G acts by the identity) and the action of any $g \in G$ on the vector $(x_1, \ldots, x_n)^T$ is given by left multiplication by a matrix A with entries in L. Then

$$Spec(L[x_1, \ldots, x_n]^G) \cong \mathbb{A}^n_{L^G}$$

where the G-superscript means the ring of fixed points. [Let $\lambda_1, \ldots, \lambda_m$ be a basis of L over L^G. By Galois theory (Exercise 34), $G = \{\sigma_1, \ldots, \sigma_m\}$ has the same number of elements and the matrix $(\sigma_i(\lambda_j))_{ij} \in GL_m(L)$. Let V be the free E-vector space on $x_1, \ldots, x_n$. Then by what we just observed, any $v \in V$ is an E-linear combination of the fixed vectors

$$v_j = \sum_{i=1}^{m} \sigma_i(\lambda_j v).$$

Pick an E-basis of V among the vectors in V^G.]

36. Prove that for algebraic varieties X, Y over an algebraically closed field k, the canonical projection

$$(X \times_{Spec(k)} Y)_m \rightarrow X_m \times Y_m$$

is a bijection. Is it a homeomorphism with respect to the induced topology of the scheme topology? How about if we replace X_m, Y_m with X_{an}, Y_{an} (i.e. the analytic topology) in the case $k = \mathbb{C}$?

37. Prove in detail the unit and inverse property in $\pi_1(X)$ for a topological space X.

38. Describe how a continuous map of spaces $f : X \rightarrow Y$ defines a functor $\pi_1(f) :$ $\pi_1(X) \rightarrow \pi_1(Y)$ and how a homotopy $h : X \times [0, 1] \rightarrow Y$ between maps $f, g :$ $X \rightarrow Y$ defines a natural transformation (hence natural isomorphism) $\pi_1(f) \cong \pi_1(g)$. Conclude that a homotopy equivalence of spaces X, Y induces an equivalence of categories between $\pi_1(X)$ and $\pi_1(Y)$. In particular, if a space X is *contractible*, i.e. homotopy equivalent to a point, then for every $x \in X$, $\pi_1(X, x) = 0$.

39. Let S^1 be the unit circle in $\mathbb{C}$ (with the topology induced by the analytic topology on $\mathbb{A}^1_{\mathbb{C}}$). Prove that $\pi_1(S^1) \cong \mathbb{Z}$. [Prove that the universal cover of S^1 is the map $\mathbb{R} \rightarrow S^1$ given by $t \mapsto e^{2\pi i t}$.]

40. Let X be a smooth proper variety over $\mathbb{C}$.
    (a) Prove that a morphism $h : Spec(K(\mathbb{A}^n_{\mathbb{C}})) \rightarrow X$ over $Spec(\mathbb{C})$ extends to a morphism $U \rightarrow X$ where $U \subseteq \mathbb{A}^n_{\mathbb{C}}$ is an open subvariety whose complement has codimension > 1 (meaning that all its irreducible components have codimension > 1). [Follow the method of Lemma 4.2.2. Consider the set of all non-empty open subvarieties of $\mathbb{A}^n_{\mathbb{C}}$ to which h extends. First show that it is non-empty. Let U be its maximal element with respect to inclusion. Suppose an irreducible component Z of its complement has codimension 1. Let P be its generic point. Apply the valuation criterion of properness for X to the discrete valuation ring $\mathcal{O}_{\mathbb{A}^n_{\mathbb{C}}, P}$ and use it to extend h to a larger open subvariety than U, thereby deriving a contradiction.]
    (b) Prove that for a rational smooth proper variety X over $\mathbb{C}$, we have $\pi_1(X_{an}) = 0$. [If X is rational, it contains an open subvariety V isomorphic to an open subvariety of $\mathbb{A}^n_{\mathbb{C}}$. Now assume a closed path α represents a non-zero element of $\pi_1(X_{an})$. Then α is homotopic to a closed path in V. However, by (a), the embedding $f : V \rightarrow X$ extends to an open subvariety $U \subseteq \mathbb{A}^n_{\mathbb{C}}$ whose complement has codimension > 1. But then $\pi_1(U_{an}) = 0$.

    (The reader can treat these statements about the fundamental group as a black box. For a deeper understanding, one needs the concept of a *manifold* from Sect. 1 of Chap. 5. For a smooth variety X over $\mathbb{C}$, X_{an} is a complex (and in particular real smooth) manifold. Now the complement Z of an open subvariety is a union of finitely many smooth submanifolds of codimension ≥ 2 (≥ 4 if Z is of codimension > 1). Note that the irreducible components of Z many not be non-singular, but the subset of points where a variety is singular, called the *singular locus*, is a proper Zariski closed subset, and thus, our statement can be proved by induction. Now a closed path in a manifold can be made smooth by a homotopy, and using the technique of *regular value*, a smooth path can be deformed by a smooth homotopy to a path which avoids a finite union of submanifolds of (real) codimension > 1. Similar comments apply to path homotopies and submanifolds of (real) codimension > 2. We briefly touch on regular values in Sect. 3.6 of Chap. 5. For details, see [20].)]

41. Consider the variety $X = Proj(\mathbb{C}[x, y, z]/(x^2z - y^3 - y^2z))$ over $\mathbb{C}$ (prove that the polynomial is irreducible). Prove that X is rational, while $\pi_1(X_{an}) \neq 0$. How is it possible in view of Exercise 40 (b)? [Construct a morphism of varieties $\mathbb{P}^1_{\mathbb{C}} \to X$ which is bijective with the exception of identifying two points. Then find a copy of $S^1 \subset X_{an}$ which is a *retract* (i.e. the inclusion has a left inverse).]

42. Calculate $\pi_1^{et}(Spec(\mathbb{Z}), P)$ for any geometrical point P.

43. Consider the étale morphism $Spec(x^2) : \mathbb{A}^1_{\mathbb{C}} \smallsetminus \{0\} \to \mathbb{A}^1_{\mathbb{C}} \smallsetminus \{0\}$. Prove that this is not a covering in the Zariski topology.

44. Prove that for varieties over $\mathbb{C}$, a finite étale morphism $f : X \to Y$ defines a finite degree covering $f_{an} : X_{an} \to Y_{an}$.

45. Prove that the category of finite étale morphisms over a scheme X has finite limits.

46. Prove that an étale morphism of Noetherian schemes is open (i.e. an image of an open set is open). [First prove that locally, an étale morphism of Noetherian schemes is given as $Spec$ of a homomorphism of rings $R \to S$ where R, S are Noetherian rings, and S is a finitely generated projective R-module; you may borrow Lemma 2.2.4 of Chap. 4. Next, prove that if $h : R \to S$ is a homomorphism of rings which makes S a finitely generated projective R-module, then $Spec(h)$ sends a distinguished open set to an open set; use the fact that for R-modules $M \oplus N = F$ and an ideal $I \subset R$, $IF \cap M = IM$.]

47. Prove that if $\phi : X \to Y$ and $\psi : Y \to Z$ are morphisms of schemes such that $\psi \circ \phi$ and ψ are étale, then ϕ is étale. [Reduce to the case of standard smooth of dimension 0, so writing schematically, ϕ is $Spec$ of a map of R-algebras $R[Y]/G(Y) \to R[X]/F(X)$ and $Y = H(X)$ (where X, Y are tuples of variables and F, G, H are tuples of functions). Then we have $F \circ H = G$. Take total differentials and use the chain rule to show that ϕ is standard smooth of dimension 0 over a point in Z, hence, by Nakayama's lemma, on local rings, and hence on open neighborhoods of any point.]

48. Using Exercises 45–47, complete the construction of the étale fundamental group as outlined in Sect. 5.6.2 by mimicking the construction of Sect. 5.6.1.

49. Let $p \in \mathbb{Z}$ be a prime number. A number field $K \subseteq \overline{\mathbb{Q}}$ is called *unramified at* p if $\mathcal{O}_K \cdot p$ is a prime ideal in $\mathcal{O}_K$ (recall that $\mathcal{O}_K$ denotes the integral closure of $\mathbb{Z}$ in K). Let S be a set of primes in $\mathbb{Z}$. Prove that $\pi_1^{et}(Spec(S^{-1}\mathbb{Z}), Spec(\overline{\mathbb{Q}}))$ is the profinite group formed by the groups $Gal(K/\mathbb{Q})$ over all fields K unramified at all primes $p \notin S$.

50. Let k be a field. Construct a monomorphism of profinite groups

$$\widehat{\mathbb{Z}} \to \pi_1^{et}(Spec(k[x, x^{-1}]), Spec(\overline{k(x)}))$$

where $\widehat{\mathbb{Z}}$ is the profinite group given by all the finite quotients of $\mathbb{Z}$.

51. Prove that the étale fundamental group $\pi_1^{et}(Spec(k), Spec(\overline{k}))$ where $\overline{k}$ is a separable closure of a field k is isomorphic to the pro-finite group $Gal(\overline{k}/k)$.

52. Calculate $\zeta(\mathbb{P}^n_{\mathbb{F}_p}, s)$.

Sheaves of Modules

4

An important concept in topology is a *vector bundle*, which is, roughly, a locally trivial parametric family of vector spaces indexed by a topological space. It is possible to think of a vector bundle as its total space with extra structure, or the sheaf of its sections. Both points of view still exist in schemes, but the sheaf-theoretical point of view is more fundamental, and reveals some additional features. In particular, taking kernels a cokernels, one gets the abelian category of *coherent sheaves*. Coherent sheaves are more general than algebraic vector bundles, including, for example, sheaves of ideals, which correspond, for Noetherian schemes, to closed subschemes. A particularly important application of sheaves of ideals is the theory of *blow-ups*, a construction which allows us, for example, to replace a point with a subscheme of codimension 1, while not disturbing (and, in fact, often even improving) smoothness.

Before getting to blow-ups, however, we will study line bundles, and the geometric concepts of *divisors*, to which they correspond in good cases. As the name suggests, divisors are important in the study of divisibility, allowing, for example, a very general characterization of unique factorization domains. From another point of view, certain divisors called *very ample* allow us to construct embeddings of varieties into projective spaces, thus giving tools for showing that a given smooth complete variety is projective. (It is, however, not true in general, as we will see in Exercise 34.)

Some new technical points will also arise. First, we will need to improve our foundations of sheaf theory. While studying line bundles, we will also encounter *first cohomology groups*, thus motivating the systematic study of cohomology in Chap. 5. We will also need still more commutative algebra, in particular more theory of regular rings, which, as it happens, will need to use the cohomological techniques of Chap. 5 for its fundamental proofs.

1 Sheaves of Modules

1.1 Presheaves and Sheaves Valued in a Category

Generally, for any category C, we can define *presheaves valued in C* (or briefly just presheaves in C) on a topological space X as functors

$$\mathcal{F} : (\text{Open sets in } X, \subseteq)^{Op} \to C. \tag{1.1.1}$$

(A partially ordered set is a category where there is one morphism $x \to y$ when $x \leq y$ and no morphism otherwise.) The functoriality incorporates the restriction axiom. Section objects are the values of the functor F on objects, and restrictions are its values on morphisms.

To define a *sheaf $\mathcal{F}$ in C* (or *sheaf valued in C*) on a topological space X, the category C needs to have limits. The gluing axiom cannot be phrased by referring to individual sections. The axiom says instead that given

$$U = \bigcup_{i \in I} U_i \tag{1.1.2}$$

open in X, then $\mathcal{F}(U)$ is the limit (equalizer) of the diagram

$$\prod_{i \in I} \mathcal{F}(U_i) \rightrightarrows \prod_{j,k \in I} \mathcal{F}(U_j \cap U_k) \tag{1.1.3}$$

where the two arrows are products of restrictions by setting $i = j$ resp. $i = k$.

Morphisms of presheaves and sheaves on X are defined as natural transformations. Explicitly, a morphism of sheaves

$$f : \mathcal{F} \to \mathcal{S}$$

consists of morphisms in C

$$f(U) : \mathcal{F}(U) \to \mathcal{S}(U)$$

and commutative diagrams in C

$$
\begin{array}{ccc}
\mathcal{F}(U) & \xrightarrow{f(U)} & \mathcal{S}(U) \\
\downarrow & & \downarrow \\
\mathcal{F}(V) & \xrightarrow{f(V)} & \mathcal{S}(V)
\end{array}
$$

where the vertical arrows are restrictions.

Assuming the category C has also directed colimits (meaning colimits over directed partially ordered sets), we may define the *stalk* $\mathcal{P}_x$ of a presheaf $\mathcal{P}$ at a point $x \in X$ as the colimit of $\mathcal{P}(U)$ over all open sets $U \ni x$ (with respect to restriction).

Denote the category of presheaves in a category C on a topological space X by pre-C-Sh_X, and the category of sheaves in a category C on X by C-Sh_X. Then by definition, there is a forgetful functor

$$U : C\text{-}Sh_X \to pre\text{-}C\text{-}Sh_X$$

given by taking a sheaf and considering it as a presheaf. (Because this is just an inclusion functor, it is often omitted from the notation.) Given a certain assumption on the category C, this functor has a left adjoint called *sheafification* and denoted by sh. For a presheaf $\mathcal{P}$ and a sheaf $\mathcal{F}$ on X, we have a natural bijection

$$C\text{-}Sh_X(sh(\mathcal{P}), \mathcal{F}) \cong pre\text{-}C\text{-}Sh_X(\mathcal{P}, U\mathcal{F}). \tag{1.1.4}$$

The assumption on the category C which we will use is:

ASSUMPTION *Directed colimits in the category C exist and commute with products and equalizers.*

Explicitly, this means that for any functor

$$F_s : I_s \to C, \quad s \in S,$$

where I_s are directed partially ordered sets, and S is any set, the canonical map

$$\operatorname*{colim}_{(i_s) \in \prod_S I_s} \prod_{s \in S} F_s(i_s) \to \prod_{s \in S} \operatorname*{colim}_{i \in I_s} F_s(i)$$

is an isomorphism, and for a directed partially ordered set I, two functors $F, G : I \to C$ and two natural transformations $h, k : F \to G$, the canonical map from the colimit over I of the equalizers of the diagrams

$$F(i) \rightrightarrows G(i)$$

into the equalizer of

$$\operatorname*{colim}_{i \in I} F(i) \rightrightarrows \operatorname*{colim}_{i \in I} G(i)$$

is an isomorphism.

This assumption is true for example for any category of universal algebras.

The functor sh is obtained as follows: One first constructs an auxiliary functor L. For a presheaf $\mathcal{P}$ in a category C on X and an open cover (U_i) of U, let

$$\mathcal{P}((U_i)) = \varprojlim \left(\prod_i \mathcal{P}(U_i) \rightrightarrows \prod_{j,k} \mathcal{P}(U_j \cap U_k) \right).$$

Then

$$LP(U)$$

is the colimit of $\mathcal{P}(U_i)$ over all open covers, where arrows are given by refinement. (A *refinement* of an open cover $(U_i)_{i \in I}$ is an open cover $(V_j)_{j \in J_i}$ where for each $j \in J$, there exists an $i \in I$ with $V_j \subseteq U_i$.)

A presheaf $\mathcal{P}$ on a topological space X is called *separated* if the canonical morphism

$$\mathcal{P}(U) \to \mathcal{P}((U_i))$$

is a monomorphism (which means injective in the case of a category of universal algebras) for every open cover (U_i) of every open set $U \subseteq X$.

1.1.1 Proposition

1. *For every presheaf $\mathcal{P}$ on X, $L\mathcal{P}$ is a separated presheaf.*
2. *For every separated presheaf $\mathcal{P}$ on X, $L\mathcal{P}$ is a sheaf.*
3. *Consequently, one can put*

$$sh = L \circ L.$$

Proof 3 is clearly a consequence of 1 and 2. To prove 1, we note that given our Assumption, a colimit of monomorphisms in C over a directed partially ordered set is a monomorphism. Additionally, a colimit over a directed poset P (also called *directed colimit*) is the same as the colimit over a *cofinal* subset Q, which means that for every element $x \in P$ there exists an element $y \in Q$ with $y \leq x$.

Now by taking intersections, refinements of a given open covering (U_i) of an open set $U \subseteq X$ are cofinal in the directed poset of coverings, which implies that the canonical morphism

$$LPL(U) \to \prod_i LPL(U_i) \qquad (1.1.5)$$

is a monomorphism. Thus,

$$LP(U) \rightarrow LP((U_i))$$

through which (1.1.5) factors, is a monomorphism, as claimed.

To prove 2, let P be a separated presheaf and let (U_i) be an open cover of an open set U. By 1, we need to prove that

$$LP(X) \rightarrow LP((U_i)) \tag{1.1.6}$$

is an isomorphism. By the Assumption, for an object $Z \in C$, a morphism

$$Z \rightarrow LP((U_i)) \tag{1.1.7}$$

for an open cover (U_i) of an open set U factors through a morphism

$$Z \rightarrow P((V_{ij}))$$

for some open cover (V_{ij}) of (U_i). These morphisms further are compatible on intersections, so they give a morphism

$$Z \rightarrow P((V_{ij})_{i,j}) \rightarrow LP(U).$$

The fact that this morphism indeed factors through the chosen morphism (1.1.7) follows from the assumption that P is separated. □

Example (The Constant Sheaf) Let us work, say, in the category of abelian sheaves (i.e. sheaves of abelian groups). (We could also work in sets, but we will need the case of abelian groups later.) The *constant sheaf* $\underline{A}$ for an abelian group A is the sheafification of the presheaf P where $P(U) = A$ for every $U \subseteq X$ open. Consider the space $X = \{1, 2\}$ with the discrete topology (i.e. where every subset is open), and let us construct the constant sheaf $\underline{\mathbb{Z}}$. We start with the presheaf P where

$$P(\emptyset) = P(\{1\}) = P(\{2\}) = P(\{1, 2\}) = \mathbb{Z}.$$

Then

$$(LP)(\emptyset) = 0$$

(using the empty cover of $\emptyset$), and

$$(LP)(\{1\}) = (LP)(\{2\}) = (LP)(\{1, 2\}) = \mathbb{Z}.$$

We see that this is not a sheaf (by considering the cover $\{1\}, \{2\}$ of $\{1, 2\}$). Note that the presheaf $\mathcal{P}$ violates the condition of being separated precisely on the empty cover of the empty set. On the other hand,

$$(LL\mathcal{P})(\emptyset) = 0,$$

$$(LL\mathcal{P})(\{1\}) = (LL\mathcal{P})(\{2\}) = \mathbb{Z},$$

$$(LL\mathcal{P})(\{1, 2\}) = \mathbb{Z} \oplus \mathbb{Z},$$

and we see that this is indeed a sheaf. Thus, we really need two steps. For a general description of the constant sheaf, see Exercise 2.

1.1.2 An alternate characterization of sheafification

For sheaves of sets (or universal algebras), there is also an alternate characterization of $sh(\mathcal{P})$: $sh\mathcal{P}(U)$ is the subset of

$$\prod_{x \in U} \mathcal{P}_x$$

consisting of all those systems $(s_x)_{x \in U}$ where for all $x \in U$, there exists an open set $V \subseteq U$ such that $x \in V$ and there is an $s_V \in \mathcal{P}(V)$ which maps to $s_x \in \mathcal{P}_x$ for every $x \in V$ by the canonical map (See Exercise 3.)

1.2   The Effect of Continuous Maps on Sheaves

Let

$$f : X \to Y$$

be a continuous map and let C be a category. Then as before, there is a functor

$$f_* : C\text{-}Sh_X \to C\text{-}Sh_Y$$

where for a C-sheaf $\mathcal{F}$ on $U \subseteq Y$ open, we put

$$(f_*\mathcal{F})(U) = \mathcal{F}(f^{-1}(U)).$$

This functor has a left adjoint, which is denoted by

$$f^{-1} : C\text{-}Sh_Y \to C\text{-}Sh_X$$

and is defined as the sheafification $sh(\mathcal{P})$ where

$$\mathcal{P}(V) = \operatorname*{colim}_{f(V)\subseteq U} \mathcal{S}(U).$$

These functors are sometimes referred to as the *pushforward* and *pullback*. To prove the adjunction, since we used sheafification, which itself is a left adjoint, it suffices to prove that with $\mathcal{P}$ as above,

$$pre\text{-}\mathcal{C}\text{-}Sh_X(\mathcal{P}, \mathcal{F}) = \mathcal{C}\text{-}Sh_Y(\mathcal{S}, f_*\mathcal{F}).$$

By the universality of the colimit, this is the same thing as producing a morphism, natural with respect to inclusions in the U and V variables,

$$\mathcal{S}(U) \to \mathcal{F}(V) \tag{1.2.1}$$

for all $U \subseteq Y$ open with $f(V) \subseteq U$, which is the same thing as $V \subseteq f^{-1}(U)$. Thus, in (1.2.1), we may as well restrict to the case $V = f^{-1}(U)$, which is the same thing as producing a morphism of sheaves $\mathcal{S} \to f_*\mathcal{F}$.

It is worth pointing out that the functor f^{-1} *preserves stalks* in the sense that (by definition), we have for $x \in X$,

$$(f^{-1}(\mathcal{F}))_c \cong \mathcal{F}_{f(x)}. \tag{1.2.2}$$

This is because a stalk is the same thing as pullback of the sheaf to a point. Because of that, the functor f^{-1} is exact (see Exercises 1 and 4). However, it does not in general preserve products (and hence, does not in general have a left adjoint). For example, let $Y = \mathbb{R}$, let

$$j_n : U_n = (-1/n, 1/n) \subset \mathbb{R}$$

and let

$$\mathcal{F} = \prod_{n\in\mathbb{N}} (j_n)_!\underline{\mathbb{Z}}_{U_n} \tag{1.2.3}$$

(where $\underline{\mathbb{Z}}_U$ denotes the constant $\mathbb{Z}$ sheaf on U and $j_!$ denotes extension by 0 along an open inclusion i; see Sect. 2.3 of Chap. 2). Let $f : \{0\} \to \mathbb{R}$ be the inclusion. Then, identifying abelian sheaves over a point with abelian groups, we have

$$f^{-1}(\mathcal{F}) = \mathcal{F}_0 = \bigoplus_{n\in\mathbb{N}} \mathbb{Z},$$

while

$$\prod_{n \in \mathbb{N}} f^{-1}(j_! \mathbb{Z}_{U_n}) = \prod_{n \in \mathbb{N}} \mathbb{Z}.$$

Recall on the other hand that if $j : U \to Y$ is an inclusion of an open subset, then $j_!$ is a left adjoint to j^{-1}.

Note that the functor f_*, which is right adjoint to f^{-1}, is therefore left exact, but is not exact in general (for $f : X \to *$ the projection to a one point set, f_* can be identified with the global sections functor—see Exercise 5). For $f = i : Z \to X$ an inclusion of a closed subset, however, we have $(i_*(\mathcal{F}))_x = \mathcal{F}_x$ for $x \in Z$, and $(i_*(\mathcal{F}))_x = 0$ for $x \notin Z$, and thus i_* is exact (see Exercise 6).

1.3 Sheaves of Modules

Let us define a category C whose objects are pairs (R, M) where R is a commutative ring and M is an R-module. Morphisms $(f, g) : (R, M) \to (S, N)$ consist of a homomorphism of rings $f : R \to S$ and a map $g : M \to N$ such that for all $x, y \in M$,

$$g(x + y) = g(x) + g(y)$$

(g is a homomorphism of abelian groups) and for all $x \in M, r \in R$,

$$g(rx) = f(r)g(x).$$

We see that the objects of C still behave the same way as universal algebras, with the difference that instead of a set, we have a pair of sets, and each operation specifies whether each input variable is in the first or the second set, and also which of the two sets the output is in. Instead of two sets, we can have a system of sets S_i indexed by $i \in I$ for some indexing set I, and again, each operation with n input variables specifies from what sets $S_{i_1}, \dots S_{i_n}$ the inputs variables are, and in which set S_i the output is in. Equations between operations can be imposed when they make sense. A structure of this type is called a *multisorted universal algebra*; in general, a homomorphism of multisorted algebras of the same type $(S_i)_{i \in I} \to (T_i)_{i \in I}$ consists of systems of maps

$$f_i : S_i \to T_i, \ i \in I$$

which preserve the operations. Categories of multisorted universal algebras also satisfy the Assumption of Sect. 1.1.

In summary, the category C of pairs (R, M) where R is a commutative ring and M is an R-module is an example of a category of multisorted universal algebras and homomorphisms. *A sheaf of modules on a topological space X is a sheaf in this category C.*

To be explicit, a sheaf of modules on X specifies for every open set $U \subseteq X$ a commutative ring $\mathcal{R}(U)$ and a $\mathcal{R}(U)$-module $\mathcal{M}(U)$ where $\mathcal{R}$ is a sheaf of commutative rings and $\mathcal{M}$ is a sheaf of abelian groups where, for an open subset $V \subseteq U$ and sections $r \in \mathcal{R}(U)$, $m \in \mathcal{M}(U)$, we have

$$rm|_V = r|_V \cdot m|_V.$$

We may then also say that $\mathcal{M}$ *is a sheaf of modules over the sheaf of rings* $\mathcal{R}$.

1.4 The Effect of Continuous Maps on Sheaves of Modules

By Sect. 1.2, for a continuous map

$$f : X \to Y \tag{1.4.1}$$

and a sheaf of modules $\mathcal{M}$ over a sheaf of commutative rings $\mathcal{R}$ on X, we have a sheaf of modules $f_*\mathcal{M}$ over the sheaf of rings $f_*\mathcal{R}$ and similarly, for a sheaf of commutative rings S on Y and a sheaf of S-modules $\mathcal{N}$, we have a sheaf of $f^{-1}S$-modules $f^{-1}\mathcal{N}$, and the functor f^{-1} is left adjoint to the functor f_*.

Typically, however, our point of view is different, i.e. X and Y are ringed spaces and the map f is a morphism of ringed spaces. (The case of a morphism of schemes is a special case.) In the case of ringed spaces, the sheaves of commutative rings on X and Y are given, and are denoted by $\mathcal{O}_X$, $\mathcal{O}_Y$. A morphism of ringed spaces then comes with a morphism of sheaves of rings

$$\phi : \mathcal{O}_Y \to f_*\mathcal{O}_X. \tag{1.4.2}$$

This means that for a sheaf of $\mathcal{O}_X$-modules $\mathcal{M}$, $f_*\mathcal{M}$ is automatically a sheaf of $\mathcal{O}_Y$-modules, using its automatic structure of a sheaf of $f_*\mathcal{O}_X$-modules, and the morphism of sheaves of rings (1.4.2).

The same does not hold for f^{-1}, however: for a sheaf of $\mathcal{O}_Y$-modules $\mathcal{N}$, $f^{-1}\mathcal{N}$ is a sheaf of modules over $f^{-1}\mathcal{O}_Y$. From (1.4.2), using the adjunction, we get a morphism of sheaves of rings

$$\psi : f^{-1}\mathcal{O}_Y \to \mathcal{O}_X \tag{1.4.3}$$

which goes "in the wrong direction." However, we can "push forward" the sheaf $f^{-1}\mathcal{N}$ from $f^{-1}\mathcal{O}_Y$ to $\mathcal{O}_X$ using the morphism (1.4.3):

Let

$$f^*\mathcal{N} = \mathcal{O}_X \otimes_{f^{-1}\mathcal{O}_Y} f^{-1}\mathcal{N} \tag{1.4.4}$$

be the sheafification of the presheaf whose set of sections on an open set $V \subseteq X$ is the $\mathcal{O}_X(V)$-module

$$\mathcal{O}_X(V) \otimes_{(f^{-1}\mathcal{O}_Y)(V)} (f^{-1}\mathcal{N})(V).$$

If we denote by $\mathcal{O}_X\text{-}Mod$ the category of sheaves of $\mathcal{O}_X$-modules, then from this point of view, the functor

$$f_* : \mathcal{O}_X\text{-}Mod \to \mathcal{O}_Y\text{-}Mod$$

is right adjoint to

$$f^* : \mathcal{O}_Y\text{-}Mod \to \mathcal{O}_X\text{-}Mod.$$

In fact, reserving the symbol f^* for this situation is the historical reason why the left adjoint of f_* for sheaves of sets (or rings) is denoted by f^{-1}.

Note that the sheafification of a sheaf of modules $(\mathcal{R}, \mathcal{M})$ where $\mathcal{R}$ is already a sheaf of rings does not change $\mathcal{R}$. We can phrase this to say that there is a sheafification functor (left adjoint to the forgetful functor) from presheaves of modules to sheaves of modules over a fixed sheaf of rings $\mathcal{R}$.

1.5 Biproduct, Tensor Product and Hom of Sheaves of Modules

Let X be a ringed space with structure sheaf $\mathcal{O}_X$ and consider again the category $\mathcal{O}_X\text{-}Mod$ of $\mathcal{O}_X$-modules. Then we have functors

$$? \oplus ? : \mathcal{O}_X\text{-}Mod \times \mathcal{O}_X\text{-}Mod \to \mathcal{O}_X\text{-}Mod,$$

$$? \otimes_{\mathcal{O}_X} ? : \mathcal{O}_X\text{-}Mod \times \mathcal{O}_X\text{-}Mod \to \mathcal{O}_X\text{-}Mod,$$

$$Hom_{\mathcal{O}_X}(?, ?) : (\mathcal{O}_X\text{-}Mod)^{Op} \times \mathcal{O}_X\text{-}Mod \to \mathcal{O}_X\text{-}Mod$$

called the *biproduct, tensor product and Hom-functor of sheaves.* The biproduct of sheaves of $\mathcal{O}_X$-modules $\mathcal{M}, \mathcal{N}$ is defined by

$$(\mathcal{M} \oplus \mathcal{N})(U) = \mathcal{M}(U) \oplus \mathcal{N}(U)$$

for an open set $U \subseteq X$.

Note that there is also a biproduct of sheaves of abelian groups (also sometimes called *abelian sheaves*). In fact, it is easily seen that both sheaves of abelian groups on a given topological space, and sheaves of modules on a given ringed space, form abelian categories (see Sect. 1.4 of Chap. 2, and also Exercise 7). Also, both categories have all limits and all colimits.

The Hom-functor of sheaves of modules is defined by setting, for an open set $U \subseteq X$,

$$(Hom_{\mathcal{O}_X}(\mathcal{M}, \mathcal{N}))(U) = Hom_{\mathcal{O}_U}(\mathcal{M}|_U, \mathcal{N}|_U), \tag{1.5.1}$$

(recall that $\mathcal{M}|_U$ denotes the restriction of the sheaf $\mathcal{M}$ to the open set U—see Exercise 8). The tensor product is defined by letting $\mathcal{M} \otimes_{\mathcal{O}_X} \mathcal{N}$ be the sheafification of the presheaf of $\mathcal{O}_X$-module $\mathcal{P}$ where for an open set $U \subseteq X$,

$$\mathcal{P}(U) = \mathcal{M}(U) \otimes_{\mathcal{O}_X(U)} \mathcal{N}(U).$$

These functors satisfy all the usual properties of the biproduct, tensor product and Hom. For example, the biproduct and tensor product are commutative, associative and unital and the tensor product is distributive under the biproduct. Also, we have for sheaves of $\mathcal{O}_X$-modules $\mathcal{L}, \mathcal{M}, \mathcal{N}$,

$$Hom_{\mathcal{O}_X}(\mathcal{L} \otimes_{\mathcal{O}_X} \mathcal{M}, \mathcal{N}) \cong Hom_{\mathcal{O}_X}(\mathcal{L}, Hom_{\mathcal{O}_X}(\mathcal{M}, \mathcal{N})) \tag{1.5.2}$$

(Exercise 9).

2 Quasicoherent and Coherent Sheaves

2.1 Invertible Sheaves, Picard Group, Locally Free Sheaves, Algebraic Vector Bundles, Algebraic K-Theory

The sheaf $\mathcal{O}_X$ is, of course, a sheaf of modules over itself, and it is the unit (neutral element) with respect to $\otimes_{\mathcal{O}_X}$. We say that a sheaf of $\mathcal{O}_X$-modules $\mathcal{L}$ is *invertible* if there exists a sheaf of $\mathcal{O}_X$-modules $\mathcal{L}^{-1}$ such that

$$\mathcal{L} \otimes_{\mathcal{O}_X} \mathcal{L}^{-1} \cong \mathcal{O}_X.$$

The set of isomorphism classes of invertible sheaves of $\mathcal{O}_X$-modules forms an abelian group with respect to the operation $\otimes_{\mathcal{O}_X}$, which is called the *Picard group* and denoted by $Pic(X)$. When $X = Spec(R)$, we shall also write $Pic(R)$ for the Picard group.

A sheaf $\mathcal{M}$ of $\mathcal{O}_X$-modules is called *locally free* if for every point $x \in X$, there exists an n and an open set $U \ni x$ such that

$$\mathcal{M}|_U \cong \underbrace{\mathcal{O}_U \oplus \cdots \oplus \mathcal{O}_U}_{n}.$$

(n can be infinite, in which case we mean the direct sum.) If n is always finite, we say that $\mathcal{M}$ is a finite-dimensional locally free sheaf. If n is independent of U, we say that $\mathcal{M}$ is a locally free sheaf of dimension n. (Dimension of a free module over a non-zero commutative ring is well-defined, see Exercises 10, 11.)

An invertible sheaf is easily seen to be the same thing as a locally free sheaf of $\mathcal{O}_X$-modules of dimension 1 (Exercise 12).

When X is a scheme, we also call a finite-dimensional locally free sheaf $\mathcal{M}$ on X an *algebraic vector bundle*. This terminology is motivated as follows: Suppose $X = Spec(R)$ is an affine scheme. Then we can also think of the ring R as the sections (in the sense of right inverses), in the category of schemes, of the projection

$$\mathbb{A}^1_X \to X. \tag{2.1.1}$$

Indeed, since we are in the category of affine schemes, the projection (2.1.1) is the *Spec* of the inclusion of constant polynomials

$$R \to R[x]. \tag{2.1.2}$$

The right inverses of (2.1.1) are then *Spec* of left inverses of (2.1.2), which are in bijective correspondence with elements of R (by where we send x). Similarly, elements of the free module R^n can be thought of as sections of the projection

$$\mathbb{A}^n_X \to X$$

in the category of schemes. From this point of view, we can think of an n-dimensional algebraic vector bundle on a scheme X as a "twisted n-dimensional affine space over X." This can be made more precise as follows: On an affine open subset $X \supseteq V_i \cong Spec(R_i)$ on which $\mathcal{M}$ is free of dimension n, (we may speak of a *coordinate neighborhood*), let

$$U_i = \mathbb{A}^n_{V_i}, \tag{2.1.3}$$

subject to choosing an isomorphism

$$\phi_i : \mathcal{M}|_{V_i} \cong \underbrace{\mathcal{O}_{U_i} \oplus \cdots \oplus \mathcal{O}_{U_i}}_{n}. \tag{2.1.4}$$

Then, letting $p_i : U_i \to V_i$ be the projection, let $U_{ij} = p_i^{-1}(V_i \cap V_j)$. Then composing $(\phi_j)|_{V_i \cap V_j}$ with $(\phi_i)^{-1}|_{V_i \cap V_j}$, and letting

$$U_{ij} = p_i^{-1}(V_i \cap V_j) = \mathbb{A}^n_{V_i \cap V_j},$$

we obtain an isomorphism of schemes

$$\phi_{ij} : U_{ij} \to U_{ji}$$

and the assumptions of Sect. 2.5 are satisfied. Therefore, we can define a scheme U as a colimit of the schemes U_i. We have a projection

$$p_{\mathcal{M}} : U_{\mathcal{M}} \to X. \tag{2.1.5}$$

Then $p_{\mathcal{M}}^{-1}(V_i)$ is (2.1.3), and the sheaf $\mathcal{M}$ is the sheaf of sections (right inverses) of restrictions of the projection p to open subsets. It is possible to be more precise about the conditions on the isomorphisms ϕ_{ij} and also about equivalence of such data, and define algebraic vector bundles in this way (see the Comment at the end of Sect. 3.1).

Since, as we already remarked, it is not difficult to show that invertible sheaves are the same thing as locally free sheaves of modules of dimension 1, they are also referred to as (algebraic) *line bundles*.

Example Perhaps the most well known example of a non-trivial line bundle is the *Möbius strip*. Its algebraic version is a line bundle over the variety

$$X = Spec(R)$$

where

$$R = \mathbb{R}[x, y]/(x^2 + y^2 - 1).$$

Then consider the open sets

$$U_1 = X \smallsetminus \{(x, y - 1)\},$$

$$U_2 = X \smallsetminus \{(x, y + 1)\}.$$

Note: one can of course think naively of the closed point $(x, y - 1)$ resp. $(x, y + 1)$ as the points $(0, 1)$ resp. $(0, -1)$ in 2-dimensional affine space. It turns out that those are not principal ideals. Nevertheless, U_1 and U_2 are distinguished open sets, since one has

$$(y - 1) = (x, y - 1)^2, \quad (y + 1) = (x, y + 1)^2,$$

so

$$U_1 = U_{(y-1)}, \quad U_2 = U_{(y+1)}. \tag{2.1.6}$$

Let, for $V \subseteq X$ open,

$$\mathcal{L}(V) = \{(s_1, s_2) \in \mathcal{O}_X(U_1 \cap V) \times \mathcal{O}_X(U_2 \cap V) \mid$$
$$s_1|_{U_1 \cap U_2 \cap V} = x s_2|_{U_1 \cap U_2 \cap V}\}. \tag{2.1.7}$$

This is obviously a sheaf of $\mathcal{O}_X$-modules. It is invertible, and in fact is inverse to itself: we have

$$\mathcal{L} \otimes_{\mathcal{O}_X} \mathcal{L}(V) = \{(s_1, s_2) \in \mathcal{O}_X(U_1 \cap V) \times \mathcal{O}_X(U_2 \cap V) \mid$$
$$s_1|_{U_1 \cap U_2 \cap V} = x^2 s_2|_{U_1 \cap U_2 \cap V}\},$$

but this is isomorphic to $\mathcal{O}_X$, by noting that

$$x^2 = (1 - y)(1 + y)$$

and $1 - y \neq 0$ on U_1, while $1 + y \neq 0$ on U_2. Therefore, we can define an isomorphism

$$\mathcal{L} \otimes_{\mathcal{O}_X} \mathcal{L}(U) \xrightarrow{\cong} \mathcal{O}_X(U)$$

by sending a pair of sections

$$(s_1 \in \mathcal{O}_X(U_1 \cap V), s_2 \in \mathcal{O}_X(U_2 \cap V))$$

where

$$s_1|_{U_1 \cap U_2 \cap V} = x^2 s_2|_{U_1 \cap U_2 \cap V}$$

to the section in $s \in \mathcal{O}_X(V)$ where

$$s|_{U_1 \cap V} = \frac{s_1}{1 - y},$$

$$s|_{U_2 \cap V} = s_2 \cdot (1 + y)$$

(which exists by the sheaf property).

On the other hand, we can see in the usual way that the Möbius strip is non-trivial, i.e. not isomorphic to $\mathcal{O}_X$. Indeed, let $s = (s_1, s_2) \in \mathcal{L}(X)$ be a global section, and consider s_1, s_2 as "naive" functions from the unit circle to $\mathbb{R}$. These functions are continuous in the analytic topology, and additionally have same signs on the $x > 0$ semicircle, and opposite signs on the $x < 0$ semicircle. Thus, it is not possible for both s_1 and s_2 to be non-zero on all of their respective semicircles (because then they would not change signs). However, if $\mathcal{L}$ were isomorphic to $\mathcal{O}_X$, then the image $s \in \mathcal{L}(X)$ of $1 \in \mathcal{O}_X(X)$ cannot be 0 at any point of the circle. This is because if $s(x_0, y_0) = 0$, then in a Zariski open neighborhood V of (x_0, y_0), $s = g \cdot t$ for some section t, and some linear function g (we can choose $g = x - x_0$ if $y_0 \neq 0$ or $g = y - y_0$ if $x_0 \neq 0$). However, this contradicts the isomorphism since $g \in \mathcal{O}_X(V)$ is not invertible.

Note that $\mathcal{L}$ does possess nonzero global sections, for example the pair

$$(x, 1) \in \mathcal{O}_X(U_1) \times \mathcal{O}_X(U_2).$$

We therefore constructed a subgroup of $Pic(X)$ isomorphic to $\mathbb{Z}/2$. It turns out that in this case, this is all of $Pic(X)$, as we shall prove later in Sect. 3.4.

The set $\mathcal{Q}_X$ of isomorphism classes of locally free sheaves of modules of finite dimension form a commutative monoid with respect to $\oplus$ (which means that the operation is commutative and associative and has a neutral element, but not necessarily an inverse). In fact, with respect to the operations $\oplus$, $\otimes_{\mathcal{O}_X}$, it forms a commutative semiring, which means an algebraic structure with two operations $+$ and $\cdot$ which both are commutative monoids, and $\cdot$ is distributive under $+$.

Now the forgetful functor from abelian groups to commutative monoids (or from commutative rings to commutative semirings) has a left adjoint which is denoted by K. Its construction resembles localization, but with respect to $+$: For a monoid M with respect to an operation $+$, the *Grothendieck group* of M is

$$K(M) = \{(m, n) \mid m, n \in M\}/\sim$$

where $\sim$ is the equivalence relation given by

$$(k, \ell) \sim (m, n)$$

if

$$k + n + u = \ell + m + u \text{ for some } u \in M.$$

Returning to the commutative semiring $\mathcal{Q}_X$ of isomorphism classes of finite-dimensional locally free sheaves on a ringed space X, if X is an affine Noetherian scheme (ie. $X = Spec(R)$ where R is a Noetherian ring), we denote

$$K(X) = K(\mathcal{Q}_X).$$

This commutative ring is called the *algebraic K-theory of* X. For a general Noetherian scheme X, a more technical definition of $K(X)$ is needed. One possibility is to take the free abelian group on all isomorphism classes of finite-dimensional locally free sheaves $\mathcal{F}$ on X, and factor out by the subgroup generated by $\mathcal{G} - \mathcal{F} - \mathcal{H}$ whenever for some inclusion $\mathcal{F} \subseteq \mathcal{G}, \mathcal{H} \cong \mathcal{G}/\mathcal{F}$.

More specifically, the group defined here is denoted by K_0. Groups K_i for $i > 0$ (and in some definitions also $i < 0$) can also be defined, but we shall not discuss them here. For more information, see [24].

Example It turns out that in the above example, the Möbius strip is also a non-trivial element of $K(X)$ (Here X is an affine scheme, so we can use the simpler definiton of $K(X)$.). In fact, we will see later that one has $K(X) \cong \mathbb{Z}/2 \oplus \mathbb{Z}$ where the second summand is given by dimension (Comment 2 in Sect. 2.2 below and Exercise 21). Let us show explicitly that, in the notation of the above example,

$$\mathcal{L} \oplus \mathcal{L} \cong 2\mathcal{O}_X. \tag{2.1.8}$$

To this end, we note that, by (2.1.6),

$$\mathcal{O}_X(U_1) = (1 - y)^{-1}R, \quad \mathcal{O}_X(U_2) = (1 + y)^{-1}R.$$

Thus, to show (2.1.8), it suffices to exhibit an invertible 2×2 matrix A in $\mathcal{O}_X(U_1)$ and an invertible 2×2 matrix B in $\mathcal{O}_X(U_2)$ such that

$$A = \begin{pmatrix} x & 0 \\ 0 & x \end{pmatrix} \cdot B. \tag{2.1.9}$$

We can put

$$A = \begin{pmatrix} x & 1-y \\ 1-y & -x \end{pmatrix}, \tag{2.1.10}$$

$$B = \begin{pmatrix} 1 & \frac{x}{1+y} \\ \frac{x}{1+y} & -1 \end{pmatrix}.$$

(Note that $det(A) = 2(y - 1)$, $det(B) = -2/(y + 1)$, so we can find inverses by the Cramer rule.)

2.2 Quasicoherent and Coherent Sheaves of Modules

The geometric interpretation of algebraic vector bundles introduced in the last subsection also works for morphisms. In other words, recalling (2.1.5), it is possible to interpret a morphism

$$\phi : \mathcal{M} \to \mathcal{N}$$

of locally free sheaves of modules on a scheme X as a morphism of schemes

$$U_\phi : U_\mathcal{M} \to U_\mathcal{N}$$

which makes a commutative diagram

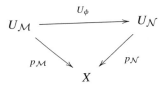

and which is "linear on fibers." By this last condition, we mean that for a coordinate neighborhood for both bundles

$$V \cong Spec(R),$$
$$p_\mathcal{M}^{-1}(V) \cong Spec(R[x_1, \ldots, x_m]),$$
$$p_\mathcal{N}^{-1}(V) \cong Spec(R[y_1, \ldots, y_n]),$$

the restriction of U_ϕ to $p_\mathcal{M}^{-1}(V)$ is $Spec$ of a homomorphism of rings

$$h : R[y_1, \ldots, y_n] \to R[x_1, \ldots, x_m]$$

where $h(y_j)$ is a *linear combination* of the x_i's with coefficients in R.

Now, however, the sheaf-theoretical point of view reveals a side of the story which may not immediately obvious from the geometrical picture, and which is particular to algebraic geometry. Namely, what can happen when ϕ is not an isomorphism?

Example Consider the free sheaf of dimension 1 on $X = \mathbb{A}^1_{\mathbb{C}} = Spec(\mathbb{C}[t])$, i.e. $\mathcal{M} = \mathcal{O}_{\mathbb{A}^1_{\mathbb{C}}}$. Then all of X is a coordinate neighborhood, and from the above discussion,

$$x \mapsto tx$$

defines a morphism

$$\phi : \mathcal{M} \to \mathcal{M}.$$

From the point of view of sheaves, on a Zariski open set $U \subseteq X$,

$$\phi(U) : \mathcal{M}(U) \to \mathcal{M}(U)$$

is defined as the multiplication by the restriction of $t \in \mathcal{O}_X(X) = \mathbb{C}[t]$. What is the kernel and cokernel of ϕ?

Geometrically, the morphism ϕ appears to be an "isomorphism on fibers except for the fiber at 0, where it is trivial."

Sheaf-theoretically, the kernel is a limit, so it can be calculated directly on sections over open sets. We see however immediately that for a distinguished open set $U_f = Spec(f^{-1}\mathbb{C}[t])$ for some polynomial $f \in \mathbb{C}[t]$, $\phi(U_f)$ is the localization of an injective map, and hence is injective. Since limits obviously preserve injective maps of abelian groups (which, after all, are the same as monomorphisms), $\phi(U)$ is injective for any Zariski open set $U \subseteq X$. Therefore, we have

$$Ker(\phi) = 0.$$

On the other hand, $Coker(\phi)$ is the sheafification of the presheaf $\mathcal{P}$ where

$$\mathcal{P}(U) = Coker(\phi(U)).$$

For a distinguished open set U_f, $f \in \mathbb{C}[t]$, we then see that

$$\mathcal{P}(U_f) = f^{-1}\mathbb{C}[t]/(t)$$

which is $\mathbb{C}$ when f is not divisible by t, and 0 otherwise. This, as it turns out, is already a sheaf, since obviously its stalk at $0 \in X$ is $\mathbb{C}$, while its stalk at any other point is 0. We therefore see that

$$Coker(\phi) = i_* \mathcal{O}_{Spec(\mathbb{C})}$$

where i is the inclusion of the closed point 0 (in other words, i is $Spec$ of the projection $\mathbb{C}[t] \to \mathbb{C}, t \mapsto 0$). More generally, i_* of a constant sheaf, where i is the inclusion of the closure of a point, (i.e. in particular i_* of any sheaf where i is the inclusion of a closed point) is called a *skyscraper sheaf*.

This example motivates looking for the smallest (in a suitable sense) abelian subcategory of the category of sheaves of modules on a ringed space X which contains all locally free sheaves. There is such a suitable category, called the category of *quasicoherent sheaves*.

2.2.1 Lemma *Let $X = Spec(R)$ be an affine scheme and let M be an R-module. Then the sheaf of $\mathcal{O}_X$-modules $M \otimes_R \mathcal{O}_X$ which is the sheafification of the presheaf $\mathcal{P}$ given by*

$$\mathcal{P}(U) = M \otimes_R (\mathcal{O}_X(U)) \tag{2.2.1}$$

satisfies

$$(M \otimes_R \mathcal{O}_X)_x = M \otimes \mathcal{O}_{X,x} \tag{2.2.2}$$

for every point $x \in X$, and

$$(M \otimes_R \mathcal{O}_X)(U_r) = M \otimes_R (\mathcal{O}_X(U_r)) = M \otimes_R (r^{-1}R) \tag{2.2.3}$$

for every $r \in R$. (The equalities in (2.2.2), (2.2.3) mean canonical isomorphisms.)
    The functor

$$? \otimes_R \mathcal{O}_X : R\text{-}Mod \to \mathcal{O}_X\text{-}Mod \tag{2.2.4}$$

is left adjoint to the functor

$$?(X) : \mathcal{O}_X\text{-}Mod \to R\text{-}Mod. \tag{2.2.5}$$

COMMENT

1. It is not true that the adjunction (2.2.4), (2.2.5) would in general be an equivalence of categories. For example, for $r \in R$ which is not a unit and not a zero divisor, considering the inclusion

$$j : Spec(r^{-1}R) = U_r \subsetneq X = Spec(R),$$

the sheaf

$$j_! \mathcal{O}_{U_r}$$

is obviously a non-zero sheaf of $\mathcal{O}_X$-modules, whose R-module of sections on X is 0. Therefore, it cannot be in the image of (2.2.4).

2. It is not in general true that the presheaf $(2.2.1)$ is a sheaf. Let, for example,

$$R = \prod_{n \in \mathbb{N}_0} \mathbb{Z},$$

and consider

$$r_i = (0, \ldots, 0, 1, 0, \ldots)$$

where the 1 is in the i'th place. Then

$$r_i^{-1} R \cong \mathbb{Z},$$

$$(r_i r_j)^{-1} R = 0,$$

so

$$U_{r_i} \cap U_{r_j} = \emptyset$$

for $i \neq j$. Thus, putting

$$U = \bigcup_{i \in \mathbb{N}_0} U_{r_i},$$

and letting, for an abelian group A,

$$M = A \otimes R,$$

the map

$$\mathcal{P}(U) \to (M \otimes_R \mathcal{O}_X)(U)$$

is the canonical inclusion

$$A \otimes R \subseteq \prod_{n \in \mathbb{N}_0} A,$$

which is not the equality for a general abelian group A.

Proof of Lemma 2.2.1 First, (2.2.2) is formal by exactness of directed colimits. Thus, by the characterization of sheafification, $(M \otimes_R \mathcal{O}_X)(U)$ is the subset of

$$\prod_{x \in U} M \otimes \mathcal{O}_{X,x}$$

which are locally of the form

$$M \otimes_R (r^{-1} R)$$

for suitable choices of r. Then the first isomorphism (2.2.3) is proved analogously as for the structure sheaf (Exercise 14).

The last statement is essentially by definition. □

2.2.2 Definition A *quasicoherent sheaf* on a scheme X is a sheaf $\mathcal{M}$ of $\mathcal{O}_X$-modules with the property that X is covered by affine open sets $U_i = Spec(R_i)$ such that for all i,

$$\mathcal{M}|_{U_i} \cong M_i \otimes_{R_i} \mathcal{O}_{U_i} \tag{2.2.6}$$

for some R_i-module M_i. A *coherent sheaf* on a Noetherian scheme X is a quasicoherent sheaf $\mathcal{M}$ where the R_i-modules M_i of (2.2.6) can be chosen to be finitely generated.

Note that because of Comment 1 after Lemma 2.2.1, we know that not every sheaf of modules on a scheme is quasicoherent. Definition 2.2.2 of quasicoherent sheaves can be also applied to locally ringed spaces.

2.2.3 Lemma

1. *A sheaf $\mathcal{M}$ of $\mathcal{O}_X$-modules over a scheme X is quasicoherent if and only if for every affine open set $U = Spec(R)$ in X,*

$$\mathcal{M}|_U \cong M \otimes_R \mathcal{O}_U \tag{2.2.7}$$

   for a suitable R-module M. If X is Noetherian, then $\mathcal{M}$ is coherent if and only if the modules M of (2.2.7) are finitely generated R-modules. In particular, the category of quasicoherent sheaves on $Spec(R)$ (resp. coherent sheaves on $Spec(R)$ with R Noetherian) is equivalent to the category of R-modules (resp. finitely generated R-modules).
2. *Quasicoherent sheaves on a scheme and coherent sheaves on a Noetherian scheme X form abelian subcategories of the category of $\mathcal{O}_X$-modules.*
3. *The adjunction of Lemma 2.2.1 gives an equivalence of categories between the category of quasicoherent sheaves on $Spec(R)$ (resp. coherent sheaves on $Spec(R)$ when R is*

Noetherian) and the category of R-modules (resp. finitely generated R-modules when R is Noetherian).

4. *If $f : X \to Y$ is a morphism of schemes, then f^* of a quasicoherent sheaf is quasicoherent. If X is Noetherian, then f_* of a quasicoherent sheaf is quasicoherent. If both X and Y are Noetherian, then f^* of a coherent sheaf is coherent.*

COMMENTS

1. It is obviously not true in general that f_* of a coherent sheaf with X, Y Noetherian would be coherent. For example, f could be the inclusion of a non-closed point. However, it can be shown to be true when f is proper.
2. The definition of a coherent sheaf given in Definition 2.2.2 could be given for a scheme X which is not Noetherian, but then coherent sheaves may not form an abelian category. There are ways to fix that by changing the definition in the general case.

Proof of Lemma 2.2.3 1. Clearly, it suffices to consider the case of $X = Spec(R)$ affine. By (2.2.3) of Lemma 2.2.1, we may assume that $Spec(R)$ is covered by finitely many distinguished open sets U_{f_i}, and there are $f_i^{-1} R$-modules M_i such that

$$\mathcal{M}|_{U_{f_i}} = M_i \otimes_{f_i^{-1} R} \mathcal{O}_{U_{f_i}}.$$

Let

$$M = \mathcal{M}(Spec(R)).$$

Then the counit of adjunction gives a morphism of sheaves of $\mathcal{O}_{Spec(R)}$-modules

$$M \otimes_R \mathcal{O}_{Spec(R)} \to \mathcal{M}. \tag{2.2.8}$$

We shall prove that (2.2.8) is an isomorphism. First, let

$$s \in g^{-1} M = M \otimes_R \mathcal{O}_{Spec(R)}(U_g).$$

By definition, then, there is an element $t \in M$ such that the image of t in $g^{-1} M$ is $g^n s$. Now assume that the restriction of s to each $U_{f_i g}$ maps to 0 in $\mathcal{M}(U_{f_i g}) = g^{-1} M_i$. Thus, by definition, for some m, the image of $g^m t$ is 0 in each M_i (and hence its restriction to each U_{f_i} is 0). Therefore, by the sheaf property of $\mathcal{M}$,

$$g^m t = 0 \in M = \mathcal{M}(Spec(R)),$$

and hence $s = 0$.

To show that (2.2.8) is onto, fix i and let $s_i \in M_i$. We need to show that for some $N >> 0$, $f_i^N s$ is a restriction of a section in $\mathcal{M}(Spec(R)) = M$. To this end, note that first of all, by the sheaf property, there exist sections $s_j \in M_j$ such that

$$s_j|_{U_{f_i f_j}} = f_i^m s$$

for some m, which can be assumed constant by taking the maximum over finitely many. By the injectivity applied to $f_i^{-1} \mathcal{M}$, however, we also know that

$$f_i^n s_j|_{U_{f_j f_k}} = f_i^n s_k|_{U_{f_j f_k}},$$

and again we may assume that n is constant by taking the maximum. Then let s be the section on $Spec(R)$ obtained by gluing the sections $f_i^n s_j$.

For 2, the main point is to check that quasicoherent sheaves have finite limits and colimits, and that the inclusion functor of quasicoherent sheaves to sheaves of modules is exact. To this end, it suffices to consider the case of X affine, and note that localization is exact.

3 follows immediately from 1 and Lemma 2.2.1, as does the statement of 4 about f^* (since there we may assume without loss of generality that both X and Y are affine) and the statement of 4 in case X and Y are affine. In the general case, we may assume that Y is affine, and since X is Noetherian, that it is covered by finitely many affine open sets U_i such that $U_i \cap U_j$ is covered by finitely many affine open sets U_{ijk}. Then $f_*(\mathcal{F})$ can be written as the kernel of a morphism of the form

$$\bigoplus_i f_*(\mathcal{F}|_{U_i}) \to \bigoplus_{i,j,k} f_*(\mathcal{F}|_{U_{ijk}}).$$

$\square$

2.2.4 Lemma *Over a commutative Noetherian ring R, finitely generated locally free modules (i.e. algebraic vector bundles over $Spec(R)$) are the same thing as finitely generated projective modules.*

Proof Over a local ring, a finitely generated projective module is free. This is because over a local ring, an $m \times n$ matrix which has a right inverse has a right inverse over the residue field, and hence has an $m \times m$ submatrix with unit determinant. Thus, finitely generated projective modules are locally free. (In fact, the assumption of being finitely generated is not needed—see Lemma 2.6.3 of Chap. 5.)

On the other hand, if M is a finitely generated locally free module, we can think about it as a coherent sheaf. In particular, there exists an epimorphism $F \to M$ where F is a free

R-module. Now the induced homomorphism

$$Hom_R(M, F) \to Hom_R(M, M) \tag{2.2.9}$$

can be thought of as a morphism of coherent sheaves, which is an epimorphism on stalks since M is locally free. On the other hand, epimorphisms of coherent sheaves are the same as epimorphisms of sheaves, i.e. can be detected on stalks (by considering skyscraper sheaves). Thus, (2.2.9) is an epimorphism of coherent sheaves and hence of R-modules, and hence is onto. An element in the inverse image of $Id : M \to M$ shows that M is a direct summand of F.

□

When the stalks of an algebraic vector bundle $\mathcal{F}$ all have the same dimension as free modules over the stalk of the structure sheaf, we call this the *rank* of the vector bundle $\mathcal{F}$.

2.2.5 Corollary *For* $X = Spec(R)$, $Pic(X)$ *is canonically isomorphic to* $Pic(R)$, *which is the group of isomorphism classes of R-modules invertible with respect to* $\otimes_R$. *Additionally, if R is Noetherian, then* $K(X)$ *is canonically isomorphic to* $K(R)$, *which is defined as the Grothendieck group of the commutative monoid of isomorphism classes of finitely generated projective R-modules with respect to* $\oplus$.

Proof The statement about the Picard group is an immediate consequence of 1 of Lemma 2.2.3. For the statement about K-theory, one uses Lemma 2.2.4.

□

COMMENT

1. For a Noetherian scheme X, we define $G(X)$ to be the free abelian group on coherent sheaves modulo the subgroup generated by $\mathcal{G} - \mathcal{F} - \mathcal{H}$ whenever we have an inclusion $\mathcal{F} \subseteq \mathcal{G}$ such that $\mathcal{H} \cong \mathcal{G}/\mathcal{F}$. We have an obvious map

$$K(X) \to G(X), \tag{2.2.10}$$

and a theorem called *dévissage* states that for regular separated Noetherian schemes, (2.2.10) is an isomorphism. (Recall that a scheme X is regular if it is locally Noetherian and the stalks $\mathcal{O}_{X,x}$ are regular local rings for all points $x \in X$; it suffices to show for closed points, since localization of a regular local ring is regular.) We see analogously to Corollary 2.2.5 that for $X = Spec(R)$ with R Noetherian, $G(X)$ is the factor of the free abelian group on all isomorphism classes of finitely generated R-modules by the subgroup generated by $M - L - N$ whenever we have an inclusion $L \subseteq M$ with $N \cong M/L$. For a proof of the dévissage theorem, see Exercises 21, 22 of Chap. 5.

2. We can now understand a little better the example in Sect. 2.1. The coordinate ring R of the unit circle over $\mathbb{R}$ is a Dedekind domain. Recall that this means that it is Noetherian,

normal and of dimension ≤ 1. It is normal because it is regular. It is regular because it is smooth over $\mathbb{R}$. For a Dedekind domain R, invertible modules are (up to isomorphism) precisely non-zero ideals. Ideals isomorphic to R as modules are precisely principal ideals. Thus, $Pic(R)$ for R a Dedekind domain is the *ideal class group*, which is the factor of the commutative monoid of ideals by the submonoid of principal ideals (with respect to $\cdot$). We will return to this point in Sect. 3.3.

In the Example of Sect. 2.1 (the Möbius band), consider the mapping

$$\phi : \mathcal{L} \to \mathcal{O}_X \qquad (2.2.11)$$

where in the notation of (2.1.7), $\mathcal{L}(V)$ is mapped to $\mathcal{O}_X(V)$ by

$$\phi : (s_1, s_2) \mapsto (s_1, x s_2)$$

(where we note that by the sheaf property, the pair of sections on the right hand side specifies a section in $\mathcal{O}_X(V)$. We see then that the image of

$$\phi(X) : \mathcal{L}(X) \to R = \mathcal{O}_X(X) \qquad (2.2.12)$$

contains the element x (which comes from the pair $(x, 1)$) and the element $(1 - y)$, which comes from the pair $(1 - y, \frac{x}{1+y})$). Therefore, the image of (2.2.12) contains the ideal $I = (x, 1 - y)$ which is obviously maximal (since the quotient R/I is a field). Also, (2.2.12) is obviously injective, since R is an integral domain. It is impossible, however, for the image of (2.2.12) to be all of R, since we already proved that the line bundle $\mathcal{L}$ is not isomorphic to $\mathcal{O}_X$. Therefore, we have proved that

$$\mathcal{L}(X) \cong (x, 1 - y). \qquad (2.2.13)$$

We can then also see where the matrices A, B or (2.1.9) come from: in the language of modules, we need to construct an isomorphism

$$2R \cong 2(x, 1 - y).$$

The isomorphism is given by the matrix A of (2.1.10) (identifying the image with $I \subset R$). In terms of sections of the Möbius band on the unit circle, we see that x is 0 only on the point $(0, -1)$ while $1 - y = \frac{x}{1+y}$ is 0 only on the point $(0, 1)$. Thus, it makes sense to choose the first basis element of the module of sections of the sum of two copies of the Möbius band to have these two sections as its coordinates (since then the resulting section of the sum is not zero anywhere). To get the second basis element, we then "rotate the first basis element section by $90°$," thus obtaining the matrix A of (2.1.10).

It is also true that for a Dedekind domain R, we have in general an isomorphism

$$K(R) \cong Pic(R) \oplus \mathbb{Z}.$$

$$(2.2.14)$$

where the second factor is dimension of the associated vector bundle (See Exercise 21.) The isomorphism from the right side of (2.2.14) to the left is given by

$$(L, n) \mapsto (L + n - 1).$$

The main point is that one checks that in $K(R)$, which is a ring with respect to the tensor product over R, for ideals L_1, L_2,

$$(L_1 - 1) \cdot (L_2 - 1) = 0.$$

$$(2.2.15)$$

The easiest way to see (2.2.15) intuitively may be to work in $G(R)$. If L is an ideal, one can represent $1 - L \in G(R)$ as the module R/L. In a Dedekind domain, an ideal can be uniquely written as a product of powers of prime ideals (we will prove that in the next Subsection), and by the Chinese remainder theorem, if

$$I = p_1^{i_1} \cdot \cdots \cdot p_n^{i_n},$$

then

$$R/I \cong R/p_1^{i_1} \otimes \ldots R/p_n^{i_n}.$$

$$(2.2.16)$$

Consequently, for a non-zero ideal $J \subseteq R$,

$$J/I \cong J/p_1^{i_1} \otimes \ldots J/p_n^{i_n}.$$

$$(2.2.17)$$

But

$$R/p_j^{i_j} \cong R_{p_j}/p_j^{i_j} \cong J \otimes R_{p_j}/p_j^{i_j} \cong J/p_j^{i_j},$$

(the second isomorphism is because R_p is a discrete valuation ring, hence a principal ideal domain), so (2.2.16) is isomorphic to (2.2.17), which therefore implies (2.2.15).

2.3 Sheaves of Ideals

A *sheaf of ideals* on a scheme X is a sub-sheaf of $\mathcal{O}_X$-modules of $\mathcal{O}_X$ (just as for a commutative ring R, an ideal is a sub-R-module of R). Let X be a scheme and $Y \subseteq X$ a closed subscheme. Then, as a part of the structure, we get a homomorphism of sheaves of rings

$$\phi : \mathcal{O}_X \to i_* \mathcal{O}_Y.$$

This makes $i_* \mathcal{O}_Y$ a sheaf of $\mathcal{O}_X$-modules. We call the sheaf of $\mathcal{O}_X$-ideals

$$Ker(\phi)$$

the *sheaf of ideals associated with the closed subscheme* $Y \subseteq X$.

2.3.1 Proposition

1. The sheaf of ideals on X associated with a closed subscheme $Y \subseteq X$ is quasicoherent, and coherent if X is Noetherian.
2. Conversely, for any sheaf of ideals $\mathcal{I}$ on X which is quasicoherent, there exists a closed subscheme $Y \subseteq X$ such that $\mathcal{I}$ is isomorphic to the sheaf of ideals associated with Y.

Proof To prove 1, clearly it suffices to assume that $X = Spec(R)$ is affine. Then a closed subscheme is of the form $Y = Spec(R/I)$ for some ideal I, and the corresponding sheaf of ideals is

$$I \otimes_R \mathcal{O}_X,$$

which is quasicoherent, and coherent if X is Noetherian.

2. By definition, for $X = Spec(R)$ affine, a quasicoherent sheaf of ideals is of the form $I \otimes_R \mathcal{O}_X$ where I is an ideal of R, so it is associated with the closed subscheme $Y = Spec(R/I)$.

Further, a closed subscheme is determined by its intersection with distinguished open sets. Therefore, for a general scheme X, if U_i, U_j are two open affine subschemes of X, and Y_i, Y_j are the closed subschemes associated with $\mathcal{I}|_{U_i}$, $\mathcal{I}|_{U_j}$, we have an isomorphism $Y_i \cap U_j \cong Y_j \cap U_i$. Thus, the closed subschemes Y_i satisfy the gluing condition Sect. 2.5 of Chap. 2. Let Y be the scheme obtained by gluing the Y_i's. By the colimit property of gluing, then, we automatically obtain a morphism of schemes

$$Y \subseteq X. \tag{2.3.1}$$

Further, $Y \cap U_i = Y_i$, so (2.3.1) is a closed immersion (since this can be verified locally). For the same reason, the sheaf of ideals associated with the closed immersion (2.3.1) is $\mathcal{I}$. $\square$

COMMENT Note that the statement of the Proposition logically implies that a quasicoherent sheaf of ideals on a Noetherian scheme is coherent. Of course, we know more generally that if X is a Noetherian scheme, then a quasicoherent sub-sheaf of modules of a coherent sheaf is coherent. (Just as for a Noetherian commutative ring R, a submodule of a finitely generated module is finitely generated.)

As an application of Proposition 2.3.1, we can give an easy proof of the following

2.3.2 Theorem *Let R be a Dedekind domain. Then every non-zero ideal $I \subseteq R$ factors, uniquely up to order of terms, as*

$$I = p_1^{i_1} \cdot \ldots \cdot p_n^{i_n}$$

where p_i are prime ideals.

Proof Let p be a maximal ideal of R. In R_p, the ideal generated by I is principal by Lemma 2.4.2 of Chap. 2. Consider the generator x. Since R is Noetherian, I is finitely generated, so x generates the ideal generated by I in $r^{-1}R$ for some $r \notin p$. Thinking geometrically, the closed subscheme $Z(I)$ of $Spec(R)$ is discrete as a topological space, and hence finite. Finite subschemes of $Spec(R)$ are of the form

$$Z(p_1^{i_1} \cdot \ldots \cdot p_n^{i_n})$$

where p_i are prime ideals (it suffices to consider subschemes with a single point p, which are also subschemes of $Spec(R_p)$; however, the only ideals in a discrete valuation ring are powers of the maximal ideal.)

Uniqueness follows from the bijective correspondence between ideals in R and closed subschemes of $Spec(R)$. $\qquad\square$

Example

1. Let us consider the example of Sect. 2.2. There, we had the closed subscheme $Spec(\mathbb{C}) \to \mathbb{A}^1_{\mathbb{C}}$ given by the homomorphism of rings

$$\mathbb{C}[x] \to \mathbb{C}, \quad x \mapsto 0,$$

and we saw that the associated sheaf of ideals is the coherent sheaf of ideals corresponding to the ideal

$$(x) \subset \mathbb{C}[x].$$

2. In Comment 2 at the end of Sect. 2.2, we constructed an embedding (as sheaves, see (2.2.11)) of the Möbius band $\mathcal{L}$ into $\mathcal{O}_X$ where

$$X = Spec(R), R = \mathbb{R}[x]/(x^2 + y^2 - 1).$$

Thus, we can view $\mathcal{L}$ as a sheaf of ideals on X, and in fact, we saw that this is a coherent sheaf of ideals corresponding to the ideal

$$(x, 1 - y) \subset R.$$

This is associated to the closed subscheme

$$Spec(\mathbb{R}) \cong Spec(R/(x, 1 - y)) \subset X, \tag{2.3.2}$$

which we can think of as the "inclusion of the point $(0, 1)$." What would happen if we took the ideal $(1 - y)$ instead of $(x, 1 - y)$? In this case, we would get the closed subscheme

$$Spec(\mathbb{R}[x]/(x^2)) \cong Spec(R/(1 - y)) \subset X. \tag{2.3.3}$$

We see that the closed subschemes (2.3.2) and (2.3.3) of X have the same points, but (2.3.3) is not reduced.

What would happen if we use (x)? In that case, we get

$$Spec(\mathbb{R}[y]/(1 - y^2))$$
$$= Spec(\mathbb{R}[y]/(1 - y)) \amalg Spec(\mathbb{R}[y]/(1 + y)) \subset X$$

by the Chinese remainder theorem. This can be interpreted as the inclusion of the disjoint union of the points $(0, 1)$ and $(0, -1)$ into X.

3. Looking at part 2 of this example, it is worth noting that we can get a different subscheme of X if we pick a different embedding (2.2.11). In fact, choose any two real numbers a, b such that

$$a^2 + b^2 = 1,$$

and assume for simplicity that

$$a, b \neq 0.$$

It is easy to check that the only zeros of

$$ay + (1 - b)x - a$$

on the unit circle are $(x, y) = (a, b)$ and $(x, y) = (0, 1)$. Similarly, the only zeros of

$$ay - (1 + b)x + a$$

on the unit circle are $(x, y) = (a, b)$ and $(x, y) = (0, -1)$. One in fact checks that in R,

$$a \cdot (ay + (1 - b)x - a) = x \cdot (1 - b) \cdot \frac{ay - (1 + b)x + a}{1 + y},$$

(one only needs to match the coefficients), so

$$(s_1, s_2) = (a \cdot (ay + (1 - b)x - a), (1 - b) \cdot \frac{ay - (1 + b)x + a}{1 + y})$$

gives a global section of $\mathcal{L}$ which only vanishes at the point $(x, y) = (a, b)$ of the unit circle. Furthermore, we see that

$$(1 - b) \cdot \frac{ay - (1 + b)x + a}{1 + y}$$

must divide both $x - a$ and $y - b$ in $(1 + y)^{-1}R$. More explicitly, realizing that

$$(b + 1)x + ay + a$$

vanishes precisely on the points $(x, y) = (-a, b)$ and $(x, y) = (0, -1)$ of the unit circle, we see that

$$((b + 1)x + ay + a) \cdot (ay - (1 + b)x + a) = (y - b) \cdot (1 + y) \cdot (2 + 2b)$$

(again, only the coefficient need to be checked). Similarly,

$$ay - (1 - b)x + a$$

is only 0 on the points $(x, y) = (a, -b)$ and $(x, y) = (0, -1)$, so we get

$$(ay - (1 - b)x + a)(ay - (1 + b)x + a) = -2a \cdot (1 + y) \cdot (x - a)$$

(again, only the coefficient need to be checked).

In any case, defining (2.2.11) by

$$(s_1, s_2) \mapsto s_2(ay - (1 + b)x + a),$$

(which is correct, since when proving that the square of $\mathcal{L}$ is isomorphic to $\mathcal{O}_X$, we proved that for two global sections (s_1, s_2) and (t_1, t_2) of $\mathcal{L}$, $s_2 t_2 (1+y) \in R$), the image of (2.2.12) will be the ideal

$$(x - a, y - b),$$

which corresponds to the closed subscheme

$$Spec(\mathbb{R}) \to X$$

into the point (a, b).

Thus, the closed subscheme corresponding to a sheaf of ideals is not preserved by isomorphisms of sheaves of ideals in the category of sheaves of modules. We shall learn more about this phenomenon in the next Section.

2.4   Quasicoherent Sheaves on *Proj* Schemes

Let a scheme X be of the form $X = Proj(R)$ where R is a graded ring (see Sect. 2.7 of Chap. 2). Let M be a graded module of R, which means that M is a graded abelian group, i.e.

$$M = \bigoplus_{n \in \mathbb{Z}} M_n$$

and multiplication by elements of R_k raises degree by k:

$$R_k \otimes M_n \to M_{n+k}.$$

(We say that an element of R_k defines a *graded homomorphism of modules of degree k*.) Then M gives rise to a quasicoherent sheaf of modules

$$\mathcal{M} = M \otimes_R \mathcal{O}_X \tag{2.4.1}$$

on X: for $f \in R_k$, on the open neighborhood $Spec((f^{-1}R)_0)$, the module of sections of $\mathcal{M}$ is

$$(f^{-1}M)_0 = (M \otimes_R f^{-1}R)_0.$$

(Recall that the subscript 0 means taking the subgroup of elements of degree 0.) This construction glues to give a sheaf of $\mathcal{O}_X$-modules.

Of course, we have, by definition,

$$\mathcal{R} = \mathcal{O}_X.$$

But now for a graded R-module M and $\ell \in \mathbb{Z}$, we have another R-module M obtained by shifting the degrees of M by ℓ:

$$M(\ell)_k = M_{\ell+k}.$$

We define

$$\mathcal{O}_X(n) = R(n) \otimes_R \mathcal{O}_X.$$

For any sheaf of $\mathcal{O}_X$-modules $\mathcal{F}$, we denote

$$\mathcal{F}(\ell) = \mathcal{F} \otimes_{\mathcal{O}_X} \mathcal{O}_X(\ell).$$

The quasicoherent sheaf $\mathcal{O}_X(1)$ is called the *twisting sheaf of Serre*.

2.4.1 Proposition *If R is generated by R_1 as an R_0-algebra, then $\mathcal{O}_X(1)$ is an invertible sheaf, and*

$$\mathcal{O}_X(k + \ell) = \mathcal{O}_X(k) \otimes_{\mathcal{O}_X} \mathcal{O}_X(\ell) \tag{2.4.2}$$

for all $k, \ell \in \mathbb{Z}$.

Proof We have an obvious morphism from the right hand side of (2.4.2) to the left hand side. Let elements $r_i \in R_1$, $i \in I$, generate R. This is actually equivalent to the augmentation ideal being

$$(r_i, i \in I),$$

which is in turn equivalent to the distinguished open sets U_{r_i} being a cover of X. Now at the affine open set U_{r_i}, we obtain the sections of M by inverting r_i and taking elements of degree 0. In particular,

$$(\mathcal{O}_X(k))(U_{r_i})$$

is identified with

$$(r_i^{-1} R)_k,$$

which in turn can be thought of as the degree k elements of the direct limit of the diagram

$$R \xrightarrow{r_i} R \xrightarrow{r_i} R \xrightarrow{r_i} \cdots ,$$

which clearly does not depend on k (in a way compatible with the morphism of (2.4.2)), since r_i has degree 1. Thus, (2.4.2) follows. This obviously implies invertibility. □

Now assume that R is a graded ring generated by R_1 as an R_0-algebra. Then for a sheaf $\mathcal{M}$ of $\mathcal{O}_X$-modules, we can define a graded R-module $\mathcal{M}_*(X)$ by

$$\mathcal{M}_*(X)_n = \mathcal{M}(n)(X). \tag{2.4.3}$$

2.4.2 Lemma *If R is a graded ring generated by R_1 as an R_0-algebra, $X = Proj(R)$, then the functor (2.4.3) from sheaves of $\mathcal{O}_X$-modules to graded R-modules is right adjoint to the functor (2.4.1) from graded R-modules to sheaves of $\mathcal{O}_X$-modules.*

Proof Let M be a graded R-module, and let $\mathcal{F}$ be a sheaf of $\mathcal{O}_X$-modules. Suppose we are given a homomorphism of graded R-modules

$$M \to \mathcal{F}_*(X). \tag{2.4.4}$$

So for every $n \in \mathbb{N}_0$, we are given a homomorphism of R_0-modules

$$M_n \to \mathcal{F}(n)(X)$$

compatibly with multiplication by R_1. Then for every $f \in R_1$, we obtain, by taking colimits under multiplication by f, a homomorphism of R_0-modules

$$(f^{-1}M)_0 \to (f^{-1}\mathcal{F}(X))_0 \cong \mathcal{F}(n)(U_f).$$

Clearly, these are compatible to give a morphism of sheaves of $\mathcal{O}_X$-modules

$$M \otimes_R \mathcal{O}_X \to \mathcal{F}. \tag{2.4.5}$$

Conversely, given (2.4.5), we have

$$M_n \to (M \otimes_R \mathcal{O}_X)(n)(X) \to \mathcal{F}(n)(X) \tag{2.4.6}$$

where the first morphism is obtained by taking, on U_f, the colimit homomorphism

$$M_n \to (f^{-1}M)_n ,$$

which obviously are compatible to give the first homomorphism (2.4.6). The second morphism is obtained by taking (2.4.5), twisting by n and taking global sections. It is easily verified that the two correspondences between (2.4.4) and (2.4.5) and vice versa are inverse to each other.

□

The functor (2.4.1), by definition, always produces a quasicoherent sheaf on X. It is important to note however that unlike the affine case, we cannot expect the unit of the adjunction of Lemma 2.4.2 to be an isomorphism in general, i.e. for an arbitrary graded R-module M, the graded R-module of twisted global sections of the associated quasicoherent sheaf $M \otimes_R \mathcal{O}_X$ of $\mathcal{O}_X$-modules to be isomorphic to M. To see this, it suffices to note that $M \otimes_R \mathcal{O}_X$ only depends on $M_{\geq N}$ for any chosen $N \in \mathbb{Z}$, so we could replace M by a graded R-module which is the same in degrees $\geq N$ but 0 in degrees $< N$, and obtain the same associated quasicoherent sheaf.

On the other hand, we have the following

2.4.3 Lemma *Suppose R is a graded ring generated by R_1 as an R_0-algebra, and suppose R_1 is a finitely generated R_0-module. Let $X = Proj(R)$. Then the counit of the adjunction of Lemma 2.4.2 is always an isomorphism, i.e. any quasicoherent sheaf of $\mathcal{O}_X$-modules is isomorphic to the sheaf of $\mathcal{O}_X$-modules associated with its graded R-module of twisted global sections.*

Proof The role of assuming that R_1 is finitely generated as an R_0-module, say, by elements $f_1, \dots f_n$, is that $f_1, \dots, f_n$ generate the augmentation ideal, i.e. U_{f_i} cover X, which is a substitute for the fact in the affine case that an affine scheme is quasicompact. At this point, we have, for every homogeneous element $h \in R$,

$$h = a_1 f_1 + \cdots + a_n f_n,$$

$a_i \in R$, and for a given h, we can replace the f_i's by an arbitrary power, assuming we are willing to replace h by a sufficiently high power. With this in mind, letting $\mathcal{F}$ be a quasicoherent sheaf on X, we can show that the counit of adjunction of Lemma 2.4.2 is onto on sections on U_h: for $s \in \mathcal{F}(U_h)$, multiplying by sufficiently high powers of h, we can find sections of $\mathcal{F}(?) \otimes_R \mathcal{O}_X$ on U_{f_i} which coincide with s on $U_h \cap U_{f_i}$. Multiplying by sufficiently high powers of h and f_i, those sections are compatible, and glue to produce an element of $\mathcal{F}(N)(X)$ for some $N >> 0$, which maps onto s on U_h. One proves similarly that the homomorphism is injective.

□

The following is essentially just a restatement:

2.4.4 Proposition *Suppose R is a graded ring generated by R_1 as an R_0-algebra, and suppose R_1 is a finitely generated R_0-module. Let $\mathcal{F}$ be a quasicoherent sheaf over*

$X = Proj(R)$. *Then there exists an onto map*

$$\bigoplus_{i \in I} \mathcal{O}_X(n_i) \to \mathcal{F}.$$

If $\mathcal{F}$ is coherent and X is Noetherian, the set I can be taken to be finite.

$\square$

Using Lemma 2.4.3, if R is finitely generated by R_1 over R_0, we can identify the category of quasicoherent sheaves on $Proj(R)$ with a full subcategory of the category of graded R-modules and morphisms preserving the degree of homogeneous elements (since, by passing to twisted sections, a morphism of quasicoherent sheaves is always induced by a homomorphism of graded modules). Additionally, the inclusion of the subcategory has a left adjoint (i.e. is a right adjoint). Such a full subcategory is called a *reflexive subcategory*. The left adjoint to the inclusion of a reflexive subcategory is called a *reflection*.

We have already encountered other examples of reflexive subcategories. For example, the category of abelian groups is a reflexive subcategory of the category of groups, the category of abelian sheaves on a topological space X is a reflexive subcategory of the category of abelian presheaves on X, the category of affine schemes is a reflexive subcategory of the category of schemes.

Example Let $R = R_0[x]$, so $Proj(R) = \mathbb{P}^0_{Spec(R_0)} = Spec(R_0)$. Then the category of quasicoherent sheaves on $Proj(R)$ is identified, by Lemma 2.4.2, with the subcategory of $R_0[x]$-modules which have the structure of $R_0[x, x^{-1}]$-modules.

If $R = R_0[x, y]$, i.e. $Proj(R) = \mathbb{P}^1_{Spec(R_0)}$, then the category of quasicoherent sheaves on $Proj(R)$ is identified with the full subcategory of graded $R_0[x, y]$-modules M such that the canonical morphism from M to the pullback

$$\lim \begin{pmatrix} & & x^{-1}M \\ & & \downarrow \\ y^{-1}M & \longrightarrow & (xy)^{-1}M \end{pmatrix} \tag{2.4.7}$$

is an isomorphism. For a general graded $R_0[x, y]$-module M, the reflection of M into the full subcategory identified with quasicoherent sheaves on $Proj(R)$ is given by (2.4.7).

For $n > 1$, a similar characterization of quasicoherent sheaves on $\mathbb{P}^n_{Spec(R_0)} = Proj(R_0[x_0, \ldots, x_n])$ also holds when we replace (2.4.7) by the diagram which contains all the canonical arrows from all the modules $x_i^{-1}M$ to all the modules $(x_i x_j)^{-1}M$, $i \neq j$.

COMMENT We will see in Sect. 4.4.2 of Chap. 6 (proof of Theorem 4.4.3 and Exercise 25 of Chap. 6) that in general when R is generated over R_0 by R_1 which is a finitely generated R_0-module, the reflexive subcategory of the category of graded R-modules which correspond to quasicoherent sheaves is the full subcategory on those graded modules M for which the local cohomology groups $H_I^0(M)$ and $H_I^1(M)$ vanish where I is the augmentation ideal of R.

The following corollary of Lemma 2.4.3 is quite important:

2.4.5 Proposition

1. *If Y is a closed subscheme of $\mathbb{P}^n_{Spec(A)} = Proj(A[x_0, \ldots, x_n])$, then there exists a homogeneous ideal*

$$I \subseteq A[x_0, \ldots, x_n]$$

   such that $Y = Z(I)$.
2. *A scheme over $Spec(A)$ is projective if and only if it is isomorphic to $Proj(R)$ where R is a graded ring with $R_0 = A$ generated by R_1 where R_1 is a finitely generated R_0-module.*

Proof To prove 1, consider the quasicoherent sheaf of ideals $\mathcal{I}$ associated with the closed subscheme Y. Then $\mathcal{I} \subseteq \mathcal{O}_X$, so $\mathcal{I}(m) \subseteq \mathcal{O}_X(m)$ since $\mathcal{O}_X(m)$ is invertible, so we can define the homogeneous ideal I in R by

$$I_m = \mathcal{I}(m)(X).$$

By Lemma 2.4.3, $Y = Z(I)$.

Recall that a scheme X is projective over $Spec(A)$ if it is isomorphic to a closed subscheme of $\mathbb{P}^n_{Spec(A)}$. By part 1 of this proposition, this happens if and only if X is isomorphic to $Proj(A[x_0, \ldots, x_n]/I)$ for some homogeneous ideal I. On the other hand, $A[x_0, \ldots, x_n]/I$ is the general ring R described in our condition for 2.

$\square$

3   Divisors

We saw from the examples in the last Section that an isomorphism of line bundles on a scheme X can be tricky to construct from first principles. Divisors can be thought of as geometric models of line bundles, (often with additional assumptions on X), where isomorphism of bundles corresponds to more manifest concepts. We will give three approaches each increasingly more concrete (and also more restrictive).

3.1   First Cohomology

This approach works in general, for all schemes X, and continues the discussion of the beginning of Sect. 2.1. A line bundle (i.e., an invertible sheaf of $\mathcal{O}_X$-modules) $\mathcal{L}$ on a scheme X admits an open cover U_i, $i \in I$ of X for which we have isomorphisms

$$\phi_i : \mathcal{L}|_{U_i} \xrightarrow{\cong} \mathcal{O}_X|_{U_i} = \mathcal{O}_{U_i}. \tag{3.1.1}$$

We have elements

$$h_{ij} \in \mathcal{O}_X(U_i \cap U_j)^\times \tag{3.1.2}$$

(recall that $R^\times$ is the group of units of a ring R) given by

$$h_{ij} = \phi_i(U_i \cap U_j) \circ \phi_j^{-1}(U_i \cap U_j). \tag{3.1.3}$$

The elements h_{ij} may be referred to as *transition functions*. Of course, (3.1.3) implies that for all $i, j, k \in I$,

$$h_{ij}|_{U_i \cap U_j \cap U_k} \cdot h_{jk}|_{U_i \cap U_j \cap U_k} = h_{ik}|_{U_i \cap U_j \cap U_k}, \tag{3.1.4}$$

but the real point is that to specify $\mathcal{L}$ up to isomorphism, it suffices to specify the open cover U_i and the elements (3.1.3), together with the condition (3.1.4) (which may be referred to as the *cocycle condition*).

It is useful here to note that since the group of units functor from commutative rings to abelian groups preserves limits (it is in fact right adjoint to a functor called the *group ring*, which is the free abelian group $\mathbb{Z}G$ with multiplication given by the multiplication in G), the abelian presheaf $\mathcal{O}_X^\times$ whose abelian group of sections on U is $(\mathcal{O}_X(U))^\times$ is in fact a sheaf. We will denote it by $\mathcal{O}_X^\times$.

The open cover $(U_i)_{i \in I}$ together with the elements (3.1.2) which satisfy the cocycle condition (3.1.4) is called a *(Čech) 1-cocycle with coefficients in $\mathcal{O}_X^\times$*. Note that abstractly, the definition would work for any abelian sheaf $\mathcal{F}$ (or even just a sheaf of groups) on an arbitrary topological space X in place of $\mathcal{O}_X^\times$ on a scheme X, so we can analogously speak of Čech 1-cocycles with coefficients in $\mathcal{F}$. Note that we think of $\mathcal{O}_X^\times$ multiplicatively, while an abelian sheaf will often be written additively, in which case $\cdot$ will become $+$, and 1 will become 0.

Note that the cocycle condition (3.1.4) implies

$$h_{ii} = 1, \quad h_{ij} = h_{ji}^{-1}.$$

On the Möbius band example, I was a two-element set, so there was only one transition function between them, which is why the cocycle condition was not needed.

A line bundle $\mathcal{L}$ can be constructed from a Čech 1-cocycle

$$((U_i)_{i \in I}, h_{ij}, i, j \in I) \tag{3.1.5}$$

with coefficients in $\mathcal{O}_X^\times$ in the same way as in the case of the Möbius band: For an open set $V \subseteq X$, put

$$\begin{aligned} \mathcal{L}(V) = \{(s_i \in \mathcal{O}_X(V \cap U_i))_{i \in I} \mid \\ (\forall i, j \in I) s_i|_{U_i \cap U_j \cap V} = h_{ij}|_{U_i \cap U_j \cap V} \cdot s_j|_{U_i \cap U_j \cap V}\}. \end{aligned} \tag{3.1.6}$$

But when is the line bundle $\mathcal{L}$ given by (3.1.5) isomorphic to another line bundle $\mathcal{M}$, given by another Čech 1-cocycle

$$((V_j)_{j \in J}, k_{ij}, i, j \in J)?$$

We see right away one awkward point: the open covers can be different. This can be dealt with by observing that if the open cover $(V_j)_{j \in J}$ is a *refinement* of the open cover $(U_i)_{i \in I}$ in the sense that for each $j \in J$, there exists an $i \in I$ such that $V_j \subseteq U_i$, then a Čech 1-cocycle defined using the covering (U_i) also specifies a Čech 1-cocycle defined using the covering (V_j). (Recall that this notion of refinement was also consider in Sect. 1.1 when we constructed the sheafification—this is related.)

For two cocycles defined on open covers $(U_i)_{i \in I}$, $(V_j)_{j \in J}$, we can then replace both by a cocycle defined on the common refinement

$$(U_i \cap V_j)_{i \in I, j \in J}.$$

Thus, it suffices to find conditions when a line bundle $\mathcal{L}$ defined by the 1-cocycle (3.1.5) is isomorphic to a line bundle $\mathcal{M}$ defined by another 1-cocycle using the same open cover:

$$((U_i)_{i \in I}, k_{ij}, i, j \in I).$$

In this case, we see that $\mathcal{L}$ and $\mathcal{M}$ will be isomorphic if and only if there exist

$$f_i \in \mathcal{O}_X^\times(U_i), i \in I \tag{3.1.7}$$

such that for all $i, j \in I$,

$$f_i|_{U_i \cap U_j} \cdot h_{ij} = k_{ij} \cdot f_j|_{U_i \cap U_j}. \tag{3.1.8}$$

For then, we can, starting from a section (3.1.6), produce a section in $\mathcal{M}(V)$ simply by replacing each s_i by $f_i \cdot s_i$ and vice versa. If, on the other hand, $\mathcal{L}$ and $\mathcal{M}$ are isomorphic, then the functions f_i must exist by looking at sections of each on U_i, and using the fact that for any commutative ring R, any isomorphism of R-modules $R \to R$ is given precisely by multiplication by an element of $R^\times$.

For an arbitrary abelian sheaf (or more generally sheaf of groups) $\mathcal{F}$, this works basically the same way except for the last part we need to argue that if the f_j's exist on a refinement (V_j), then setting $U_{(i,j)} = U_i \cap V_j$, and denoting by $f_{(i,j)}$ the restrictions of f_j to $U_{(i,j)}$ and by $h_{(i,j),(i',j')}$ the restrictions of $h_{ii'}$ to $U_{(i,j)} \cap U_{(i',j')}$, we have

$$h_{(i,j),(i,j')} = k_{(i,j),(i,j')} = 1,$$

so

$$f_{(i,j)}|_{U_{(i,j)} \cap U_{(i,j')}} = f_{(i,j')}|_{U_{(i,j)} \cap U_{(i,j')}},$$

and the elements $f_{(i,j)}$ can therefore be glued by the sheaf property to give f_i.

In any case, if (3.1.8) holds, the Čech 1-cocycles h_{ij}, k_{ij} are called *cohomologous*, and the set of equivalence classes of cocycles with respect to the relation of cohomology is denoted by

$$\check{H}^1(X, \mathcal{F}) \tag{3.1.9}$$

and called *first Čech cohomology of X with coefficients in $\mathcal{F}$.* If $\mathcal{F}$ is an abelian sheaf, then of course 1-cocycles with coefficients in $\mathcal{F}$ form an abelian group, and two 1-cocycles $(h_{ij})_{i,j\in I}$, $(k_{ij})_{i,j\in I}$ defined on the same open cover are cohomologous if and only if (writing multiplicatively), $(h_{ij}^{-1} k_{ij})_{i,j\in I}$ is cohomologous to 1. It is worth writing explicitly what it means that a 1-cocycle $(k_{ij})_{i,j\in I}$ is cohomologous to 1: it means that there exists $f_i \in \mathcal{F}(U_i)$ such that

$$k_{ij} = f_i|_{U_i \cap U_j} f_j^{-1}|_{U_i \cap U_j}. \tag{3.1.10}$$

In this case, we say that that the 1-cocycle $(k_{ij})_{i,j\in I}$ is a *coboundary* of $(f_i)_{i\in I}$, and write

$$(k_{ij})_{i,j\in I} = \delta(f_i)_{i\in I}.$$

When $\mathcal{F}$ is an abelian sheaf, the set of all coboundaries forms an abelian subgroup of the set of all cocycles, thus giving (3.1.9) the structure of an abelian group. We speak of the *first Čech cohomology group with coefficients in the abelian sheaf $\mathcal{F}$.*

Thus, we proved in particular the following

3.1.1 Theorem *For any scheme X, we have a canonical isomorphism*

$$Pic(X) \cong \check{H}^1(X, \mathcal{O}_X^\times).$$

(3.1.11)

$\square$

COMMENT It is customary to write H^1 instead of $\check{H}^1$ in (3.1.11). This is because there is another definition of cohomology with coefficients in an abelian sheaf which coincides with Čech cohomology in dimension 1. We shall study cohomology in more detail in Chap. 5. In fact, in any abelian category $\mathcal{A}$, recall that an *exact sequence* is a sequence of morphisms (finite or infinite)

$$\ldots A_k \xrightarrow{\ f_k\ } A_{k-1} \xrightarrow{\ f_{k-1}\ } A_{k-2} \longrightarrow \ldots$$

such that

$$Ker(f_{k-1}) = Im(f_k)$$

whenever applicable. In particular, a *short exact sequence* is an exact sequence of the form

$$0 \longrightarrow A \xrightarrow{\ f\ } B \xrightarrow{\ g\ } C \longrightarrow 0.$$

(3.1.12)

This then just means that f is a monomorphism, g is an epimorphism and

$$Ker(g) = Im(f).$$

If $\mathcal{A}$ is, say, the category of abelian groups or R-modules, this means that up to isomorphism $C \cong B/A$ where f is the inclusion of A into B and g is the projection from B to B/A.

Now for fixed objects $A, C \in Obj(\mathcal{A})$, a short exact sequence (3.1.12) is called an *extension* of C by A. We denote by

$$Ext_{\mathcal{A}}^1(C, A)$$

(3.1.13)

the set of isomorphism classes of all extensions (3.1.12). By an isomorphism of extensions, we mean a diagram of the form

$$
\begin{array}{ccccccccc}
0 & \longrightarrow & A & \longrightarrow & B & \longrightarrow & C & \longrightarrow & 0 \\
& & \Big\downarrow{\scriptstyle Id} & & \Big\downarrow{\scriptstyle \cong} & & \Big\downarrow{\scriptstyle Id} & & \\
0 & \longrightarrow & A & \longrightarrow & B' & \longrightarrow & C & \longrightarrow & 0.
\end{array}
$$

One notes that (3.1.13) is actually an abelian group: Addition of two extensions is obtained by adding them using $\oplus$, and then pushing forward via the codiagonal morphism

$$\nabla : A \oplus A \to A,$$

i.e. making the following diagram so the left hand square is a pushout

$$
\begin{array}{ccccccccc}
0 & \longrightarrow & A \oplus A & \longrightarrow & B \oplus B' & \longrightarrow & C \oplus C & \longrightarrow & 0 \\
& & \Big\downarrow{\scriptstyle \nabla} & & \Big\downarrow & & \Big\downarrow{\scriptstyle Id} & & \\
0 & \longrightarrow & A & \longrightarrow & B_0 & \longrightarrow & C \oplus C & \longrightarrow & 0
\end{array}
$$

and then puling back via the diagonal

$$\Delta : C \to C \oplus C,$$

i.e. making the following diagram so the right hand square is a pullback

$$
\begin{array}{ccccccccc}
0 & \longrightarrow & A & \longrightarrow & B_1 & \longrightarrow & C & \longrightarrow & 0 \\
& & \Big\downarrow{\scriptstyle Id} & & \Big\downarrow & & \Big\downarrow{\scriptstyle \Delta} & & \\
0 & \longrightarrow & A & \longrightarrow & B_0 & \longrightarrow & C \oplus C & \longrightarrow & 0.
\end{array}
$$

(Recall that $\oplus$ in an abelian category is both the product and the coproduct; the codiagonal of a coproduct is the morphism from the coproduct obtained by sending each copy of A to A via the identity, and the diagonal of a product is again obtained by sending C by Id to both copies of C.)

The inverse of an extension (3.1.12) is obtained by reversing the sign of f or g. The neutral element is the extension

$$0 \longrightarrow A \overset{f}{\longrightarrow} A \oplus C \overset{g}{\longrightarrow} C \longrightarrow 0 \qquad (3.1.14)$$

where f is a coproduct injection and g is a product projection. (3.1.14) is also called a *split short exact sequence*.

(Note: one problem with this definition in an arbitrary abelian category is that we haven't shown that $Ext^1(C, A)$ is actually a set—it could be a proper class; but this will not come up in the examples we are interested in.)

Now if X is a topological space, and $\mathcal{F}$ is an abelian sheaf on X, one defines

$$H^1(X, \mathcal{F}) = Ext^1_{Ab-sh(X)}(\underline{\mathbb{Z}}, \mathcal{F})$$

where $Ab\text{-}sh(X)$ is the category of abelian sheaves on X and $\underline{\mathbb{Z}}$ is the constant sheaf with values in $\mathbb{Z}$, i.e. $f^{-1}(\mathbb{Z})$ where $f : X \to *$ is the unique continuous map to a point and the sheaf $\underline{\mathbb{Z}}$ on a point takes the value $\mathbb{Z}$ on the set containing the point, and the value 0 on the empty set.

3.1.2 Theorem   *For any space X and any abelian sheaf $\mathcal{F}$, we have a canonical isomorphism*

$$H^1(X, \mathcal{F}) \cong \check{H}^1(X, \mathcal{F}). \qquad (3.1.15)$$

Proof   Consider an extension

$$0 \longrightarrow \mathcal{F} \overset{f}{\longrightarrow} \tilde{\mathcal{F}} \overset{g}{\longrightarrow} \underline{\mathbb{Z}} \longrightarrow 0 \qquad (3.1.16)$$

of abelian sheaves on X. The stalk of $\underline{\mathbb{Z}}$ at every point is $\mathbb{Z}$, so we have a short exact sequence

$$0 \to \mathcal{F}_x \to \tilde{\mathcal{F}}_x \to \mathbb{Z} \to 0.$$

This means that there exists an open set $V \ni x$ and an element $s \in \tilde{\mathcal{F}}(V)$ such that $(g(V))(s)$ represents

$$1 \in \underline{\mathbb{Z}}_x = \mathbb{Z}.$$

Note carefully that then $(g(V))(s) - 1$ represents 0 in $\underline{\mathbb{Z}}_x = \mathbb{Z}$, so by definition there exists an open set U, $x \in U \subseteq V$ such that the restriction of $(g(V))(s) - 1$ to U, which is $(g(U))s|_U - 1$, is 0, so on U', there is a section in $\tilde{\mathcal{F}}(U)$ (namely $s|_U$) which maps to $1 \in \underline{\mathbb{Z}}(U)$ by $g(U)$. (Note: All this is needed because $\underline{\mathbb{Z}}(U)$ is actually not $\mathbb{Z}$ if U is not connected.)

Thus, we have an open cover (U_i) of X together with sections $s_i \in \tilde{\mathcal{F}}(U_i)$ where $(g(U_i))(s_i) = 1$. Then for any i, j, the image of

$$s_i|_{U_i \cap U_j} - s_j|_{U_i \cap U_j} \tag{3.1.17}$$

under $g(U_i \cap U_j)$ is $1 - 1 = 0$, so (3.1.17) is the image, under $f(U_i \cap U_j)$, of an element

$$h_{ij} \in \mathcal{F}(U_i \cap U_j),$$

which is a Čech 1-cocycle. This procedure can obviously be reversed to construct, from a Čech 1-cocycle valued in $\mathcal{F}$, an extension of abelian sheaves (3.1.16). This gives inverse isomorphisms between the two sides of (3.1.15). (Note that this classification of extensions of $\mathbb{Z}$ up to isomorphism is in fact analogous with our discussion on classification of line bundles up to isomorphism.) □

COMMENT An analogue of Theorem 3.1.1, along with its proof, also works essentially unchanged for bundles of dimension n. In this case, we consider the sheaf $M_n(\mathcal{O}_X)$ of $n \times n$ matrices in $\mathcal{O}_X$. This is, of course, a sheaf of $\mathcal{O}_X$-modules (a sum of n^2 copies of $\mathcal{O}_X$), but matrices can be multiplied, and so $M_n(\mathcal{O}_X)$ can also be considered as a sheaf of (non-commutative) rings. Furthermore, the sheaf $M_n(\mathcal{O}_X)^\times$ is denoted by $GL_n(\mathcal{O}_X)$. (Recall that the group of units of the ring $M_n(R)$ of $n \times n$ matrices over a commutative ring R is denoted by $GL_n(R)$ and called a *general linear group of R*.)

In any case, essentially the same discussion as the proof of Theorem 3.1.1 leads to a proof of the fact that the set of isomorphism classes of n-bundles on a scheme X is bijective to the first Čech cohomology

$$\check{H}^1(X, GL_n(\mathcal{O}_X)).$$

Note that in general this is not (i.e. does not have a canonical structure of) a group, since $GL_n(\mathcal{O}_X)$ is a sheaf of groups, not of abelian groups.

3.2 Cartier Divisors

If X is an integral scheme, then it has a well-defined field of rational functions $K(X)$ (unrelated, despite the notation, to algebraic K-theory), which is the field of fractions of $\mathcal{O}_X(U)$ for every non-empty affine open set $U \subseteq X$. We have therefore a canonical embedding

$$\mathcal{O}_X(U) \subseteq K(X),$$

which passes to an embedding on units:

$$\iota_U : \mathcal{O}_X^\times(U) \subseteq K(X)^\times. \tag{3.2.1}$$

For a line bundle $\mathcal{L}$ on X given by a cocycle

$$((U_i)_{i \in I}, h_{ij}, i, j, \in I),$$

therefore, if we choose a particular $j_0 \in I$, we can define a section of the sheaf

$$\underline{K(X)^\times / \mathcal{O}_X^\times} \tag{3.2.2}$$

(recall that the underline means a constant sheaf) by gluing the sections

$$\iota_{U_i \cap U_{j_0}}(h_{i,j_0}) \in (\underline{K(X)^\times / \mathcal{O}_X^\times})(U_i). \tag{3.2.3}$$

The cocycle condition assures that these sections glue to a global section of the sheaf (3.2.2) is called a *Cartier divisor* on X.

Conversely, every Cartier divisor s gives rise to a line bundle: On some open cover $(U_i)_{i \in I}$, each $s|_{U_i}$ comes from a section of $\underline{K(X)^\times}(U_i)$, i.e. an element of $s_i \in K(X)^\times$. We can then simply put

$$h_{ij} = s_i|_{U_i \cap U_j} \cdot s_j^{-1}|_{U_i \cap U_j}.$$

This correspondence between line bundles and Cartier divisors is obviously not bijective (for example, when associating a Cartier divisor with a line bundle, we made a choice of $j_0 \in I$). In fact, even more generally, we could have multiplied the Cartier divisor (3.2.3) by any element of $f \in K(X)^\times$, i.e. a global section of

$$\underline{K(X)^\times}. \tag{3.2.4}$$

Such a Cartier divisor, i.e. one which is the projection of a section of (3.2.4), is denoted by (f), and called a *principal Cartier divisor*. We see by definition that two Cartier divisors whose fraction is a principal Cartier divisor give rise to isomorphic line bundles. Such Cartier divisors are called *linearly equivalent*.

Conversely, if two Cartier divisors s, t (i.e. global sections of (3.2.2)) give rise to isomorphic line bundles, and on an open cover U_i, the isomorphism is given by

$$\phi_i \in \mathcal{O}_X^\times(U_i),$$

then the image of ϕ_{j_0} in $K(X)^\times$ defines a principal Cartier divisor which is the ratio of the associated Cartier divisors. Thus, we proved

3.2.1 Theorem *For every integral scheme X, the factor group of the (abelian) group of Cartier divisors by the subgroup of principal Cartier divisors is canonically isomorphic to Pic(X).*

$\square$

Example Considering again the unit circle

$$X = Spec(R), \quad R = \mathbb{R}[x, y]/(x^2 + y^2 - 1),$$

and the Möbius band whose global sections are pairs of sections

$$(s_1 \in (1 - y)^{-1} R, s_2 \in (1 + y)^{-1} R)$$

such that $s_1 = xs_2$, we find a corresponding Cartier divisor by gluing, say, the section $1 \in K(X)^\times / \mathcal{O}_X^\times(U_{(1+y)})$ with the section $x \in K(X)^\times / \mathcal{O}_X^\times(U_{(1-y)})$. So this basically does not give anything new, although from this point of view we see that there is nothing special about the fact that x has a zero at the point $(0, -1)$: Considering the rational function

$$f = \frac{ay - (1 + b)x + a}{1 + y} \in K(X)^\times,$$

(noting that the numerator has zeros at $(0, -1)$ and (a, b)), we may take the linearly equivalent Cartier divisor obtained by gluing

$$f \in K(X)^\times / \mathcal{O}_X^\times(U_{(1+y)})$$

and

$$fx \in K(X)^\times / \mathcal{O}_X^\times(U_{(1-y)}),$$

which has "a zero at the point (a, b)." We will see this point more clearly in the next section from the point of view of Weil divisors.

The concept of a Cartier divisor can be extended to an arbitrary scheme X by replacing the sheaf $K(X)$ by the sheaf $\mathcal{K}$ which is the sheafification of the presheaf $\mathcal{P}$ where $\mathcal{P}(U)$ is the localization of $\mathcal{O}_X(U)$ obtained by inverting the multiplicative set consisting of all sections which are non-zero divisors in every local ring $\mathcal{O}_{X,x}$ for $x \in U$. Since sheafification preserves stalks, note that the natural morphism of sheaves

$$\mathcal{O}_X \to \mathcal{K}$$

is injective. A Cartier divisor in this more general context is a section of $\mathcal{K}^\times/\mathcal{O}_X^\times$, and a principal Cartier divisor is one which is the projection of a section of $\mathcal{K}^\times$. The (abelian) group which is the factor of the group of Cartier divisors by the subgroup of principal Cartier divisors is called the *Cartier divisor class group* and denoted by $Ca\text{-}Cl(X)$. In this case, an analogue of the above construction still gives an injective map

$$Ca\text{-}Cl(X) \xrightarrow{\;\subseteq\;} Pic(X) \tag{3.2.5}$$

The map (3.2.5) is constructed the same way as in the case of integral schemes: lift generators s_i of sections of a Cartier divisor to $\mathcal{K}^\times$ locally, and then record the Čech 1-cocycle of their differences in $\mathcal{O}_X$, thus giving a cocycle for a line bundle. Moreover, note that the inverses s_i^{-1} represent this line bundle as a subsheaf of $\mathcal{K}$. The map (3.2.5) is also injective for the same reason: if two invertible subsheaves of $\mathcal{K}$ are isomorphic via an abstract isomorphism of $\mathcal{O}_X$-modules, then this isomorphism, considered as a section of $\mathcal{K}^\times$, gives a principal Cartier divisor which is the ratio of the Cartier divisors corresponding to the subsheaves.

We saw however that by construction, the image of (3.2.5) lands in the subgroup of $Pic(X)$ of line bundles generated by invertible subsheaves of $\mathcal{K}$, which, as it turns out, does not have to include all line bundles. Thus, the map (3.2.5) is not onto in general.

COMMENT For any topological space X, and any abelian sheaf $\mathcal{F}$, one can write

$$\check{H}^0(X, \mathcal{F}) = \mathcal{F}(X) \cong Hom(\underline{\mathbb{Z}}, \mathcal{F}) = H^0(X, \mathcal{F}).$$

(Here *Hom* denotes the abelian group of homomorphisms in the category of abelian sheaves.) Then for any short exact sequence of abelian sheaves

$$0 \to \mathcal{E} \to \mathcal{F} \to \mathcal{G} \to 0 \tag{3.2.6}$$

on X, we obtain an exact sequence

$$
\begin{array}{ccccc}
0 \longrightarrow & H^0(X, \mathcal{E}) & \longrightarrow & H^0(X, \mathcal{F}) & \longrightarrow & H^0(X, \mathcal{G}) \\
& & & & & \downarrow \\
& H^1(X, \mathcal{G}) & \longleftarrow & H^1(X, \mathcal{F}) & \longleftarrow & H^1(X, \mathcal{E}).
\end{array}
\tag{3.2.7}
$$

The vertical homomorphism (also referred to as the *connecting map*) can be constructed by taking a morphism of sheaves

$$f : \underline{\mathbb{Z}} \to \mathcal{G}$$

to the extension which is the top row of the diagram

$$
\begin{array}{ccccccccc}
0 & \longrightarrow & \mathcal{E} & \longrightarrow & \mathcal{Q} & \longrightarrow & \mathbb{Z} & \longrightarrow & 0 \\
& & \downarrow{\scriptstyle Id} & & \downarrow & & \downarrow{\scriptstyle f} & & \\
0 & \longrightarrow & \mathcal{E} & \longrightarrow & \mathcal{F} & \longrightarrow & \mathcal{G} & \longrightarrow & 0
\end{array}
$$

where the right square is a pullback in the category of abelian sheaves. The other morphisms (3.2.7) all come by functoriality from the morphisms (3.2.6). It is a little tedious but easy to show that the sequence (3.2.7) is indeed exact. In fact, higher cohomology can be defined (analogously to our present definitions), and the exact sequence (3.2.7) then continues analogously through the higher cohomology groups. This exact sequence is then referred to as the *long exact sequence in cohomology*. We shall study it in detail in Chap. 5 (Theorem 2.3.7).

The inclusion (3.2.5) then can be interpreted as the homomorphism in (3.2.7) from the cokernel of the last arrow of the top row to the last term of the bottom row, where (3.2.6) is the short exact sequence

$$
0 \to \mathcal{O}_X^\times \to \mathcal{K}^\times \to \mathcal{K}^\times/\mathcal{O}_X^\times \to 0. \tag{3.2.8}
$$

In the case when X is an integral scheme, we have $\mathcal{K} = \underline{K(X)}$ so $\mathcal{K}^\times = \underline{K(X)^\times}$, and furthermore, since X is irreducible,

$$
\underline{K(X)^\times}(U) = \begin{array}{ll} K(X)^\times & \text{if } U \neq \emptyset \\ 0 & \text{if } U = \emptyset. \end{array}
$$

This means that the restriction maps of $\underline{K(X)^\times}$ are onto. A sheaf $\mathcal{F}$ whose restriction maps are onto is called *flasque*. We will see in Chap. 5 that all higher cohomology (i.e. all cohomology except possibly H^0) of a flasque sheaf is 0. It is easy to check directly that for a flasque sheaf $\mathcal{F}$,

$$
\check{H}^1(X, \mathcal{F}) = 0. \tag{3.2.9}
$$

Consider a Čech 1-cocycle $((U_i)_{i \in I}, h_{ij})$ in $\mathcal{F}$. Without loss of generality, I is an ordinal number. We will construct, by transfinite induction on $i \in I$, an $s_i \in \mathcal{F}(U_i)$ such that (writing additively),

$$
s_j|_{U_j \cap U_k} - s_k|_{U_j \cap U_k} = h_{jk} \text{ for } j, k \leq i.
$$

Assume that s_j has been constructed for all $j < i$. Then the sections

$$h_{ij} + s_j|_{U_i \cap U_j}, \ j < i$$

glue (by the cocycle condition) to a section

$$\tilde{s}_i \in \mathcal{F}(U_i \cap (\bigcup_{j<i} U_j)).$$

By $\mathcal{F}$ being flasque, this section is the restriction of a section

$$s_i \in \mathcal{F}(U_i),$$

completing the induction hypothesis, and thus proving (3.2.9).

Returning to the case of $\mathcal{F} = \underline{K(X)}^\times$ for an integral scheme X, considering the long exact sequence (3.2.7) associated with the short exact sequence of sheaves (3.2.8), we have

$$H^1(X, \underline{K(X)}^\times) = 0,$$

so we get an exact sequence

$$0 \longrightarrow \mathcal{O}_X(X)^\times \longrightarrow K(X)^\times \longrightarrow (\underline{K(X)}^\times/\mathcal{O}_X^\times)(X)$$

$$0 \longleftarrow Pic(X),$$

which is another way to express Theorem 3.2.1.

3.3   Weil Divisors

In the example of the Möbius band $\mathcal{L}$, we were able to construct an isomorphism of $\mathcal{L}$ with a coherent sheaf of ideals, thereby establishing a connection between $\mathcal{L}$ and a closed subscheme. This is attractive, since closed subschemes are very geometric. We may ask two questions: (1) Which closed subschemes do we encounter? (2) Does this work for every line bundle?

Even before making precise definitions, note that the answer to 1 appears to be that we encounter closed subschemes of *codimension* 1, since the closed subscheme, at least locally, arises as the 0 locus of a regular function. Recall that in view of the example in Exercise 38 of Chap. 2, "codimension 1" may not be the same thing as "dimension 1 lower than $dim(X)$," and we must therefore use the precise definition of codimension introduced there.

To help answer question 2, let us consider another basic example:

Example Consider the scheme

$$X = \mathbb{P}^n_{Spec(k)} = Proj(k[x_0, \ldots, x_n])$$

for a field k. Then by Proposition 2.4.1, $\mathcal{O}_X(1)$ is a line bundle. In fact, we have an injective morphism of line bundles

$$x_i : \mathcal{O}_X \to \mathcal{O}_X(1)$$

for any $i = 0, \ldots, n$, in fact, generating the vector space of global sections of $\mathcal{O}_X(1)$, which is

$$k[x_0, \ldots, x_n]_1 = k^{n+1}.$$

On the other hand, the vector space of morphisms of line bundles

$$\mathcal{O}_X(1) \to \mathcal{O}_X$$

is isomorphic to the vector space of global sections of $\mathcal{O}_X(-1)$, which is

$$k[x_0, \ldots, x_n]_{-1} = 0.$$

In particular, $\mathcal{O}_X(1)$ is not isomorphic, as a line bundle, to any sheaf of ideals, and therefore cannot correspond to a closed subscheme of X (which would be the zero locus of the ideal). On the other hand, $\mathcal{O}_X(-1)$ is isomorphic to the sheaf of ideals (x_i) for any $i = 0, \ldots, n$, and thus, there is an associated closed subscheme, namely any of the hyperplanes $Z(x_i)$. Denoting the inclusion of the hyperplane by

$$i : \mathbb{P}^{n-1}_k \to \mathbb{P}^n_k,$$

we get, in fact, a short exact sequence of coherent sheaves

$$0 \to \mathcal{O}_{\mathbb{P}^n_k}(-1) \to \mathcal{O}_{\mathbb{P}^n_k} \to i_*\mathcal{O}_{\mathbb{P}^{n-1}_k} \to 0 \tag{3.3.1}$$

This example suggests that to get $\mathcal{O}_X(1)$, we should take "the hyperplane $Z(x_i)$ with coefficient -1" (to correspond to the inverse of $\mathcal{O}_X(-1)$).

Actually, the choice of sign is a matter of convention (although must remain consistent throughout geometry), and as it turns out, the usual choice of signs is the opposite, assigning the $+$ sign to the zero locus of a global section of a line bundle, and the sign $-$ to the subscheme corresponding to an isomorphic sheaf of ideals. So $\mathcal{O}_X(1)$ will be assigned the hyperplane $Z(x_i)$, and $\mathcal{O}_X(-1)$ will be assigned its negative. In any case, this discussion motivates the following definition.

3.3.1 Definition Suppose that X is a Noetherian integral separated scheme such that all of the local rings $\mathcal{O}_{X,x}$ (i.e. stalks of the structure sheaf) which are of dimension 1 are regular. Then a *prime Weil divisor* on X is an integral closed subscheme Y of codimension 1 (i.e. such that the local ring $\mathcal{O}_{X,y}$ at the generic point y of Y has dimension 1, see Exercise 15). A *Weil divisor* on X is an element of the free abelian group whose free generators are prime Weil divisors. Thus, a Weil divisor can be viewed as a formal $\mathbb{Z}$-valued linear combination of prime Weil divisors. A divisor is called *effective* if all the coefficients of the linear combination are non-negative.

COMMENT Note that the condition of Definition 3.3.1 is always satisfied when X is an integral Noetherian separated scheme that is also normal, which implies that all its local rings are normal integral domains. (Recall that a Noetherian normal local ring of dimension 1 is always regular by Lemma 2.4.2, 2 of Chap. 3; see also Exercise 42 of Chap. 1).

3.3.2 Lemma *Let X be a scheme which satisfies the assumptions of Definition 3.3.1, and let $f \in K(X)^\times$ (where $K(X)$ is the field of rational functions on X). Then there exists only finitely many integral closed subschemes $Y \subset X$ of codimension 1 such that*

$$f \notin \mathcal{O}_{X,y}^\times \tag{3.3.2}$$

where y is the generic point of Y.

Proof Since X is Noetherian and therefore quasicompact, it suffices to assume that $X = Spec(R)$ is affine. Then

$$f = \frac{g}{h}$$

where g and h are non-zero regular functions. A point y where we have (3.3.2) must be in

$$Z(g) \cup Z(h). \tag{3.3.3}$$

Since X is Noetherian, (3.3.3) is a union of finitely many irreducible closed sets $Y_1, \ldots, Y_m$, $Y_i = Spec(R/y_i)$ where y_i is a prime ideal. Now suppose $y \in Y_i$. Then $y_i \subseteq y$, which, since $dim(R_y) = 1$, means $y_i = y$ or $y_i = (0)$. The latter is impossible since $g, h \neq 0$. Thus, there are only finitely many choices of y. Note that Y is the closure of y, and hence is determined by y. □

3.3.3 Definition Let $f \in K(X)^{\times}$, and let (3.3.2) where y is a generic point of a closed subscheme of codimension 1. Then $\mathcal{O}_{X,y}$ is a regular local ring of dimension 1, hence a discrete valuation ring with valuation v_y. Define

$$(f) = \sum_y v_y(f) \cdot Y \qquad (3.3.4)$$

which, by Lemma 3.3.2, is a finite sum. We call (f) the *principal Weil divisor* associated with f. Again, two Weil divisors whose difference is a principal Weil divisor are called *linearly equivalent*. Clearly, principal Weil divisors form a subgroup of the (abelian) group of all Weil divisors on X. The factor group is denoted by $Cl(X)$ and called the *(Weil) divisor class group* of X.

COMMENT The theory of Weil divisors can, in fact, be extended beyond assuming that the scheme X is regular in codimension 1, by using concepts of intersection theory, namely the intersection of a Weil and a Cartier divisor. However, we shall not develop this generalization here. (For information on intersection theory, see [7].)

The term "divisor" is motivated by the following fact.

3.3.4 Theorem *A Noetherian integral domain R is a unique factorization domain if and only if it is normal and satisfies*

$$Cl(Spec(R)) = 0.$$

The proof of this theorem relies on some commutative algebra which we developed in Chap. 1. Prime Weil divisors in $Spec(R)$ are closed subschemes of the form R/p where p is a prime ideal of height 1.

3.3.5 Lemma *A Noetherian integral domain R is a unique factorization domain if and only if every height 1 prime ideal in R is principal.*

Proof Suppose R is a unique factorization domain and suppose p is a prime ideal of height 1. Let $0 \neq x \in p$, and let $x = a_1 \cdot \ldots \cdot a_n$ be a factorization into irreducibles. Then there exists an i such that $a_i \in p$, so $(a_i) \subseteq p$. However, (a_i) must be prime by uniqueness of factorization, and since p is height 1, $(a_i) = p$.

Suppose, on the other hand, that every prime ideal p of R of height 1 is principal. In a Noetherian ring, every element has a factorization into irreducibles. For uniqueness of factorization, we need to prove that if x is irreducible, then (x) is prime. Let x

be irreducible, and let p be a minimal prime containing x. Krull's Hauptidealsatz (see Theorem 5.3.7 of Chap. 1 and the subsequent Comment) says that in any Noetherian ring, a minimal prime ideal containing an element which is neither a 0 divisor nor a unit has height 1. Thus, p has height 1. By assumption, then, $p = (y)$ for some element $y \in R$. Therefore, $x \in (y)$, which means that y is a factor of x, and since x is irreducible, $p = (y) = (x)$. □

3.3.6 Lemma *Let R be a normal Noetherian integral domain, and let K be its field of fractions. Then R is equal to the intersection $A \subseteq K$ of all localizations R_p over prime ideals p of height 1.*

Proof Obviously, $R \subseteq A$. Suppose then that $x, a \in R$ and x is not divisible by a. Our job is to show that

$$\frac{x}{a} \notin A. \tag{3.3.5}$$

Consider the ideal

$$I = \{z \in R \mid xz \text{ is a multiple of } a\}.$$

By assumption, $I \neq R$. On the other hand, note that without loss of generality, we may assume that I is prime. If not, we would have a multiple xzt of a where xz and xt are not multiples of a. Thus, we could replace x by xz or xt and repeat this procedure until we get a prime ideal. (The procedure would have to terminate after finitely many steps because R is Noetherian.) Thus, assume I is prime. We will prove that in fact I is of height 1, which proves (3.3.5), since

$$\frac{x}{a} \notin R_I.$$

(Note that if $x/a = u/b$, then $xb = au$, so $b \in I$.) In fact, without loss of generality, we may as well now assume that R is local with maximal ideal I, since we can replace R by the localization R_I. Recall now the definition (2.4.4) of Chap. 3. To prove that I is of height 1, by part 1 of Lemma 2.4.2 of Chap. 3, it suffices to prove that

$$II^{-1} = R. \tag{3.3.6}$$

Of course, again, $II^{-1} \subseteq R$, so if (3.3.6) fails, we must have

$$II^{-1} = I.$$

As in the proof of part 2 of Lemma 2.4.2 of Chap. 3, this implies that $I(I^{-1})^n = I$ and hence all elements of I^{-1} are integral over R, which, since R is normal, implies $I^{-1} = R$.

This is, however, a contradiction with (3.3.5), since

$$\frac{x}{a} \in I^{-1}.$$

$\square$

Proof of Theorem 3.3.4 By Lemma 3.3.5, all we need to prove is that for a normal Noetherian integral domain R,

$$Cl(Spec(R)) = 0 \tag{3.3.7}$$

if and only if every height 1 prime ideal in R is principal. Of course, height 1 prime ideals in R correspond bijectively to prime Weil divisors in $Spec(R)$, so we see right away that if every height 1 prime ideal in R is principal, then (3.3.7) holds.

Suppose, conversely, that (3.3.7) holds. Then for every prime ideal $p \subset R$ of height 1, there exists an element f of the field of fractions $K \supseteq R$ such that

$$v_p(f) = 1, \; v_q(f) = 0 \text{ for other height 1 prime ideals } R.$$

By Lemma 3.3.6, however, we know that $f \in R$, since otherwise we would have $v_q(f) < 0$ for some height 1 prime ideal q. But on the other hand, if $g \in p$ is any element, then

$$v_q(\frac{g}{f}) \geq 0$$

for all height 1 prime ideals q, so again by Lemma 3.3.6,

$$\frac{g}{f} \in R,$$

$\square$

so $p = (f)$ and p is principal, thus proving (3.3.7).

Example Consider, say, on $\mathbb{A}^3_{\mathbb{C}} = Spec(\mathbb{C}[x, y, z])$, the Weil divisor

$$D = 5Z(x^2 + y^2 - 1) - 3Z(x - y - z).$$

Since $\mathbb{C}[x, y, z]$ is a unique factorization domain, we know that any divisor on $\mathbb{A}^3_{\mathbb{C}}$ must be principal. It is, of course, easy to see directly that

$$D = (f)$$

where

$$f = \frac{(x^2 + y^2 - 1)^5}{(x - y - z)^3}.$$

Now assume a scheme X satisfies the assumptions of Definition 3.3.1. Then in particular X is an integral scheme, so

$$Pic(X) \cong Ca\text{-}Cl(X).$$

On the other hand, every Cartier divisor on X determines a Weil divisor: send a global section $f \in \underline{K(X)^\times / \mathcal{O}_X^\times(X)}$ to

$$\sum_Y v_y(f) Y \tag{3.3.8}$$

where the sum is over all codimension 1 integral closed subschemes $Y \subset X$ and y is the generic point of Y. Note that the sum (3.3.8) is well-defined since multiplication by a section of $\mathcal{O}_X^\times$ on an open set containing the given generic point does not affect the valuation. Note also that the sum (3.3.8) is finite because X is Noetherian (see Lemma 3.3.2). Under this correspondence, principal Cartier divisors clearly go to principal Weil divisors, so we obtain an inclusion

$$Ca\text{-}Cl(X) \subseteq Cl(X). \tag{3.3.9}$$

On the other hand, by definition, any Weil divisor Z which comes from a Cartier divisor clearly has the property that every point of X has an open neighborhood U such that the restriction $Z|_U$ (defined by intersecting all codimension 1 integral subschemes Y with U) is principal. A Weil divisor with this property is called *locally principal*. Note that principal Weil divisors are locally principal, so being locally principal is really a property of an element of $Cl(X)$. We can summarize this discussion in the following

3.3.7 Proposition *There is a canonical embedding (3.3.9) whose image coincides with the classes of locally principal Weil divisors.*

$\square$

Example Let $X = Spec(R)$, $R = k[x, y, z]/(z^2 - xy)$ where k is a field of characteristic not equal to 2 (a quadratic cone). Then X is normal. To show this, we need to show that R is integrally closed in its field of fractions K. Because of the equation for z^2, as an $k[x, y]$-module, R is free with basis $1, z$ and similarly K is a free $k(x, y)$-module with basis $1, z$. Therefore, every element of R is integral over $k[x, y]$. Now if an element $t = u + zv$,

$u, v \in k(x, y)$, is integral over R, it is therefore also integral over $k[x, y]$. If $t \in k(x, y)$, we are done, since $k[x, y]$ is normal. Otherwise, over $k(x, y)$, t is a zero of the polynomial

$$(t - u)^2 - xyv^2 = t^2 - 2tu + (u^2 - xyv^2), \tag{3.3.10}$$

and hence this must be the minimal polynomial of t over $k(x, y)$. Hence, since t is integral over $k[x, y]$, (3.3.10) must be a factor of a polynomial in t with coefficients in $k[x, y]$, and hence by the Gauss lemma, its coefficients must be in $k[x, y]$. Thus, $u \in k[x, y]$, and also

$$u^2 - xyv^2 \in k[x, y],$$

which implies $xyv^2 \in k[x, y]$. Since x, y are relatively prime irreducibles in the unique factorization domain $k[x, y]$, we also have $v \in k[x, y]$ and hence $t \in R$.

Then $Z(y, z) \cong Spec(k[x])$ is the line $y = z = 0$ on the cone, hence an integral subscheme of codimension 1. (Note that in the local ring $R_{(y,z)}$, x is invertible, so the maximal ideal $(y, z) \subset R_{(y,z)}$ is principal, generated by z, and hence of height 1.) Thus, $Z(y, z)$ defines a Weil divisor on X. However, note that (y, z) is not a principal ideal at $R_{(x,y,z)}$, since x, y, z are K-linearly independent in maximal ideal m/m^2 (since the polynomial $z^2 - xy$ is homogeneous of degree 2). Therefore, $(y, z)/m^2$ has dimension 2 over k, and hence (y, z) is not a principal ideal in $R_{(x,y,z)}$, and hence cannot be locally principal. This is an example of a Weil divisor which is not a Cartier divisor.

We see immediately that this is essentially the only difficulty that can happen: if X is *locally factorial*, which means that X is integral, Noetherian, separated and all the local rings $\mathcal{O}_{X,x}$ are unique factorization domains (note that then automatically X is normal), then every Weil divisor is principal in every local ring $\mathcal{O}_{X,x}$, and hence in some open neighborhood of x, and hence it is locally principal. Thus, we have the following

3.3.8 Proposition *For a locally factorial scheme, the inclusion (3.3.9) is an isomorphism. This happens in particular if X is Noetherian regular separated, since all regular local rings are unique factorization domains (See Exercise 18.)*

□

Thus, for Noetherian regular integral separated schemes, the divisor class group and the Picard group are isomorphic (Exercise 20).

3.3.9   Example: Divisors on Dedekind Domains

Suppose R is a Dedekind domain. Denote by $\Gamma(R)$ the set of nonzero ideals on R. This is a commutative monoid with respect to the product of ideals. We have a homomorphism of commutative monoids

$$\Gamma(R) \to Cl(Spec(R)) \tag{3.3.11}$$

by sending

$$p_1^{i_1} \cdot \cdots \cdot p_n^{i_n} \mapsto i_1 Z(p_1) + \cdots + i_n Z(p_n).$$

(Recall Theorem 2.3.2. Note that the operation is written multiplicatively in the source and additively in the target.) We will show that (3.3.11) is, in fact, onto. This is the same as showing that every Weil divisor on $Spec(R)$ is linearly equivalent to an effective divisor. It follows from the fact that the Weil divisor $-Z(p)$ for a maximal ideal $p \subsetneq R$ is linearly equivalent to $(r) - Z(p)$ for any element $r \in p$, which is effective.

Thus, (3.3.11) is onto. Additionally, (3.3.11) clearly sends principal ideals to 0 (since they are sent to the corresponding principal divisor). In fact, we see that if two ideals I, J are sent to the same divisor class, there exist elements $r, s \in R \setminus \{\emptyset\}$ such that

$$I \cdot (s) = J \cdot (r). \tag{3.3.12}$$

Thus, if we denote the equivalence relation (3.3.12) by $\sim$ and call it *equivalence of ideals*, then the set $Cl(R)$ of equivalence classes $\Gamma(R)/\sim$ is, by (3.3.11) bijective to

$$Cl(Spec(R)) \cong Pic(R).$$

The bijection with $Cl(Spec(R))$ gives $Cl(R)$ a structure of an abelian group, which is called the *ideal class group* of R. The operation is, of course, still given by multiplication of ideals. By Theorem 3.3.4, a Dedekind domain R is a unique factorization domain if and only if $Cl(R) = 0$, which happens if and only if R is a principal ideal domain (which means an integral domain in which every ideal is principal).

In number theory, a *number field* is a finite extension K of $\mathbb{Q}$. The integral closure of $\mathbb{Z}$ in K is then a Dedekind domain, which is denoted (somewhat conflictingly) by $\mathcal{O}_K$. The ideal class groups $Cl(\mathcal{O}_K)$ is not known in general, and is of major interest in number theory. A major known result is that $Cl(\mathcal{O}_K)$ is always finite (using the techniques built up in the Exercises of Chap. 3, we are able to prove this, see Exercise 25). However, the ideal class group of a general Dedekind domain can be any abelian group. (We will see examples in Sect. 3.4 below where it is not finitely generated.)

Example Take for example *quadratic fields* which are of the form $\mathbb{Q}(\sqrt{D})$ where D is a *square free integer*, which means that it is not divisible by any n^2 for an integer $n > 1$. One has

$$\mathbb{Q}(\sqrt{D}) = \{a + b\sqrt{D} \mid a, b \in \mathbb{Q}\}.$$

(We have

$$\frac{a+b\sqrt{d}}{c+d\sqrt{D}} = \frac{(a+b\sqrt{D})(c-d\sqrt{D})}{c^2 - Dd^2}$$

where the denominator is non-zero because D is not a square.) To calculate $\mathcal{O}_{\mathbb{Q}(\sqrt{D})}$, note that for $b \neq 0$, the minimal polynomial of $u = a + b\sqrt{D}$, $a, b \in \mathbb{Q}$, is

$$(u-a)^2 - b^2 D = u^2 - 2au + (a^2 - b^2 D),$$

so if u is integral over $\mathbb{Z}$, then by Gauss' lemma,

$$2a \in \mathbb{Z}, \ a^2 - b^2 D \in \mathbb{Z}. \tag{3.3.13}$$

If D is not congruent to 1 modulo 4, then we have to have $a \in \mathbb{Z}$, and it then follows that $b \in \mathbb{Z}$ because D is square free. If $D \equiv 1 \mod 4$, then we can have either $a, b \in \mathbb{Z}$ or $a, b \in \frac{1}{2} + \mathbb{Z}$. Thus,

$$\mathcal{O}_{\mathbb{Q}(\sqrt{D})} = \begin{cases} \mathbb{Z}[\dfrac{1+\sqrt{D}}{2}] & \text{if } D \equiv 1 \mod 4 \\ \mathbb{Z}[\sqrt{D}] & \text{otherwise.} \end{cases}$$

Even for quadratic fields $\mathbb{Q}(\sqrt{D})$, the ideal class group $Cl(\mathcal{O}_{\mathbb{Q}(\sqrt{D})})$ (or even its order) is not known in general. For $D < 0$ square free, it is known that $Cl(\mathcal{O}_{\mathbb{Q}(\sqrt{D})}) = 0$ (i.e. $\mathcal{O}_{\mathbb{Q}(\sqrt{D})}$ is a unique factorization domain, or equivalently prime ideal domain) for precisely 9 different values of D: $-1, -2, -3, -7, -11, -19, -43, -67, -163$. It is not known if there are infinitely many $D > 0$ square free for which

$$Cl(\mathcal{O}_{\mathbb{Q}(\sqrt{D})}) = 0,$$

although there appears to be a positive proportion of them.

Example Consider again the Dedekind domain

$$R = \mathbb{R}[x, y]/(x^2 + y^2 - 1).$$

We know that the Weil divisor on $Spec(R)$ corresponding to the Möbius band cannot be principal (and hence, R cannot be a unique factorization domain). From the point of view of the ideal class group, we already saw that the Möbius band, as an invertible sheaf, is isomorphic to the non-principal ideal $(x, 1 + y)$. Let us see it in another way using the passage from Cartier to Weil divisors. We know that the Cartier divisor is x on $U_{(1-y)}$ and

1 on $U_{(1+y)}$. Therefore, the Weil divisor will be some coefficient times $Z(x, 1 + y)$. (We also know that any other point of the circle with residue field $\mathbb{R}$ could be chosen instead.) To calculate the coefficient from the point of view of valuations, we must work in the local ring

$$R_{(x,1+y)}. \tag{3.3.14}$$

The key observation is that in (3.3.14), $1 - y$ is invertible, so the maximal ideal is generated by x (since, again, $(1 + y) = x^2/(1 - y)$). Thus, x has valuation 1 and therefore the Weil divisor of the Möbius band is $1 \cdot Z(x, 1 + y)$.

3.4   Examples and Calculations

In this section, we will use our techniques to give examples of calculations of divisor class groups and Picard groups.

Example Let k be a field, $n > 0$. Consider $X = \mathbb{P}^n_k = Proj(R)$, $R = k[x_0, \dots, x_n]$, which is a regular scheme. The Weil divisor associated with $\mathcal{O}_X(1)$ is any hyperplane, say, $H = Z(x_0)$, since $\mathcal{O}_X(1)$ can be represented as a Cartier divisor by its global section $s = x_0$, which has $v_H(s) = 1$, and $v_Y(s) = 0$ for any other integral subscheme $Y \neq H$ of codimension 1. Now any codimension 1 integral subscheme of X corresponds to a height 1 homogeneous prime ideal I of R (by Proposition 2.4.5 part 1), which is principal. Let $I = (g)$ where g is a homogeneous irreducible polynomial of degree d. We see that the prime Weil divisor $Z(g)$ is linearly equivalent to $d(H)$ by multiplying by the rational function x_0^d/g. This defines a homomorphism of abelian groups

$$\mathbb{Z} \to Pic(\mathbb{P}^n_k)$$

which is onto. We can show that it is also injective by noting that a Weil divisor

$$a_1 Z(g_1) + \cdots + a_m Z(g_m) \tag{3.4.1}$$

where g_i are irreducible homogeneous polynomials of degrees d_i can only be principal if

$$a_1 d_1 + \cdots + a_n d_n = 0. \tag{3.4.2}$$

This is because if (3.4.1) is the principal divisor associated with a rational function f on $\mathbb{P}^n_k$ then we have $f = g/h$ where g, h are homogeneous polynomials of equal degrees, so if we factor $g = g_1^{k_1} \dots g_m^{k_m}, h = h_1^{\ell_1} \dots h_p^{\ell_p}$, where g_i are irreducible of degrees d_i and h_i

are irreducible of degree e_i, we have

$$k_1 d_1 + \cdots + k_m d_m - \ell_1 e_1 - \cdots - \ell_p e_p = 0,$$

which is (3.4.2). Thus, we used Weil divisors to prove that

$$Pic(\mathbb{P}_k^n) \cong \mathbb{Z}.$$

3.4.1 Proposition *Suppose X is a Noetherian integral separated scheme whose all local rings of dimension 1 are regular. Let $Z \subsetneq X$ be a closed subset. Then intersecting with $U = X \smallsetminus Z$ defines an onto homomorphism*

$$\phi : Cl(X) \to Cl(U).$$

Further, this is an isomorphism if Z is of codimension ≥ 2 (i.e. if all generic points of irreducible components of Z are of height ≥ 2). If Z is irreducible of codimension 1, we have an exact sequence

$$\mathbb{Z} \xrightarrow{\iota} Cl(X) \xrightarrow{\phi} Cl(U) \longrightarrow 0 \tag{3.4.3}$$

where $\iota(1) = Z$.

Proof The fact that ϕ is onto follows from the fact that the closure of a prime divisor in U is a prime divisor in X. If Z is of codimension ≥ 2, the height 1 generic points (which correspond to prime Weil divisors) are the same in X and U. Also, rational functions on U are the same as on X. Thus, ϕ is an isomorphism. If Z is irreducible of codimension 1, then it contains a unique height 1 point, namely the generic point of Z. Thus, the free abelian group on prime Weil divisors of X has the same generators as the group of prime Weil divisors on U, plus the additional generator Z. Noting again that the rational functions on U and X are the same, and by factoring them out, we obtain exactly (3.4.3). (Note that ι does not have to be injective because some multiple of Z can be a principal divisor.) $\square$

3.4.2 Proposition *Suppose X is a Noetherian integral separated scheme whose every local ring of height 1 is regular. Then the same holds for $\mathbb{A}_X^1 = X \times_{Spec(\mathbb{Z})} Spec(\mathbb{Z}[t])$, and we have a canonical isomorphism*

$$\phi : Cl(X) \xrightarrow{\cong} Cl(\mathbb{A}_X^1). \tag{3.4.4}$$

Proof Clearly, the condition we are studying is local, so to verify the condition, it suffices to assume that $X = Spec(R)$ is affine. The fact that if R is an integral domain, so is $R[t]$, is trivial. If p is a prime ideal of $R[t]$ of height 1, then $q = p \cap R$ has height ≤ 1 in R, so R_q is regular, therefore so is $R_q[t]$. Thus, $R[t]_p = (R_q[t])_p$ is regular (See Corollary 2.6.8 of Chap. 5.)

Returning to the case of a general scheme X satisfying our hypotheses, the isomorphism ϕ of (3.4.4) is constructed by sending a prime Weil divisor $Y \subset X$ to $\pi^{-1}(Y)$ where $\pi : \mathbb{A}_X^1 \to X$ is the projection. Clearly, principal divisors are sent to principal divisors (for a rational function f on X, the principal divisor (f) is sent to (f), considered as a rational function on $\mathbb{A}_X^1$). In fact, we can show in the same way that ϕ is injective: Denote the field of rational functions on X by K. Suppose $f \in K(t)$ such that

$$(f) \in Im(\phi). \tag{3.4.5}$$

If $f \notin K$, then factor $f = g/h$ where $g, h \in K[t]$ are relatively prime. Then g, h are not both constant, so at least one of them has a zero in the subset $\mathbb{A}_{Spec(K)}^1$ of $\mathbb{A}_X^1$, which means that (f) has a non-zero coefficient at a height 1 point which projects to the generic point of X, which contradicts the assumption (3.4.5). Thus, $f \in K$, and hence $(f) = \phi((f))$. Thus, ϕ is injective.

To prove that ϕ is onto, consider an arbitrary divisor D on $\mathbb{A}_X^1$, and consider the sum D_0 of all its terms which are divisors on $\mathbb{A}_{Spec(K)}^1$. Since $K(t)$ is a principal ideal domain, there exists a rational function $f \in K(t)$ such that $(f) = D_0$ in $\mathbb{A}_{Spec(K)}^1$. Therefore $D - (f)$ is in the image of ϕ in $\mathbb{A}_X^1$. $\qquad\square$

Remark This statement (and proof) generalize to prove that the projection $p_{\mathcal{M}}$ induces an isomorphism

$$Cl(U_{\mathcal{M}}) \cong Cl(X) \tag{3.4.6}$$

for every locally free sheaf of modules $\mathcal{M}$ on X. (See (2.1.5) of Sect. 2.1 and Exercise 28.)

Example We can now finally put to rest the Möbius band example by calculating completely $Pic(X)$ where

$$X = Spec(R), \quad R = \mathbb{R}[x, y]/(x^2 + y^2 - 1).$$

Since X is regular (it is smooth over $Spec(\mathbb{R})$), we can use Weil divisors. This can be done pretty much directly, but there is a nice way using Proposition 3.4.1: Let $Z = Z(x, y - 1)$. Then, as remarked before,

$$X \smallsetminus Z = U_{(1-y)} \cong Spec((1 - y)^{-1}R).$$

Now we have an isomorphism of rings

$$(1 - y)^{-1} R = (1 - y)^{-1} \mathbb{R}[x, y]/(x^2 + y^2 - 1) \cong (1 + t^2)^{-1} \mathbb{R}[t] \qquad (3.4.7)$$

given by the Euler parametrization of the circle:

$$t = \frac{y + 1}{x}, \ \ (\text{note that } t^2 + 1 = \frac{2}{1 - y})$$

and conversely

$$x = \frac{2t}{t^2 + 1}, \ y = \frac{t^2 - 1}{t^2 + 1}, \ \ (\text{note that } 1 - y = \frac{2}{t^2 + 1}).$$

We have $Cl(\mathbb{A}^1_{\mathbb{R}}) = 0$ by Proposition 3.4.2, so by Proposition 3.4.1 (applied to $Z = Z(t^2 + 1)$), we have $Cl(U_{(1-y)}) = 0$. Thus, by Proposition 3.4.1 again now applied to our X and $Z = Z(x, 1 - y)$, We have an onto homomorphism

$$\mathbb{Z} \to Cl(X) = Pic(X),$$

where $1 \in \mathbb{Z}$ maps to $(x, 1 - y)$. But we already saw that this element of $Pic(X)$ is non-trivial, represented by the Möbius band. We also saw that the second power of the Möbius band is $0 \in Pic(X)$, so we have proved that

$$Pic(X) \cong \mathbb{Z}/2.$$

Example Let $X = Spec(R)$ where

$$R = \mathbb{R}[x, y, z]/(x^2 + y^2 + z^2 - 1).$$

This ring has dimension 2, so it is not a Dedekind domain. It is, nevertheless, smooth over $\mathbb{R}$ and hence regular, and it is a rational variety, i.e. its field of rational functions is isomorphic to the field of rational functions on an affine space. The rational isomorphism is given by the *stereographic projection*, i.e. projecting points on the sphere $x^2 + y^2 + z^2 = 1$ onto the plane $z = 0$ using rays originating in the point $(0, 0, 1)$. More specifically, the stereographic projection is an isomorphism

$$(1 - z)^{-1} R \cong (u^2 + v^2 + 1)^{-1} \mathbb{R}[u, v] \qquad (3.4.8)$$

given by

$$u = \frac{2x}{1 - z}, \ v = \frac{2y}{1 - z}, \qquad (3.4.9)$$

using the similarity between the triangle with vertices

$$(0, 0, 1), \quad (0, 0, -1), \quad (u, v, -1)$$

and the triangle with vertices

$$(x, y, z), \quad (x, y, -1), \quad (u, v, -1).$$

The inverse of the homomorphism (3.4.9) is

$$x = \frac{2u}{u^2 + v^2 + 1}, \quad y = \frac{2v}{u^2 + v^2 + 1}, \quad z = \frac{u^2 + v^2 - 1}{u^2 + v^2 + 1} \tag{3.4.10}$$

(discovered by computing $u^2 + v^2$ from (3.4.9), using $x^2 + y^2 + z^2 = 1$ and canceling $1 - z$ in the numerator and denominator, thus obtaining a fractional linear equation for z). In any case, set $U = X \smallsetminus Z$ where

$$Z = Z(1 - z), \tag{3.4.11}$$

and observe that Z is an integral closed subscheme, since it is isomorphic to

$$Spec(\mathbb{R}[x, y]/(x^2 + y^2)),$$

and $x^2 + y^2$ is an irreducible polynomial over $\mathbb{R}$. Now we have $Cl(U) = 0$ by (3.4.8), and we can use the exact sequence (3.4.3) to construct an onto homomorphism

$$\mathbb{Z} \to Cl(X) \to 0. \tag{3.4.12}$$

The generator goes again to the divisor $1 \cdot Z$. This time, however, observe that this divisor is principal by (3.4.11) (since $(1 - z)$ is a principal ideal). Thus, the generator of $\mathbb{Z}$ in (3.4.12) goes to 0, and we conclude that $Cl(X) = 0$, and hence R is a unique factorization domain.

COMMENT Note that the stereographic projection works in any dimension. (The Euler parametrization is the stereographic projection in dimension 1.) One can use this to prove that for any $n > 1$,

$$Cl(Spec(\mathbb{R}[x_1, \ldots, x_n]/(x_1^2 + \cdots + x_n^2 - 1))) = 0. \tag{3.4.13}$$

(Exercise 29.)

Example (The Hyperboloid in One Sheet) Let k be a field of characteristic $\neq 2$. Consider the smooth affine scheme $X = Spec(R)$ over $Spec(k)$ given by

$$R = k[x, y, z]/(x^2 + y^2 - z^2 - 1). \qquad (3.4.14)$$

Let $Z = Z(1 - y) \subseteq X$. In R, we have

$$x^2 - z^2 = 1 - y^2,$$

so

$$(x - z)^{-1} R \cong v^{-1} k[y, v]$$

where $v = x - z$. Thus, putting $U = X \smallsetminus Z$, we have

$$Cl(U) = 0.$$

However, Z is not irreducible. Rather, we see immediately that its irreducible components are

$$Z_1 = Z(1 + y, x - z), \quad Z_2 = Z(1 - y, x - z).$$

We cannot therefore apply Proposition 3.4.1 as stated, but the same argument gives an onto homomorphism

$$\mathbb{Z} \oplus \mathbb{Z} \to Cl(X) \to 0 \qquad (3.4.15)$$

where

$$(a, b) \mapsto a \cdot Z_1 + b \cdot Z_2.$$

Observe that we can calculate the kernel of (3.4.15) as well: If

$$a \cdot Z_1 + b \cdot Z_2 = (f)$$

for a rational function f, then f must be an invertible regular function on U. But we saw that $U \cong Spec(v^{-1}k[y, v])$, and we have

$$(v^{-1}k[y, v])^{\times} \cong \mathbb{Z} \oplus k^{\times}$$

where the $k^\times$ comes from constant functions and the generator of the $\mathbb{Z}$ is $v = x - z$, and we have

$$(x - z) = Z_1 + Z_2.$$

Therefore, we conclude that

$$Pic(X) = Cl(X) \cong \mathbb{Z}.$$

Note that for $k \cong \mathbb{R}$, the circle is isomorphic to the closed subscheme $Z(z)$ of X, and the generator of $Pic(X)$ pulls back to the Möbius band. Note also that for $k = \mathbb{C}$, the hyperboloid in one sheet is isomorphic to

$$Spec(\mathbb{C}[x, y, z]/(x^2 + y^2 + z^2 - 1)).$$

In particular, in the previous example, if we replaced the field $\mathbb{R}$ with the field $\mathbb{C}$, the Picard group would be $\mathbb{Z}$, not 0.

Remark Note that for a Noetherian separated integral scheme X whose local rings $\mathcal{O}_{X,x}$ of dimension 1 are regular, and a reduced closed subscheme Z of codimension 1, if we denote by S the set of irreducible components of Z of codimension 1 in X, and by $\mathbb{Z}(S)$ the free abelian group on S, the argument of the previous example gives an exact sequence

$$0 \to \mathcal{O}_X^\times \to \mathcal{O}_U^\times \to \mathbb{Z}(S) \to Cl(X) \to Cl(U) \to 0, \tag{3.4.16}$$

which is a generalization of Proposition 3.4.1.

It is tempting to try to compare (3.4.16) with the long exact sequence in cohomology associated with a short exact sequence of sheaves (see the end of Sect. 3.2). But it is not as straightforward as one may think. If we denote by $j : U \subseteq X$ the inclusion, we have a short exact sequence of sheaves

$$0 \to \mathcal{O}_X^\times \to j_*\mathcal{O}_U^\times \to \mathcal{Z} \to 0 \tag{3.4.17}$$

where $\mathcal{Z}$ is the sheaf of locally principal Weil divisors with non-zero coefficients only on irreducible components of Z. (We speak of divisors with *support* on Z.) The long exact sequence (3.2.7) in cohomology associated with (3.4.17) is

$$0 \longrightarrow \mathcal{O}_X^\times \longrightarrow \mathcal{O}_U^\times \longrightarrow \mathcal{Z}(X) \longrightarrow Pic(X)$$

$$\downarrow$$

$$H^1(X, j_*\mathcal{O}_U^\times) \qquad\qquad (3.4.18)$$

$$\downarrow$$

$$H^1(X, \mathcal{Z}).$$

It maps into the long exact sequence (3.4.16) (by which we mean there are morphisms on corresponding terms which form commutative diagrams with the morphisms in the sequences). In effect, the terms before $\mathcal{Z}(X)$ coincide, $\mathcal{Z}(X)$ is clearly a subgroup of the group of all Weil divisors on X with support in Z, and $Pic(X) \subseteq Cl(X)$ similarly. There are mappings

$$Pic(X) \to H^1(X, j_*\mathcal{O}_U^\times) \to Pic(U) \qquad\qquad (3.4.19)$$

given by restriction of sheaves. Hence, the composition is given by restriction of line bundles. This means that there is a canonical map of the long exact sequence (3.4.18) to the long exact sequence (3.4.16). However, either of the maps (3.4.19) may fail to be onto, as we shall see in the following two examples.

Example Let us revisit the cone $X = Spec(R)$,

$$R = k[x, y, z]/(z^2 - xy)$$

where k is a field of characteristic $\neq 2$. We saw that the Weil divisor $Z(y, z)$ on X is not locally principal, and hence does not come from a Cartier divisor. On the other hand, the principal divisor (y) is equal to $2 \cdot Z(y, z)$, and we have

$$X \smallsetminus Z(y, z) = U_{(y)} \cong Spec(y^{-1}k[y, z]),$$

so by Proposition 3.4.1,

$$Cl(X) \cong \mathbb{Z}/2.$$

On the other hand, Now consider the scheme

$$V = X \smallsetminus Z(x, y, z).$$

Then V is a regular scheme, so by Proposition 3.4.1 (applied to $Z(x, y, z) \subset X$), we have

$$Pic(V) = Cl(V) = Cl(X) \cong \mathbb{Z}/2.$$

In this case, however, we can show that

$$H^1(X, \mathcal{Z}) = 0. \tag{3.4.20}$$

Therefore, the first map (3.4.19) is onto by the long exact sequence (3.4.18), and hence the second map (3.4.19) is not onto (since the composition is not onto). Of course, we already know that $Pic(X) = 0$ (since it is a subgroup of $Cl(X)$ and is not equal to it). To show (3.4.20), consider the short exact sequence

$$0 \to \mathcal{Z} \to i_*\underline{\mathbb{Z}}_{Z(y,z)} \to \mathcal{F} \to 0 \tag{3.4.21}$$

where $i : Z(y, z) \to X$ is the closed immersion (and recall that $\underline{\mathbb{Z}}$ denotes the constant sheaf). Then from the above discussion it follows that

$$\mathcal{F} = \iota_*\underline{\mathbb{Z}}_{Z(x,y,z)}$$

where $\iota : Z(x, y, z) \to X$ is again the closed immersion. Thus, the map in 0'th cohomology induced by the second morphism (3.4.21) is $Id : \mathbb{Z} \to \mathbb{Z}$, while $i_*\underline{\mathbb{Z}}_{Z(y,z)}$ is flasque, so (3.4.20) is implied by the long exact sequence in cohomology associated with the short exact sequence (3.4.21).

Example Consider $X = Spec(R)$,

$$R = \mathbb{C}[x, y, z]/(z^2(1 - z)^2 + x^2 - y^2). \tag{3.4.22}$$

One shows that R is normal by the same method as for the ring in the previous example (which was treated in the previous subsection), by showing that it contains the integral closure of $\mathbb{C}(y, z)$ in $\mathbb{C}(y, z)[x]/(z^2(1 - z)^2 + x^2 - y^2)$. Let

$$Z = Z(z(1 - z) - y, x), \quad U = X \smallsetminus Z$$

and let $j : U \to X$ be the open immersion. It is easy to compute $Cl(X)$ by the same method as we used in the previous example (Exercise 30), but that is not what we are after. Instead, we will focus on the sheaf $\mathcal{Z}$. Denoting again by $i : Z \to X$ the closed immersion, this time, one has a short exact sequence

$$0 \to \mathcal{Z} \to i_*\underline{\mathbb{Z}}_Z \to (i_0)_*\underline{\mathbb{Z}}_{Z(x,y,z)} \oplus (i_1)_*\underline{\mathbb{Z}}_{Z(x,y,z-1)} \to 0 \tag{3.4.23}$$

where $i_0 : Z(x, y, z) \to X, i_1 : Z(x, y, z - 1) \to X$ again are the closed immersions. Therefore, there is a non-zero element

$$\alpha \in H^1(X, \mathcal{Z})$$

represented by a Čech cocycle defined on the open cover

$$U_{(z)}, U_{(z-1)}$$

of X and the locally principal divisor

$$Z_{01} = Z(Z \cap U_{(z)} \cap U_{(z-1)})$$

on $U_{(z)} \cap U_{(z-1)}$. However,

$$U_{(z)} \cap U_{(z-1)} = Spec(z^{-1}(z - 1)^{-1}\mathbb{C}[x, y, z]/(z^2(z - 1)^2 + x^2 - y^2))$$
$$Spec(z^{-1}(z - 1)^{-1}u^{-1}\mathbb{C}[z, u])$$

(by substituting $u = x - y$, so $Cl(U_{(z)} \cap U_{(z-1)}) = 0$ and hence the divisor $Z_{01} = (f)$ is principal. The function f then lifts the element α to $H^1(X, j_*\mathcal{O}_U^\times)$, thus showing that the first map (3.4.19) is not onto in this case.

Example Let X be a complete smooth curve over an algebraically closed field k. Define the *degree* of a Weil divisor $\sum n_p Z(p)$ to be $\sum n_p$. The degree of a principal divisor is always 0: Consider a rational function $f \in K(X)^\times$. If $f \in k^\times$, the divisor is 0 and there is nothing to prove. Otherwise, we can think of f as a dominant morphism $f : X \to \mathbb{P}^1$. Then by Theorem 4.2.6 of Chap. 3, considering the points $0, \infty \in \mathbb{P}^1_k$,

$$deg((f)) =$$

$$\sum_{q \in f^{-1}(0)} v_q(t) - \sum_{q \in f^{-1}(\infty)} v_q(t) = deg(f) - deg(f) = 0.$$

Thus, we obtain an onto homomorphism

$$Cl(X) \xrightarrow{\ deg\ } \mathbb{Z} \longrightarrow 0. \tag{3.4.24}$$

We already know that it is an isomorphism for $X = \mathbb{P}^1_k$ by the first example of this subsection. We will next see an example where it is not an isomorphism. The kernel of

(3.4.24) is usually denoted by $Pic_0(X)$. Thus, we will next give an example of a complete smooth curve X for which $Pic_0(X) \neq 0$.

Example Let E be an elliptic curve, i.e. the complete smooth curve over an algebraically closed field of characteristic 0 rationally equivalent to (4.2.4) in the example in Sect. 4.2 of Chap. 3. (Again, our arguments work equally well for (4.2.5).) In other words,

$$E = Proj(R), \quad R = k[x, y, z]/(x^2z - y^3 - yz^2).$$

By the previous example, degree gives an onto homomorphism (3.4.24) for $X = E$, but now we will compute the kernel of deg.

We saw in the Example in Sect. 4.2 of Chap. 3 that E is not rational. This means that for two closed points $p \neq q \in E$,

$$Z(p) - Z(q) \neq 0 \in Cl(E) = Pic(E). \tag{3.4.25}$$

Indeed, if the Weil divisor $Z(p) - Z(q)$ were principal, this would give a degree 1 morphism $E \to \mathbb{P}^1_k$, which by definition would mean that E is rational. Thus, choosing a point q_0 (usually, one chooses $q_0 = [0:0:1]$),

$$p \mapsto Z(p) - Z(q_0)$$

gives an injective map

$$E_m \to Pic_0(E). \tag{3.4.26}$$

(Recall that E_m is the set of closed points of E.) We will see that (3.4.26) is a bijection, thus giving E_m a structure of a group!

So all we need to show is that (3.4.26) is onto. Consider any line

$$ax + by + cz = 0 \tag{3.4.27}$$

in $\mathbb{P}^2_k$, and the associated rational function

$$\frac{ax + by + cz}{z} \tag{3.4.28}$$

on E. We will assume that a and b are not both 0, so we can think of (3.4.28) as a dominant morphism $E \to \mathbb{P}^1_k$. If the point q_0 is not on the line (3.4.27) then (3.4.28) is of degree 3. In this case, (3.4.27) has three zeros on E counting multiplicities, and Theorem 4.2.6 of

Chap. 3 tells us that the principal divisor of (3.4.28) is of the form

$$Z(p_1) + Z(p_2) + Z(p_3) - 3Z(q_0) \tag{3.4.29}$$

where p_1, p_2, p_3 are three (not necessarily distinct) points of E, all distinct from q_0. If the line (3.4.27) contains q_0 but is not equal to the line $z = 0$, then (3.4.28) is of degree 2 and the principal divisor is of the form

$$Z(q_1) + Z(q_2) - 2Z(q_0). \tag{3.4.30}$$

Note that we can choose q_1 in (3.4.30). Thus, choosing $q_1 = p_3$ and subtracting from (3.4.29), we get

$$Z(p_1) + Z(p_2) - Z(q_2) - Z(q_0). \tag{3.4.31}$$

We can add integral multiples of (3.4.29) and (3.4.30) to show that any Weil divisor on E of degree 0 is equivalent to $Z(p) - Z(q_0)$ for some point $p \in E$, which proves our claim. To see this, note that if a divisor D has non-zero coefficients at two different points $\neq q_0$, we can make all such coefficients positive by adding multiples of the form (3.4.30), picking lines (3.4.27) through points with negative coefficients and q_0. If all of the coefficients at points other than q_0 are positive, we can subtract positive multiples of (3.4.31) where p_1 and p_2 are two of the points (note that they do not have to be distinct, if we want them to be the same, pick a line tangent to E) to reduce the sum of coefficients. Repeat the reduction until it is 1.

Note that thinking of the point q_0 as 0, (3.4.31) can be thought of as a formula for a group operation on E_m itself, simply defining q_3 the sum of p_1 and p_2. Similarly, (3.4.30) makes q_2 the additive inverse of q_1. It is worth noting that if E is defined over a field k which is not necessarily algebraically closed, one can show that the formulas given by (3.4.30) and (3.4.31) produce a group law on E itself, not just its pullback to $Spec(\bar{k})$. More precisely, one proves that the composition $E \times E \to E$ unit $Spec(k) \to E$ and inverse $E \to E$ are morphisms of schemes, thus making an *group scheme*. A group scheme which is also a variety is called an *algebraic group*. Thus, an elliptic curve is an algebraic group projective over $Spec(k)$.

Example The *cusp curve* E is the closure in $\mathbb{P}^2_k$ (assume still k algebraically closed) of

$$Spec(R), \quad R = k[x, y]/(x^2 - y^3).$$

It can be also written as

$$Proj\,(k[x, y, z]/(x^2 z - y^3)).$$

One sees easily that the $\mathcal{O}_{E,q_\infty} = R_{q_\infty}$ is not normal for

$$q_\infty = [0:0:1].$$

Therefore, $Cl(E)$ is not defined, but E is an integral scheme, so we can use Cartier divisors to work out $Pic(E)$. One checks that $U = E \setminus Z(q_\infty)$ is smooth over k, in fact isomorphic to $\mathbb{A}_k^1$. To see this, note that $Z(q_\infty)$ is the reduced closed subscheme associated with the intersection of E with $Z(x)$ in $\mathbb{P}_k^2$. Thus, U is simply the intersection of E with $U_{(x)} \subset \mathbb{P}_k^2$, which is

$$Spec(S), \quad S = k[y, z]/(z - y^3).$$

Clearly, this scheme is isomorphic to $\mathbb{A}_k^1$.

Now every Cartier divisor is linearly equivalent to a Cartier divisor D which is 1 in an open neighborhood of any chosen point q. (If in a neighborhood of q, the divisor is equal to $f \in K(X)$, simply divide it by f globally.) Choosing $q = q_\infty$, we see that our Cartier divisor D which is invertible in an open neighborhood of q_∞ then determines and is determined by a Weil divisor on U. However, it does not follow that two linearly equivalent divisors on U would necessarily come from linearly equivalent Cartier divisors D on E. In fact, linear equivalence on U is sort of irrelevant (recall that $Pic(\mathbb{A}_k^1) = 0$), and we have to work out the E-linear equivalence of Weil divisors on U (which we will denote by $\sim_E$) from other facts.

First, let us discuss degree. Note that since E is not a smooth curve over k, degree of Cartier divisors on E has not been defined. However, note that we have a morphism

$$g : \mathbb{P}^1 \to E$$

defined by extending the composition $\mathbb{A}_k^1 \cong U \subset E$ to $\mathbb{P}^1$ by Lemma 4.2.2 of Chap. 3. Then defining the E-degree of a Weil divisor $\sum n_i Z(p_i)$ on U as $\sum n_i$, the degree of an E-principal divisor (f) we have

$$deg((f)) = \sum_{q \in (f \circ g)^{-1}(0)} v_q(t) - \sum_{q \in (f \circ g)^{-1}(\infty)} v_q(t) =$$
$$deg(f \circ g) - deg(f \circ g) = 0.$$

Next, note that for points $p \neq q \in U_m$,

$$Z(p) \sim_E Z(q). \tag{3.4.32}$$

To prove (3.4.32), assume that $Z(p)$ and $Z(q)$ are linearly equivalent on E. Then there exists a dominant morphism $f : E \to \mathbb{P}^1$ (where $p \mapsto 0, q \mapsto \infty$) which induces an isomorphism on fields of fractions. Localizing at q_∞, this gives an embedding of the discrete valuation ring $\mathcal{O}_{\mathbb{P}^1_k, f(q_\infty)}$ to R_{q_∞} which induces an isomorphism on fields of fractions. Since discrete valuation rings are maximal, we would have to have

$$\mathcal{O}_{\mathbb{P}^1_k, f(q_\infty)} \cong R_{q_\infty}$$

which is impossible, since R_{q_∞} is not normal. This proves (3.4.32). On the other hand, the discussion of lines (3.4.30) and corresponding rational functions (3.4.31) is precisely the same as in the previous example, with the exception that we want to avoid any lines containing the point q_∞. This happens naturally, however, noting that any line through the point q_∞ contains at most one other point of E, with multiplicity 1. Thus, connecting two points of U_m (including the case of a point with itself) never creates a line which would contain q_∞.

It is also worth noting that the group law on U coincides with the ordinary group law on $\mathbb{A}^1_k$ (addition). This is because the points (u, u^3), (v, v^3), $(-u - v, -(u+v)^3)$ actually lie on a straight line in the vector space k^2.

The algebraic group $\mathbb{A}^1_k$ with respect to addition is called the *additive group* and denoted by $\mathbb{G}_a$ (although the disadvantage of this notation is that it does not display k). In any case, we proved that

$$Pic(E) \cong \mathbb{Z} \oplus (\mathbb{G}_a)_m.$$

Example The *nodal curve* E is the closure in $\mathbb{P}^2_k$ (k algebraically closed of characteristic $\neq 2$) of

$$Spec(R), \quad R = k[x, y]/(x^2 - y^3 - y^2),$$

or

$$Proj(k[x, y, z]/(x^2 z - y^3 - y^2 z)).$$

In this case, we factor the projective equation

$$x^2 z - y^3 - y^2 z = 0$$

as

$$(x + y)(x - y)z = y^3.$$

It is convenient to substitute

$$u = \frac{x+y}{2}, \quad v = \frac{x-y}{2},$$

so then we have

$$4uvz = (u-v)^3.$$

We see that a line in $\mathbb{P}_k^2$ which intersects E only in the point

$$q_\infty = [0:0:1]$$

is $Z(u)$. In the complement of this line, then, the equation becomes

$$4vz = (1-v)^3.$$

We see that this is isomorphic to $Spec(k[v, v^{-1}]) \cong \mathbb{A}_k^1 \smallsetminus \{0\}$. This is a group under multiplication. In fact, it is an algebraic group, called the *multiplicative group* and denoted by $\mathbb{G}_m$. Thus, we have

$$Pic(E) \cong \mathbb{Z} \oplus (\mathbb{G}_m)_m,$$

(It is unfortunate that the two subscripts in the above formula have different meanings, the first one indicating the multiplicative group, the second one closed points.)

To show that the group law is the same as in the multiplicative group, one checks that the points

$$(r, \frac{(1-r)^3}{4r}), \quad (s, \frac{(1-s)^3}{4s}), \quad (1/rs, \frac{(1-1/rs)^3}{4/rs}),$$

$r, s \neq 0$, lie on the same line in k^2.

3.5   Very Ample Line Bundles

Let $f : X \to A$ be a morphism of schemes. One is often interested in determining if X is a projective scheme over A. Recall that this means that X is isomorphic, over A, to a closed subscheme of some $\mathbb{P}_A^n$. More generally, X is called *quasiprojective over A* if X is isomorphic, over A, to an open subscheme V of a closed subscheme of some $\mathbb{P}_A^n$. The composition

$$X \xrightarrow[\cong]{} V \xrightarrow[\subseteq]{open} Z \xrightarrow[\subseteq]{closed} \mathbb{P}_A^n \tag{3.5.1}$$

over A is then called an *immersion*

$$g : X \to \mathbb{P}^n_A. \tag{3.5.2}$$

Recall, of course, that f is called a closed immersion if V is a closed subscheme of $\mathbb{P}^n_A$, i.e. if $V = Z$. This always happens when X is proper over A (by Theorem 2.6.2 of Chap. 2).

Denoting the composition (3.5.1) by g, we may study the line bundle

$$\mathcal{L} = g^*\mathcal{O}_A(1). \tag{3.5.3}$$

If a line bundle $\mathcal{L}$ on X comes from an immersion g via (3.5.3), we say that the line bundle $\mathcal{L}$ is *very ample*.

Note of course that $\mathcal{L}$ may depend on the choice of the immersion g. For example, on $\mathbb{P}^n_A$ itself, the line bundle $\mathcal{O}_A(d)$ for any $n \geq 1$ is very ample, corresponding to the embedding

$$\mathbb{P}^n_A \to \mathbb{P}^{\binom{n+d}{d}-1}_A$$

given by all the possible polynomials of degree d in the $n+1$ projective coordinates of $\mathbb{P}^n_A$ (the *Veronese embedding*, see Exercise 6 of Chap. 1).

One is interested in criteria determining whether a given line bundle is very ample. We start with a simpler question, namely when a line bundle can be expressed in the form (3.5.3) for some (3.5.2), regardless of whether (3.5.2) is an immersion. We say that a line bundle $\mathcal{L}$ is *generated by global sections* $s_0, \ldots, s_n \in \mathcal{L}(X)$ if every section $s \in \mathcal{L}(U)$ is an $\mathcal{O}_X(U)$-linear combination of $s_0|_U, \ldots, s_n|_U$. The same definition clearly can be made for any sheaf of modules $\mathcal{L}$. It is obviously equivalent to requiring that the morphism of sheaves of modules from the free sheaf of modules on a set of $n+1$ elements to $\mathcal{L}$ given by the sections $s_0, \ldots, s_n$ be onto.

3.5.1 Proposition *A morphism (3.5.2) over A such that (3.5.3) holds exists if and only if $\mathcal{L}$ is generated by $n+1$ global sections.*

Proof We already commented in the beginning of Sect. 3.3 that $\mathcal{O}_A(1)$ is generated by $n+1$ global sections (the projective coordinates), so necessity follows.

By the same token, it also implies that the morphism (3.5.2) can be given by the ratio

$$[s_0 : \cdots : s_n]. \tag{3.5.4}$$

Note that while the s_i do not make good sense as coordinates, their ratios do. We have to be careful, however, about the fact that some of the sections s_i can be 0. But X is covered by open sets on each of which one of the s_i's is non-zero (if all were zero simultaneously at a point, they would not generate the line bundle sections). Then dividing by this section gives a well-defined morphism on that open set, and those morphisms are compatible. This proves sufficiency.

$\square$

Note that the assumption of Proposition 3.5.1 does not really depend on A, but then again, neither does the conclusion: the statement says the same thing for $A = Spec(\mathbb{Z})$.

Example We can now understand better why the Picard group of the hyperboloid in one sheet H is $\mathbb{Z}$: The line bundle, say, $(x - z, 1 - y)$ is generated by two global sections, and hence defines a morphism g to $\mathbb{P}^1_k$. Explicitly, it is defined by the ratio

$$[x - z : 1 - y].$$

As in the proof above, when both coordinates are 0, we simply note that it is also equal to the ratio

$$[1 + y : x + z].$$

The essential point is however that one now sees that the morphism g defines an isomorphism

$$H \cong U_{\mathcal{O}_{\mathbb{P}^1_k}(1)} \tag{3.5.5}$$

(see (2.1.5)). To see (3.5.5), if we consider the morphism

$$f : H \to \mathbb{A}^2_k = Spec(k[u, v])$$

given by $u = x - z$, $v = 1 - y$, then

$$f \times g : H \to \mathbb{A}^2_k \times \mathbb{P}^1_k$$

is a closed immersion whose image is identified with $U_{\mathcal{O}_{\mathbb{P}^1_k}(1)}$. Therefore, its Picard group is the same as the Picard group of $\mathbb{P}^1_k$ (by the Remark under Proposition 3.4.2), which we already saw is $\mathbb{Z}$.

Now suppose we already have constructed a morphism of schemes of the form (3.5.2). How can we tell whether it is an immersion? We will give one criterion now.

3.5.2 Proposition *Let X be a projective scheme over an algebraically closed field k, let $\mathcal{L}$ be a line bundle generated by global sections $s_0, \ldots, s_n$ and let*

$$\phi : X \to \mathbb{P}^n_k$$

be the corresponding morphism of schemes given by Proposition 3.5.1. Denote by $V \subseteq \mathcal{L}(X)$ the k-vector space spanned by $s_0, \ldots, s_n$.

Then ϕ is an immersion (hence a closed immersion) if and only if the following conditions hold:

(1) V separates points, i.e. for every two closed points $x \neq y \in X$, there exists an $s \in V$ such that $s \in m_x \mathcal{L}_x$, $s \notin m_y \mathcal{L}_y$ where $m_x \subset \mathcal{O}_{X,x}$ denotes the maximal ideal.
(2) V separates tangent vectors, i.e. for every closed point $x \in X$ the set of all $v \in V \cap m_x \mathcal{L}_x$ generates $m_x \mathcal{L}_x / m_x^2 \mathcal{L}_x$. (Here we identify a global section with its image in the corresponding local ring.)
(3) $\phi_* \mathcal{O}_X$ is a coherent sheaf on $\mathbb{P}^n_k$.

COMMENT As we already remarked, one can prove that condition (3) follows just from X being proper over $Spec(k)$, and hence over $\mathbb{P}^n_k$, see Exercise 5 of Chap. 3).

Proof Necessity: If ϕ is a closed immersion, then by Proposition 2.3.1, the kernel of the onto morphism of sheaves

$$\mathcal{O}_{\mathbb{P}^n_k} \to \phi_* \mathcal{O}_X$$

is a quasicoherent (hence coherent, since $\mathbb{P}^n_k$ is a Noetherian scheme) sheaf of ideals $\mathcal{I}$. Therefore,

$$\phi_* \mathcal{O}_X \cong \mathcal{O}_{\mathbb{P}^n_k} / \mathcal{I},$$

which is a coherent sheaf. This proves (3). For (1) and (2), it suffices to consider the case when $X = \mathbb{P}^n_k$ and ϕ is the identity. (Note: Here we use the fact that on local rings, morphisms of schemes by definition induce morphisms of local rings on stalks, and for a closed immersion, those morphisms are onto.) In the case $\phi = Id$, condition (1) says that in homogeneous coordinates $[x_0 : \cdots : x_n]$, for two different closed points of $\mathbb{P}^n_k$, there exists a linear function

$$a_0 x_0 + \cdots + a_n x_n \tag{3.5.6}$$

which is zero on one of them and not the other, which is obvious. (Note that since k is algebraically closed, closed points can be represented by actual choices of ratios of coordinates in k.) For condition (2), for a closed point $P \in \mathbb{P}^n_k$, $m_P \mathcal{O}_{\mathbb{P}^n_k}(1)$ is precisely generated by those linear functions (3.5.6) which vanish on P, which is what we need.

Sufficiency: Since X is proper over k, ϕ is a closed map on points. Condition (1) shows that the morphism ϕ is injective on closed points and hence on points (since any point is characterized by the set of closed points in its closure, as X is projective). Since ϕ is continuous, closed and injective, it is a homeomorphism onto its image, which is a closed subset of $\mathbb{P}^n_k$. Therefore, it suffices to show that ϕ induces onto homomorphisms on stalks. By localization, it suffices to consider closed points. By our assumptions, this follows from the following result. $\qquad\square$

3.5.3 Lemma *Let* $f : (A, m) \to (B, n)$ *be a morphism of local Noetherian rings such that* f *induces an isomorphism on residue fields,* $f(m)$ *maps onto* n/n^2 *and* B *is a finitely generated* A-module. Then f is onto.

Proof Let a be the ideal in B generated by $f(m)$. Then by assumption, $a \subseteq n$, and its projection to n/n^2 is onto. Consider the B-module $M = n/a$. We have

$$a + n^2 = n,$$

so n^2 projects onto M, in other words, $nM = M$. Since n is the maximal ideal, by Nakayama's lemma, $M = 0$ and

$$a = n.$$

Now consider the finitely generated A-module $Q = B/f(A)$. Then, since $f(m) \cdot B = n$, and $n + f(A) = B$ by the assumption on residue fields, we conclude that $mQ = Q$, and hence by Nakayama's lemma, $Q = 0$ since m is the maximal ideal of A. $\qquad\square$

3.6 Blow-ups

We begin with generalizing the construction of the scheme $Proj(S)$ over $Spec(S_0)$ for a graded ring S to the situation when $Spec(S_0)$ is replaced by an arbitrary, not necessarily affine, scheme X. Recall (Proposition 2.4.5) that $Proj(S)$ is projective over $Spec(S_0)$ when S is generated, as an S_0-algebra, by S_1, which in turn is a finitely generated S_0-module. We will be interested in generalizing this case.

Suppose we have a Noetherian scheme X, and coherent sheaves S_n on X, $n \in \mathbb{N}_0$, such that

$$S_0 = \mathcal{O}_X,$$

together with a commutative associative unital system of "multiplication" morphisms of sheaves of $\mathcal{O}_X$-modules

$$\mu : S_k \otimes_{\mathcal{O}_X} S_\ell \to S_{k+\ell}.$$

We also assume that the iterated multiplication morphisms

$$\mathcal{S}_1 \otimes_{\mathcal{O}_X} \cdots \otimes_{\mathcal{O}_X} \mathcal{S}_1 \to \mathcal{S}_n$$

are onto. Then we can define a scheme $Proj(\mathcal{S})$ and a projection

$$\pi_\mathcal{S} : Proj(\mathcal{S}) \to X \qquad (3.6.1)$$

by letting, for an affine open subset $U \subseteq X$,

$$Proj(\mathcal{S}|_U) = Proj\left(\bigoplus_{n \in \mathbb{N}_0} \mathcal{S}_n(U)\right) \qquad (3.6.2)$$

and taking the colimit over $U \subset X$ affine open. (In more detail, one sees that for $U \subseteq X$ affine open and $V \subseteq U$ open, $\pi_{\mathcal{S}|_U}^{-1}(V)$ is the colimit of $Proj(\mathcal{S}|_W)$ over $W \subseteq V$ affine open, so we can use the technique of Sect. 2.5 of Chap. 2.)

Note that we have a line bundle

$$\mathcal{O}_{Proj(\mathcal{S})}(1) \qquad (3.6.3)$$

obtained by gluing the line bundles

$$\mathcal{O}_{Proj(\bigoplus_{n \in \mathbb{N}_0} \mathcal{S}_n(U))}(1)$$

for affine open subsets $U \subseteq X$.

3.6.1 Theorem *Let X be a projective scheme over a Noetherian affine scheme Y, and let $\mathcal{S}$ be as above. Then $Proj(\mathcal{S})$ is projective over Y.*

Proof Denote by $f : X \to \mathbb{P}^n_Y$ the closed immersion over Y. Let

$$U_i = f^{-1}(U_{(x_i)}).$$

Then $U_i \subseteq X$ form an open cover of X. Let x_i be a choice of homogeneous coordinates for $\mathbb{P}^n$, and let $\mathcal{I}_i$ be the sheaf of ideals defining the reduced closed subscheme $Z_i = Proj(\mathcal{S}) \smallsetminus \pi_\mathcal{S}^{-1}(U_i)$. Then the sheaf

$$f_* \pi_{\mathcal{S}*}(\mathcal{O}_{Proj(\mathcal{S})}(1) \otimes_{\mathcal{O}_{Proj(\mathcal{S})}} \mathcal{I}_i) \qquad (3.6.4)$$

is quasicoherent by Lemma 2.2.3.

The sheaf

$$\mathcal{O}_{Proj(S)}(1)|_{\pi_S^{-1}(U_i)} = (\mathcal{O}_{Proj(S)}(1) \otimes_{\mathcal{O}_{Proj(S)}} \mathcal{I}_i)|_{\pi_S^{-1}(U_i)}$$

is very ample by Proposition 2.4.5, in particular, it is generated by finitely many sections

$$r_{i1}, \ldots r_{i,n_i}. \tag{3.6.5}$$

By adjunction, we can think of (3.6.5) as sections of (3.6.4) on $U_{(x_i)}$, and hence by Lemma 2.4.3, there exists an $N \in \mathbb{N}_0$ such that all

$$x_i^N r_{ij}$$

extend to global sections t_{ij} on $Proj(S)$ (See also Exercise 26.) Since

$$\mathcal{O}_{Proj(S)}(1) \otimes_{\mathcal{O}_{Proj(S)}} \mathcal{I}_i$$

is a subsheaf of $\mathcal{O}_{Proj(S)}(1)$, this means that the global sections t_{ij} generate

$$\mathcal{O}_{Proj(S)}(1) \otimes_{\mathcal{O}_{Proj(S)}} \pi_S^*(\mathcal{O}_X(N)). \tag{3.6.6}$$

We claim in fact that (3.6.6) is very ample. Let

$$\phi : Proj(S) \to \mathbb{P}_Y^N$$

be the morphism defined by the global sections t_{ij} and let x_{ij} be the corresponding homogeneous coordinates on $\mathbb{P}_Y^N$. Since we tensored with the sheaves of ideals $\mathcal{I}_i$, we know that for a fixed i,

$$\phi^{-1}(U_{(x_{i1}, \ldots, x_{i,n_i})}) = \pi_S^{-1}(U_i).$$

On the other hand, by construction, the restriction of ϕ to $\pi_S^{-1}(U_i)$ is an immersion of $\pi_S^{-1}(U_i)$ into $U_{(x_{i1}, \ldots, x_{i,n_i})}$. This shows that ϕ is an immersion. $\qquad\square$

Now let again X be a Noetherian scheme. Suppose $\mathcal{I} = \mathcal{I}_Z$ is a quasicoherent (hence coherent) sheaf of ideals in $\mathcal{O}_X$ corresponding to a closed subscheme Z. Then putting

$$(S_Z)_n = \mathcal{I}^n, \quad (S_Z)_0 = \mathcal{O}_X, \tag{3.6.7}$$

we call $Proj(S_Z)$ the *blow-up of the scheme X at the closed subscheme Z*. It is sometimes also denoted by $Bl_Z(X)$.

To illustrate what the blow-up looks like, first note that of course, the restriction of $\pi_{\mathcal{S}_Z}$ defines an isomorphism

$$\pi_{\mathcal{S}_Z}^{-1}(X \smallsetminus Z) \to X \smallsetminus Z.$$

Next, note that for any morphism of schemes $f : Y \to X$ and a closed subscheme $Z \subseteq X$, the quasicoherent sheaf of ideals $\mathcal{I}_{f^{-1}(Z)}$ on Y defining the closed subscheme $f^{-1}(Z)$ (defined as a pullback) is, by construction of the pullback of schemes, the image of $f^*(\mathcal{I}_Z)$ in $\mathcal{O}_Y$.

3.6.2 Proposition *The sheaf of ideals $\mathcal{I}_{\pi_{\mathcal{S}_Z}^{-1}(Z)}$ on $Proj(\mathcal{S}_Z)$ is invertible. (The divisor $\pi_{\mathcal{S}_Z}^{-1}(Z)$ is called the* exceptional divisor *of the blow-up.)*
Further, $Proj(\mathcal{S}_Z)$ is universal *in the sense that for every morphism of schemes $f : Y \to X$ such that $\mathcal{I}_{f^{-1}(Z)}$ is an invertible sheaf on Y factors uniquely through $Proj(\mathcal{S}_Z)$, i.e. there exists a unique morphism of schemes g completing the following diagram:*

$$
\begin{array}{ccc}
Y & \overset{g}{\dashrightarrow} & Proj(\mathcal{S}_Z) \\
& \searrow_{f} & \downarrow_{\pi_{\mathcal{S}_Z}} \\
& & X.
\end{array}
\tag{3.6.8}
$$

Proof For both statements, we can assume that both $X = Spec(A)$ and $Y = Spec(B)$ are affine (in case of the second statement, this is because of the uniqueness). Now in the affine case, the first statement follows from Proposition 2.4.1 since

$$\mathcal{I}_{\pi_{\mathcal{S}_Z}^{-1}(Z)} = \mathcal{O}_{Proj(\mathcal{S}_Z)}(1)$$

where $\mathcal{S}_Z$ is the graded ring of global sections of $\mathcal{S}_Z$ (i.e., its graded components are powers of the ideal I_Z defining Z).

For the second statement, let $f = Spec(h)$ for a homomorphism of rings $h : A \to B$ and let I be the ideal in A defining the closed subscheme Z. If the ideal $J = I \cdot B$ in B defining $f^{-1}(Z)$ is an invertible B-module, without loss of generality, we may assume that it is isomorphic to B as a B-module (by passing, if necessary, to another open cover). Thus, we have $J = (b)$ for some $b \in B$ which is not a zero divisor. Further, we have a canonical morphism over $Spec(B)$

$$Proj\left(\bigoplus_{r \in \mathbb{N}_0} J^n\right) \to Proj\left(\bigoplus_{n \in \mathbb{N}_0} I^n\right) \times_{Spec(A)} Spec(B),$$

so for existence in the case of X affine, we may assume without loss of generality that $A = B$ and h is the identity. But the open set

$$U_{(b)} \subseteq Proj\left(\bigoplus_{n \in \mathbb{N}_0} J^n\right)$$

is the $Spec$ of

$$(b^{-1} Proj\left(\bigoplus_{n \in \mathbb{N}_0} J^n\right))_0 = B,$$

and thus the identical map from $Spec(B)$ to $U_{(b)}$ satisfies our requirements.

For uniqueness in the affine case, to avoid confusion, for $u \in I^n \subseteq A$, denote by $u(n)$ its copy in the n'th summand I^n of

$$\bigoplus_{n \in \mathbb{N}_0} I^n.$$

By further localizing, we may then assume that g maps $Spec(B)$ to a distinguished open set

$$U_{(a(1))} = Spec((a(1))^{-1} \bigoplus_{n \in \mathbb{N}_0} I^n)_0)$$

for some $a \in I$. Then, by assumption, $h(a) = b \cdot c$ for some $c \in B$. Thus, we have a homomorphism of rings

$$\tilde{h} : (a(1))^{-1} \bigoplus_{n \in \mathbb{N}_0} I^n)_0 \to B$$

extending h. By assumption, there exist $c_1, \ldots, c_n \in B$ and $x_1, \ldots, x_n \in I$ such that

$$c_1 h(x_1) + \cdots + c_n h(x_n) = b.$$

We compute

$$\begin{aligned} &h(a) \cdot (c_1 \tilde{h}(a(1)^{-1} x_1(1)) + \cdots + c_n \tilde{h}(a(1)^{-1} x_n(1))) \\ &= c_1 \tilde{h}(a(1)^{-1} a x_1(1)) + \cdots + c_n \tilde{h}(a(1)^{-1} a x_n(1)) \\ &= c_1 h(x_1) + \cdots + c_n h(x_n) = b. \end{aligned}$$

Hence, $h(a)$ is a factor of b, and hence is a non-zero divisor in B. But now for any $x \in I$,

$$h(a)\tilde{h}(a(1)^{-1} x(1)) = \tilde{h}(a(1)^{-1} a x(1)) = h(x),$$

which means that $\widetilde{h}(a(1)^{-1}x(1)) \in B$ is uniquely determined. However, those elements generate the ring

$$(a(1)^{-1} \bigoplus_{n \in \mathbb{N}_0} I^n)_0.$$

□

Note that this immediately implies functoriality: if $f : X \to Y$ is a morphism of Noetherian schemes and $Z \subseteq Y$ a closed subscheme then we have a unique diagram

$$\begin{array}{ccc} Proj(\mathcal{S}_{f^{-1}(Z)}) & \xrightarrow{\quad g \quad} & Proj(\mathcal{S}_Z) \\ \downarrow & & \downarrow \\ X & \xrightarrow{\quad f \quad} & Y \end{array} \qquad (3.6.9)$$

and moreover if f is a closed immersion, so is g. When f is a closed immersion, the closed subscheme of $Proj(\mathcal{S}_Z)$ associated with g is called the *strict transform* of X under the blowing up morphism $\pi_{\mathcal{S}_Z}$.

In particular, when X is a quasiprojective variety over a field k and Z is a closed subscheme, it follows from Theorem 3.6.1 that $Proj(\mathcal{S}_Z)$ is a quasiprojective variety over k and the morphism $\pi_{\mathcal{S}_Z}$ is birational, projective and onto. (Note: the reason we can generalize to X quasi-projective is the functoriality of the blow-up—we can always pass to the closure of X.)

The main point of this discussion is that a converse also holds:

3.6.3 Theorem *Let $f : T \to X$ be a birational projective onto morphism where X is a quasiprojective variety over a field k. Then f is isomorphic to the blow-up projection $\pi_{\mathcal{S}_Z}$ for some closed subscheme $Z \subseteq X$.*

The proof of this theorem is technical and we omit it. It may be found in [11], Theorem 7.17.

4   Exercises

1. (Godement) Let $\mathcal{C}$ be a category satisfying the Assumption of Sect. 1.1, and let X be a topological space. Let X_0 be the set X with the discrete topology, and let $f : X_0 \to X$ be the identity. For a sheaf $\mathcal{F}$ on X valued in $\mathcal{C}$, define

$$M\mathcal{F} = f_* f^{-1} \mathcal{F}.$$

This is a functor $C\text{-}sh(X) \to C\text{-}sh(X)$.

(a) Construct natural transformations $Id \to M$ and $MM \to M$ which are associative and unital with respect to composition of functors. (This makes M a *monad*; In general, a monad arises from composing a right adjoint with its corresponding left adjoint.

(b) Using the Assumption, prove that a C-valued sheaf $\mathcal{F}$ is the equalizer of the sheaves

$$M\mathcal{F} \rightrightarrows MM\mathcal{F}.$$

(c) Using (b), prove that under the Assumption, a morphism of C-valued sheaves is an isomorphism if and only if it induces an isomorphism on stalks, and that two morphisms of sheaves $\mathcal{F} \rightrightarrows \mathcal{G}$ coincide if and only if they induce the same morphisms on stalks.

2. Prove that for the constant sheaf $\underline{A}$ on a topological space X, for an open set $U \subseteq X$, $\underline{A}(U)$ is the set of continuous functions $U \to A$ with the discrete topology on A (also known as *locally constant functions* on U).

3. Prove the alternate characterization of sheafification for universal algebras as described in Sect. 1.1.2.

4. Prove that for a continuous map $f : X \to Y$ and for a category C satisfying our Assumption, the functor $f^{-1} : C\text{-}Sh_Y \to C\text{-}Sh_X$ preserves stalks and is left exact. (Note that it also preserves colimits since it is a left adjoint.)

5. Give an example where the global section functor on abelian sheaves, (resp. sheaves of sets), is not right exact.

6. Prove in detail that the pushforward functor of abelian sheaves (resp. sheaves of sets) for a closed inclusion is right exact.

7. Prove that abelian sheaves on a space, as well as sheaves of modules over a ringed space, form abelian categories.

8. Prove that formula (1.5.1) defines a sheaf.

9. Prove that the tensor product and Hom functor of sheaves of modules satisfy formula (1.5.2).

10. Prove that if R is a non-zero commutative ring, then free R-modules on bases of different cardinalities are not isomorphic. [If m is a maximal ideal of R and F is a free R-module, then F/mF is a free R/m-module.]

11. Prove that if V is an infinite dimensional vector space over a field F, and R is the (non-commutative) ring of endomorphisms of V (i.e. homomorphisms $V \to V$), then the free left R-module on 1 generator is isomorphic to the free left R-module on 2 generators. (Recall that a left R-module M has distributive associative unital multiplication $rm \in M$ for $r \in R, m \in M$.)

12. Prove that an invertible sheaf over a scheme X is the same thing as a locally free sheaf of $\mathcal{O}_X$-modules of dimension 1.

13. Suppose that we complexified the Möbius strip in the example of Sect. 2.1 by replacing $\mathbb{R}$ with $\mathbb{C}$ in the definition of the ring R. Will the corresponding algebraic line bundle be trivial? [Show that $(x, y - 1)$ then is a principal ideal.]

14. Complete the proof of Lemma 2.2.1. [Model the proof on the proof of Lemma 2.2.2 of Chap. 2.]

15. Prove that the definition of codimension given in Definition 3.3.1 agrees with the definition given in Exercise 38 of Chap. 2.

16. Recall that for a module M over a commutative ring R, we have the *tensor algebra* (or free associative algebra) $T_R M$ which is a graded (usually non-commutative) ring where $(T_R M)_n$ is a tensor product over R of n copies of M.

The *exterior algebra* $\wedge_R M$ is the quotient of the tensor algebra $T_R M$ by the relations $x^2 = 0$ for all $x \in M$. Thus, $\wedge_R M$ is a graded non-commutative R-algebra, where $\wedge_R^n M = (\wedge_R M)_n$ is generated by products $x_1 \ldots x_n$ (which is often written as $x_1 \wedge \cdots \wedge x_n$). In fact, the ring $\wedge_R M$ is *graded-commutative* which means that for $x \in \wedge_R^m M$, $y \in \wedge_R^n M$, we have

$$x \wedge y = (-1)^{mn} y \wedge x.$$

Prove that for a Noetherian ring R, there is a *determinant* homomorphism of abelian groups

$$Det : K(R) \to Pic(R)$$

which maps $[M]$ to $[\wedge_R^n M]$ for a locally free (i.e. projective) R-module M of dimension n.

17. Let R be a regular local ring and $f \in R$ a non-nilpotent element. Prove that $Pic(f^{-1}R) = 0$. [For an invertible $f^{-1}R$-module L, $L = f^{-1}M$ where M is a finitely generated R-module. You may borrow from Chap. 5, Sect. 2.6, and Exercise 21 of Chap. 5 the result that the homomorphism from K to G is an isomorphism for R and $f^{-1}R$. Thus, since $KR = \mathbb{Z}$, $[L] = n$ for some $n \in \mathbb{Z}$. Now use the determinant homomorphism of Exercise 16 to prove that $[L] = 1 \in Pic(f^{-1}R)$.]

18. Prove the *Auslander-Buchsbaum Theorem* which says that every regular local ring is a UFD. [It suffices to prove that every height 1 prime p is principal. Use induction on dimension. Let R be a regular local ring of dimension n and let $p \subset R$ be a height 1 prime. Let x be one of the n generators of the maximal ideal of R. Without loss of generality, $x \notin p$. Then $x^{-1}R$ is regular of dimension $< n$. By the induction hypothesis, the ideal $x^{-1}p \subset x^{-1}R$ is locally free of dimension 1, and thus is invertible. By Exercise 17, it is principal. Let $x^{-1}p = (f/x^m)$ for some $f \in R$. Now factor $f = a_1 \ldots a_s$ where $a_i \in R$ are irreducible. Since p is prime, we have $a_i \in p$ for some i. Since a_i divides the generator of $x^{-1}p$, which is prime, it generates $x^{-1}p$, and thus is prime in $x^{-1}R$. We claim that $a_i \in R$ is prime (and hence, $(a_i) = p$ since p is minimal). To prove that a_i is prime, let $a_i \mid bc$ with $b, c \in R$. Since a_i is

prime in $x^{-1}R$, a_i divides $x^n b$ or $x^n c$ in R for some $n \in \mathbb{N}_0$. Let, say, $a_i u = x^n b$, $u \in R$. (Prove that x is prime by showing that the associated graded ring of $R/(x)$ is the symmetric algebra over R/m on $m/(m^2 + (x))$.) But since by assumption $x \nmid a_i$, we have $x \mid u$, so choosing n minimal gives $n = 0$.]

19. Prove that every regular ring is normal. [Use Exercise 18, Exercise 43 of Chap. 1, and Exercise 6 of Chap. 3.]

20. Prove that if X is a Noetherian regular integral separated scheme, then $Pic(X) \cong Cl(X)$. [Prove that every Weil divisor is a Cartier divisor. Reduce to the affine case and use Exercises 19, 18, and Exercise 43 of Chap. 1.]

21. Let R be a Dedekind domain.

(a) Let $I, J \subseteq R$ be non-zero ideals which are relatively prime, meaning that their prime factors are different. (See Exercise 8 of Chap. 3.) Then $I \oplus J \cong R \oplus IJ$ as R-modules. [Use the short exact sequence

$$0 \to I \cap J \to I \oplus J \to R \to 0.$$

Prove that $I \cap J = IJ$.]

(b) Let $I \subset R$ be a non-zero ideal. Use the Chinese remainder theorem to prove that there exists an element $x \in I$ such that $x R_p = I R_p$ for every prime p containing I. Then $(x) \subseteq I$, so there exists an ideal J with $IJ = (x)$. Prove that I, J are relatively prime and that $J \cong I^{-1}$ as R-modules. Using (a), conclude that $I \oplus I^{-1} \cong R \oplus R$. In particular, ideals of R are projective modules.

(c) Let p be a non-zero prime in R, and let $0 < m \le n$ be integers. Using the method of (b), prove that $p^n \oplus p^{-m} \cong R \oplus p^{n-m}$ and hence, multiplying by p^m, $p^m \oplus p^n \cong R \oplus p^{m+n}$. Conclude from (a) and (c) that for any non-zero ideals $I_1, \ldots, I_n \subseteq R$,

$$I_1 \oplus \cdots \oplus I_n \cong (n-1)R \oplus I_1 \ldots I_n.$$

(d) Prove that every finitely generated projective R-module M is isomorphic to a direct sum of finitely many ideals. [If $M \subseteq R^n$, proceed by induction on n, using the short exact sequence

$$0 \to M \cap R^{n-1} \to M \to I \to 0$$

where I is the projection of M onto the last coordinate. Note that we have not used the fact tat M is a direct summand, thus having proved that every submodule of R^n is projective.]

(d) Recall the exterior algebra from Exercise 16. By considering $\wedge^n M$, $\wedge^n N$, prove that when $M \cong N$ and $M = (n-1)R \oplus I$, $N = (m-1)R \oplus J$ with non-zero ideals I, J, then $m = n$ and $I \cong J$.

(e) Conclude that every finitely generated projective R-module M is isomorphic to $R^{n-1} \oplus I$ for some ideal I where $n \in \mathbb{N}$ and the class of I in $Pic(R)$ are uniquely determined. Also conclude that $K(R) \cong \mathbb{Z} \oplus Pic(R)$.

22. (Grothendieck) Prove that every algebraic vector bundle over $\mathbb{P}^1_k$ for a field k isomorphic to a direct sum of bundles of the form $\mathcal{O}(\ell_i)$, $\ell_i \in \mathbb{Z}$, where the numbers ℓ_i are uniquely determined up to order. [(Proof by Hazewinkel and Martin) First note that isomorphism classes of the bundles in question are bijective to equivalence classes of square matrices A over $k[x, x^{-1}]$ with determinant x^m for some $m \in \mathbb{Z}$ modulo the relation $A \sim BAC$ whether B resp. C is an invertible matrix over $k[x^{-1}]$ resp. $k[x]$. We want to prove that each equivalence class is uniquely represented by a diagonal matrix with diagonal entries $x^{r_1}, \ldots x^{r_n}$, $r_1 \geq \cdots \geq r_n$. To prove uniqueness, suppose another diagonal matrix with $r'_1 \geq \cdots \geq r'_n$ is in the same class. By multiplying by C^{-1} from the right and considering the formula for $m \times m$ submatrices of the product of two $n \times n$ matrices, prove that $r_1 + \cdots + r_m \geq r'_1 + \cdots + r'_m$ for all m. Reversing the roles of the matrices, equality arises, which implies the claim.

For existence, proceed by induction on n. Multiplying by x^N with some $N \in \mathbb{N}_0$, we can assume that the entries of A are polynomials. By performing column operations over $k[x]$ to find gcd of the first row (which must be a power of x) and using the induction hypothesis, show that each class contains a matrix of the form

$$\begin{pmatrix} x^{k_1} & 0 & \ldots & 0 \\ q_2 & x^{k_2} & \ldots & 0 \\ \ldots & 0 & \ldots & \ldots \\ q_n & & & x^{k_n} \end{pmatrix}.$$

Now if k_1 is maximal possible (which exists since its degree is bounded by the degree of $det(A)$), then it is $\geq k_i$ for all i. Indeed, suppose $k_1 < k_i$. Without loss of generality, q_i has no terms of degree $\leq k_1$. Thus, all terms of q_i are divisible by x^{k_1+1}. Then switching the first and i'th row and performing the Euclidean algorithm in the first row using column operations again, we can increase k_1—contradiction. Finish the proof from this point.]

23. Generalize Exercise 21 of Chap. 3 to prove that for a number field K and a fractional ideal I in K (i.e. a finitely generated $\mathcal{O}_K$-submodule of K), there exists an element $x \in I$ with

$$N(x) \leq \frac{n!}{n^n} \left(\frac{4}{\pi} \right)^s vol(K_{\mathbb{R}}/I).$$

[The same method works.]

24. Prove that for a number field K, every element of the ideal class group $Cl(\mathcal{O}_K)$ is represented by an ideal $I \subseteq \mathcal{O}_K$ with

$$|\mathcal{O}_K/I| \leq \frac{n!}{n^n}\left(\frac{4}{\pi}\right)^s vol(K_\mathbb{R}/\mathcal{O}_K).$$

[By Exercise 23, for any fractional ideal J, there is an element $x \in J^{-1}$ of norm less or equal to

$$N(x) \leq \frac{n!}{n^n}\left(\frac{4}{\pi}\right)^s vol(K_\mathbb{R}/J^{-1}).$$

Now take $I = Jx$ and prove that $Jx \subseteq \mathcal{O}_K$. Use the fact that $vol(K_\mathbb{R}/x^{-1}J^{-1}) = vol(K_\mathbb{R}/\mathcal{O}_K)/|\mathcal{O}_K/Jx|$.]

25. Prove that $Cl(\mathcal{O}_K)$ is finite. [By Exercise 24, we have a natural number N such that every element of $Cl(\mathcal{O}_K)$ is represented by an ideal $I \subseteq \mathcal{O}_K$ with $\mathcal{O}_K \subseteq I \subseteq N\mathcal{O}_K$.]

26. Let X be a quasicompact separated scheme. Let L be an invertible sheaf of $\mathcal{O}_X$-modules, and let f be a section of L with zero set Z. Let $\mathcal{F}$ be a quasicoherent sheaf on X and let s be a section of $\mathcal{F}$ on $X \smallsetminus Z$. Prove that there exists an $n \in \mathbb{N}$ such that the section $s \otimes f^n$ of $L^{\otimes n} \otimes_{\mathcal{O}_X} \mathcal{F}$ on $X \smallsetminus Z$ extends to X. [Cover X by open affine subsets on which L is trivial. Note that you need to tensor with another power of L to glue the sections constructed. Pass again to an affine open cover, and use the definition of localization.]

27. (Kleiman) Prove that if X is a regular separated scheme, and $\mathcal{M}$ is a coherent sheaf, then there exists an epimorphism $\mathcal{F} \to \mathcal{M}$ where $\mathcal{F}$ is a locally free sheaf. [Without loss of generality, X is connected. Let U_i be an affine open cover of X. Then $X \smallsetminus U_i$ is the zero set of a section of an invertible sheaf $\mathcal{L}_i$ on X. Use Exercise 26.]

28. Suppose X is a Noetherian integral separated scheme whose every local ring of height 1 is regular. Let M be a locally free sheaf on X. Prove that then the same holds for the corresponding scheme $\mathcal{U}_M$, and we have an isomorphism $Cl(X) \cong Cl(\mathcal{U}_M)$.

29. Prove formula (3.4.13).

30. Compute $Cl(R)$ where R is the ring given by formula (3.4.22).

31. (a) Let $n, N \in \mathbb{N}$, $1 \leq n \leq N$. For a subset $S \subseteq \{1, \ldots, N\}$ of cardinality n, let $U_S = \mathbb{A}_\mathbb{Z}^{n(N-n)}$ be identified with the space of $S \times \{1, \ldots, N\}$)-matrices with a unit $S \times S$ submatrix. Let $U_{S,T}$ be the open subscheme of such matrices whose $(S \times T)$-submatrix $M_{S,T}$ is invertible (make that precise). Define a scheme by gluing the U_S's along the open subschemes $U_{S,T}$ as in Sect. 2.5 of Chap. 2, identifying $U_{S,T} \cong U_{T,S}$ by switching implicit and explicit coordinates (write down the precise formula). This is the *Grassmannian* $Gr_n(N) = Gr_n(N)_\mathbb{Z}$. For any scheme X, $Gr_n(N)_X$ is the pullback to X, as usual.

(b) Construct an algebraic line bundle *Det* on $Gr_n(N)$ by gluing the trivial line bundles on U_S by the transition function $det(M_{S,T})$. Note that notation must be fixed to distinguish Det from its inverse. Prove that with the correct choice (which can be derived from the case $n = 1$), the line bundle Det is very ample, and conclude that $Gr_n(N)$ is a projective scheme.

32. Define a function $v : \mathbb{C}(x, y)^\times \to \mathbb{Z}$ by letting, for a polynomial $f(x, y)$, $v(f)$ be the order of vanishing of $\phi(x) = f(x, e^x)$ at $x = 0$, i.e. the maximum n such that $\phi(0) = \phi'(0) = \cdots = \phi^{(n-1)}(0) = 0$, and letting, for a rational function g/h where $g, h \in \mathbb{C}[x, y]$, $v(g/h) = v(g) - v(h)$.

   (a) Prove that v is a well-defined discrete valuation on $\mathbb{C}(x, y)$. Let R be the corresponding discrete valuation ring.

   (b) Prove that there does not exist a prime Weil divisor on $\mathbb{P}^2_{\mathbb{C}}$ with generic point p such that $v = v_p$ (where we identify $\mathbb{C}(x, y)$ with the field of rational functions on $\mathbb{P}^2_{\mathbb{C}}$ by letting x, y be two affine coordinates).

   (c) Letting $\gamma : Spec(\mathbb{C}(x, y)) \to \mathbb{P}^2_{\mathbb{C}}$ be the inclusion of the generic point. Since $\mathbb{P}^2_{\mathbb{C}}$ is proper, we know that γ extends to a morphism of schemes $Spec(R) \to \mathbb{P}^2_{\mathbb{C}}$. What is the closure Y of the closed point of $Spec(R)$?

   (d) Blow up Y and describe the extension of γ to $Spec(R) \to Bl_Y(\mathbb{P}^2_{\mathbb{C}})$. What happens when we iterate this procedure of blowing up the closure of the image of the closed point of $Spec(R)$ infinitely many times, and take the limit?

33. Let $f : Y \to X$ be a morphism of Noetherian schemes. Let $Z \subseteq X$ be a closed subscheme and let $g : Bl_Z(X) \to X$ be the projection. Is it always true that

$$Bl_{(f\times g)^{-1}Z}((Bl_Z X) \times Y) \cong Bl_{f^{-1}(Z)}Y?$$

34. (Hironaka's example) Consider the images Y, Z of $\mathbb{P}^1_{\mathbb{C}}$ in $\mathbb{P}^3_{\mathbb{C}}$ with coordinates $[x : 0 : y : 0]$ and $[x^2 : xy : y^2 : 0]$ where x, y are the projective coordinates in $\mathbb{P}^1$. Then the intersection $Y \cap Z$ consists of two closed points A, B with projective coordinates $[1 : 0 : 0 : 0]$ and $[0 : 0 : 1 : 0]$.

   (a) Consider the scheme U obtained by blowing up $\mathbb{P}^3_{\mathbb{C}} \setminus \{B\}$ first at Y and then at Z, and the scheme V obtained by blowing up $\mathbb{P}^3_{\mathbb{C}} \setminus \{A\}$ first at Z and then at Y. Then $W = \mathbb{P}^3_{\mathbb{C}} \setminus \{A, B\}$ is an open subscheme of both U and V. Let X be the scheme obtained as a pushout of U and V along W. Prove that X is a smooth proper variety over $\mathbb{C}$.

   (b) Prove that the inverse image of the point A (resp. B) under the map $X \to \mathbb{P}^3_{\mathbb{C}}$ consists of a union of two subvarieties A_1, A_2 (resp. B_1, B_2) isomorphic to $\mathbb{P}^1_{\mathbb{C}}$ where A_1, B_1 result from the first blow-up and A_2, B_2 from the second. Additionally, observe that the inverse image of Y is a surface $\widetilde{Y}$ containing the subvarieties A_1, A_2, B_2 and the inverse image of Z is the surface $\widetilde{Z}$ containing the subvarieties B_1, B_2, A_2.

   (c) Prove that in $\widetilde{Y}$, the (Weil) divisor $(A_1) + (A_2)$ is equivalent to the divisor (B_2) (i.e. their difference is a principal divisor). and in $\widetilde{Z}$, the divisor $(B_1) + (B_2)$ is equivalent to the divisor (A_2).

(d) Now suppose we have an embedding $X \to \mathbb{P}_{\mathbb{C}}^N$. Denote by a_1, a_2, b_1, b_2 the degrees of the restrictions to A_1, A_2, B_1, B_2. Use (c) to prove that $a_1 + a_2 = b_2$, $b_1 + b_2 = a_2$. Derive a contradiction.

35. *The Wonderful Compactification I* (DeConcini, Procesi, Fulton, MacPherson) We shall describe here a special case of an important construction using blow-ups. Its significance will become clearer in Chap. 6 (Exercise 15). Let k be an algebraically closed field and let $X = \mathbb{A}_k^m$. Let, for $S \subset \{1, \ldots, n\}$, D_S be the subvariety of X^n given by the equations $x_i = x_j$ with $i, j \in S$ where $x_i, i = 1, \ldots, n$ are the coordinates of a point in X^n.

Perform a sequence of blow-ups to construct a variety $X[n]$ along with projections $p_n : X[n] \to X^n$ and prime divisors $\Xi_S^n \subset X[n]$ projecting to D_S for $S \subseteq \{1, \ldots, n\}$, $|S| > 1$. Put $X[1] = X$. Suppose $X[n]$ has been constructed together with a projection $p_n : X[n] \to X^n$. Construct a sequence of varieties $X[n+1, i]$ together with divisors $\Xi_S^{n+1,i} \subset X[n+1, i]$ for $|S| > 1, i = n+1, n, \ldots, 1$, and

$$S \subseteq \{1, \ldots, n\}$$

or

$$S \subseteq \{1, \ldots, n+1\}, \ n+1 \in S \text{ and } |S| > i,$$

and closed subvarieties $\Xi_S \subseteq X[n+1, i]$ for $|S| > 1$ for

$$S \subseteq \{1, \ldots, n+1\}, \ n+1 \in S, \ |S| \le i.$$

as follows: $X[n+1, n+1] = X[n] \times X$ and let, for $S \subseteq \{1, \ldots, n\}, |S| > 1$,

$$\Xi_S^{n+1,n+1} = \Xi_S^n \times X.$$

For $S = \{i, n+1\}, \in \{1, \ldots, n\}$, put

$$\Xi_S^{n+1,n+1} = (p_n \times Id)^{-1}(D_S).$$

For $S \subseteq \{1, \ldots, n\}, |S| \ge 2$, put

$$\Xi_{S \cup \{n+1\}}^{n+1,n+1} = (\Xi_S^n \times X) \times_{D_S \times X} D_{S \cup \{n+1\}}.$$

Given $X[n+1, i+1]$, we put

$$X[n+1, i] = Bl_{Z[i]}(X[n+1, i+1])$$

where $Z[i]$ is the (as it turns out, disjoint) union of $\Xi_S^{n+1,i+1}$ where $S \subseteq \{1, \ldots, n+1\}$ runs over all subsets of cardinality $i+1$ which contain $n+1$. Let $\Xi_S^{n+1,i} \subset X[n+1, i]$ be the exceptional divisor of $\Xi_S \subset X[n+1, i+1]$ for $S \subseteq \{1, \ldots, n+1\}$ of cardinality $i+1$ which contain $n+1$, and the strict transform of $\Xi_S^{n+1,i+1} \subset X[n+1, i+1]$ for all other applicable sets S. Let $\Xi_S^{n+1} = \Xi_S^{n+1,1}$.

By an S-*screen*, where $S \subseteq \{1, \ldots, n\}$, $|S| > 1$ we mean an unordered set of at least two distinct closed points in $X = \mathbb{A}_k^m$, which are labeled by disjoint non-empty sets whose union is S. Two S-screens are considered the same when one can be transformed to the other by a translation followed by a multiplication by a scalar $\lambda \in k^\times$.

By an (n, i)-*tree* T, we mean an unordered collection of distinct closed points of X decorated by non-empty disjoint subsets of $\{1, \ldots, n\}$ whose union is $\{1, \ldots, n\}$ (without any identifications), together with a minimal collection of screens satisfying the following conditions: For a subset $S \subseteq \{1, \ldots, n\}$, if $|S| > 1$ and $n \notin S$, or $n \in S$ and $|S| > i$ and T contains (in the initial configuration or any of its screens) a point with label S, then an S-screen is present. If $n \in S$ and $2 < |S| \le i$ and a point with label S is present, then an $(S \setminus \{n\})$-screen is present.

Prove, by induction on n and i, the following statements:

(a) The closed points of $X[n, i]$ can be bijectively identified with (n, i)-trees in such a way that a point corresponding to an (n, i)-tree T is contained in $\Xi_S^{n,i}$ if and only if one of the following conditions arises:

1. $S \subseteq \{1, \ldots, n-1\}$ or $n \in S \subseteq \{1, \ldots, n\}$ and $|S| > i$, and T contains an S-screen.

2. $n \in S \subseteq \{1, \ldots, n\}$, $1 < |S| \le i$, and T contains (possibly in one of its screens) a point labeled by a subset S' such that $S \subseteq S'$ and $|S'| \le i$.

(b) Prove that in $X[n]$, a collection of subvarieties $p_n^{-1}(D_{S_i})$ for subsets S_i of $\{1, \ldots, n\}$ of cardinality > 1 has an empty intersection unless for all pairs i, j, S_i, S_j are either disjoint or one contains the other.

36. Let $S = \mathbb{C}[x, y]$ and let R be the subring of S generated by x^5, xy^3, x^2y, y^5. Denote by I the ideal $(x^5, xy^3, x^2y, y^5) \subset R$, let $\mathcal{I}$ be the associated sheaf of ideals on $Spec(R)$. Let $\mathcal{J}$ be sheaf of ideals on $Spec(S)$ associated to the ideal (x, y).

(a) Prove that I is a maximal ideal in R, and that the ring R_I is not regular.

(b) Prove that $Bl_{\mathcal{I}}(Spec(R))$ is a smooth variety over $Spec(\mathbb{C})$ (and hence regular).

(c) Does there exist a morphism of schemes $Bl_{\mathcal{J}}(Spec(S)) \to Bl_{\mathcal{I}}(Spec(R))$ over the morphism $Spec(S) \to Spec(R)$ coming from the inclusion of rings? Reconcile your answer with universality of blow-ups.

Introduction to Cohomology

5

It is now time to study the subject of cohomology in detail. In Chap. 4, we already encountered its special cases in degrees 0 and 1. This motivates understanding the general machinery which lets us set up cohomology groups in any degree. Even more importantly, a careful reader noticed that there are very important facts about regular rings (for example the fact that a localization of a regular ring is regular) which we have not proved so far, and deferred to when we can characterize regular rings cohomologically. For this, we will certainly need cohomology in higher degrees. Without filling this gap, we would not even be able to use Weil divisors to calculate Picard groups rigorously in general examples such as those of Chap. 4. We will complete those proofs in the present chapter.

The two topics mentioned in the last paragraph (i.e. introducing higher sheaf cohomology and characterizing regular rings cohomologically) already reveal sufficiently different aspects of the idea of cohomology to necessitate a more comprehensive discussion. But there is more. As many of the concepts we already discussed, cohomology originated in topology, where it appears together with homology, which, from the topological point of view, is even more fundamental. In analysis, a very direct approach to cohomology arises from differential forms.

In this chapter, we will study all these facets of (co)homology. We will also introduce the computational tool of spectral sequences. In Chap. 6, we will then see some powerful applications and further methods our general treatment of cohomology leads to in algebraic geometry.

© The Author(s), under exclusive license to Springer Nature Switzerland AG 2021
I. Kriz, S. Kriz, *Introduction to Algebraic Geometry*,
https://doi.org/10.1007/978-3-030-62644-0_5

1   De Rham Cohomology in Analysis

With all the different aspects of cohomology, where should we start? Perhaps the most elementary and geometrically intuitive topic is de Rham cohomology in differential geometry. The concept of a differential form is very easily geometrically motivated by the theory of integration.

Giving a brief introduction to differentiable manifolds, we will experience 'Grothendieck's approach to geometry,' which we used to introduce schemes, in action in a different field. Thus, we learn that this approach is more general. The reader may find it fun to see differentiable manifolds treated in this light, which is not typical for analysis textbooks. (For a more traditional approach, see [19].)

There is another benefit of describing manifolds from scratch: There is in fact an analogue of the concept of de Rham cohomology in schemes, and in Chap. 6, Sect. 2, we will see that sometimes they coincide. It is important to understand the subtle but profound contextual difference of both definitions in order to appreciate the deep fact of their agreement.

1.1   Smooth and Complex Manifolds

Smooth and complex manifolds can be defined in an analogous way as we defined schemes in Sect. 1.3 of Chap. 2. In general, this approach is sometimes referred to as the *Grothendieck approach to geometry*. On an open subset $U \subseteq \mathbb{R}^n$ in the analytic topology, a function $f : U \to \mathbb{R}$ is called *smooth* (in the sense of real analysis) if it is continuous and has all higher partial derivatives, and all are continuous. Similarly, on an open subset $U \subseteq \mathbb{C}^n$ in the analytic topology, a function $f : U \to \mathbb{C}$ is called *holomorphic* if it has the first complex partial derivatives on U. (Then it has also all higher complex partial derivatives, and all are continuous. For an introduction to complex analysis in one variable, we recommend [1].) Smooth functions on an open set $U \subseteq \mathbb{R}^n$ form a sheaf $\mathcal{C}_U^\infty$, and holomorphic functions on an open subset $U \subseteq \mathbb{C}^n$ form a sheaf $\mathcal{O}_U^{an}$, where restriction is given by restriction of functions.

A *smooth (resp. complex) manifold* of dimension n can be defined as a locally ringed space M such that for every $x \in M$, there exists an open neighborhood V which, with the restriction of the structure sheaf on M, is isomorphic, as a locally ringed space, to $(U, \mathcal{C}_U^\infty)$ for some open set $U \subseteq \mathbb{R}^n$, (resp. $(U, \mathcal{O}_U^{an})$ for some open set $U \subseteq \mathbb{C}^n$). We will denote the structure sheaf on a smooth (resp. complex) manifold M by $\mathcal{C}_M^\infty$ (resp. $\mathcal{O}_M^{an}$), and similarly we denote by $\mathcal{C}_V^\infty$ (resp. $\mathcal{O}_V^{an}$) its restriction to an open subset V. The global sections of $\mathcal{C}_M^\infty$ (resp. $\mathcal{O}_M^{an}$) will be denoted by $C^\infty(M)$ (resp. $Hol(M)$). Their elements will be called *smooth* (resp. *holomorphic*) functions on M.

We will also assume (as one usually does) that M is Hausdorff and the connected components (i.e. maximal connected subsets) of M have countable bases of topology. This is equivalent to M being Hausdorff and satisfying another property which will be also useful later, namely that M is *paracompact*. For an open cover (U_i) of a topological space X, a *refinement* is an open cover (V_j) such that for every j, there exists an i with $V_j \subseteq U_i$. We say that a cover (V_j) is *locally finite* if for every point $x \in X$, there exists an open set $U \ni x$ such that there are only finitely many j with $V_j \cap U \neq \emptyset$. A topological space X is *paracompact* if every open cover of X has a locally finite refinement.

A smooth (resp. complex) manifold M of dimension n can also be called a real (resp. complex) n-*manifold*. Of course, a complex n-manifold automatically has the structure of a smooth $2n$-manifold.

A morphism of smooth (resp. complex) manifolds is a morphism of locally ringed spaces which are smooth (resp. complex) manifolds. Often, those morphisms are referred to simply as smooth (resp. holomorphic) maps. An isomorphism of smooth manifolds is called a *diffeomorphism*. An isomorphism of complex manifolds is called a *holomorphic diffeomorphism*. When there exists a diffeomorphism (resp. holomorphic diffeomorphism) between smooth (resp. complex) manifolds, we call them *diffeomorphic* (resp. *holomorphically diffeomorphic*).

Two comments are in order. First, in the definition of a smooth manifold, the open set U can be equivalently replaced with $\mathbb{R}^n$. This is because an open ball is diffeomorphic to $\mathbb{R}^n$. An analogous statement is not true however about the definition of a complex manifold, since a bounded open disk in $\mathbb{C}^n$ is not holomorphically diffeomorphic to $\mathbb{C}^n$ (see [1]). The second comment is that the words "locally" can be omitted from the definitions of smooth and complex manifolds, because all points are closed, and therefore, morphisms of ringed spaces between smooth or complex manifolds are automatically morphisms of locally ringed spaces.

An *atlas* is an open cover (V_i) of a smooth (resp. complex) manifold M by sets $(V_i, C_{V_i}^\infty)$ (resp. $(V_i, \mathcal{O}_{V_i}^{an})$) together with isomorphisms of ringed spaces $h_i : V_i \to U_i$ into open subsets of $\mathbb{R}^n$ (resp. $\mathbb{C}^n$) with their respective structure sheaves. The sets V_i are called *charts* and the isomorphisms h_i are called *coordinates*. It is important to realize however that the choice of an atlas is not unique. By gluing of sheaves, an atlas (V_i, h_i) specifies a smooth (resp. complex) manifold provided that $h_i \circ h_j^{-1}$ are smooth (resp. holomorphic) on $h_j(V_i \cap V_j)$.

1.1.1 Example   The n-*sphere*

$$S^n = \{(x_0, \ldots, x_n) \in \mathbb{R}^{n+1} \mid \sum x_i^2 = 1\}$$

is a smooth manifold. The atlas can be chosen as the sets

$$V_{2i} = \{(x_0, \ldots, x_n) \in S^n \mid x_i > 0\},$$

$$V_{2i+1} = \{(x_0, \ldots, x_n) \in S^n \mid x_i < 0\}.$$

One can then let h_{2i}, h_{2i+1} be defined by

$$(x_0, \ldots, x_n) \mapsto (x_0, \ldots, \widehat{x_i}, \ldots, x_n)$$

(the hat means omitting the coordinate). The sphere S^2 also has a structure of a complex manifold because it is $\mathbb{P}^1_{\mathbb{C}}$ with the analytic topology (see the next example). Obviously, odd-dimensional spheres cannot be complex manifolds, and it is also known that S^4 is not diffeomorphic to a complex manifold. In general, whether S^n is a complex manifold is still an open problem. It is also interesting to note that for many higher n, there exist two or more non-diffeomorphic smooth manifolds with the topology of S^n.

1.1.2 Example A smooth abstract variety V over $\mathbb{C}$ gives rise to a complex manifold. The basic point is that locally, the closed points of V can be identified with the set of solutions of a system of equations

$$f_1(z_1, \ldots, z_n) = 0$$
$$\vdots$$
$$f_m(z_1, \ldots, z_n) = 0$$

(1.1.1)

where $f_1, \ldots, f_m$ are rational functions with denominators non-zero on some Zariski (hence analytically) open set $U \subseteq \mathbb{C}^n$. Additionally, $n \geq m$ and we require that the ideal of $\mathcal{O}_U$ (see formula (1.4.6) of Sect. 1.3.2 of Chap. 1) generated by determinants of $m \times m$ submatrices M of the Jacobi matrix

$$\begin{pmatrix} \frac{\partial f_1}{\partial z_1} & \cdots & \frac{\partial f_1}{\partial z_n} \\ \cdots & \cdots & \cdots \\ \frac{\partial f_m}{\partial z_1} & \cdots & \frac{\partial f_m}{\partial z_n} \end{pmatrix}$$

(1.1.2)

contains $1 \in \mathcal{O}_U$. But this means that at no point $(z_1^0, \ldots, z_n^0)$ can all the $m \times m$ determinants $det(M)$ simultaneously be 0, since then their every linear combination with coefficients in $\mathcal{O}_U$ would be 0, which is not true for the constant 1 function. Suppose in a neighborhood V of a point $(z_1^0, \ldots, z_n^0) \in V$, the determinant with columns $i_1 < \cdots < i_m$ is non-zero. Thus, we see from the implicit function theorem that the solutions of (1.1.1) are the graph of a complex-differentiable function of the $n - m$ variables z_i, $i \neq i_j$, which is a complex manifold. This manifold will be denoted by V_{an} and referred to as V *with the analytic topology.* (Note that V_{an} is Hausdorff because by definition V is separated and has a countable basis of topology because V is of finite type.)

1.2   Differential Forms

Let M be a smooth manifold and let $x \in M$. A *tangent vector* of the manifold M at the point x is defined to be a function

$$\partial : C^\infty(M) \to \mathbb{R}$$

which satisfies the following axioms:

$$\partial c = 0 \text{ for } c \text{ constant} \tag{1.2.1}$$

$$\partial(f + g) = \partial f + \partial g \tag{1.2.2}$$

$$\partial(fg) = f(x)\partial g + g(x)\partial f. \tag{1.2.3}$$

Clearly, tangent vectors to M at x form an $\mathbb{R}$-vector space, which we will denote by $T M_x$ and call the *tangent space* to M at x.

For a smooth map (i.e. morphism) of smooth manifolds $\phi : M \to N$, and a point $x \in M$, we clearly have a linear map

$$D\phi_x : T M_x \to T N_{\phi(x)}$$

given, for $\partial \in T M_x$ and $f : N \to \mathbb{R}$, by

$$(D\phi_x(\partial))(f) = \partial(f \circ \phi).$$

The linear map $D\phi_x$ is called the *total differential* of the smooth map ϕ at the point x.

1.2.1 Lemma *Suppose M is a smooth manifold, $f \in C^\infty(M)$, and $\partial \in T M_x$. If there exists $U \subseteq M$ open such that $x \in U$ and $f|_U = 0$, then $\partial f = 0$.*

Proof There exists an open subset $V \subseteq U$ with $x \in V$ and a smooth function $g : M \to \mathbb{R}$ such that $g|_V = 0$, $g|_{M \smallsetminus U} = 1$. We have $f = f \cdot g$. Thus,

$$\partial f = \partial(f \cdot g) = f(x)\partial g + g(x)\partial f = 0.$$

$\square$

So, we have a canonical isomorphism (given by restriction) $T M_x \cong T U_x$ for any open subset U containing x. In particular, we can choose U to be a chart, and it suffices to understand tangent vectors at x on an open subset of $\mathbb{R}^n$. For a vector $v = (a_1, \ldots, a_n)^T \in$

$\mathbb{R}^n$ (here the superscript $?^T$ denotes the transposed, i.e. column, vector), clearly,

$$\partial_v = a_1 \frac{\partial}{\partial x_1} + \cdots + a_n \frac{\partial}{\partial x_n} \tag{1.2.4}$$

is a tangent vector at x.

1.2.2 Lemma *All tangent vectors to an open subset $U \subseteq \mathbb{R}^n$ are of the form (1.2.4).*

Proof Without loss of generality, $x = 0$. It suffices to show that if for a tangent vector ∂, we have

$$\partial f = 0 \tag{1.2.5}$$

for every linear function f, then we have (1.2.5) for every $f \in C^\infty(U)$. Suppose then that (1.2.5) holds for every linear function f. Let $f \in C^\infty(U)$. Let $a = f(0)$. Then

$$g = \frac{f - a}{Df_0}$$

(more precisely the function defined by this formula on non-zero points, and as 1 at 0) is smooth. Thus,

$$\partial f = \partial g \cdot Df_0 + \partial a = g(0)\partial(Df_0) + \partial g \cdot Df_0(0) = 0.$$

$\square$

We have seen that if $M = U$ is an open subset of $\mathbb{R}^n$, then a basis of TM_x is

$$\frac{\partial}{\partial x_1}, \ldots, \frac{\partial}{\partial x_n}. \tag{1.2.6}$$

For an open set $V \subseteq \mathbb{R}^m$ and a map $\phi : U \to V$ given by $\phi = (f_1, \ldots, f_m)$, the matrix of $D\phi_x$ with respect to the basis (1.2.6) of TU_x and the basis

$$\frac{\partial}{\partial y_1}, \ldots, \frac{\partial}{\partial y_m}$$

of $TV_{\phi(x)}$ where $y_1, \ldots, y_m$ are the coordinates of $\mathbb{R}^m$, is the Jacobi matrix (1.1.2).

Note that for $x, y \in U$, we have a canonical isomorphism $TU_x \cong TU_y$. A *vector field* on U is of the form

$$v = g_1 \frac{\partial}{\partial x_1} + \ldots g_n \frac{\partial}{\partial x_n}$$

where $g_i \in C^\infty(U)$. Thus, more precisely, the space $Vect(U)$ of vector fields on U is the free $C^\infty(U)$-module on the basis (1.2.6).

It is important to note that the concept of a smooth vector field on an open subset of $\mathbb{R}^n$ is independent of the choice of coordinates. This is the same thing as observing that for a diffeomorphism $\phi : U \to V$ to an open subset $V \subseteq \mathbb{R}^n$, and a smooth vector field v on U, we obtain a smooth vector field $\phi_*(v)$ on V by the formula

$$y \mapsto D\phi_{\phi^{-1}(y)}(v(\phi^{-1}(y))). \tag{1.2.7}$$

Since the formula (1.2.7) does not in an obvious way generalize to arbitrary smooth maps ϕ, we do not have functoriality of vector fields with respect to general smooth maps.

Consider also the dual space $TM_x^* = Hom_{\mathbb{R}}(TM_x, \mathbb{R})$ to the vector space TM_x (called the *cotangent space*). When $U \subseteq \mathbb{R}^n$, we have a dual basis

$$dx_1, \ldots, dx_n$$

to the basis (1.2.6). Note that these basis elements can indeed be identified with the total differentials of the coordinate functions $x_1, \ldots, x_n$. We also have the k-th exterior power $\Lambda^k TM_x^*$, which has basis

$$dx_{i_1} \wedge \cdots \wedge dx_{i_k}, \ 1 \le i_1 < \cdots < i_k \le n. \tag{1.2.8}$$

We extend this notation by letting, for a permutation σ,

$$dx_{i_{\sigma(1)}} \wedge \cdots \wedge dx_{i_{\sigma(k)}}$$

be the multiple of (1.2.8) by the sign of σ, and letting any product

$$dx_{j_1} \wedge \cdots \wedge dx_{j_k}$$

be 0 if two of the j_i's coincide. Note that this is a special case of the *exterior algebra* construction, which we saw in Exercise 16 of Chap. 4.

A *differential k-form* (briefly, k-form) on U is of the form

$$\sum g_{i_1,\ldots,i_k} dx_{i_1} \wedge \cdots \wedge dx_{i_k}$$

where $g_{i_1,\ldots,i_k} \in C^\infty(U)$. Thus, more precisely, the space $\Omega^k(U)$ of k-forms on U is the free $C^\infty(U)$-module on (1.2.8).

Observe that for an open set $V \subseteq \mathbb{R}^m$, any smooth map $\phi : U \to V$, and a 1-form ω on V, we get a 1-form $\phi^*(\omega)$ on U by using, for $x \in U$ and $v \in TM_x$, the formula

$$((\phi^*(\omega))(x))(v) = (\omega(\phi(x)))(D\phi_x(v)). \tag{1.2.9}$$

Thus, 1-forms are contravariantly functorial with respect to smooth maps of open sets of Euclidean spaces. In particular, they are functorial in either direction with respect to diffeomorphisms of open subsets of $\mathbb{R}^n$, as are vector fields. A similar argument also applies to k-forms.

Obviously, we have sheaves

$$\underline{Vect}_U, \underline{\Omega}^k_U$$

where $\underline{Vect}_U(U) = Vect(U)$, $\underline{\Omega}^k_U(U) = \Omega^k(U)$. By gluing sheaves (using the functoriality with respect to diffeomorphisms of open subsets of $\mathbb{R}^n$), we therefore have sheaves $\underline{Vect}_M$, $\underline{\Omega}^k_M$ on every smooth manifold M, and we also denote their global sections by $Vect(M)$, $\Omega^k(M)$, respectively. The sections of those sheaves are called *vector fields (resp. k-forms) on M*.

We have $\underline{\Omega}^0_M = \mathcal{C}^\infty_M$, which is a sheaf of commutative rings. We see that $\underline{\Omega}^k_M$ are sheaves of modules, and that

$$\underline{\Omega}^1_M = \underline{Hom}_{\mathcal{C}^\infty_M}(\underline{Vect}_M, \mathcal{C}^\infty_M), \ \underline{\Omega}^k_M = \Lambda^k_{\mathcal{C}^\infty_M} \underline{\Omega}^1_M.$$

(The definition of the exterior algebra extends to sheaves of modules in the obvious way.) This is, in fact, a better definition of differential forms, since it is coordinate-free. We additionally see that we have a product

$$\wedge : \underline{\Omega}^k_M \otimes_{\mathcal{C}^\infty_M} \underline{\Omega}^\ell_M \to \underline{\Omega}^{k+\ell}_M,$$

making $\bigoplus_k \underline{\Omega}^k_M$ a sheaf of $\mathcal{C}^\infty_M$-algebras.

1.3 De Rham Cohomology

We begin with the following result:

1.3.1 Lemma *Let M be a smooth manifold. Then there exists a unique homomorphism of abelian sheaves*

$$d : \underline{\Omega}^k_M \to \underline{\Omega}^{k+1}_M$$

which satisfies the following axioms:

1. For a section $f \in \mathcal{C}^\infty(U)$, a point $x \in U$ and a tangent vector $v \in TM_x$,

$$\langle df_x, v \rangle = v(f).$$

2. *For sections* $\omega \in \Omega^k(U), \eta \in \Omega^\ell(U)$, *we have*

$$d(\omega \wedge \eta) = (d\omega) \wedge \eta + (-1)^k \omega \wedge d\eta.$$

3. *We have*

$$d \circ d = 0.$$

COMMENT   In Axiom 1, recall that a differential 1-form can be evaluated at a point $x \in M$ to produce an element of TM_x^*. The notation $\langle ?, ? \rangle$ means the evaluation product between an $\mathbb{R}$-vector space and its dual.

Proof We prove uniqueness first. It suffices to prove for $U \subseteq \mathbb{R}^n$ open. Then by Axiom 1, we have

$$\left\langle df_x, \frac{\partial}{\partial x_i} \right\rangle = \left. \frac{\partial f}{\partial x_i} \right|_x,$$

which implies

$$df_x = \sum_{i=1}^{n} \left. \frac{\partial f}{\partial x_i} \right|_x \cdot dx_i.$$

From Axioms 2, 3, for a smooth function $h : U \to \mathbb{R}$, we must have

$$d(hdx_{i_1} \wedge \cdots \wedge dx_{i_k}) = \sum_{i=1}^{n} \frac{\partial h}{\partial x_i} \cdot dx_i \wedge dx_{i_1} \wedge \cdots \wedge dx_{i_k}.$$

To prove existence, by the uniqueness, it suffices to consider the case of $M = U \subseteq \mathbb{R}^n$ open. To prove 1, it suffices to consider

$$v = \frac{\partial}{\partial x_i},$$

in which case it is by definition. Axiom 2 follows from the Leibniz rule. Axiom 3 follows from the commutation of partial derivatives:

$$\frac{\partial^2 h}{\partial x_i \partial x_j} = \frac{\partial^2 h}{\partial x_j \partial x_i}.$$

$\square$

A *chain complex* is a sequence C of abelian groups (more generally, objects of an abelian category) $(C_k)_k \in \mathbb{Z}$ together with a homomorphism of abelian groups (called *differential*) $d_k : C_k \to C_{k-1}$ such that $d_{k-1} \circ d_k = 0$. One then defines the k-*th homology* of C by

$$H_k(C) = Ker(d_k)/Im(d_{k+1}).$$

For a chain complex, one often denotes $Z_k = Ker(d_k)$ (resp. $B_k = Im(d_{k+1})$) and when dealing with an abelian category where the concept of elements makes sense, calls its elements *cycles* (resp. *boundaries*). We often write simply d for d_k. A *cochain complex* is a sequence C of abelian groups $(C^k)_{k \in \mathbb{Z}}$ together with a differential $d = d^k : C^k \to C^{k+1}$ which satisfies $d^{k+1} \circ d^k = 0$. We define the k-*th cohomology* of C by

$$H^k(C) = Ker(d^k)/Im(d^{k-1}).$$

The symbols Z^k, B^k, and concepts of *cocycles* and *coboundaries* for cochain complexes are defined symmetrically to chain complexes.

Note that, in fact, a chain complex can be made into a cochain complex (and vice versa) by putting

$$C^k = C_{-k}. \tag{1.3.1}$$

Chain complexes (hence also cochain complexes) form an abelian category with respect to *chain maps* where for chain complexes C, D a chain map $f : C \to D$ consists of homomorphisms $f_k : C_k \to D_k$ together with a commutative diagram

$$
\begin{array}{ccc}
C_k & \xrightarrow{f_k} & D_k \\
d \downarrow & & \downarrow d \\
C_{k-1} & \xrightarrow{f_{k-1}} & D_{k-1}.
\end{array}
$$

Cochain maps are defined analogously using the identification (1.3.1). Often, chain complexes are given by defining C_k only for an interval of integers k. In that case, we set the remaining groups C_k to 0. Similarly for cochain complexes. Homology (resp. cohomology) is a functor from chain complexes and chain maps (resp. cochain complexes and cochain maps) to abelian groups. Similar statements hold in any abelian category.

1.3.2 Definition Let M be a smooth manifold. The *de Rham cohomology* $H_{DR}^k(M)$ is defined as the k-th cohomology of the cochain complex

$$\Omega^0(M) \xrightarrow{\ d\ } \Omega^1(M) \xrightarrow{\ d\ } \cdots \xrightarrow{\ d\ } \Omega^n(M),$$

which is called the *de Rham complex* and denoted by $\Omega^*(M)$. It also has an obvious sheaf version, which we denote by $\underline{\Omega}_M^*$ (but note that it is not a cochain complex of C_M^∞-modules).

COMMENT The definition of the differential immediately implies that for a smooth map $f : M \to N$,

$$f^* d = d f^*, \tag{1.3.2}$$

and hence H_{DR}^k is a contravariant functor from the category of smooth manifolds to the category of $\mathbb{R}$-vector spaces.

1.3.3   The Difference Between Vector Fields and 1-Forms

We already noted that vector fields are functorial with respect to diffeomorphisms, while 1-forms are (contravariantly) functorial with respect to smooth maps. It turns out that the two concepts really are quite different. In some sense, any two non-zero vector fields are locally isomorphic, and we can use this to see that there cannot be any natural analogue of the de Rham differential d which would involve vector fields.

1.3.4 Proposition *Let $U \subseteq \mathbb{R}^n$ be open, and let $a \in U$. Consider a vector field*

$$v(x) = f_1(x)\frac{\partial}{\partial x_1} + \cdots + f_n(x)\frac{\partial}{\partial x_n}$$

on U with $v(a) \neq 0$. Then there exits an open set $V \subseteq U$, $a \in V$, and a diffeomorphism

$$\phi : V \xrightarrow{\ \cong\ } W$$

with $W \subseteq \mathbb{R}^n$ open such that

$$\phi_* v = \frac{\partial}{\partial y_1}$$

(where we write $y_1, \ldots, y_n$ for the coordinates in the target).

Proof Without loss of generality, $a = 0$, $v(0) = \frac{\partial}{\partial x_1}$. Choose constants $y_2, \ldots, y_n$ with $(0, y_2, \ldots, y_n)^T \in U$. Consider the following differential equation for a function $x =$

$x_{y_2,\ldots,y_n} : \mathbb{R} \to \mathbb{R}^n$:

$$x'(t) = v(x(t)),$$

with initial condition

$$x(0) = (0, y_2, \ldots, y_n)^T. \tag{1.3.3}$$

We then know that in an open neighborhood of 0, there is a unique solution. Put

$$(x_1, \ldots, x_n)^T = x_{y_2,\ldots,y_n}(y_1).$$

Considering now all y_i as variables, we have

$$Dx|_{(y_1,\ldots,y_n)^T}\left(\frac{\partial}{\partial y_1}\right) = x'_{y_2,\ldots,y_n}(y_1) = v((x_1, \ldots, x_n)^T).$$

Also,

$$Dx|_0 = Id.$$

Thus, by the inverse function theorem, the function defined by the formula (1.3.3) has an inverse in an open neighborhood of 0, which can be taken as ϕ. □

1.3.5 Example There does not exist any diffeomorphism $\phi : U \to V$ of non-empty open subsets of $\mathbb{R}^2$ with

$$\phi^*(x_2 dx_1) = dx_1.$$

This is because we would have, by (1.3.2),

$$d\phi^*(\omega) = \phi^*(d\omega),$$

and $d(dx_1) = 0$, while $d(x_2 dx_1) = -dx_1 \wedge dx_2 \neq 0$, which is a contradiction.

1.4 The de Rham Complex of a Complex Manifold

The reader may wonder why we so far ignored the case of complex manifolds in our discussion of differential forms. In the complex case, in fact, much more can be said, and in some sense, the discussion gets us closer to the algebraic situation (see Sect. 1 of Chap. 6 below). Let us start with the very basic facts here. If M is a complex manifold, it is advantageous to consider functions into $\mathbb{C}$. We can put

$$C^\infty_{M,\mathbb{C}} = C^\infty_M \otimes_{\mathbb{R}} \mathbb{C}, \ \Omega^k_{M,\mathbb{C}} = \Omega^k_M \otimes_{\mathbb{R}} \mathbb{C}, \ \underline{Vect}_{M,\mathbb{C}} = \underline{Vect}_M \otimes_{\mathbb{R}} \mathbb{C}.$$

From the real point of view, $\Omega^k_{M,\mathbb{C}}$ is, of course, just a direct sum of two copies of Ω^k_M, but it is a sheaf of $\mathbb{C}$-modules (in fact, of $C^\infty_{M,\mathbb{C}}$-modules). Similarly for $\underline{Vect}_{M,\mathbb{C}}$.

Now it is advantageous to introduce different $C^\infty_{U,\mathbb{C}}$-bases of the $\mathbb{C}$-vector spaces of sections of these sheaves on open subsets of $U \subseteq \mathbb{C}^n$. On $\mathbb{C}$, letting the complex coordinate be $z = x + iy$, we put

$$\frac{\partial}{\partial z} = \frac{1}{2}\left(\frac{\partial}{\partial x} - i\frac{\partial}{\partial y}\right),$$

$$\frac{\partial}{\partial \bar{z}} = \frac{1}{2}\left(\frac{\partial}{\partial x} + i\frac{\partial}{\partial y}\right).$$

These vectors are chosen in such a way that they are of the same length (with respect to the usual dot product), one has

$$\frac{\partial h(z)}{\partial \bar{z}} = 0$$

for a holomorphic function $h(z)$, and

$$\frac{\partial z}{\partial z} = 1.$$

Similarly, a basis of $TU_z \otimes_{\mathbb{R}} \mathbb{C}$ for U an open subset of $\mathbb{C}^n$ is

$$\frac{\partial}{\partial z_1}, \ldots, \frac{\partial}{\partial z_n}, \frac{\partial}{\partial \bar{z}_1}, \ldots, \frac{\partial}{\partial \bar{z}_n},$$

and the dual basis is

$$dz_1, \ldots, dz_n, d\bar{z}_1, \ldots, d\bar{z}_n.$$

One lets $\Omega^{k,\ell}(U; \mathbb{C})$ be the set of all

$$\sum_{\substack{1 \le i_1 < \cdots < i_k \le n \\ 1 \le j_1 < \cdots < j_\ell \le n}} h_{i_1,\ldots,i_k; j_1,\ldots,j_\ell} dz_{i_1} \wedge \cdots \wedge dz_{i_k} \wedge d\bar{z}_{j_1} \wedge \cdots \wedge d\bar{z}_{j_\ell}$$

where $h_{i_1,\ldots,i_k;j_1,\ldots,j_\ell}$ are smooth functions. Then one has

$$\Omega^n(U;\mathbb{C}) = \bigoplus_{k+\ell=n} \Omega^{k,\ell}(U;\mathbb{C}),$$

and the summands are preserved by holomorphic diffeomorphisms (in fact, contravariantly functorial with respect to holomorphic maps). Thus, one can speak of sheaves

$$\underline{\Omega}_M^{k,\ell}$$

for a complex manifold M, and one has

$$\underline{\Omega}_{M,\mathbb{C}}^n = \bigoplus_{k+\ell=n} \underline{\Omega}_M^{k,\ell}.$$

Further, all these summands are contravariantly functorial with respect to holomorphic maps. The global sections of these sheaves are denoted by $\Omega^{k,\ell}(M)$, $\Omega^n(M;\mathbb{C})$. The differential on $d : \Omega^n(M;\mathbb{C}) \to \Omega^{n+1}(M;\mathbb{C})$ is defined by tensoring the differential on $\Omega^*(M)$ over $\mathbb{R}$ with $\mathbb{C}$. The n'th cohomology of $\Omega^n(M;\mathbb{C})$ is denoted by $H_{DR}^n(M;\mathbb{C})$, and called the n'th complex de Rham cohomology of the complex manifold M, or the n'th de Rham cohomology of the complex manifold M with coefficients in $\mathbb{C}$.

Similarly as in the case of smooth manifolds, we can define differentials

$$\partial : \underline{\Omega}_M^{k,\ell} \to \underline{\Omega}_M^{k+1,\ell},$$

$$\overline{\partial} : \underline{\Omega}_M^{k,\ell} \to \underline{\Omega}_M^{k,\ell+1}$$

by defining them on open sets of $\mathbb{C}^n$ by

$$\partial(h dz_{i_1} \wedge \cdots \wedge dz_{i_k} \wedge d\overline{z}_{j_1} \wedge \cdots \wedge d\overline{z}_{j_\ell})$$
$$= \sum \tfrac{\partial h}{\partial z_i} dz_i \wedge dz_{i_1} \wedge \cdots \wedge dz_{i_k} \wedge d\overline{z}_{j_1} \wedge \cdots \wedge d\overline{z}_{j_\ell},$$

$$\overline{\partial}(h d\overline{z}_{i_1} \wedge \cdots \wedge dz_{i_k} \wedge d\overline{z}_{j_1} \wedge \cdots \wedge d\overline{z}_{j_\ell})$$
$$= \sum \tfrac{\partial h}{\partial \overline{z}_i} d\overline{z}_i \wedge dz_{i_1} \wedge \cdots \wedge dz_{i_k} \wedge d\overline{z}_{j_1} \wedge \cdots \wedge d\overline{z}_{j_\ell}.$$

We have

$$d = \partial + \overline{\partial}.$$

Again, these operators automatically pass to global sections, and are contravariantly functorial with respect to holomorphic maps.

This is an example of a more general construction which is also useful elsewhere. A *double chain complex* is a system of abelian groups (or, more generally, the objects of an abelian category) $(C_{k,\ell})$ together with homomorphisms

$$\partial : C_{k,\ell} \to C_{k-1,\ell}, \quad \overline{\partial} : C_{k,\ell} \to C_{k,\ell-1}$$

such that we have

$$\partial \overline{\partial} + \overline{\partial} \partial = 0, \quad \partial \partial = \overline{\partial} \overline{\partial} = 0. \tag{1.4.1}$$

A *double chain map* between double chain complexes C, D is a collection of homomorphisms $C_{k,\ell} \to D_{k,\ell}$ commuting with both differentials. Again, a double cochain complex is defined as $(C^{k,\ell})$ where $C^{k,\ell} = C_{-k,-\ell}$ is a double chain complex. For a double chain complex $(C_{k,\ell})$, we can obtain a chain complex $|C|$ called its *totalization* (often, the absolute value signs are omitted) by putting

$$|C|_n = \bigoplus_{k+\ell=n} C_{k,\ell}, \quad d = \partial + \overline{\partial}.$$

This is a functor from the category of double chain complexes and double chain maps to the category of chain complexes and chain maps. The $\mathbb{C}$-valued de Rham complex of a complex manifolds is thereby an example of a double chain complex, both on the level of sheaves and global sections, and both are contravariantly functorial with respect to holomorphic maps.

Another example of a double chain complex, which will be useful later, is the *tensor product of chain complexes* C, D of abelian groups, where $(C \otimes D)_{m,n} = C_m \otimes D_n$,

$$\partial = d_C \otimes Id,$$

and for $x \otimes y \in C_m \otimes D_n$,

$$\delta(x \otimes y) = (-1)^m x \otimes d_D(y).$$

This construction generalizes analogously to a tensor product in chain complexes C, D of R-modules over a commutative ring R, which is denoted by $C \otimes_R D$.

Note that by property (2) of Lemma 1.3.1, on a smooth manifold M,

$$\omega \otimes \eta \mapsto \omega \wedge \eta$$

defines a cochain map

$$\Omega^*(M) \otimes_{\mathbb{R}} \Omega^*(M) \to \Omega^*(M) \tag{1.4.2}$$

(and similarly over $\mathbb{C}$ if M is a complex manifold).

1.4.1 The Holomorphic de Rham Complex

On a complex manifold M, one also has the *holomorphic de Rham complex* $\underline{\Omega}^k_{M,Hol}$ (with global sections $\Omega^k_{Hol}(M)$) which, in coordinates for open subsets of $\mathbb{C}^n$, has sections

$$\sum h_{i_1,\ldots,i_k} dz_1 \wedge \cdots \wedge dz_k$$

with $h_{i_1,\ldots,i_k}$ holomorphic. It is again contravariantly functorial with respect to holomorphic maps both in the sheaf and global section versions. Note that one has

$$\underline{\Omega}^k_{M,Hol} = Ker(\overline{\partial} : \underline{\Omega}^{k,0}_M \to \underline{\Omega}^{k,1}_M).$$

This concept brings us closer to the context of algebraic geometry in the sense that holomorphic functions are a better approximation of regular functions on an algebraic variety over $\mathbb{C}$ than smooth functions. Note however that the holomorphic de Rham complex of a complex manifold of dimension n only goes up to degree n, while the de Rham complex goes up to degree $2n$. We will see that for example for $\mathbb{P}^1_{\mathbb{C}}$, the second de Rham cohomology is in fact non-zero. Thus, the holomorphic de Rham complex, while it may appear to be a more natural definition, does not have the same cohomology as the full de Rham complex. We will be able understand these concepts better in Sect. 1 of Chap. 6, after having gone through the prerequisites covered in the rest of this chapter.

2 Derived Categories and Sheaf Cohomology

We will now introduce some concepts which will ultimately help us understand how de Rham cohomology relates to other types of cohomology. The most immediate connection is with *sheaf cohomology*, which, in turn, is a special case of the more general concept of *derived functors*. Derived functors occur on *derived categories*, which we introduce first.

We will study derived categories in general, then specialize to the case of derived categories of abelian categories and derived functors on them, of which sheaf cohomology is an example. To this end, we will also treat the foundations of abelian categories in more detail.

We will conclude the present section by studying in more detail Tor and Ext groups over a commutative ring, and use them to obtain a cohomological criterion for regularity of rings.

2.1 Derived Categories

Sheaf cohomology groups can be defined as groups of homomorphisms in certain derived categories, and essentially the concept is as hard as the general case of the derived category of any abelian category. This is why we begin with a general discussion of derived categories. In the beginning, even restricting to abelian categories is not needed.

Let C be a category and let $E \subseteq Mor(C)$ be a class of morphisms which we will refer to as *equivalences*, and denote by $\sim$. We will assume that E contains all isomorphisms, and also the *two out of three (briefly 2/3) property*, which means that for morphisms $f : X \to Y, g : Y \to Z$, if two out of the three morphisms $f, g, g \circ f$ are in E, so is the third.

A *derived category*, if one exists, is a category $DC = D_E C$ together with a functor $\Phi : C \to DC$ which is the identity on objects, for every equivalence $f \in E, \Phi(f)$ is an equivalence, and the functor Φ is universal with respect to that property. More precisely, for every functor $F : C \to D$ which is the identity on objects such that for every equivalence $f \in E, F(f)$ is an isomorphism, there exists a unique functor $DF : DC \to D$ such that $DF \circ \Phi = F$, i.e. the following diagram of functors strictly commutes:

$$
\begin{array}{ccc}
C & \xrightarrow{\;F\;} & D \\
{\scriptstyle \Phi} \downarrow & \nearrow & \\
DC. & {\scriptstyle DF} &
\end{array}
\tag{2.1.1}
$$

The universality implies that if a derived category exists, it is unique up to isomorphism of categories which is the identity on objects (Exercise 6). Usually, the functors F we consider are not actually identity on objects. This is why the following result is useful:

2.1.1 Lemma *If $F : C \to D$ is any functor such that for $f \in E, F(f)$ is an isomorphism, then there exists a functor $DF : DC \to D$ such that the diagram (2.1.1) commutes up to natural isomorphism, i.e. there exists a natural isomorphism $\eta : F \to DF \circ \Phi$. Moreover, DF is unique in the sense that for another such $\eta' : F \to D'F \circ \Phi$, there exists a unique natural isomorphism $\kappa : DF \to D'F$ such that $(\kappa \circ \Phi) \circ \eta = \eta'$.*

Proof Up to equivalence of categories, we can always introduce new isomorphic objects. Using that, we can assume that F is injective on objects. Now we can replace D with its full subcategory on $F(Obj(C))$. $\qquad \square$

The reader may ask why the existence of a derived category is even in doubt, since it appears to be analogous to factoring a universal algebra by a relation. However, arguing along those lines directly would involve considering a class of proper classes, which we do not allow. This can be remedied by introducing certain general assumptions. However, often, the following more concrete construction, which has some specific advantages, is available.

We say that an object $X \in Obj(C)$ is *E-local* if for all $f : Y \to Z, f \in E$,

$$C(f, X) : C(Z, X) \to C(Y, X)$$

is a bijection. We say that X is *E-colocal* if for all $f : Y \to Z, f \in E$,

$$C(X, f) : C(X, Y) \to C(X, Z)$$

is a bijection.

Let B be a class of local objects. We say that an object X has *localization in B* if there exists an object $X' \in B$ and an equivalence

$$(\gamma_X : X \to X') \in E.$$

If B is a class of colocal objects, we say that an object X has *colocalization in B* if there exists an object $X' \in B$ and an equivalence

$$(\gamma_X : X' \to X) \in E.$$

We say that a category C with the given class of equivalences E has *localization in B* (resp. *colocalization in B*) if every object of B is local (resp. colocal) and every object of C has localization in B (resp. *colocalization in B*). We say that the category C *has localization (resp. colocalization)* when there exists a class of objects B such that C has localization (resp. colocalization) in B.

2.1.2 Proposition *If C has localization or colocalization in B, then a derived category $DC = D_E C$ exists and is equivalent to the full subcategory of C on B.*

Proof We will treat the case of localization (the case of colocalization then follows by passing to opposite categories). The category $D = D_E C$ must have the same objects as C. Choose for every object $X \in Obj(C)$ a localization by B

$$\gamma_X : X \to X'.$$

Define

$$D(X, Y) = C(X', Y').$$

This is automatically a category. The functor Φ must be the identity on objects, and on morphisms is defined by letting, for $f : X \to Y$, $\Phi(f)$ be the morphism g obtained by completing the diagram

$$
\begin{array}{ccc}
X & \xrightarrow{\ f\ } & Y \\
{\scriptstyle \gamma_X}\Big\downarrow{\scriptstyle \sim} & & {\scriptstyle \sim}\Big\downarrow{\scriptstyle \gamma_Y} \\
X' & \dashrightarrow{\ g\ } & Y',
\end{array}
\tag{2.1.2}
$$

which can be done uniquely by the locality of Y'. The uniqueness implies that Φ preserves identity and composition.

To show that for $f \in E$, $\Phi(f)$ is an isomorphism, consider the diagram (2.1.2) with $f \in E$. Then $g \in E$ by the 2/3 property. By locality of X',

$$
C(g, X') : C(Y', X') \to C(X', X')
$$

is now a bijection, so when $h \mapsto Id_{X'}$, $g \circ h = Id_{X'}$. By the same argument, h has a right inverse k, and we have

$$
g = g \circ h \circ k = k.
$$

For universality, let $F : C \to D'$ be a functor which is Id on objects, such that for all $f \in E$, $F(f)$ is an isomorphism. Let $f \in D$ be represented by a morphism $g : X' \to Y'$. Then we may define $DF(f) \in D'(X, Y)$ by

$$
D(F)(f) = F(\gamma_Y)^{-1} \circ F(g) \circ F(\gamma_X).
$$

To prove uniqueness, since F takes every γ_X to an isomorphism, it suffices to prove on objects of B. But for objects X, Y of B, it follows from the fact that for $f \in D(X, Y)$ represented by $g \in C(X', Y')$, there exists a unique $f \in C(X, Y)$ completing the diagram (2.1.2), by the locality of Y. $\qquad\square$

2.2 Properties of Abelian Categories

We shall now fill in some of the proofs of the properties of abelian categories which we noted in Sect. 1.4 of Chap. 2. Recall that an abelian category is defined simply as a category which has finite limits and colimits, has a zero object (i.e. the unique morphism from the initial to the terminal object is an isomorphism), such that every epimorphism is a cokernel, and every monomorphism is a kernel.

2.2.1 Lemma *In an abelian category, a morphism which is both an epimorphism and a monomorphism is an isomorphism.*

Proof An epimorphism $f : A \to B$ is, by definition, the cokernel of some morphism $g : C \to A$. But then, by definition, $f \circ g = f \circ 0$, so if f is a monomorphism, then $g = 0$, and the cokernel of 0 is an isomorphism.

$\square$

2.2.2 Lemma *Suppose in an abelian category, $f : A \to C$, $g : C \to B$ are morphisms.*

(1) If f is a monomorphism and g is the cokernel of f, then f is the kernel of g.
(2) If g is an epimorphism and f is the kernel of g, then g is the cokernel of f.

Proof Clearly, the statements are symmetrical, so it suffices to prove (1). Let $h : K \to C$ be the kernel of g. On the other hand, by assumption, f is a kernel of some morphism $k : C \to D$, and by the universality of cokernel, there exists a $\beta : B \to D$ such that $k = \beta \circ g$. By naturality of limits, we then get a morphism $\alpha : K \to A$ such that $f \circ \alpha = h$.

On the other hand, by universality of kernel, we also have a $\gamma : A \to K$ such that $h \circ \gamma = f$. Since both f and h are monomorphisms, α and γ must be inverse to each other. (Note that in any category, equalizers are monomorphisms.)

$\square$

2.2.3 Lemma *In an abelian category, if $Ker(f) = 0$ then f is a monomorphism. If $Coker(f) = 0$, then f is an epimorphism.*

Proof Clearly, the statements are symmetrical, so it suffices to prove the first one. Suppose in an abelian category, $f : A \to B$ and $Ker(f) = 0$. Suppose $g, h : C \to A$ such that $f \circ g = f \circ h = \phi : C \to B$. Let $\alpha : C' \to C \amalg C$ be the kernel of the codiagonal $\nabla : C \amalg C \to C$. Then we have $f \circ (g \amalg h) = \phi \circ \nabla$ and thus $f \circ (g \amalg h) \circ \alpha = 0$. Thus, $(g \amalg h) \circ \alpha = 0$ and thus, by Lemma 2.2.2, there exists a $\beta : C \to A$ such that $\beta \circ \nabla = g \amalg h$, which we were trying to prove.

$\square$

Note that by Lemmas 2.2.2, 2.2.3, the concept of a short exact sequence is unambiguous: We can write

$$0 \longrightarrow A \overset{f}{\longrightarrow} B \overset{g}{\longrightarrow} C \longrightarrow 0$$

to mean that f is a monomorphism and g is its cokernel, or that g is an epimorphism, and f is its kernel. Those two statements are equivalent. Also note that by Lemma 2.2.3, a morphism whose kernel and cokernel are 0 is an isomorphism.

2.2.4 Lemma *In an abelian category, suppose we have a diagram*

$$
\begin{array}{ccccccccc}
0 & \longrightarrow & A & \longrightarrow & C & \longrightarrow & B & \longrightarrow & 0 \\
 & & \downarrow{\scriptstyle Id} & & \downarrow{\scriptstyle f} & & \downarrow{\scriptstyle Id} & & \\
0 & \longrightarrow & A & \longrightarrow & C' & \longrightarrow & B & \longrightarrow & 0.
\end{array}
\qquad (2.2.1)
$$

whose rows are exact. Then f is an isomorphism.

Proof One shows that both the kernel and cokernel of f are 0. This is left as an exercise. (Exercise 7). □

2.2.5 Lemma *In an abelian category, the canonical morphism*

$$
A \amalg B \to A \sqcap B
$$

is an isomorphism. (Thus, we have a biproduct, *which is denoted by $\oplus$.)*

Proof One observes that we have a short exact sequence

$$
0 \to A \to A \amalg B \to B \to 0
\qquad (2.2.2)
$$

where the first morphism is the coproduct injection. To this end, the second morphism is defined as the cokernel of the injection, and it is formal that the target is B. To show that the injection is a monomorphism, note that it is right inverse to $Id \amalg 0 : A \amalg B \to A \amalg 0 = A$.

   Now symmetrically, we get a short exact sequence

$$
0 \to A \to A \sqcap B \to B \to 0,
\qquad (2.2.3)
$$

and a morphism from (2.2.2) to (2.2.3), thus resulting in a diagram of the form (2.2.1). □

   With a biproduct, we can add two morphisms $f, g : A \to B$ by forming the composition

$$
A \xrightarrow{\ \Delta\ } A \oplus A \xrightarrow{\ f \oplus g\ } B \oplus B \xrightarrow{\ \nabla\ } B.
$$

It remains to show how to *subtract* morphisms. To this end, we consider the diagram

$$
\begin{array}{ccccccccc}
0 & \longrightarrow & C & \longrightarrow & C \oplus C & \longrightarrow & C & \longrightarrow & 0 \\
 & & \downarrow{\scriptstyle Id} & & \downarrow{\scriptstyle \tau} & & \downarrow{\scriptstyle Id} & & \\
0 & \longrightarrow & C & \longrightarrow & C \oplus C & \longrightarrow & C & \longrightarrow & 0
\end{array}
$$

where the horizontal morphisms are injections to the first coordinate and projections to the second coordinate (and hence, the rows are exact), and τ is given by the matrix

$$
\begin{pmatrix} Id & Id \\ 0 & Id \end{pmatrix}.
$$

Then by Lemma 2.2.4, τ is an isomorphism. We see then that its inverse matrix must be of the form

$$
\begin{pmatrix} Id & -Id \\ 0 & Id \end{pmatrix},
$$

thereby providing a definition of $-Id$.

From this point on, it is not difficult to prove all "usual additive properties of the category Ab of abelian groups" in any abelian category. The reader is referred to [3] for further details.

2.3 The Derived Category of an Abelian Category

Let A be an abelian category. We will learn here how to construct the derived category of the category A-Chain of chain complexes in A where the equivalences E are *quasiisomorphisms*, i.e. chain maps which induce an isomorphism on homology. This is sometimes referred to as the *derived category* DA *of the abelian category* A. We will also consider full subcategories of DA on chain complexes C which are *bounded below* or *bounded above* which means that there exists a constant N such that $C_n = 0$ for $n < N$ (resp. $C_n = 0$ for $n > N$). The notations A-Chain$_+$, A-Chain$_-$, DA_+, DA_- mean full subcategories on bounded below (resp. bounded above) chain complexes.

We will work in chain complexes throughout this subsection. The statements proved are, of course, automatically valid for cochain complexes, with "bounded below" and "bounded above" interchanged.

2.3.1 Projective and Injective Objects

An object P of an abelian category $\mathcal{A}$ is called *projective* if for every epimorphism $A \to B$, every morphism $P \to B$ factors through A:

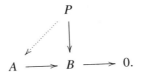

An abelian category $\mathcal{A}$ is said to have *enough projectives* if, for every object A, there exists an epimorphism

$$P \to A \to 0.$$

Dually, an object Q is *injective* if for every monomorphism $A \to B$, every morphism $A \to Q$ factors through B:

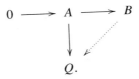

An abelian category $\mathcal{A}$ is said to have *enough injectives* if for every object A there exists a monomorphism

$$0 \to A \to Q$$

with Q injective.

2.3.2 The Homotopy Category of Chain Complexes

A *chain homotopy* between two chain maps $f : C \to D$ in $\mathcal{A}$-Chain is a sequence of $\mathcal{A}$-homomorphisms $h = h_n : C_n \to D_{n+1}, n \in \mathbb{Z}$, such that

$$f_n - g_n = dh_n + h_{n-1}d.$$

Two chain maps are called *chain homotopic* if there exists a chain homotopy between them. We write $h : f \simeq g$, or just $f \simeq g$. Clearly, this is an equivalence relation, and the equivalence classes are called *chain homotopy classes*. The concept of chain homotopy is compatible with composition, and thus, we have a category, called the *homotopy category of chain complexes* $h\mathcal{A}$-Chain whose objects are $\mathcal{A}$-chain complexes, and morphisms are chain-homotopy classes of chain maps. An isomorphism of chain complexes C, D in $h\mathcal{A}$-Chain is called a *chain homotopy equivalence*. If a chain homotopy equivalence exists, the chain complexes C, D are called *chain homotopy equivalent* and one writes $C \simeq D$.

It is immediate that chain-homotopic chain maps induce the same homomorphisms in homology, and thus the concept of a quasiisomorphism passes to the homotopy category. We will also consider the full subcategories $h\mathcal{A}$-Chain$_+$, $h\mathcal{A}$-Chain$_-$ on bounded below (resp. bounded above) chain complexes.

2.3.3 Lemma *If a derived category of an abelian category $\mathcal{A}$ exists, the functor $\Phi : \mathcal{A}$-Chain$\rightarrow D\mathcal{A}$ factors through the canonical functor $\mathcal{A}$-Chain$\rightarrow h\mathcal{A}$-Chain.*

Proof We consider the chain complex of abelian groups I where $I_0 = \mathbb{Z} \oplus \mathbb{Z}$, $I_1 = \mathbb{Z}$, $I_n = 0$ for $n \neq 0, 1$, and $d : I_1 \rightarrow I_0$ is given by the matrix $(1, -1)^T$. Considering an object A of an abelian category as a chain complex which is A in degree 0 and 0 in other degrees, we have quasiisomorphisms $i_1, i_2 : \mathbb{Z} \rightarrow I$ given by the matrices $(1, 0)^T$, $(0, 1)^T$ in degree 0 (and, of course, by 0 elsewhere), and a quasiisomorphism $\epsilon : I \rightarrow \mathbb{Z}$ given in degree 0 by the matrix $(1, 1)$ (and, again, necessarily by 0 elsewhere).

Now we have a well defined tensor product of an object of an abelian category with a finitely generated abelian group. For a chain complex C in an abelian category $\mathcal{A}$, it follows from more general facts, but one can also verify directly that

$$C \otimes i_1, C \otimes i_2 : C \rightarrow C \otimes I \qquad (2.3.1)$$

are quasiisomorphisms (since cycles in $C \otimes I$ are spanned by $c \otimes \alpha_i$ where $c \in C_n$ is a cycle and α_1, α_2 are the generators of the of the two $\mathbb{Z}$-summands of I_0; Exercise 8.)

Now for chain maps $f, g : C \rightarrow D$, consider the diagram

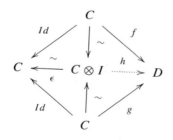

where the vertical arrows are $C \otimes i_1$, $C \otimes i_2$. One verifies that a chain homotopy between f and g is equivalent to filling the dotted arrow in the diagram. Upon applying the functor Φ into the derived category, ϵ becomes an isomorphism, so $f = h \circ \epsilon^{-1} = g$.

$\square$

Now the main result about derived categories of abelian categories is the following

2.3.4 Theorem *1. If an abelian category A has coproducts and enough projectives, then hA-Chain has colocalization.*

2. If an abelian category A has enough projectives, then hA-Chain$_+$ has colocalization by bounded below chain complexes of projective objects.

3. If an abelian category A has products and enough injectives, then hA-Chain has localization.

4. If an abelian category A has enough injectives, then hA-Chain$_-$ has localization by bounded above chain complexes of injective objects.

The derived categories of hA-Chain, hA-Chain$_+$, hA-Chain$_-$ are usually denoted by DA, DA_+, DA_- and called the *derived category (resp. bounded below derived category, resp. bounded above derived category)* of the abelian category A.

2.3.5 Introduction to homological algebra

We shall now start developing the tools needed to prove Theorem 2.3.4. This will take several sections. The proof of (1) and (2) will be finished in 2.3.11. The proof of (3) and (4) will be done in Sect. 2.3.12, using some technical facts the proof of which will follow.

For an A-chain map $f : C \to D$, the *mapping cone* is the totalization of the double chain complex Cf where $Cf_{n,0} = D_n$, $Cf_{n,1} = C_n$ and $Cf_{n,p} = 0$ for $p \neq 0, 1$, ∂ coincides with the differential on C, D and $\delta : Cf_{n,1} \to Cf_{n,0}$ is $(-1)^n f_n$. As usual, one omits the totalization signs. Then we get canonical chain maps $i : D \to Cf$, $j : Cf \to C[1]$ where for a chain complex Q, $Q[n]$ denotes shift up by n, i.e. the chain complex with $Q[n]_k = Q_{k-n}$. We see that from this point of view, the mapping cone construction is self-dual up to a shift: we can call $C[-1]$ the *mapping co-cone*. A special case of the mapping cone is the *cone* $CX = C(Id : X \to X)$.

One also sees that we have a canonical chain map

$$Cf \to D/Im(f) \tag{2.3.2}$$

and a short exact sequence of chain complexes

$$0 \longrightarrow D \stackrel{i}{\longrightarrow} Cf \stackrel{j}{\longrightarrow} C[1] \longrightarrow 0. \tag{2.3.3}$$

It is interesting to note that if we denote by $[C, D]$ the abelian group of chain-homotopy classes of chain maps from C to D (i.e. morphisms in hA-Chain), then for any chain complex X, we have an exact sequence

$$[Cf, X] \stackrel{[i,X]}{\longrightarrow} [D, X] \stackrel{[f,X]}{\longrightarrow} [C, X]. \tag{2.3.4}$$

The map $i \circ f$ is chain-homotopic to 0, which shows that the composition of the two maps (2.3.4) is 0. On the other hand, Cf is isomorphic to the colimit of the diagram

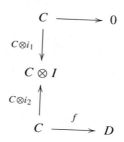

and thus, a chain map $D \to X$ chain-homotopic to 0 when composed with f extends to Cf.

It is also interesting to continue the procedure of taking mapping cones: Continuing to denote by $i : D \to Cf$ the canonical map, using (2.3.3) and (2.3.2) applied to i instead of f, we obtain a canonical map

$$Ci \to C[1]$$

which one can check to be a chain-homotopy equivalence (i.e. an isomorphism in $h\mathcal{A}$-Chain). Repeating this also for the canonical monomorphism $k : Cf \to Ci$, one obtains a commutative diagram of the form

$$
\begin{array}{ccc}
 & Ci \longrightarrow Cj \\
\nearrow & \downarrow{\scriptstyle\sim} \qquad \downarrow{\scriptstyle\sim} \\
C \xrightarrow{f} D \xrightarrow{i} Cf \xrightarrow{j} C[1] \xrightarrow{-f[1]} D[1].
\end{array}
\qquad (2.3.5)
$$

Given the self-duality of the mapping cone (up to shift), one has analogous results upon turning around arrows. Combining (2.3.4) and (2.3.5), one then gets the following

2.3.6 Theorem *For any chain map $f : C \to D$ of $\mathcal{A}$-chain complexes, we have canonical long exact sequences*

$$\cdots \longrightarrow [C[n+1], X] \longrightarrow [Cf[n], X] \xrightarrow{[i[n],X]} [D[n], X] \xrightarrow{[f[n],X]} [C[n], X] \longrightarrow \cdots$$

$$\cdots \longrightarrow [X, C[n]] \xrightarrow{[X,f[n]]} [X, D[n]] \xrightarrow{[X,i[n]]} [X, Cf[n]] \longrightarrow [X, C[n+1]] \longrightarrow \cdots$$

$$(2.3.6)$$

$$(2.3.7)$$

$\square$

Constructions of this type are often referred to as *homological algebra*. Perhaps the best-known fact of homological algebra is the following

2.3.7 Theorem *For a short exact sequence of chain complexes*

$$0 \longrightarrow C' \overset{i}{\longrightarrow} C \overset{j}{\longrightarrow} C'' \longrightarrow 0,$$

there is a long exact sequence

$$\cdots \longrightarrow H_n C' \overset{H_n i}{\longrightarrow} H_n C \overset{H_n j}{\longrightarrow} H_n C'' \overset{\Delta}{\longrightarrow} H_{n-1} C' \longrightarrow \cdots$$

where Δ is functorial in the category of short exact complexes of chain complexes, and double chain maps.

Proof To construct the morphism Δ, one observes that we also have a canonical chain map

$$C'[1] \to Cj \tag{2.3.8}$$

using the chain map i to map C'_n into $(Cj)_{n,1}$, utilizing the fact that $j \circ i = 0$. Now one shows that (2.3.8) induces an isomorphism in homology. (This can be checked directly.) Thus, Δ can be induced just by the canonical chain map $C'' \to Cj$. The rest of the proof, which is just a series of applications of the properties of the kernel and image in abelian categories, is left as an exercise (Exercise 9). □

Another standard fact of homological algebra is the following statement often used in conjunction with Theorem 2.3.7. Again, a proof follows by examining the properties of the image and kernel in an abelian category (Exercise 10).

2.3.8 Lemma (The 5-Lemma) *Consider a diagram in an abelian category*

$$
\begin{array}{ccccccccc}
A' & \longrightarrow & B' & \longrightarrow & C' & \longrightarrow & D' & \longrightarrow & E' \\
\downarrow{\scriptstyle a} & & \downarrow{\scriptstyle b} & & \downarrow{\scriptstyle c} & & \downarrow{\scriptstyle d} & & \downarrow{\scriptstyle e} \\
A & \longrightarrow & B & \longrightarrow & C & \longrightarrow & D & \longrightarrow & E.
\end{array}
$$

Assume that a is an epimorphism, e is monomorphism, and b, d are isomorphisms. Then c is an isomorphism. □

While Theorems 2.3.6 and 2.3.7 may seem similar, it is important to note that they are different results for an arbitrary abelian category $\mathcal{A}$. For example, (2.3.6), (2.3.7) are long exact sequences of abelian groups, while in Theorem 2.3.7, we are dealing with exact sequences in $\mathcal{A}$. There is, nevertheless, also an approach to Theorem 2.3.7 along the lines of Theorem 2.3.6, interpreting, for a finitely generated abelian group H and an object A of an abelian category $\mathcal{A}$, $Hom(H, A)$ as an object of $\mathcal{A}$.

2.3.9 Cell Chain Complexes

Now a *cell $\mathcal{A}$-chain complex C* is of the form

$$C = \operatorname{colim} C_{(m)}$$

where $C_{(-1)} = 0$ and for all $m \in \mathbb{N}_0$, $C_{(m)}$ is the mapping cone of a chain map

$$i_m : P_{(m)} \to C_{(m-1)}$$

where $P_{(m)}$ is a chain complex of $\mathcal{A}$-projective objects with 0 differential.

To prove that cell chain complexes are colocal, in view of the equivalence (2.3.8), Theorem 2.3.7, and (2.3.7), it suffices to prove the following

2.3.10 Lemma *If C, X are chain complexes in $\mathcal{A}$ where C is cell and $H_m(X) = 0$ for all $m \in \mathbb{Z}$, then*

$$[C, X] = 0.$$

Proof Let $f : C \to X$ be a chain map. We shall produce, by induction, chain maps $CC_{(n)} \to X$ (recall that CY denotes the cone on a chain complex Y) extending each other as well as restrictions of the chain map f. Their colimit will then be a map $CC \to X$ which is equivalent to the required homotopy. For $n = -1$, there is nothing to prove. Assuming the homotopy has been constructed for a given n, let $Q_{(n)}$ be the pushout of the diagram

$$
\begin{array}{ccc}
C_{(n)} & \longrightarrow & C_{(n+1)} \\
\downarrow {\scriptstyle C_{(n)} \otimes i_1} & & \\
CC_{(n)} & &
\end{array}
$$

Then restriction of the map f to $C_{(n+1)}$, together with the already constructed chain map $CC_{(n)} \to X$ produce a chain map

$$Q_{(n)} \to X.$$

On the other hand, we have a canonical colimit chain map

$$Q_{(n)} \to CC_{(n+1)}$$

and, in fact, we have a chain map $g_n : P_{(n+1)}[1] \to Q_{(n)}$ and an isomorphism

$$CC_{(n+1)} \cong Cg_n.$$

By Theorem 2.3.6, it suffices to prove that for a projective object P of $\mathcal{A}$ and an $\mathcal{A}$-chain complex X with 0 homology,

$$[P[m], X] = 0. \tag{2.3.9}$$

But this is trivial: a chain map (2.3.9) is the same thing as a morphism $P \to Z_m$, and by assumption, we have $B_m = Z_m$. Then apply projectivity to the epimorphism $X_{m+1} \to B_m \to 0$. □

2.3.11 Proof of (1) and (2) of Theorem 2.3.4

For (1), we just proved that cell objects are colocal. Now to prove colocalization, we need to produce, for an arbitrary $\mathcal{A}$-chain complex X, an equivalence $\gamma : C \to X$ where C is cell. To this end, keeping in mind that $\mathcal{A}$ has enough projectives, and using the notation of Sect. 2.3.9, simply select an epimorphism

$$P_{(0)}[1] \to H_* X$$

(one uses $H_* X$ for the graded $\mathcal{A}$-object $(H_n(X))$), and then inductively an epimorphism

$$P_{(m)} \to Ker(H_* C_{(m-1)} \to H_* X),$$

as inductively, we also produce chain maps $C_{(m)} \to X$. The fact that the colimit chain map is an equivalence follows from commutation of homology with colimits of sequences.

For (2) of Theorem 2.3.4, we note that when X is bounded below, a better construction is available. One notes that in fact, a bounded below chain complex of projective objects is automatically cell. To construct the colocalization in $h\mathcal{A}$-Chain$_+$, consider first the case when $X = M$ is just an object. Then from the assumption that there are enough projectives, we can produce a short exact sequence

$$0 \to M_1 \to P_0 \to M \to 0$$

and, inductively, short exact sequences

$$0 \to M_{n+1} \to P_n \to M_n \to 0.$$

Using the connecting maps, we then have produced an exact sequence

$$\cdots \to P_2 \to P_1 \to P_0 \to M \to 0$$

or, equivalently, a chain complex

$$P = (\cdots \to P_2 \to P_1 \to P_0)$$

and an equivalence $P \to M$. The chain complex P is called a *projective resolution* of M.

Now when X is bounded below, assume, without loss of generality, that $X_n = 0$ for $n < 0$. one can produce, inductively on n, a cell chain complex of projective objects $C_{(n)}$ where $(C_{(n)})_m$ is non-zero only for $0 \le m \le n$ and chain maps

$$\gamma_n : C_{(n)} \to X$$

which induce isomorphisms in homology in degrees $0 \le i \le n-1$, and an epimorphism in degree $i = n$. Such a chain map is sometimes called an *n-equivalence*. Then the mapping cone $C\gamma_n$ has 0 homology in degrees $< n + 1$, and choosing an epimorphism $P_{n+1} \to H_{n+1}C\gamma_n$, we may interpret it as a chain map $P_{n+1}[n + 1] \to C\gamma_n$. Thus, we obtain a chain map $P_{n+1}[n] \to C_{(n)}$, and if we take $C_{(n+1)}$ to be its mapping cone, we may extend the chain map γ_n to an $(n + 1)$-equivalence $\gamma_{n+1} : C_{(n+1)} \to X$. Observe carefully that restricting attention to this case, we do not need to require that $\mathcal{A}$ have arbitrary coproducts, as long as it has enough projectives.

2.3.12 Proof of (3) and (4) of Theorem 2.3.4

It is tempting to say that the proof of these cases of Theorem 2.3.4 is the precise dual of what we just did. This is, in fact, almost correct. The case (4) is, in fact, dual in this fashion. Notably, we have a concept of an *injective resolution* of an object M, which is a chain complex of the form

$$Q = (Q_0 \to Q_{-1} \to Q_{-2} \to \ldots)$$

with Q_n injective, and an isomorphism

$$M \cong H_0(Q),$$

while $H_i Q = 0$ for $i \ne 0$. Note that it is more natural in this case to think of Q as a cochain complex.

Now most of the proof of (3) is indeed a precise dual of (1). In particular, we have a notion of a *co-cell chain complex*, which the reader is encouraged to write down as a precise dual of the notion of a cell chain complex.

The one snag in this duality is that for a sequence

$$\cdots \to C_{(2)} \to C_{(1)} \to C_{(0)}$$

of chain complexes, it is in general *false* that the canonical map

$$H_* \lim C_{(n)} \to \lim H_* C_{(n)}$$

would be an isomorphism. It is, however, true under certain special conditions (see Proposition 2.3.14), which apply in the case of a localization by co-cell chain complexes which is precisely dual to our construction of co-localization by cell chain complexes. The relevant story is explained below.

Let $\cdots \to A_2 \to A_1 \to A_0$ be a sequence of objects in an abelian category $\mathcal{A}$. We say that the sequence satisfies the *Mittag-Leffler condition* if for each $n \in \mathbb{N}_0$, the sequence of images of A_m in A_n, $m \geq n$, is eventually constant in m.

Consider now a sequence of short exact sequences:

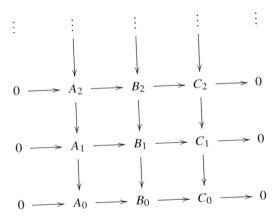

The key point is the following fact, which is an easy consequence of the properties of the image (Exercise 11):

2.3.13 Lemma

1. *If the sequence* (A_n) *satisfies the Mittag-Leffler condition, then the induced sequence*

$$0 \to \lim A_n \to \lim B_n \to \lim C_n \to 0$$

is exact. (Note: The first four terms are always exact.)

2. *If the sequences (A_n), (C_n) satisfy the Mittag-Leffler condition, so does the sequence (B_n).*

□

2.3.14 Proposition *Let $\mathcal{A}$ be an abelian category with products. Let*

$$\cdots \to {}_2C \to {}_1C \to {}_0C$$

be a sequence of $\mathcal{A}$-chain complexes, and let C be its limit. If for each $n \in \mathbb{Z}$, the sequences

$$\cdots \to {}_2C_n \to {}_1C_n \to {}_0C_n,$$

$$\cdots \to H_n({}_2C) \to H_n({}_1C) \to H_n({}_0C)$$

satisfy the Mittag-Leffler condition, then for each $n \in \mathbb{Z}$, the canonical map

$$H_n C \to \lim_k H_n({}_kC)$$

is an isomorphism.

Proof Recall that in a chain complex C, one sometimes denotes $Z_n = Ker(d : C_n \to C_{n-1})$ (resp. $B_n = Im(d : C_{n+1} \to C_n)$) and these the subobjects of cycles, resp. boundaries. (Of course, speaking of elements does not in general make sense in an abelian category.) We have short exact sequences

$$0 \to B_n \to Z_n \to H_n(C) \to 0, \tag{2.3.10}$$

$$0 \to Z_n \to C_n \to B_{n-1} \to 0. \tag{2.3.11}$$

Now applying this to ${}_kC$, (2.3.10) and (2.3.11) become sequences of short exact sequences in the variable k. If the sequence $({}_kC_n)$ satisfies the Mittag-Leffler condition for each n, so does the sequence $({}_kB_n)$ by properties of the image. Assuming also $({}_kH_nC)$ satisfies the Mittag-Leffler condition, so does $({}_kZ_n)$ by (2.3.10) and Lemma 2.3.13 (2). Thus, the inverse limits of both sequences (2.3.10), (2.3.11) over $C = {}_kC$ are exact and since the canonical morphism

$$Ker(\lim d : \lim {}_kC_n \to \lim {}_kC_{n-1}) \to \lim {}_kZ_n$$

is an isomorphism by commutation of limits, our statement follows.

□

2.3.15 Derived Functors

Let $\Phi : C \to C'$ be a functor where C, C' are small categories. Let, for a category D, $Funct(C, D)$ denote the category of functors from C to D and natural transformations. Then we have a "forgetful functor"

$$\Phi^* : Funct(C', D) \to Funct(C, D)$$

given by composition with Φ. This functor has a left (resp. right) adjoint $\Phi_\sharp$ (resp. Φ_*) called the *left (resp. right) Kan extension*. The construction of Kan extensions is, in fact, a direct generalization of their construction in the case of group actions in Sect. 5.2 of Chap. 3 above.

For a functor $F : C \to D$, $L_\Phi F = LF = \Phi_* F$ (resp. $R_\Phi F = RF = \Phi_\sharp F$) is called the *left (resp. right) derived functor* of F along Φ. (Note the switch in terminology of "left" and "right" between Kan extensions and derived functors; this is due to historical reasons.) The unit and counit of adjunction now give canonical natural transformations

$$\epsilon : \Phi^* LF \to F,$$

$$\eta : F \to \Phi^* RF,$$

and an object of C is called *F-colocal* (resp. *F-local*) (with respect to Φ) if η_X (resp. ϵ_X) is an isomorphism.

Removing the assumption of smallness on the categories C, C', the trouble is that $Funct(C, D)$, $Funct(C', D)$ now become illegal objects of set theory. Nevertheless, the above definitions still can be written down as conditions on the specific functors and objects, which are not guaranteed to exist, but are unique up to natural isomorphism if they do. A functor for which a left (resp. right) derived functor exists is called *left derivable* (resp. *right derivable*).

Concretely, we say that $LF = L_\Phi F$ is a left derived functor of F along Φ if it is a universal functor with a natural transformation $\epsilon : LF \circ \Phi \to F$:

$$
\begin{array}{ccc}
 & C & \\
\Phi \Big\downarrow & \overset{\nearrow \epsilon}{} & \searrow^{F} \\
C' & \cdots\cdots\!\!\!\!> & D, \\
 & LF &
\end{array}
$$

i.e. for any functor $F' : C' \to D$ with a natural transformation $\zeta : F' \circ \Phi \to F$, there exists a unique natural transformation $\kappa : F' \to LF$ such that

$$\zeta = \epsilon \circ (\kappa \circ \Phi).$$

An object $X \in Obj(C)$ is called F-colocal if ϵ_X is an isomorphism.

The concept of the right derived functor $RF = R_\Phi F$ is dual, i.e. we have a natural transformation

$$\eta : F \to RF \circ \Phi$$

such that for any functor $F' : C' \to D$ with a natural transformation $\mu : F \to F' \circ \Phi$, there is a unique natural transformation $\lambda : RF \to F'$ such that

$$\mu = (\lambda \circ \Phi) \circ \eta.$$

An object X of C is called F-local with respect to Φ if η_X is an isomorphism.

Often the interesting case is when $C' = DC$ is the derived category and $\Phi : C \to DC$ is the universal functor. In that case, essentially from the definitions, we obtain the following

2.3.16 Proposition *Let $\Phi : C \to DC$ be the universal functor where DC is the derived category of C with respect to a class of equivalences E which includes all isomorphism, and satisfies 2/3.*

1. *If C has localization by a class B of objects, then a right derived functor $RF = R_\Phi F$ exists and is given by $RF(X) = F(X')$ for $X' \in B$ with an equivalence $\gamma_X : X \to X'$, and*

$$\eta_X = F(\gamma_X).$$

   In particular, a local object is F-local.
2. *If C has colocalization by a class of objects B, then a left derived functor $LF = L_\Phi F$ exists and is given by $LF(X) = F(X')$ for $X' \in B$ with an equivalence $\gamma_X : X' \to X$, and*

$$\epsilon_X = F(\gamma_X).$$

In particular, a colocal object is F-colocal.

$\square$

2.3.17 Proposition *Suppose C is a category with a class of equivalences E containing all isomorphisms and satisfying 2/3, and a derived category DC exists. Then for an object $X \in Obj(C)$, with respect to Φ, the functor $C(?, X) : C \to Sets^{Op}$ is left derivable, and* *its left derived functor is $C(\Phi(?), \Phi(X))$, and the functor $C(X, ?) : C \to Sets$ is right derivable, and its right derived functor is $C(\Phi(X), \Phi(?))$.*

□

2.3.18 Derived Functors in an Abelian Category

Let us now consider the case when $\mathcal{A}$, $\mathcal{B}$ are abelian categories, and $F : \mathcal{A} \to \mathcal{B}$ is an additive functor. Then F canonically extends to a functor from $h\mathcal{A}$-Chain to $h\mathcal{B}$-Chain, which is usually denoted also by F. If $\mathcal{A}$ has enough projectives (resp. injectives), it follows from Proposition 2.3.16 that the functor F is automatically left- (resp. right-) derivable with respect to the universal functor Φ from $h\mathcal{A}$-Chain to its derived category. The derived functors LF, RF are often referred to as the *total left (resp. right) derived functor*. One also denotes $L_n F = H_n(LF) = L(H_n \circ F)$, $R^n F = H_{-n}(RF) = R(H_{-n} \circ F)$ and calls them the *n'th left (resp. right) derived functor*. For $X, Y \in Obj(\mathcal{A})$, one denotes by $Ext^n_{\mathcal{A}}(X, Y)$ the value on X of the n-th right derived functor of

$$Hom_{\mathcal{A}}(?, Y) : \mathcal{A}^{Op} \to Ab,$$

or equivalently the value on Y of the n-th right derived functor of

$$Hom_{\mathcal{A}}(X, ?) : \mathcal{A} \to Ab.$$

2.3.19 Proposition *Let $\mathcal{A}$ be an abelian category. let*

$$0 \to M \to N \to P \to 0 \qquad (2.3.12)$$

be a short exact sequence, and let $F : \mathcal{A} \to \mathcal{B}$ be an additive functor where $\mathcal{B}$ is another abelian category.

1. *If $\mathcal{A}$ has enough projectives, then (2.3.12) gives rise to a long exact sequence of the form*

$$\dots L_n F M \to L_n F N \to L_n F P \to L_{n-1} F P \to \dots$$

   where for $n < 0$, we have $L_n F = 0$. If F is right exact, then $L^0 F \cong F$.
2. *If $\mathcal{A}$ has enough injectives, then (2.3.12) gives rise to a long exact sequence of the form*

$$\dots R^n F M \to R^n F N \to R^n F P \to R^{n+1} F M \to \dots$$

   where for $n < 0$, we have $R^n F = 0$. If F is left exact, then $R^0 F \cong F$.

Proof We shall prove (1). If C is a projective resolution of M and D is a projective resolution of N, then we have a morphism $f : C \to D$ which on homology induces the first morphism (2.3.12). Then the sequence

$$C \xrightarrow{\ f\ } D \xrightarrow{\ i\ } Cf$$

induces (2.3.12) on homology, and consists of colocal objects. Applying the functor F preserves mapping cones, so the exactness statement follows from Theorem 2.3.7. The statement about L_0 follows from the definition of right exactness: If

$$\cdots \to P_1 \to P_0$$

is a projective resolution of M, then we have an induced exact sequence

$$F P_1 \to F P_0 \to F M \to 0.$$

The proof of (2) is symmetrical.

$\square$

2.3.20 Proposition *Let A be an abelian category with enough injectives and let $F : A \to B$ be an additive functor to another abelian category which is left exact. Then an object X of A is F-local if and only if*

$$R^n F(X) = 0 \text{ for all } n > 0.$$

An analogous result holds if we replace R with L, injectives with projectives and local with colocal.

Proof Let Q be an injective resolution of X. Then we have the quasiisomorphism

$$\epsilon : X \to Q. \tag{2.3.13}$$

Since F is left exact, however, $R^0 F\epsilon$ is an isomorphism. Now consider the diagram

$$
\begin{array}{ccc}
FX & \xrightarrow{\ \eta\ } & RFX \\
\downarrow & & \downarrow{\scriptstyle\sim} \\
FQ & \xrightarrow{\ \sim\ } & RFQ.
\end{array}
$$

By what we just said, the left vertical arrow is a quasiisomorphism if and only if $R^i F X = 0$ for $i > 0$. Thus, our statement follows. □

The following result is often useful in practical calculations:

2.3.21 Proposition

1. Let $\mathcal{A}$ be an abelian category with enough projectives, and let $F : \mathcal{A} \to \mathcal{B}$ be an additive functor where $\mathcal{B}$ is another abelian category. Consider a double chain complex of the form

$$C = ((C_{m,n}), \partial, \delta)$$

for some $k, \ell \in \mathbb{Z}$, $C_{m,n} = 0$ if $m < k$ or $n < \ell$, and the chain complexes $(C_{m,*}, \delta)$ are F-colocal for each $m \in \mathbb{Z}$. Then the totalization of the double chain complex C is F-colocal. In particular, a bounded below chain complex of F-colocal objects of $\mathcal{A}$ is F-colocal.

2. Let $\mathcal{A}$ be an abelian category with enough injectives, and let $F : \mathcal{A} \to \mathcal{B}$ be an additive functor where $\mathcal{B}$ is another abelian category. Consider a double chain complex of the form

$$C = ((C_{m,n}), \partial, \delta)$$

for some $k, \ell \in \mathbb{Z}$, $C_{m,n} = 0$ if $m > k$ or $n > \ell$, and the chain complexes $(C_{m,*}, \delta)$ are F-local for each $m \in \mathbb{Z}$. Then the totalization of the double chain complex C is F-local. In particular, a bounded above chain complex of F-local objects of $\mathcal{A}$ is F-local.

Proof The two proofs are completely symmetrical. We shall prove (1). First, if we also assume there is a constant K such that $C_{m,n} = 0$ for $m > K$, then C can be obtained from the chain complexes $C_{m,*}$ by repeated application of mapping cones. Thus, to prove that case, it suffices to show that if $f : C \to D$ is a chain map and C, D are F-local, then so is Cf. To this end, by properties of the mapping cone, we may construct a diagram

$$\begin{array}{ccccc}
C' & \xrightarrow{f'} & D' & \xrightarrow{i'} & Cf' \\
{\scriptstyle \gamma_C}\downarrow{\scriptstyle \sim} & & {\scriptstyle \gamma_D}\downarrow{\scriptstyle \sim} & & \downarrow{\scriptstyle g} \\
C & \xrightarrow{f} & D & \xrightarrow{i} & Cf
\end{array}$$

commutative up to chain homotopy (i.e. commutative in $h\mathcal{A}$-Chain) where C', D' are cell. Then Cf' is also cell. We are also assuming that $F(\gamma_C)$, $F(\gamma_D)$ are quasiisomorphisms.

Since, by definition, F preserves chain homotopy and mapping cones, it suffices to prove that g and $F(g)$ are quasiisomorphisms. In view of Theorem 2.3.7, these statements follow from the 5-lemma (Lemma 2.3.8).

Now to remove the assumption about the existence of K, note that we can form a double chain complex $_K C$ where $_K C_{m,n} = C_{m,n}$ for $m \leq K$, and $_K C_{m,n} = 0$ for $m > K$. Then we have a canonical double chain map $_K C \to C$ which is an isomorphism in degrees (m, n) with $m \leq K$. In particular, upon totalization, we obtain an isomorphism in degrees $\leq K + \ell$. Now by the proof of (2) of Theorem 2.3.4, we may choose $\gamma_C : C' \to C$, $\gamma_{_K C} : _K C' \to _K C$ (omitting totalization signs) so that we have a chain map $_K C' \to C'$ which is an isomorphism in degrees $\leq K + \ell$. Then the same will be true after applying F, and upon applying homology, we obtain an isomorphism in degrees $< K + \ell$. Thus, $F \epsilon_C$ is an isomorphism in degrees $< K + \ell$ which is arbitrarily large, thus proving our statement.

$\square$

2.4 Examples of Abelian Categories

2.4.1 Abelian Groups and Modules

For a commutative ring R, the category R-Mod of R-modules (and homomorphisms) has enough projectives, namely free R-modules: For any R-module M, we can take the free R-module RM on M, and from the universal property (see Sect. 1.4 of Chap. 2) obtain an epimorphism

$$RM \to M.$$

For $R = \mathbb{Z}$, then, we obtain the fact that the category Ab of abelian groups has enough projectives. We will now see that these categories also have enough injectives.

2.4.2 Lemma

1. *An abelian group Q is injective if and only if it is divisible which means that for every $x \in Q$ and $n \in \mathbb{N}$ there exists a $y \in Q$ with $ny = x$.*
2. *For a commutative ring R, the category R-Mod has enough injectives.*
3. *For a topological space X, the category Ab-Sh_X of abelian sheaves on X (and more generally R-Mod-Sh_X of sheaves of R-modules) has enough injectives.*

Proof For (1), clearly, the divisibility condition tests injectivity for the monomorphism

$$n : \mathbb{Z} \to \mathbb{Z},$$

and hence is necessary for injectivity.

To prove that it is sufficient, we will use Zorn's lemma.

Suppose an abelian group Q is divisible. Suppose we have an inclusion of abelian groups

$$A \subseteq B$$

and a homomorphism $F : A \to Q$. By Zorn's lemma, there exists a maximal subgroup C, with respect to inclusion, of B containing A such that h extends to C. If $C = B$, we are done. Assume there exists an element $x \in B \smallsetminus C$. Let $C' = \langle C, x \rangle$ be the subgroup of B generated by C and x. If for all $n \in \mathbb{N}$, $nx \notin C$, then $C' \cong C \oplus \mathbb{Z}$, so clearly we can extend f to C' by using 0 on $\mathbb{Z}$. Otherwise, let $n \in \mathbb{N}$ be minimal such that $nx \in C$. Then C' is the pushout of a diagram

so f can be extended to C' by the divisibility condition. In either case, we obtain a contradiction with the maximality of C.

To prove (2), first consider the category $\mathbb{Z}\text{-Mod} = Ab$. Let A be an abelian group. Then for every $x \in A$, the cyclic subgroup of A embeds into either $\mathbb{Z}$ or $\mathbb{Q}/\mathbb{Z}$, and the embedding extends to a homomorphism on A by injectivity. The product of these homomorphisms is then injective, and a product of injective objects is injective.

Now for a commutative ring R, the right adjoint (right Kan extension)

$$A \mapsto Hom_{\mathbb{Z}}(R, A)$$

to the forgetful functor from R-Mod to Ab preserves injectives (since it is a right adjoint, and its left adjoint preserves monomorphisms), so the statement follows.

The statement (3) is proved similarly. We will only consider the case of $Ab\text{-}Sh_X$ (the general case is analogous). When X is discrete then $Ab\text{-}Sh_X$ is just a product of copies of the category Ab, so it has enough injectives. Let X_{disc} be the set X with the discrete topology and let $f : X_{disc} \to X$ be the identity map. Then f is continuous, and for an injective sheaf $\mathcal{F}$ on X_{disc}, $f_*\mathcal{F}$ is injective, and clearly, by adjunction, every abelian sheaf on X injects (i.e. admits a monomorphism) into such a sheaf. $\qquad\square$

2.4.3 Definition of $Tor_R(M, N)$

Let R be a commutative ring. Then $\otimes_R$, considered as a functor in one variable, with the other variable fixed, is right exact. The n'th left derived functor is denoted by $Tor_n^R(?, ?)$. This does not depend on which variable we take the derived functor in. Letting C, D

be projective resolutions of R-modules M, N, this follows from the fact that we have a diagram

$$C \otimes_R N \xleftarrow{\sim} C \otimes_R D \xrightarrow{\sim} M \otimes_R D,$$

(totalization understood), which in turn follows from the following

2.4.4 Lemma *If C is a cell R-module, then $C \otimes_R ?$ preserves quasiisomorphisms.*

Proof By Theorem 2.3.7, it suffices to prove that for a cell chain complex of R-modules C and any chain complex of R-modules D with $H_* D = 0$ we have

$$H_*(C \otimes_R D) = 0.$$

For C a graded free R-module with 0 differential, this follows from distributivity of $\otimes_R$ under direct sums. A projective R-module is a direct sum of a free one, thus implying the statement for projective chain complexes with 0 differential. Using Theorem 2.3.7 and commutation of homology with colimits of sequences, the statement follows for a cell chain complex C.

$\square$

An important application of this fact is the following fact:

2.4.5 Theorem (Hilbert Syzygy Theorem) *Let k be a field and let M be a finitely generated graded $R = k[x_1, \ldots, x_n]$-module (where each x_i has degree 1). Then there exists a free R-resolution*

$$F_n \to \cdots \to F_1 \to F_0$$

of M where each F_i is a finitely generated R-module.

Proof We shall build inductively a graded free R-resolution of M

$$\cdots \to F_{i+1} \to F_i \to \cdots \to F_0$$

of M. Suppose we have defined a complex

$$F_i \to \cdots \to F_0$$

which has homology M in degree 0 and no homology in degrees $1, \ldots, i-1$, such that all F_j, $j = 1, \ldots, i$ are finitely generated graded R-modules, and the differentials are graded homomorphisms. Then $K_i = Ker(d : F_i \to F_{i-1})$ is a finitely generated graded

R-module since R is Noetherian (Hilbert basis theorem). Let S_i be a set of homogeneous generators of K_i with the smallest possible number of elements, and let $C_{i+1} = RS_i$ be the free R-module on S_i. Letting for $s \in S_i$,

$$ds = \sum_{t \in S_{i-1}} a_{ts} t \tag{2.4.1}$$

where $a_{ts} \in R$ are homogeneous polynomials and the right hand side of (2.4.1) represents $s \in F_i$, then none of the polynomials a_{ts} have degree 0 (since then $S_{i-1} \smallsetminus \{t\}$ would be a set of generators of K_{i-1}). Thus, we have $d_i \otimes_R k = 0$ (where x_i act by 0 on k). In other words,

$$F_i \otimes_R k \cong Tor_i^R(M, k).$$

By Lemma 2.4.4, this is isomorphic to $Tor_i^R(k, M)$. But as a free R-resolution of k, we can take the tensor product, over k, of the complexes $k[x_i] \xrightarrow{\ x_i\ } k[x_i]$. Thus,

$$Tor_i^R(k, M) = 0 \text{ for } i > n.$$

$\square$

We shall discuss more results in this direction in Sect. 2.6 below.

2.5   Sheaf Cohomology

We now turn to sheaves.

2.5.1 Definition For an abelian sheaf $\mathcal{F}$ on a topological space X, we write

$$H^n(X, \mathcal{F}) = Ext^n_{Ab\text{-}Sh_X}(\mathbb{Z}, \mathcal{F}) \tag{2.5.1}$$

(where $\mathbb{Z}$ is the constant sheaf on X, i.e. $\pi^{-1}(\mathbb{Z})$ where π is the map from X to a point), and call this the *n'th cohomology group of X with coefficients in the sheaf $\mathcal{F}$*. By the remarks in the last subsection, (2.5.1) is the same as the n'th derived functor of global sections, as well as of the functor $Ab\text{-}Sh_X(\mathbb{Z}, ?)$, applied to the sheaf $\mathcal{F}$.

From our point of view, this concept is then canonically extended to when $\mathcal{F}$ is replaced by a chain complex $\underline{C}$ of sheaves on X, i.e. one writes

$$H^n(X, \underline{C}) = R^n Hom(\mathbb{Z}, \underline{C}) \tag{2.5.2}$$

where the *Hom* is understood in the category of chain complexes of abelian sheaves on X (or, equivalently in their homotopy category). All derived functors are considered with respect to the canonical functor Φ into the derived category. The construction (2.5.2) is often referred to as *hypercohomology* with coefficients in $\underline{C}$. The corresponding total derived functor can also be referred to as the hypercohomology complex.

2.5.2 Functoriality of Sheaf Cohomology

For a morphism of abelian sheaves $\phi : \mathcal{F} \to \mathcal{G}$ on a space X, we obviously have an induced homomorphism

$$\phi_* : H^n(X, \mathcal{F}) \to H^n(X, \mathcal{G}). \tag{2.5.3}$$

This makes cohomology with coefficients in a varying abelian sheaf into a covariant functor in the sheaf.

For a continuous map $f : X \to Y$ between topological spaces and an abelian sheaf $\mathcal{F}$ on Y, we also have a canonical homomorphism

$$f^* : H^n(Y, \mathcal{F}) \to H^n(X, f^{-1}(\mathcal{F})). \tag{2.5.4}$$

The reason is that if

$$\mathcal{I}_0 \to \mathcal{I}_1 \to \dots$$

is an injective resolution of $\mathcal{F}$, then the cochain complex

$$f^{-1}(\mathcal{I}_0) \to f^{-1}(\mathcal{I}_1) \to \dots \tag{2.5.5}$$

is exact except in degree 0 where it has cohomology $f^{-1}(\mathcal{F})$, by the fact that f^{-1} is an exact functor. Thus, (2.5.5) is canonically quasiisomorphic to $f^{-1}(\mathcal{F})$ and hence has a cochain map, unique up to chain homotopy, to an injective resolution of $f^{-1}(\mathcal{F})$ inducing an isomorphism in 0'th cohomology.

The functor f_* is in general only left exact, but there is a canonical homomorphism

$$H^m(Y, f_*(\mathcal{F})) \to H^m(X, \mathcal{F}) \tag{2.5.6}$$

given as a composition of the canonical homomorphisms

$$H^m(Y, f_*(\mathcal{F})) \to H^m(X, f^{-1}f_*(\mathcal{F})) \to H^m(X, \mathcal{F})$$

where the first arrow is (2.5.4), and the second arrow is induced by the counit of adjunction

$$f^{-1}f_*(\mathcal{F}) \to \mathcal{F}. \tag{2.5.7}$$

On stalks, one sees that when f is an inclusion of a subset with the induced topology, (2.5.7) is, in fact, an isomorphism.

When $f = i$ is the inclusion of a closed subset with the induced topology, then f_* is exact, and also preserves injectives (since it is a right adjoint to a functor f^{-1} which preserves monomorphisms). Letting $\mathcal{I}$ be an injective resolution of $\mathcal{F}$, then $f_*(\mathcal{I})$ is an injective resolution of $f_*(\mathcal{F})$, while $f^{-1}f_*(\mathcal{I}) \cong \mathcal{I}$. Thus, in this case, we conclude that (2.5.6) is an isomorphism.

Similar comments apply to hypercohomology with coefficients in bounded above chain complexes (indexed homologically).

2.5.3 Flasque and Soft Sheaves

It is advantageous to identify certain classes of sheaves which are local with respect to the functor of taking global sections. Recall that an abelian sheaf $\mathcal{F}$ on a space X is called *flasque* (or *flabby*) if for every open set $U \subseteq X$, the restriction

$$\mathcal{F}(X) \to \mathcal{F}(U)$$

is an epimorphism. A sheaf $\mathcal{F}$ on X is called *soft* if for every closed set $i : Z \subseteq X$, the canonical map

$$\mathcal{F}(X) \to (i^*\mathcal{F})(Z)$$

is an epimorphism. It is immediate that every flasque sheaf is soft.

2.5.4 Proposition

1. *Flasque sheaves are Γ-local where Γ denotes the global section functor.*
2. *On paracompact Hausdorff spaces, soft sheaves are Γ-local.*

(Both statements are understood with respect to the canonical functor Φ into the derived category.)

The proof rests on the following two lemmas.

2.5.5 Lemma *Suppose that*

$$0 \to \mathcal{F}' \to \mathcal{F} \to \mathcal{F}'' \to 0 \qquad (2.5.8)$$

is a short exact sequence of sheaves on a space X. If either $\mathcal{F}'$ is flasque or both $\mathcal{F}'$ is soft and X is paracompact Hausdorff, then applying the global section functor to (2.5.8) *produces a short exact sequence:*

$$0 \to \mathcal{F}'(X) \to \mathcal{F}(X) \to \mathcal{F}''(X) \to 0. \tag{2.5.9}$$

Proof The global sections functor is left exact, so it suffices to show that the last map (2.5.9) is onto. Let $s \in \mathcal{F}''(X)$. First consider the case of $\mathcal{F}'$ flasque. Consider the partially ordered set $\mathcal{P}$ of pairs (V, t) where $V \subseteq X$ is open, and $t \in \mathcal{F}(V)$ lifts the restriction of s to V. Then by Zorn's Lemma, $\mathcal{P}$ has a maximal element (V, t). We claim that $V = X$. Otherwise, let $x \in X \smallsetminus V$. Then by the exactness of (2.5.8), the restriction of s to $\mathcal{F}''_x$ lifts to $\mathcal{F}_x$. By definition, then, for some $W \ni x$ open, the restriction $s|_W$ lifts to a section $q \in \mathcal{F}(W)$. Now

$$t_{V \cap W} - q_{V \cap W}$$

maps to 0 in $\mathcal{F}''(V \cap W)$, and hence comes from $\mathcal{F}'(V \cap W)$, which, since $\mathcal{F}'$ is flasque, is a restriction of a global section $r \in \mathcal{F}'(W)$. Thus,

$$t|_{V \cap W} = (q + r)|_{V \cap W},$$

and thus t and $q + r$ can be glued to a section of $\mathcal{F}(V \cup W)$ which lifts the restriction of s, thus contradicting the maximality of (V, T).

To adapt this proof to the soft case, by definition, for every $x \in X$, there exists an open neighborhood U_x such that $s|_{U_x}$ lifts to $\mathcal{F}(U_x)$. In a paracompact Hausdorff space, we may further select an open set V_x such that $x \in V_x$ and the closure of V_x is contained in U_x. Now since X is paracompact, the open cover $(V_x)_{x \in X}$ has a locally finite refinement $(W_i)_{i \in I}$. Then by local finiteness, for any $J \subseteq I$, the union of closures

$$\bigcup_{j \in J} \overline{W_j}$$

is closed. On the other hand, for each inclusion $\iota_i : \overline{W_i} \to X$, we have a lift of the restriction of s to $\iota_i^*(\mathcal{F}'')(W_i)$ to $\iota_i^*(\mathcal{F})(W_i)$. Thus, we can repeat the proof in the flasque case if we let $\mathcal{P}$ be the set of pairs (J, t) where $J \subseteq I$ and t is a lift of the restriction of s to

$$\iota^* \mathcal{F}\left(\bigcup_{j \in J} \overline{W_j}\right)$$

where $\iota : \bigcup_{j \in J} \overline{W_j} \to X$ is the inclusion. $\qquad \square$

2.5.6 Lemma *Suppose we have an exact sequence of sheaves (2.5.8) on a space X.*

1. *If $\mathcal{F}'$, $\mathcal{F}$ are flasque, then $\mathcal{F}''$ is flasque.*
2. *If X is paracompact Hausdorff and $\mathcal{F}'$, $\mathcal{F}$ are soft, then $\mathcal{F}''$ is soft.*

Proof We shall prove (1). Suppose $\mathcal{F}'$, $\mathcal{F}$ are flasque. Let $i : U \subseteq X$ be open, and let $s \in \mathcal{F}''(U)$. Clearly, the sheaf $i^*\mathcal{F}'$ is also flasque, so by (2.5.9) with X replaced by U, s lifts to a section $t \in \mathcal{F}(U)$. Since $\mathcal{F}$ is flasque, s is a restriction of a global section of $\mathcal{F}$. Projecting that global section to $\mathcal{F}''(X)$, we get the required global section whose restriction to U is s.

Now (2) is completely analogous when upon replacing "open" by "closed." $\qquad\square$

2.5.7   Proof of Proposition 2.5.4

We shall prove (1). By Proposition 2.3.20, it suffices to prove that

$$H^n(X, \mathcal{F}) = 0 \text{ for } n > 0. \tag{2.5.10}$$

Since we have enough injectives, there is a short exact sequence

$$0 \to \mathcal{F} \to \mathcal{Q} \to \mathcal{F}_1 \to 0$$

where $\mathcal{Q}$ is injective, and hence flasque (since the property of being flasque is a special case of the condition for injectivity with respect to the inclusion $i_!\mathbb{Z}_U \to \mathbb{Z}$ for an inclusion of an open set $i : U \to X$). Therefore, by Lemma 2.5.5, $\mathcal{F}_1$ is also flasque. Now by Theorem 2.3.7, we have a long exact sequence

$$\cdots \to H^n(X, \mathcal{F}) \to H^n(X, \mathcal{Q}) \to H^n(X, \mathcal{F}_1) \to H^{n+1}(X, \mathcal{F}) \to \cdots. \tag{2.5.11}$$

Now since $\mathcal{Q}$ is injective, we have $H^n(X, \mathcal{Q}) = 0$ for $n > 0$. On the other hand, the map

$$H^0(X, \mathcal{Q}) \to H^0(X, \mathcal{F}_1)$$

of (2.5.11) is an epimorphism by Lemma 2.5.5, and thus, by (2.5.11), $H^1(X, \mathcal{F}) = 0$. On the other hand, also by (2.5.11),

$$H^n(X, \mathcal{F}) \cong H^{n-1}(X, \mathcal{F}_1)$$

for $n > 1$, and thus, since $\mathcal{F}_1$ is also flasque, (2.5.10) follows by induction. $\qquad\square$

The case of (2) is now completely analogous.

We can now prove that de Rham cohomology is a special case of sheaf cohomology.

2.5.8 Lemma (Poincaré Lemma)   *We have*

$$H^0_{DR}(\mathbb{R}^n) = \mathbb{R},$$

where the cycles are constant functions, and

$$H^i_{DR}(\mathbb{R}^n) = 0$$

for $i > 0$.

Proof Induction on n. For $n = 0$, the result is obvious. Now consider the projection

$$\pi : \mathbb{R}^n \to \mathbb{R}^{n-1}, \ (x_1, \ldots, x_n) \mapsto (x_1, \ldots, x_{n-1})$$

and the inclusion

$$i : \mathbb{R}^{n-1} \to \mathbb{R}^n, \ (x_1, \ldots, x_{n-1}) \mapsto (x_1, \ldots, x_{n-1}, 0).$$

For the induction step, we define a chain homotopy h between the identity and $i \circ \pi$ on $\Omega^*(\mathbb{R}^n)$ given for $1 \le i_1 < \cdots < i_k \le n$ by

$$h(g dx_{i_1} \wedge \cdots \wedge dx_{i_k}) =$$

$$(-1)^{k+1} \left(\int_0^{x_n} g(x_1, \ldots, x_{n-1}, t) dt \right) dx_{i_1} \wedge \cdots \wedge dx_{i_{k-1}} \qquad (2.5.12)$$

when $i_k = n$ and

$$h(g dx_{i_1} \wedge \cdots \wedge dx_{i_k}) = 0$$

otherwise.

$\square$

2.5.9 Corollary *For a smooth manifold M, the cohomology of the chain complex $\underline{\Omega}^*_M$ in the category of abelian sheaves on M is the constant sheaf $\underline{\mathbb{R}}$ (concentrated in degree 0).*

Proof Since we have a map $\underline{\mathbb{R}} \to \underline{\Omega}^0_M$ given by constant functions, it suffices to prove the statement on the level of stalks at a point $x \in M$. To this end, without loss of generality, $M = \mathbb{R}^n$, $x = 0$. Apply the Poincaré Lemma to balls with center 0 and radius $1/m$, and pass to the colimit over m.

$\square$

2.5.10 Theorem *Let M be a smooth manifold. Then we have a canonical natural isomorphism*

$$H_{DR}^n(M) \cong H^n(M, \mathbb{R}).$$

Proof It is proved in analysis (using a partition of unity) that the abelian sheaf C_M^∞, and thus also the abelian sheaves $\underline{\Omega}_M^k$ are soft. Thus, by Corollary 2.5.9, the de Rham complex $\underline{\Omega}_M^*$ is a resolution of $\mathbb{R}$ by soft sheaves, which are Γ-local (where Γ denotes global sections) by Proposition 2.5.4. Thus, by Proposition 2.3.21, $\underline{\Omega}_M^*$ is Γ-local. The statement follows by definition of Γ-locality. $\qquad\square$

We have, of course, an analogous statement for complex manifolds:

$$H_{DR}^n(M; \mathbb{C}) \cong H^n(M, \underline{\mathbb{C}}).$$

2.6   A Cohomological Criterion for Regular Local Rings

In this subsection, we will prove what is known as the Auslander-Buchsbaum-Serre criterion of regularity. For a ring R, the *global dimension* is defined as the maximum possible number d such that $Ext_R^d(M, N) \neq 0$ for any R-modules M, N. If no such maximum exists, we say that the global dimension is infinity: $gl(R) = \infty$. The supremum of n such that $Ext_R^n(M, N) \neq 0$ for a given module M is called the *cohomological dimension* of M and denoted by $cd(M) = cd_R(M)$. (Again, the cohomological dimension can be infinite.)

2.6.1 Theorem (Auslander-Buchsbaum-Serre) *Let R be a local Noetherian ring. Then $gl(R) < \infty$ if and only if R is regular, in which case we have $gl(R) = dim(R)$.*

To prove Theorem 2.6.1, we will need some observations from commutative as well as homological algebra. First of all, we note that when defining global dimension, equivalently, we can restrict to the case when the module M is finitely generated (or, for that matter, cyclic, which means generated by one element). This is delicate, since it is *not true* that $cd(M)$ would be in general equal to the supremum of the numbers $cd(N)$ with $N \subseteq M$ finitely generated. (See Exercise 14.) However, it follows from the next lemma (since by Zorn's lemma, we can choose $M_{\alpha+1}/M_\alpha$ cyclic).

2.6.2 Lemma *Let β be an ordinal number and let $M_\alpha \subseteq M$, $\alpha \leq \beta$ be submodules such that $M_0 = 0$, for a limit ordinal $\gamma \leq \beta$,*

$$M_\gamma = \bigcup_{\alpha < \gamma} M_\alpha$$

and such that $cd(M_{\alpha+1}/M_\alpha) \le d$ *for all* $\alpha < \beta$. *Then* $cd(M) \le d$.

Proof Induction on d. For $d = 0$, since all the $M_{\alpha+1}/M_\alpha$'s have cohomological dimension 0, they are (by definition) projective, and thus, M is their direct sum. Therefore, M is projective, and hence has cohomological dimension 0.

To prove the statement for a given $d > 0$, using Zorn's lemma, we can construct free modules K_α, $\alpha < \beta$ and an onto homomorphism of modules

$$h : \bigoplus_{\alpha<\beta} K_\alpha \to M$$

such that for all $\gamma \le \beta$,

$$h(\bigoplus_{\alpha<\gamma} K_\alpha) = M_\gamma.$$

In particular, if we let $N = Ker(h)$,

$$N_\alpha = Ker(h) \cap (\bigoplus_{\alpha<\beta} K_\alpha),$$

using the long exact sequence in Ext, we have $cd(N_{\alpha+1}/N_\alpha) \le d - 1$ by the induction hypothesis, and thus, $cd(N) \le d - 1$ by the induction hypothesis, and thus, using the long exact sequence in Ext again, $cd(M) \le d$.

$\square$

2.6.3 Lemma *Let R be a local ring. Then every projective R-module is free.*

Proof Let $F = P \oplus Q$ be R-modules where F is free. Our job is to prove that P is free. Let F be the free R-module RS on a set S.

Step 1: For any $x \in P$, there exists a countable subset $T \subseteq S$ such that $x \in RT$ and

$$P = (P \cap RT) + (P \cap R(S \smallsetminus T)). \tag{2.6.1}$$

To see this, we produce a sequence of finite subsets $T_i \subseteq S$: Let T_0 be a set of elements of S whose linear combination is x. Given T_i, let $T_{i+1} \subseteq S$ be a set of elements containing T_i, as well as elements of S whose linear combinations are the projections of all elements of T_i to P. Then we may put $T = \bigcup T_i$.

Step 2: P is a direct sum of countably generated (automatically projective) R-modules. To prove this, by Step 1, we can use Zorn's lemma on the set of all pairs $(T, \mathcal{D})$ where $T \subseteq S$ is subset for which (2.6.1) holds, and $\mathcal{D}$ is a decomposition of $P \cap RT$ into a direct sum of countably generated R-modules. (There is an obvious notion of "inclusion" of decompositions.) The maximal element then has to have $T = S$, for otherwise we could use Step 1 with S replaced by $S \smallsetminus T$.

Step 3: Every element of P is contained in a finitely generated direct summand P' of P which is a free R-module. To prove this, let us assume we have selected a basis S such that

$$x = \sum_{i=1}^{n} a_i e_i$$

with $a_i \in R$, $e_i \in S$, $i = 1, \ldots n$. Let $e_i = y_i + z_i$, $y_i \in P$, $z_i \in Q$ such that n is the smallest possible. Then we have

$$a_i \notin (a_j \mid j \neq i), \tag{2.6.2}$$

for if

$$a_i = \sum_{j \neq i} c_j a_j, \ c_j \in R,$$

we could eliminate the basis element e_i by replacing e_j with $e_j + c_j e_i$ for $j \neq i$, contradicting the minimality of n.

Now let $y_i = \sum b_{ij} e_j + t_i$ where $t_i \in S \setminus \{e_1, \ldots, e_n\}$. Then by projecting to P, $a_i = \sum b_{ij} a_j$. Now (2.6.2) implies that the elements $1 - b_{ii}$ and b_{ij}, $i \neq j$ are not units of R. Since R is local, b_{ii} are units, and thus the matrix (b_{ij}) is invertible (since its determinant is a unit). Thus, we can construct another basis of F by replacing $e_1, \ldots e_n$ with $y_1, \ldots, y_n$, and put $P' = R\{y_1, \ldots, y_n\}$.

Now to conclude the argument, by Step 2, without loss of generality, P is countably generated with generators, say, $x_1, \ldots, x_n \ldots$. We will produce, by inductions, free summands F_i of P on bases S_i such that $x_1, \ldots, x_i \in F_i$, and for $i < j$, $S_i \subseteq S_j$. Suppose F_i is given. Then apply Step 3 to P replaced with P/F_i, where x is the projection of x_{i+1}. Since x_i are generators, $\bigcup S_i$ is a basis of P. $\square$

Let R be a Noetherian local ring with maximal ideal m and residue field k, and let M be a finitely generated R-module. We saw that we have a projective resolution

$$\ldots \xrightarrow{d} P_n \xrightarrow{d} P_{n-1} \xrightarrow{d} \ldots \xrightarrow{d} P_0 \tag{2.6.3}$$

of M. By Lemma 2.6.3, P_n is a free R-module on a set S_n. The resolution (2.6.3) is called *minimal* when $dP_n \subseteq mP_{n-1}$. Note that a minimal resolution always exists when M is a finitely generated R-module. For this, it suffices to show that for any R-module M there is an onto homomorphism of R-modules

$$FS \to M \tag{2.6.4}$$

whose kernel is in mS and such that S is finite. However, to show that, it suffices to choose S so that kS maps isomorphically to the k-module M/mM (since then the cokernel C of (2.6.4) satisfies $mC = C$, and thus is 0 by Nakayama's lemma).

Now we see that $gl(R)$ is equal to the supremum of the lengths of minimal projective resolutions of finitely generated R-modules M (where *length* is defined as the supremum of d such that $P_d \neq 0$). Indeed, if M has a minimal resolution of length d, then obviously $Ext_R^n(M, N) = 0$ for $n > d$ for any R-module N. On the other hand, however, $Ext_R^d(M, k) \neq 0$, since after Hom-ing a minimal resolution to k, the differentials (by definition) become 0. This also shows that when calculating the cohomological dimension, it suffices to specialize to $N = k$.

Next, we will need the following

2.6.4 Lemma (Serre) *Let R be a local Noetherian ring and let the dimension of the $k = R/m$-vector space m/m^2 be n. Then*

$$Ext_R^n(k, k) \neq 0. \tag{2.6.5}$$

Proof Consider the initial map of rings $f : \mathbb{Z} \to R$. We distinguish two cases.

Case 1: $f(f^{-1}(m)) \subseteq m^2$. Let $u_1, \ldots, u_n$ be lifts to R of generators of the k-vector space m/m^2. Then let $A = \mathbb{Z}[x_1, \ldots, x_n]$. There is a homomorphism of rings

$$A \to R$$

sending $x_i \mapsto u_i$. By universality, this induces a functor on derived categories $DR \to DA$, and thus we have a homomorphism of (non-commutative) rings

$$Ext_R^*(k, k) \to Ext_A^*(k, k) \tag{2.6.6}$$

(where multiplication is by composition), and the superscript $*$ means sum over all $i \in \mathbb{N}_0$. Additionally, considering the long exact sequence on $Ext_R(?, k)$ corresponding to the short exact sequence of R-modules

$$0 \to m/m^2 \to R/m^2 \to k \to 0,$$

the image in $Ext_R^1(k, k)$ under the connecting map of the basis of $Hom_R(m/m^2, k) = Hom_k(m/m^2, k)$ dual to the k-basis $x_1, \ldots, x_n$ of m/m^2 maps under (2.6.6) to elements $\alpha_1, \ldots, \alpha_n$ of

$$Ext_A^1(k, k) \tag{2.6.7}$$

which are given by images under the connecting map of the $Ext_A(?, k)$ long exact sequence associated with the short exact sequence of A-modules

$$0 \to k\{x_1, \ldots, x_n\} \to A/(x_1, \ldots, x_n)^2 \otimes k \to k \to 0$$

of homomorphisms of A-modules $k\{x_1, \ldots, x_n\} \to k$ which is Id_k tensored with the dual basis of $x_1, \ldots, x_n$. But Ext_A is easy to compute. In particular,

$$\alpha_1 \ldots \alpha_n \neq 0 \in Ext_A^n(k, k) \tag{2.6.8}$$

(see Exercise 17).

Case 2: $f^{-1}(m) = (p)$ where $p \in \mathbb{Z}$ is a prime number, whose image under f projects to a non-trivial element $u \in m/m^2$. In particular, k has characteristic p. Then choose a k-basis $u_1, \ldots, u_n$ of m/m^2 such that $u = u_n$. Now let $A = \mathbb{Z}[x_1, \ldots, x_{n-1}] \to R$ map $x_i \mapsto u_i$. Again, we have (2.6.6). Now consider the long exact sequence in $Ext_R(?, k)$ associated with the short exact sequence of R-modules

$$0 \to m/(m^2, p) \to R/(m^2, p) \to k \to 0$$

Consider the images, under the connecting map, of homomorphisms of R-modules $m/(m^2, p) \to k$ which form a dual basis to the basis $u_1, \ldots, u_{n-1}$. Under (2.6.6), these map to elements $\alpha_1, \ldots, \alpha_{n-1}$ of (2.6.7) which are images under the connecting map of the long exact sequence in $Ext_A(?, k)$ associated with the short exact sequence

$$0 \to k\{x_1, \ldots, x_{n-1}\} \to A/(x_1 \ldots, x_{n-1})^2 \otimes k \to k \to 0$$

of homomorphisms of A-modules $k\{x_1, \ldots, x_{n-1}\} \to k$ which are Id_k tensored with the dual basis of $x_1, \ldots, x_{n-1}$. Let, additionally, $\alpha_n \in Ext_A^1(k, k)$ be the image of the element of $Ext_R^1(k, k)$ associated with the short exact sequence of R-modules

$$0 \to k \to R/(m^2, p^2, u_1, \ldots, u_{n-1}) \to k \to 0$$

(which, note, is a sum of copies of the short exact sequence of A-modules

$$0 \to \mathbb{Z}/p \to \mathbb{Z}/p^2 \to \mathbb{Z}/p \to 0$$

where $x_1, \ldots, x_{n-1}$ act trivially). Again, Ext_A is not mysterious, and we have (2.6.8). (Exercise 17.) □

Now we have the following fact

2.6.5 Lemma *Suppose R is a Noetherian local ring with maximal ideal m where every element of m is a zero divisor. Then there exists a $0 \neq z \in R$ such that $zm = 0$.*

Proof Consider a minimal primary decomposition

$$(0) = J_1 \cap \cdots \cap J_n.$$

We claim that

$$m \subseteq \bigcup_{i=1}^{n} \sqrt{J_i}. \tag{2.6.9}$$

In fact, otherwise, we have a $y \in m$ none of whose powers is contained in any of the J_i's. But by assumption, there is a $0 \neq x \in R$ with $xy = 0$. Thus, since the J_i's are primary, we have $x \in J_i$ for all $i = 1, \ldots, n$, and thus, $x = 0$.

Now (2.6.9) implies that $m = \sqrt{J_i}$ for some i. By Proposition 5.1.3 of Chap. 1,

$$m = \sqrt{\{y \in R \mid xy = 0\}}$$

for some $x \in R$. This means that if $m = (y_1, \ldots, y_s)$, then there is an N such that $y_i^N x = 0$ for all $i = 1, \ldots, s$. Let z be a multiple of x by a maximal monomial in the y_i's which is non-zero. $\square$

2.6.6 Lemma *Let R be a Noetherian local ring with maximal ideal m such that every element of m is a zero divisor. Let M be a finitely generated R-module such that $cd_R(M) \geq 1$. Then $cd_R(M) = \infty$.*

Proof Suppose that $n = cd_R(M)$ is finite. Considering a minimal projective resolution (2.6.3) of M, we have $P_1 \neq 0$. Consider the exact sequence

$$0 \longrightarrow P_n \xrightarrow{\ d\ } P_{n-1}.$$

Then P_n, P_{n-1} are free R-modules, and $d(P_n) \subseteq m P_{n-1}$. If $a \in P_n$ is one of the free generators and z is as in Lemma 2.6.5, then $d(az) = 0$, which contradicts injectivity. $\square$

2.6.7   Proof of Theorem 2.6.1

Now suppose R is a Noetherian local ring with maximal ideal m and suppose $x \in m$ is a regular element (a non-zero divisor). Let M be finitely generated R-module with $cd_R(M) \geq 1$ and with a minimal resolution (2.6.3). Then we have a short exact sequence of the form

$$0 \to N \to P_0 \to M \to 0.$$

Now for any submodule of a free R-module $Q \subseteq F$, we have

$$Tor_1^R(Q, R/(x)) = 0. \tag{2.6.10}$$

Indeed, the long exact sequence in Tor gives

$$0 \to Tor_2^R(F/Q, R/(x)) \to Tor_1^R(Q, R/(x)) \to Tor_1^R(F, R/(x))$$

where the first and last term are 0. Thus, tensoring the minimal resolution (2.6.3) with the last term omitted with $R/(x)$ over R, we obtain a minimal $R/(x)$-resolution

$$\to P_n/(x) \to \cdots \to P_1/(x)$$

of $N/(x)$, and thus,

$$cd_{R/(x)}(N/(x)) = cd_R(N) - 1. \tag{2.6.11}$$

Proof of Theorem 2.6.1 Let m be the maximal ideal of R. Denote the residue field $k = R/m$. By the argument we made in Sect. 2.6.7 and by Lemma 2.6.6, if M is finitely generated R-module with $cd_R(M) = d < \infty$, there must exist a regular sequence of length d in m. This implies $dim(R) \geq d$. However, if R is not a regular local ring, then, by Lemma 2.6.4, there exists a finitely generated R-module M with $cd_R(M) > dim(R)$, and thus, $cd_R(M) = \infty$. Thus, R being regular is a necessary condition.

On the other hand, when R is regular local, then generators

$$x_1, \ldots, x_{dim(R)}$$

of m form a regular sequence, and thus, by Sect. 2.6.7 again, for any finitely generated R-module M, we have $cd_R(M) \leq dim(R)$. On the other hand, using the long exact sequence in Ext, by induction on i, $Ext_R^s(R/(x_1, \ldots, x_i), k), k)$ is 0 for $s > i$ and k for $s = i$, so $cd_R(k) = dim(R)$, which proves the last statement of the Theorem. □

One important application of Theorem 2.6.1 is the following

2.6.8 Corollary *Suppose R is a regular ring. Then the ring of polynomials $R[x]$ is regular.*

Proof Suppose R is regular. Now consider a prime ideal $p \subset R[x]$. Then $q = p \cap R$ is prime in R, and thus, $gl(R_q) = d < \infty$. Thus, we have $gl(R_q[x]) \leq d + 1$ (Exercise 15).

Now putting $q' = q \cdot R[x]$, note that $R[x]_p$ is a localization of $R[x]_{q'} = R_q[x]$. Since localization preserves projective resolutions, $gl(R[x]_p) \leq gl(R_q[x]) \leq d + 1$, and thus, $R[x]_p$ is regular local.

$\square$

3   Singular Homology and Cohomology

We still have not computed any actual cohomology groups of manifolds. While it is possible to compute directly with de Rham cohomology, the story would not be complete if we did not mention singular homology and cohomology, which are the main computational tools for basic examples. It also allows computations with coefficients in $\mathbb{Z}$ rather than just $\mathbb{R}$, thus including torsion information. Therefore, this method is a refinement of de Rham cohomology.

3.1   The Singular Chain and Cochain Complex

When discussing singular homology and cohomology, we will restrict attention to *Hausdorff spaces*. The *standard n-simplex* is

$$\Delta^n = \{(t_0, \ldots, t_n) \in \mathbb{R}^{n+1} \mid t_i \geq 0, \ \sum t_i = 1\}, \tag{3.1.1}$$

with the induced topology from $\mathbb{R}^{n+1}$. The coordinates $(t_0, \ldots, t_n)$ (the symbol $[t_0, \ldots, t_n]$ is also used) are called *barycentric coordinates*. Actually, the inequalities $t_i \geq 0$ serve a purely aesthetic purpose; if we removed them, we would obtain an equivalent theory. For a topological space X, a *singular n-simplex in X* is a continuous mapping

$$\sigma : \Delta^n \to X.$$

The set of all singular n-simplices in X is denoted by $S_n X$. The abelian group $C_n X$ of *singular n-chains in X* is the free abelian group on $S_n X$:

$$C_n X = \mathbb{Z} S_n X.$$

It is also useful to consider the category *Pair* of *pairs* of topological spaces, whose objects are pairs of spaces (X, Y) where $Y \subseteq X$ has the induced topology, and morphisms $f : (X, Y) \to (X', Y')$ are continuous maps $f : X \to X'$ such that $f(Y) \subseteq Y'$. We can then also define the abelian group of *relative singular n-chains on a pair (X, Y)* by

$$C_n(X, Y) = C_n X / C_n Y.$$

observe that this is also a free abelian group.

The point of introducing the standard simplex is that we have canonical maps

$$\partial_i : \Delta^{n-1} \to \Delta^n, \; i = 0, \ldots n,$$

given by

$$(t_0, \ldots, t_{n-1}) \mapsto (t_0, \ldots, t_{i-1}, 0, t_i, \ldots, t_n).$$

We therefore have

$$\partial_i \circ \partial_j = \partial_{j+1} \circ \partial_i \text{ for } 0 \le i \le j \le n - 1. \tag{3.1.2}$$

We define

$$d : C_n X \to C_{n-1} X$$

by setting, for a singular simplex σ on X,

$$d(\sigma) = \sum_{i=0}^{n} (-1)^i \sigma \circ \partial_i. \tag{3.1.3}$$

The relation (3.1.2) implies

$$0 = d \circ d : C_n X \to C_{n-2} X,$$

(Exercise 26) and thus the system $CX = C_* X = ((C_n X), d)$ becomes a chain complex called the *singular chain complex of* X. (As usual for chain complexes, when not defined, i.e. here for n negative, we set $C_n X = 0$.) Similarly on pairs, we have a chain complex $C(X, Y) = C_*(X, Y) = (C_n(X, Y), d)$, called the *relative singular chain complex of the pair* (X, Y) and we have, in fact, a short exact sequence of chain complexes

$$0 \to CY \to CX \to C(X, Y) \to 0. \tag{3.1.4}$$

It is immediate that for a continuous map $f : X \to X'$, we get a chain map $Cf : CX \to CX'$ by sending a singular n-simplex σ in X to $f \circ \sigma$. Thus, we have functors C from the category of topological spaces (resp. the category of pairs) to the category of chain complexes. Composing with the homology functors H_n, we obtain functors

$$H_n : Top \to Ab, \; H_n : Pair \to Ab$$

which are called *the n'th singular homology* and *the n'th relative singular homology*.
It is useful to generalize this somewhat. Let A be any abelian group, sometimes referred to as the group of *coefficients*. Then we have a chain complex

$$C(X; A) = CX \otimes A$$

where $C(X; A)_n = C_n X \otimes A$ and the differential is $d \otimes Id_A$, and a cochain complex

$$C^*(X; A) = Hom(CX, A)$$

where $Hom(CX, A)^n = Hom(C_n X, A)$ and the differential

$$Hom(C_n X, A) \to Hom(C_{n+1} X, A)$$

is

$$f \mapsto f \circ d.$$

One defines $C(X, Y; A) = C_*(X, Y; A)$ and $C^*(X, Y; A)$ analogously. The corresponding (co)homology groups are referred to as *singular homology and cohomology with coefficients in A*:

$$H_n(X; A) = H_n C(X; A), \ H_n(X, Y; A) = H_n C(X, Y; A),$$
$$H^n(X; A) = H^n C^*(X; A), \ H^n(X, Y; A) = H^n C^*(X, Y; A).$$

As above, $H_n(?; A)$ are functors from Top or $Pair$ to Ab, $H^n(?; A)$ are functors from Top^{Op}, $Pair^{Op}$ to Ab. It is important to observe that for a pair (X, Y), we still have short exact sequences of (co)chain complexes

$$0 \to C(Y; A) \to C(X; A) \to C(X, Y; A) \to 0 \tag{3.1.5}$$

$$0 \to C^*(X, Y; A) \to C^*(X; A) \to C^*(Y; A) \to 0, \tag{3.1.6}$$

since the abelian groups $CX, C(X, Y)$ are free.

3.2   Eilenberg-Steenrod Axioms

Singular homology and cohomology satisfy certain properties which make them calculable. For example, Theorem 2.3.6 together with the short exact sequences (3.1.5), (3.1.6) give long exact sequences of the form

$$\cdots \to H_n(Y; A) \to H_n(X; A) \to H_n(X, Y; A) \to H_{n-1}(Y; A) \to \cdots \tag{3.2.1}$$

and

$$\cdots \to H^n(X, Y; A) \to H^n(X; A) \to H^n(Y; A) \to H^{n+1}(X, Y; A) \to \cdots . \tag{3.2.2}$$

This is referred to as the *exactness axiom*.

Another easy property is that if we denote by $*$ the one point topological space, then we have

$$H_0(*; A) = H^0(*; A) = A$$

and

$$H_n(*; A) = H^n(*; A) = 0$$

for $n \neq 0$. This is called the *dimension axiom*. It follows immediately from the definitions.

Recall from Sect. 5.5 of Chap. 3 that a *homotopy* between continuous maps of topological spaces $f, g : X \to Y$ is a continuous map $h : X \times [0, 1] \to Y$ (where $[0, 1]$ has the induced topology from the analytic topology on $\mathbb{R}$) such that $h(x, 0) = f(x)$, $h(x, 1) = g(x)$. Again, we write $h : f \simeq g$, or just $f \simeq g$, and say that f, g are *homotopic*. Again, this is compatible with composition, so we have the *homotopy category of topological spaces* $hTop$ whose objects are topological spaces and morphisms are homotopy classes of continuous maps. Similarly we form the homotopy category of pairs $hPair$ where a homotopy between two morphisms of pairs $f, g : (X, Y) \to (Z, T)$ is a homotopy $h : X \times [0, 1] \to Z$ such that for every $t \in [0, 1]$ and every $y \in Y, h(y, t) \in T$. We then have canonical functors

$$Top \to hTop, \quad Pair \to hPair \tag{3.2.3}$$

which are the identity on objects, and send a morphism to its homotopy class. An isomorphism of topological spaces X, Y in $hTop$ is called a *homotopy equivalence*. If a homotopy equivalence exists, the spaces X, Y are called *homotopy equivalent* and one writes $X \simeq Y$. Similarly for pairs.

The *homotopy axiom* states that the functors $H_n(?; A)$ and the contravariant functors $H^n(?; A)$ factor through the functors (3.2.3), or, in other words, that homotopic maps of spaces or pairs induce the same map on homology or cohomology. The reader may be reminded of the same fact about chain complexes in Sect. 2.3.2 above. However, the proof for spaces is harder, since we have to find a way to pass from homotopy of spaces to chain homotopy. To make our treatment self-contained, we include a proof in the next section. More details can be found in [20].

The last Eilenberg-Steenrod axiom, which is also proved in the following subsection, states that whenever we have a pair (X, Y) and a subset $Z \subseteq X$ whose closure is contained in the interior of Y

$$\overline{Z} \subseteq Y^\circ,$$

then, for the inclusion of pairs

$$i : (X \smallsetminus Z, Y \smallsetminus Z) \to (X, Y),$$

$$H_n(i; A) : H_n(X \smallsetminus Z, Y \smallsetminus Z; A) \to H_n(X, Y; A)$$

and

$$H^n(i; A) : H^n(X, Y; A) \to H^n(X \smallsetminus Z, Y \smallsetminus Z; A)$$

are isomorphisms. This is called the *excision axiom*.

In the process of the proof, one uses a lemma which is useful on its own. Let $\mathcal{U} = (Z_i)_{i \in I}$ be a system of subsets of X. We can then form a variant $C^{\mathcal{U}} X$ of the singular chain complex where we only consider singular simplices whose image is in Z_i for some $i \in I$. By repeating the above construction, we then have also chain complexes $C^{\mathcal{U}}(X; A)$.

3.2.1 Lemma *Suppose that the interiors (Z_i°) form an open cover of X. Then the canonical chain map*

$$C^{\mathcal{U}}(X; A) \to C(X; A)$$

induces an isomorphism in homology (and hence is a chain-homotopy equivalence).

$\square$

3.3   Proof of the Homotopy and Excision Axioms

The homotopy and excision axioms form the technical core of the singular homology method, and for this reason, we include their proofs here, even though they really belong to the field of algebraic topology. We shall only discuss the case of homology with coefficients in $\mathbb{Z}$. Homology with coefficients in a general abelian group and cohomology can be treated analogously, or alternately, one can use universal coefficients, which we will discuss in Sect. 3.5.

3.3.1 Theorem *Two homotopic maps of pairs induce the same map in singular homology.*

We first prove the following partial result.

3.3.2 Lemma *If X is a contractible space, then $X \to *$ induces an isomorphism in homology. (In particular, $H_0(X) = \mathbb{Z}$, $H_n(X) = 0$ for $n > 0$.)*

Proof If X is contractible, we have a homotopy $k : X \times [0, 1] \to X$ with $k(x, 0) = x$, $k(x, 1) = *$ for some $* \in X$ (independent of x). Let a chain map $\epsilon : CX \to CX$ (where CX is the singular chain complex of X) be defined on a singular simplex $\sigma : \Delta^n \to X$ by $\epsilon(\sigma) = 0$ for $n > 0$, and $\epsilon(\sigma) = *$ for $n = 0$ (identifying singular 0-simplices with points). We shall exhibit a chain homotopy

$$h : Id_{CX} \simeq \epsilon,$$

which implies the statement (by considering what ϵ induces on homology).

In effect, we may define, for $\sigma : \Delta^n \to X$,

$$h(\sigma) : \Delta^{n+1} \to X$$

on barycentric coordinates by

$$[t_0, \ldots, t_{n+1}] \mapsto k(\sigma[\frac{t_1}{1 - t_0}, \ldots, \frac{t_n}{1 - t_0}], t_0)$$

for $0 \leq t_0 < 1$, and

$$[0, \ldots, 0, 1] \mapsto *.$$

One sees that

$$dh(\sigma) = \sigma - \sum_{i=0}^{n}(-1)^i h(\sigma \circ \partial_i)$$

except for $n = 0$, where

$$dh(\sigma) = \sigma - *,$$

(identifying singular 0-simplices with points). Thus, h is the required chain homotopy. □

3.3.3 Lemma *Let $\iota_t : X \to X \times [0, 1]$ be given by $x \mapsto (x, t)$. Then there is a natural chain homotopy $h : C(\iota_0) \simeq C(\iota_1)$. Naturality means that for a continuous map $f : X \to$*

Y, *the following diagram commutes:*

$$
\begin{array}{ccc}
C_n(X) & \xrightarrow{\;h_n\;} & C_{n+1}(X \times [0, 1]) \\
\Big\downarrow{\scriptstyle f_*} & & \Big\downarrow{\scriptstyle (f \times [0,1])_*} \\
C_n(Y) & \xrightarrow{\;h_n\;} & C_{n+1}(Y \times [0, 1]).
\end{array}
\tag{3.3.1}
$$

Proof We shall construct h_n simultaneously for all topological spaces X by induction on n. For $n = 0$, choose an affine map $\iota : \Delta^1 \to [0, 1]$ which sends $(0, 1) \mapsto 0$, $(1, 0) \mapsto 1$. Then for a singular 0-simplex in X, which can be identified with a point $x \in X$, let

$$
h(x) = const_x \times \iota.
$$

Now suppose h_{n-1} is defined and natural in the sense of (3.3.1) (with n replaced by $n-1$). We will first define $h(\kappa_n)$ where $\kappa_n = Id : \Delta^n \to \Delta^n$. Then for any singular n-simplex $\sigma : \Delta^n \to X$ for any topological space X, we have

$$
\sigma = \sigma_*(\kappa_n).
$$

Thus, to satisfy naturality (3.3.1), we can (and must) put

$$
h(\sigma) = \sigma_*(h(\kappa_n)).
$$

Now to construct $\lambda = h(\kappa_n)$, we must solve the equation

$$
d\lambda + h(d\kappa_n) = (\iota_0)_* \kappa_n - (\iota_1)_* \kappa_n,
$$

or

$$
d\lambda = (\iota_0)_* \kappa_n - (\iota_1)_* \kappa_n - h(d\kappa_n).
\tag{3.3.2}
$$

Let us verify that the right hand side of (3.3.2) is a cycle. We have

$$
\begin{aligned}
d((\iota_0)_* \kappa_n - (\iota_1)_* \kappa_n - h(d\kappa_n)) = \\
(\iota_0)_* d\kappa_n - (\iota_1)_* \kappa_n - dhd\kappa_n = \\
dhd\kappa_n - hdd\kappa_n - dhd\kappa_n = 0,
\end{aligned}
\tag{3.3.3}
$$

as required. Now for $n > 0$, the right hand side of (3.3.2) being a cycle, it is also a boundary, since $\Delta^n \times [0, 1]$ is contractible, and thus,

$$
H_n(\Delta^n \times [0, 1]) = 0
$$

by Lemma 3.3.2. Thus, λ exists, and the induction step is complete. $\qquad\square$

COMMENTS

1. The method used in this proof is known as the *method of acyclic models*. In [20], a general categorical version of this method is treated.
2. The above proof demonstrates the difference between "natural" and "canonical": The homotopy constructed is natural in the sense of (3.3.1), but is by no means canonical, in that we have no preferred choice for the class λ in the induction step.

Now Theorem 3.3.1 follows from Lemma 3.3.3: A homotopy of pairs is a morphism of pairs of the form $k : (X_1 \times [0, 1], X_2 \times [0, 1]) \to (Y_1, Y_2)$. The chain homotopy of Lemma 3.3.3 gives a chain homotopy between the chain maps

$$(X_1, X_2) \to (X_1 \times [0, 1], X_2 \times [0, 1])$$

given by

$$(x_1, x_2) \mapsto ((x_1, t), (x_2, t))$$

with $t = 0, 1$. Compose this with

$$k_* : C((X_1 \times [0, 1], X_2 \times [0, 1])) \to C(Y_1, Y_2),$$

and take homology.

Next, we prove the excision axiom.

3.3.4 Theorem *Let Z and Y be subsets of a topological space X such that the closure of Z in X is contained in the interior of Y in X. Then the inclusion of pairs $(X \setminus Z, Y \setminus Z) \to (X, Y)$ induces an isomorphism on singular homology.*

First, we shall show how to deduce the theorem from Lemma 3.2.1.

Proof of Theorem 3.3.4 Let $\mathcal{U} = \{Y, X \setminus Z\}$. We have a diagram of chain complexes where the rows are induced by inclusions, and exact:

$$
\begin{array}{ccccccccc}
0 & \longrightarrow & C(Y) & \longrightarrow & C^{\mathcal{U}}(X) & \longrightarrow & C(X \setminus Z, Y \setminus Z) & \longrightarrow & 0 \\
 & & \downarrow{\scriptstyle Id} & & \downarrow{\scriptstyle \subseteq} & & \downarrow{\scriptstyle \subseteq} & & \\
0 & \longrightarrow & C(Y) & \longrightarrow & C(X) & \longrightarrow & C(X, Y) & \longrightarrow & 0.
\end{array}
$$

Apply homology, and then use Lemma 3.2.1 and the 5-lemma. □

The first step toward proving Lemma 3.2.1 is defining a natural *barycentric subdivision* chain map

$$sd : CX \to CX.$$

Let

$$\alpha : \{0, \ldots, n\} \to \{0, \ldots, n\} \tag{3.3.4}$$

be a permutation. We will define a singular n-simplex

$$\lambda_\alpha : \Delta^n \to \Delta^n.$$

First define λ_{Id} as the affine map which has

$$\lambda_{Id}([1, 0, \ldots, 0] = [\frac{1}{n+1}, \frac{1}{n+1}, \cdots \frac{1}{n+1}],$$

$$\lambda_{Id}([0, 1, 0, \ldots, 0] = [0, \frac{1}{n}, \frac{1}{n}, \ldots, \frac{1}{n}],$$

$$\cdots$$

$$\lambda_{Id}([0, 0, \ldots, 1] = [0, \ldots, 0, 1],$$

Then define

$$\lambda_\alpha([t_0, \ldots, t_n]) = \lambda_{Id}([t_{\alpha^{-1}(0)}, \ldots, t_{\alpha^{-1}(n)}]).$$

Finally, put, for a singular n-simplex $\sigma : \Delta^n \to X$,

$$sd(\sigma) = \sum_\alpha sign(\alpha)\sigma \circ \lambda_\alpha \tag{3.3.5}$$

where the sum is over all permutations (3.3.4). To see that sd is a chain map, taking the 0'th face on the right hand side of (3.3.5) is $sd(d\sigma)$, the remaining terms cancel in pairs, taking the $(i-1)$'st resp. i'th face for two permutations σ, $\tau \circ \sigma$ where τ is the 2-cycle permutation $(i-1, i)$.

3.3.5 Lemma *There exists a natural chain homotopy*

$$h : sd \simeq Id.$$

Proof We use the method of acyclic models. We have $s d_0 = I d_0$, so we can put $h_0 = 0$. Now suppose that h_{n-1} is constructed. We shall construct

$$h_n(\kappa_n)$$

where $\kappa_n = Id : \Delta^n \to \Delta^n$. Then we can, again, represent any singular n-simplex $\sigma : \Delta^n \to X$ as $\sigma_* \kappa_n$, and thus we can (and must) put

$$h_n(\sigma) = \sigma_* h_n(\kappa_n).$$

To find $\lambda = h_n(\kappa_n)$, we have, again, the equation

$$d\lambda = sd(\kappa_n) - \kappa_n - h_{n-1}(d\kappa_n).$$

We find that the right hand side is a cycle in $C_n X$ by a calculation identical to (3.3.3). Thus, it is a boundary by Lemma 3.3.2. Thus, we can solve for λ, completing the induction step. $\qquad \square$

Proof of Lemma 3.2.1 Consider the short exact sequence of chain complexes

$$0 \to C^{\mathcal{U}}(X) \to C(X) \to C(X)/C^{\mathcal{U}}(X) \to 0.$$

By the long exact sequence in homology, it suffices to show that the last term has homology 0. A cycle in $C(X)/C^{\mathcal{U}}(X)$ is represented by a chain $c \in C(X)$ such that

$$d(c) \in C^{\mathcal{U}}(X). \tag{3.3.6}$$

By the Lebesgue number theorem, however, there exists an $n \in \mathbb{N}$ such that $sd^n(c) \in C^{\mathcal{U}}(X)$. Now by Lemma 3.3.5 (and induction), there exists a chain homotopy

$$k : sd^n \simeq Id,$$

i.e.

$$dk(c) + k(dc) = sd^n(c) - c,$$

or

$$c + dk(c) = sd^n(c) - k(dc).$$

Observe that by (3.3.6), the right hand side is in $C^{\mathcal{U}}(X)$. Thus,

$$c \in Im(d) + C^{\mathcal{U}}(X),$$

or, in other words, c is a boundary in $C(X)/C^{\mathcal{U}}(X)$, as required.

$\square$

3.4   The Homology of Spheres

A good example of the use of the Eilenberg-Steenrod axioms is the proof of the following

3.4.1 Proposition *For all* $k, n \in \mathbb{N}_0$, *we have*

$$H_k(S^n; A) \cong H^k(S^n; A).$$

This group is isomorphic to $A \oplus A$ *if* $k = n = 0$, *to* A *if* $k = n \neq 0$, *and to* 0 *otherwise.*

The proof goes a lot easier if we introduce *reduced (co)homology* $\widetilde{H}_n(X)$. For a non-empty space X, the map $\epsilon : X \to *$ has a right inverse. One denotes

$$\widetilde{H}_n(X; A) = Ker(H_n(\epsilon; A) : H_n(X; A) \to A),$$

$$\widetilde{H}^n(X; A) = Coker(H^n(\epsilon; A) : A \to H^n(X; A)).$$

The existence of a right inverse to ϵ gives isomorphisms

$$H_0(X; A) \cong \widetilde{H}_0(X; A) \oplus A, \quad H^0(X; A) \cong \widetilde{H}^0(X; A) \oplus A,$$

and

$$H_n(X; A) \cong \widetilde{H}_n(X; A), \quad H^n(X; A) \cong \widetilde{H}^n(X; A) \text{ for } n \neq 0.$$

There is no reduced homology or cohomology of pairs, but the exactness axiom has a reduced analogue

$$\cdots \to \widetilde{H}_n(Y; A) \to \widetilde{H}_n(X; A) \to H_n(X, Y; A) \to \widetilde{H}_{n-1}(Y; A) \to \cdots \qquad (3.4.1)$$

and

$$\cdots \to H^n(X, Y; A) \to \widetilde{H}^n(X; A) \to \widetilde{H}^n(Y; A) \to H^{n+1}(X, Y; A) \to \cdots . \qquad (3.4.2)$$

This can be proved directly from the Eilenberg-Steenrod axioms, but it is easier to introduce the *reduced singular chain complex*

$$\widetilde{C}X = Ker(C\epsilon : CX \to C*).$$

This is a chain complex of free abelian groups, so we can extend this to the case of coefficients and cochains using $? \otimes A$, $Hom(?, A)$, and then observe that for a pair (X, Y) of non-empty spaces, we have short exact sequences of chain complexes

$$0 \to \widetilde{C}(Y; A) \to \widetilde{C}(X; A) \to C(X, Y; A) \to 0$$

$$0 \to C^*(X, Y; A) \to \widetilde{C}^*(X; A) \to \widetilde{C}^*(Y; A) \to 0.$$

Proof of Proposition 3.4.1 We prove the statement for homology with coefficients in $\mathbb{Z}$. The other cases are completely analogous. Observe that in terms of reduced homology, the statement just says that $\widetilde{H}_n S^k$ is isomorphic to $\mathbb{Z}$ for $k = n$, and is 0 otherwise. This will be proved by induction on n. For $n = 0$, it follows directly from the dimension axiom and the observation that homology takes disjoint unions to direct sums. (By the way, cohomology takes disjoint unions to products.) Now for $n > 0$, consider $\mathbb{R}^n$ as a subset of $\mathbb{R}^{n+1}$ consisting of those points whose last coordinate is 0. Accordingly, S^{n-1} becomes a subset of S^n of those points whose last coordinate is 0. Let also

$$S^n_+ = \{(x_0, \dots, x_n) \in S^n \mid x_n \geq 0\}.$$

This space is homotopy equivalent to $*$ (we say *contractible*), and thus

$$\widetilde{H}_k S^n_+ = 0$$

for all k. Now by the long exact sequence in reduced homology of the pair (S^n, S^n_+), we get

$$\widetilde{H}_k(S^n) \cong H_k(S^n, S^n_+).$$

Now denoting $v = (0, \dots, 0, 1)$, by excision, we have

$$H_k(S^n, S^n_+) \cong H_k(S^n \smallsetminus \{v\}, S^n_+ \smallsetminus \{v\}).$$

On the other hand, the pair $S^n \smallsetminus \{v\}$ is contractible, so by the long exact sequence in reduced homology of the pair $(S^n \smallsetminus \{v\}, S^n_+ \smallsetminus \{v\})$, we have

$$H_k(S^n \smallsetminus \{v\}, S^n_+ \smallsetminus \{v\}) \cong \widetilde{H}_{k-1}(S^n_+ \smallsetminus \{v\}).$$

But the space $S^n_+ \setminus \{v\}$ is homotopy equivalent to S^{n-1}, so our statement follows from the induction hypothesis.

$\square$

3.5 Universal Coefficients and Künneth Theorem

Actually, homology or cohomology with any coefficients can be readily calculated from the homology with coefficients in $\mathbb{Z}$. More generally, let R be a principal ideal domain and let C be a chain complex of R-modules. Then, recalling our notation $Z_n = Ker(d : C_n \to C_{n-1})$, $B_n = Im(d : C_{n+1} \to C_n)$, $H_n = H_n C = Z_n/B_n$, we have short exact sequences

$$0 \to B_n \to Z_n \to H_n \to 0, \tag{3.5.1}$$

$$0 \to Z_n \to C_n \to B_{n-1} \to 0. \tag{3.5.2}$$

Since R is a PID, submodules of free modules are free, so in particular B_n and Z_n are free R-modules, and the chain complex

$$\mathcal{H}_n : B_n \to Z_n$$

(with Z_n set in degree 0 and B_n in degree 1) is a projective resolution of H_n. Also, the short exact sequence (3.5.2) splits. Denoting by $s_n : B_{n-1} \to C_n$ the splitting, the isomorphisms

$$\subset \oplus s_n : Z_n \oplus B_{n-1} \to C_n$$

actually define an *isomorphism of chain complexes*

$$\bigoplus_{n \in \mathbb{Z}} \mathcal{H}_n[n] \cong C. \tag{3.5.3}$$

The isomorphism (3.5.3) then gives, for an R-module A, isomorphisms

$$H_n(C \otimes_R A) \cong (H_n(C) \otimes_R A) \oplus Tor_1^R(H_{n-1}(C), A), \tag{3.5.4}$$

$$H^n(Hom_R(C, A)) \cong Hom_R(H^n(C), A) \oplus Ext_R^1(H_{n-1}(C), A). \tag{3.5.5}$$

Note that, if R is a field, the Tor_1 and Ext^1 terms, of course, vanish. For $R = \mathbb{Z}$ and a topological space X, we get, in particular,

$$H_n(X; A) \cong (H_n(X) \otimes A) \oplus Tor_1^{\mathbb{Z}}(H_{n-1}(X), A), \tag{3.5.6}$$

$$H^n(X; A) \cong Hom(H_n(X), A) \oplus Ext_{\mathbb{Z}}^1(H_{n-1}(X), A). \tag{3.5.7}$$

These facts are referred to as the *Universal coefficient theorem*. One caveat emptor is that the choice of the splittings was not canonical, and thus, the isomorphisms (3.5.4), (3.5.5), (3.5.6), (3.5.7) are not canonical (and, as it turns out, not even natural). In (3.5.4), (3.5.5), one observes carefully that there are in fact short exact sequences of the form

$$0 \to H_n(C) \otimes_R A \to H_n(C \otimes A) \to Tor_1^R(H_{n-1}(C), A) \to 0 \qquad (3.5.8)$$

$$0 \to Ext_R^1(H_{n-1}(C), A) \to H^n(Hom_R(C, A)) \to Hom_R(H_n(C), A) \to 0, \qquad (3.5.9)$$

which are independent of the splitting (and hence canonical and natural), and which split (non-canonically) to give (3.5.4), (3.5.5). Let us see this for the case of (3.5.8) (the other cases are analogous). To this end, the first map (3.5.8) follows simply from that fact that for $z \in Z_n, a \in A, z \otimes a$ is a cycle in $C \otimes_R A$. To construct the second map, consider the canonical map

$$\phi : C_n \otimes A \to B_{n-1} \otimes_R A$$

given by $d \otimes A$. If c is a cycle in $C \otimes A$, then $\phi(c)$ becomes 0 when we apply the map

$$\subseteq \otimes A : B_{n-1} \otimes_R A \to C_{n-1} \otimes_R A.$$

But we claim that

$$\subseteq \otimes A : Z_{n-1} \otimes_R A \to C_{n-1} \otimes_R A \qquad (3.5.10)$$

is injective. Thus, $\phi(c)$ vanishes when we apply

$$\subseteq \otimes A : B_{n-1} \otimes_R A \to Z_{n-1} \otimes_R A,$$

and thus defines an element in $Tor_1^R(H_{n-1}(C), A)$, using the projective resolution $\mathcal{H}_n$. To see (3.5.10), apply (1) of Proposition 2.3.19 to the short exact sequence (3.5.2) (with n replaced by $n - 1$) and the functor $? \otimes_R A$, recalling that since B_{n-1} is a free R-module, we have $Tor_1^R(B_{n-2}, A) = 0$.

Note that a completely similar discussion applies to (the totalization of) a tensor product of chain complexes $C \otimes_R D$ assuming R is a PID and C, D are chain complexes of free R-modules. We obtain an isomorphism

$$H_n(C \otimes_R D) = \bigoplus_{k+\ell=n} H_k(C) \otimes_R H_\ell(D) \oplus \bigoplus_{k+\ell=n-1} Tor_1^R(H_k(C), H_\ell(D)),$$

with a canonical natural short exact sequence. This is known as the *Künneth Theorem*. For example, for $R = \mathbb{R}$ (where, of course, the Tor_1-term goes away), (1.4.2) therefore gives a homomorphism (sometimes called the *cup product*)

$$\cup : H_{DR}^k(M) \otimes_{\mathbb{R}} H_{DR}^\ell(M) \to H_{DR}^{k+\ell}(M)$$

(and similarly with complex coefficients for complex manifolds M). Thinking of $H_{DR}^*(M)$ as a graded $\mathbb{R}$-module, the properties of differential forms then easily imply that $\cup$ turns $H_{DR}^*(M)$ into a graded-commutative (associative unital) ring (in fact, $\mathbb{R}$-algebra), which means that for homogeneous elements x, y of degrees k, ℓ,

$$x \cup y = (-1)^{k\ell} y \cup x.$$

(The $\cup$ symbol is pretty but a bit slow, and is often omitted.)

This brings up a natural question as to whether a similar discussion also applies to spaces, i.e. whether singular cohomology with coefficients in a commutative ring R has a cup product, which would make it a graded-commutative R-algebra. The answer to both questions turns out to be positive, the key observation being the *Eilenberg-Zilber theorem*, which states that for spaces X, Y, we have a natural chain-homotopy equivalence

$$\psi : C(X \times Y) \simeq CX \otimes CY,$$

unique in the category $h\,Ab$-Chain subject to the condition that it induces the canonical identification in degree 0. (See Exercise 27.) A detailed discussion, again, can be found in [20]. Letting $\Delta : X \to X \times X$ be the diagonal, i.e. $\Delta(x) = (x, x)$, we have a composition

$$CX \xrightarrow{\Delta} C(X \times X) \xrightarrow{\psi} CX \otimes CX,$$

and various dualizations give chain maps

$$C^*(X; \mathbb{Z}) \otimes C^*(X; \mathbb{Z}) \to C^*(X; \mathbb{Z}), \tag{3.5.11}$$

$$C^*(X; \mathbb{Z}) \otimes C_*(X; \mathbb{Z}) \to C_*(X; \mathbb{Z}) \tag{3.5.12}$$

(and similarly with coefficients in a commutative ring). Now for any two chain complexes C, D, we have a canonical homomorphism

$$H_k C \otimes H_\ell D \to H_{k+\ell}(C \otimes D)$$

by taking the class represented by the tensor product of representative cycles of given classes. Using this, (3.5.11) and (3.5.12) give homomorphisms

$$\cup : H^k(M; \mathbb{Z}) \otimes H^\ell(M; \mathbb{Z}) \to H^{k+\ell}(M; \mathbb{Z}),$$

$$\cap : H^k(M; \mathbb{Z}) \otimes H_\ell(M; \mathbb{Z}) \to H_{\ell-k}(M; \mathbb{Z})$$

(and similarly with coefficients in any commutative ring). Then $\cup$ makes $H^*(M; \mathbb{Z})$ into a graded-commutative ring. Sign conventions on $\cap$ vary, but it is possible to pick the signs in such a way that $H_*(M; \mathbb{Z})$ becomes a left $H^*(M; \mathbb{Z})$-module, i.e. that we have

$$(\alpha \cup \beta) \cap \gamma = \alpha \cap (\beta \cap \gamma). \tag{3.5.13}$$

3.6   CW-Homology

Let

$$D^n = \{(x_1, \dots, x_n) \in \mathbb{R}^n \mid \sum x_i^2 \le 1\}.$$

Thus, $S^{n-1} \subset D^n$. A *CW-complex* is a topological space X which is a union of subspaces

$$\emptyset = X_{-1} \subseteq X_0 \subseteq X_1 \subseteq \dots,$$

with the colimit topology, together with sets I_n (called *sets of n-cells*) maps

$$f_n : S^{n-1} \times I_n \to X_{n-1}$$

(called the *attaching map*) such that X_n is the pushout of the diagram

$$
\begin{array}{ccc}
S^{n-1} \times I_n & \xrightarrow{\ f_n\ } & X_{n-1} \\
{\scriptstyle \subset}\downarrow & & \\
D^n \times I_n. & &
\end{array}
$$

A *CW pair* (X, Y) is defined in the same way, except that $X_{-1} = Y$.

Note the similarity with the notion of cell chain complex of Sect. 2.3.9. In addition to moving from the category of chain complexes to the category of spaces, in a CW complex, we require that "a cell be attached to cells of lower degrees only." This requirement can in fact be added to the notion of a cell chain complex to obtain a notion of a *CW-chain complex*. Conversely, the requirement can be dropped from the definition of a CW-complex to obtain the notion of a *cell space*. These notions happen to be of lesser immediate importance to our discussion. In a CW-complex X, the subspace X_n is often referred to as the *n-skeleton*.

COMMENT In fact, the analogy with chain complexes extends further. For a space X and a point $x \in X$, we can define, for all $n \in \mathbb{N}$, $\pi_n(X, x)$ as the set of based homotopy classes of continuous maps $S^n \to X$ which send a chosen base point to x. One proves that $\pi_n(X, x)$ are groups similarly as in the case $n = 1$ (see Sect. 5.5 of Chap. 3); in fact, these groups are abelian for $n \geq 2$, and are called the *homotopy groups of the space* X. Unlike homology, the calculation of homotopy groups of a given space is a difficult, and in general unsolved, problem. They are, however, useful from a foundational point of view. To this end, it is helpful to also introduce $\pi_0(X)$, by which we mean the set of *path-components* of X, i.e. equivalence classes of points $x \in X$ where $x \sim y$ if there exists a path in X from x to y. (This is a set without a natural group structure.) The groups $\pi_n(X, x)$ and the set $\pi_0(X)$ have the expected functorialities.

Now one says that a continuous map $f : X \to Y$ is a *weak equivalence* when $\pi_0(f)$ is a bijection and for all $n \in \mathbb{N}$ and all $x \in X$,

$$\pi_n(f) : \pi_n(X, x) \to \pi_n(Y, f(x))$$

is an isomorphism.

It turns out that analogously to the case of chain complexes, the category $hTop$ with respect to the subcategory E of weak equivalences has colocalization in the class B consisting of all CW-complexes. This is sometimes referred to as the *Whitehead Theorem*. The proof is not necessarily harder than in the case of chain complexes, but it involves methods of homotopy theory of spaces. Details can be found in [13]. Thus, a derived category $DTop = DhTop$ in the sense of Sect. 2.1 exists. It is known as the *derived category of spaces*. Weak equivalences can be proved to induce isomorphisms on singular homology and cohomology, which can therefore be considered as functors on $DTop$.

3.6.1 Example The space $(\mathbb{P}^n_{\mathbb{C}})_{an}$ consisting of closed points of $\mathbb{P}^n_{\mathbb{C}}$ with the analytic topology is, in the context of topology, often denoted by $\mathbb{C}P^n$. It is, as remarked in Example 1.1.2, canonically a complex manifold. We will now also give it the structure of a CW-complex. Indeed, $\mathbb{C}P^n$ is identified with the set of complex lines through the origin in the vector space $\mathbb{C}^{n+1}$ (which is a linear algebra name for $\mathbb{A}^{n+1}_{\mathbb{C}}$). It is actually useful to mention another concept here which is important in algebraic topology: We have an obvious projection $p_n : \mathbb{C}^{n+1} \setminus \{0\} \to \mathbb{C}P^n$ by

$$(z_0, \ldots z_n) \mapsto [z_0 : \cdots : z_n],$$

and given the induced topology on $\mathbb{C}^{n+1} \setminus \{0\}$ from $\mathbb{C}^{n+1}$, the topology on $\mathbb{C}P^n$ is the universal topology which makes the projection p_n continuous. This is called the quotient topology. In general, for a topological space X (not necessarily Hausdorff) and any map $f : X \to Y$ where Y is a set, the *quotient topology* has $U \subseteq Y$ open if and only if $f^{-1}(U) \subseteq X$ is open. The quotient topology is useful in defining colimits. For example, the pushout of a diagram of topological spaces

$$
\begin{array}{ccc}
X & \xrightarrow{\ f\ } & Y \\
{\scriptstyle g}\downarrow & & \\
Z & &
\end{array}
$$

can be constructed as the quotient of $Y \amalg Z$ under the smallest equivalence relation $\sim$ which has $f(x) \sim g(x)$ for all $x \in X$, with the quotient topology.

Now using the usual embedding $\mathbb{C}^n \subset \mathbb{C}^{n+1}$ by

$$
(z_1, \ldots, z_n) \mapsto (z_1, \ldots, z_n, 0),
$$

we obtain a continuous embedding $\mathbb{C}P^{n-1} \subset \mathbb{C}P^n$. We claim that we can take $\mathbb{C}P^k$ to be the $2k$-skeleton of $\mathbb{C}P^n$ with $0 \le k \le n$, and no odd-dimensional cells. To this end, by induction on n, we just need to show how $\mathbb{C}P^n$ is obtained from $\mathbb{C}P^{n-1}$ by attaching a single $2n$-cell. To this end, consider the space

$$
P_n = \{(z_0, \ldots, z_n) \in \mathbb{C}^{n+1} \mid \sum |z_k|^2 = 1\}
$$

with the induced topology from $\mathbb{C}^{n+1}$. (Of course, $P_n \cong S^{2n+1}$.) Then consider the subspace

$$
Q_n = \{(z_0, \ldots, z_n) \in P_n \mid z_n \in \mathbb{R}, \ z_n \ge 0\}.
$$

Then we have $P_{n-1} \subset Q_n$, and in fact the pair (Q_n, P_{n-1}) is homeomorphic to (D^{2n}, S^{2n-1}). Letting $f_n : P_{n-1} \to \mathbb{C}P^{n-1}$ be the obvious projection

$$
(x_0, \ldots, x_{n-1}, 0) \mapsto [x_0 : \cdots : x_{n-1}],
$$

we then obtain a canonical homomorphism from the pushout of the diagram of topological spaces

$$
\begin{array}{ccc}
P_{n-1} & \xrightarrow{\ f_n\ } & \mathbb{C}P^{n-1} \\
{\scriptstyle \subset}\downarrow & & \\
Q_n & &
\end{array}
$$

to $\mathbb{C}P^n$, which is bijective, and hence by Theorem 2.1.2 of Chap. 2, is a homeomorphism.

Similarly, one can show that the real projective space $\mathbb{R}P^n$ of lines through the origin in $\mathbb{R}^{n+1}$ with the quotient topology from $\mathbb{R}^{n+1} \setminus \{0\}$ is a CW-complex with one cell in each dimension $0 \le k \le n$, with skeleta $\mathbb{R}P^k$. The pair (Q_n, P_{n-1}) is replaced by (S^n_+, S^{n-1}). This is a good example in algebraic topology. We should note however that $\mathbb{R}P^n$ is, in our discussion so far, "not an object of algebraic geometry" in the sense that it is the set of closed points of $\mathbb{P}^n_\mathbb{R}$ of residue field $\mathbb{R}$ only; the variety $\mathbb{P}^n_\mathbb{R}$ also has closed points with residue field $\mathbb{C}$, and from our point of view should therefore be regarded as $\mathbb{C}P^n$ with $Gal(\mathbb{C}/\mathbb{R})$-action by complex conjugation.

3.6.2 Lemma *Let X be a CW-complex. Then for $n \ge 0$, the group $H_k(X_n, X_{n-1})$ is canonically isomorphic to the free abelian group $\mathbb{Z}I_n$ when $k = n$, and to 0 otherwise.*

Proof Let

$$D^n_\circ = D^n \setminus \{0\}, \quad B^n = D^n \setminus S^{n-1}, \quad B^n_\circ = D^n_\circ \setminus S^{n-1}.$$

Let $X_{n,\circ}$ be the pushout of the diagram

$$
\begin{array}{ccc}
S^{n-1} \times I_n & \xrightarrow{f_n} & X_{n-1} \\
\cap \downarrow & & \\
D^n_\circ \times I_n. & &
\end{array}
$$

The pair (X_n, X_{n-1}) is homotopy equivalent to the pair $(X_n, X_{n,\circ})$ whose homology is, by excision, isomorphic to the homology of the pair

$$\coprod_{I_n} (B^n, B^n_\circ).$$

But the pair $(B^n, B^n_\circ)$ is homotopy equivalent to the pair (D^n, S^{n-1}), whose homology was calculated in our proof of Proposition 3.4.1. $\qquad\square$

Now define, for a CW-complex X, the chain complex $C^{CW}X$ where

$$C^{CW}_n X = \mathbb{Z}I_n,$$

and using the identification of Lemma 3.6.2, the differential $d^{CW} = d^{CW}_n$ is the composition

$$H_n(X_n, X_{n-1}) \to H_{n-1}(X_{n-1}) \to H_{n-1}(X_{n-1}, X_{n-2}) \qquad (3.6.1)$$

which is the composition of the maps involved in the long exact sequences in homology of the pairs (X_n, X_{n-1}), (X_{n-1}, X_{n-2}). Note that the composition of two consecutive maps in (3.6.1) contains a composition of two consecutive maps in the long exact sequence in homology of the pair (X_{n-1}, X_{n-2}), and thus is 0. We denote

$$H_n^{CW}(X) = H_n(C^{CW}X),$$

and call this the *CW-homology of X*. CW homology with coefficients and cohomology is defined analogously, using the chain complexes $C^{CW}(X; A) = C^{CW}X \otimes A$, $C^*_{CW}(X; A) = Hom(C^{CW}X, A)$ (which also satisfy a version of Lemma 3.6.2). Similarly, we also have CW-homology and cohomology of CW pairs. Now the central fact about CW homology and cohomology is that it is, in fact, the same as singular homology and cohomology.

3.6.3 Theorem *For a CW-complex X, we have canonical isomorphisms*

$$H_n(X; A) \cong H_n^{CW}(X; A), \quad H^n(X; A) \cong H^n_{CW}(X; A),$$

and similarly for CW-pairs.

This fact can be proved directly, but instead, we shall deduce it in the next subsection as an easy application of the concept of a spectral sequence, which is useful more generally in the context of (co)homology.

Meanwhile, to use Theorem 3.6.3 to compute (co)homology, we need an approach to calculating

$$d^{cell} : \mathbb{Z}I_n \to \mathbb{Z}I_{n-1}.$$

This can be considered as an "$I_{n-1} \times I_n$-matrix," which can be called the *incidence matrix*. Of course, these sets can be infinite, but only finitely many entries in each column can be non-zero. To the (i, j)-entry with $i \in I_{n-1}$, $j \in I_n$, we note that it is, in fact, induced in homology by a map

$$\phi : S^{n-1} \to S^{n-1}, \tag{3.6.2}$$

which is the composition

$$S^{n-1} \to X_{n-1} \to X_{n-1}/X_{n-2} \to D^{n-1}/S^{n-2} \cong S^{n-1}$$

where the first map is the restriction of the attaching map f_n to $S^{n-1} \times \{j\}$, X/Y is obtained from X by identifying all the points of Y (and taking the quotient topology), the penultimate map projects onto the i'th cell modulo its boundary, collapsing all the other $(n - 1)$-cells to the base point, and the last homeomorphism is chosen arbitrarily, but the same for all pairs $(i, j) \in I_{n-1} \times I_n$. Thus, we just need to compute what a map (3.6.2) induces in $(n - 1)$'st homology.

Before tackling that question, let us mention a little detail: note that this method is only correct for $n > 1$. For $n = 1$, we note that we have a canonical bijection $I_0 \cong X_0$, and the j'th column of the incidence matrix simply has 1 in the $f_1(1, j)$'th row, and -1 in the $f_1(-1, j)$'th row.

To find out what a map $\phi : S^m \to S^m$ induces in m'th homology for $m > 0$ (this is called the *degree* $deg(\phi) \in \mathbb{Z}$ of the map ϕ, it is useful to identify $S^m \cong \mathbb{R}^m \cup \{\infty\}$ (thus determining the topology on the target) via the projection from the point $(0, \ldots 0, 1)$ to $\mathbb{R}^m \subset \mathbb{R}^{m+1}$, i.e.

$$(x_0, \ldots, x_m) \mapsto (\frac{x_0}{1 - x_m}, \cdots \frac{x_{m-1}}{1 - x_m}) \tag{3.6.3}$$

which has inverse

$$(t_1, \ldots, t_{m-1}) \mapsto$$

$$\left(\frac{2t_1}{1 + t_1^2 + \cdots + t_{m-1}^2}, \cdots, \frac{2t_{m-1}}{1 + t_1^2 + \cdots + t_{m-1}^2}, \frac{t_1^2 + \cdots + t_{m-1}^2 - 1}{1 + t_1^2 + \cdots + t_{m-1}^2} \right).$$

Now every map

$$\phi : \mathbb{R}^m \cup \{\infty\} \to \mathbb{R}^m \cup \{\infty\} \tag{3.6.4}$$

is homotopic to one which sends $\infty \mapsto \infty$ and has a *regular value* which means a point $y \in \mathbb{R}^m$ such that $\phi^{-1}(y)$ is finite, and for all $x \in \phi^{-1}(y)$ there exists an open set $U \ni x$ such that $\phi|U$ is smooth and $det(D\phi|_x) \neq 0$. Then one can show that

$$deg(\phi) = \sum_{y \in \phi^{-1}(x)} sign(det(D\phi|_x)). \tag{3.6.5}$$

Details can be found in [16].

3.6.4   Example 3.6.1 Continued

We see that the cell chain complex $C^{CW}\mathbb{C}P^n$ with the CW-structure we described is of the form

$$\mathbb{Z} \to 0 \to \mathbb{Z} \to \cdots \to \mathbb{Z} \to 0 \to \mathbb{Z},$$

and thus, there is no possibility of differentials. Thus, we can conclude that $H_k(\mathbb{C}P^n) = \mathbb{Z}$ for $0 \leq k \leq 2n$ even, and 0 else.

In contrast, for $\mathbb{R}P^n$, the cell chain complex is of the form

$$\mathbb{Z} \to \mathbb{Z} \to \cdots \to \mathbb{Z} \to \mathbb{Z},$$

and thus, we must compute the differentials using the degree formula. The attaching map

$$S^{n-1} \to S^{n-1}_+ / S^{n-2}$$

sends $(x_1, \ldots, x_n) \mapsto (x_1, \ldots, x_n)$ for $x_n \geq 0$, and $(x_1, \ldots, x_n) \mapsto (-x_1, \ldots, -x_n)$ for $x_n \leq 0$ (note that when $x_n = 0$, the point goes to the base point). Conjugating the map

$$(x_1, \ldots, x_n) \mapsto (-x_1, \ldots, -x_n)$$

by (3.6.3), we get the map

$$(t_1, \ldots, t_{n-1}) \mapsto \left(\frac{-t_1}{t_1^2 + \cdots + t_{n-1}^2}, \ldots, \frac{-t_{n-1}}{t_1^2 + \cdots + t_{n-1}^2} \right),$$

whose Jacobi matrix is the $1/(t_1^2 + \cdots + t_{n-1}^2)^2$-multiple of

$$-(t_1^2 + \cdots + t_{n-1}^2) I_{n-1} + 2(t_1, \ldots, t_{n-1})^T (t_1, \ldots, t_{n-1}).$$

Thus, it has one eigenvalue $1/(t_1^2 + \cdots + t_{n-1}^2)$ and $m-2$ eigenvalues $-1/(t_1^2 + \cdots + t_{n-1}^2)^2$, so its determinant is positive when n is even and negative when n is odd.

Choosing any point $y \in \mathbb{R}^{n-1} \setminus \{0\}$ as a regular value, we therefore see that $\phi^{-1}(y)$ has two elements, and that the two summands (3.6.5) have the same signs when n is even and opposite signs when n is odd. Thus, $C^{CW}(\mathbb{R}P^n)$ is of the form

$$\mathbb{Z} \longrightarrow \cdots \longrightarrow \mathbb{Z} \xrightarrow{\ 2\ } \mathbb{Z} \xrightarrow{\ 0\ } \mathbb{Z}, \qquad (3.6.6)$$

thereby showing that $H_k(\mathbb{R}P^n)$ is $\mathbb{Z}$ when $k = 0$ or $k = n$ odd, $\mathbb{Z}/2$ when $0 < k < n$ odd, and 0 otherwise. It is also worth noting that for cohomology, the arrows (3.6.6) are reversed, so $H^k(\mathbb{R}P^n; \mathbb{Z})$ is $\mathbb{Z}$ when $k = 0$ or $k = n$ odd, $\mathbb{Z}/2$ when $0 < k \leq n$ even, and 0 otherwise. Alternately, we can also see that by the universal coefficient theorem for the last statement.

It can be shown that every complete algebraic variety over $\mathbb{C}$ with the analytic topology has a structure of a finite CW-complex (meaning with finitely many cells), and any algebraic variety over $\mathbb{C}$ is homotopy equivalent to a finite CW-complex, so at least theoretically, CW-homology can be used to calculate singular homology and cohomology.

3.7   Spectral Sequences

Spectral sequences are perhaps the most advanced tool of homological algebra. There is a homological and cohomological version of the concept of a spectral sequence, which differ by interchanging subscripts and superscripts, and reversing the signs of all gradings. We will mostly discuss the homological spectral sequence version in this subsection, only pointing out in a few places what the cohomological situation looks like, for the sake of familiarity of notation.

A homological *exact couple* consists of $\mathbb{Z}$-graded abelian groups (or more generally, objects of an abelian category) D, E, together with graded homomorphisms which form a long exact sequence,

$$
\begin{array}{ccc}
D & \xrightarrow{\quad i \quad} & D \\
& {\scriptstyle k}\nwarrow \quad \swarrow {\scriptstyle j} & \\
& E. &
\end{array}
\tag{3.7.1}
$$

where i, j are of degree 0, and k is of degree -1. Also, note that E then becomes a chain complex with the differential

$$
d = j \circ k.
$$

The main point of this concept is that we can define a *derived exact couple*

$$
\begin{array}{ccc}
D' & \xrightarrow{\quad i' \quad} & D' \\
& {\scriptstyle k'}\nwarrow \quad \swarrow {\scriptstyle j'} & \\
& E' &
\end{array}
\tag{3.7.2}
$$

where $D' = Im(i)$, $E' = H_*(E, d)$ where the homomorphisms i', k' are just induced from the homomorphisms i, k, while j' is "induced by $j \circ i^{-1}$." More precisely, for the definition of j', one observes that, (using exactness), one also has

$$
E' = k^{-1} Im(i)/j(Ker(i)).
\tag{3.7.3}
$$

Using the short exact sequence

$$
0 \to Ker(i) \to D \to Im(i) \to 0,
$$

the morphism $j : D \to E$ then induces a morphism $D \to E'$ which becomes 0 when composed with the inclusion $Ker(i) \subseteq D$, and thus induces a morphism $j' : Im(i) \to E'$. The fact that (3.7.2) is again an exact couple is readily verified by hand. (Exercise 28.)

In fact, by induction, one can show that when iterating this procedure, the m'th derived exact couple

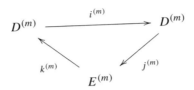

can also be described directly: One has

$$D^{(m)} = Im(i^m), \tag{3.7.4}$$

$$E^{(m)} = k^{-1} Im(i^m)/j(Ker(i^m)), \tag{3.7.5}$$

the morphisms $i^{(m)}$, $k^{(m)}$ are induced from i, k, and the morphism $j^{(m)}$ is "induced by $j \circ i^{-m}$" in precisely the same sense as for $m = 1$ (Exercise 29).

Terminology varies somewhat, but for our purposes, we can say that a *homological spectral sequence* arises from a sequence of exact couples

$$\begin{array}{ccc} D^r & \xrightarrow{i^r} & D^r \\ & \nwarrow_{k^r} \quad \swarrow_{j^r} & \\ & E^r, & \end{array} \tag{3.7.6}$$

$r \geq 1$, where the exact couple (3.7.6) for $r + 1$ is the derived exact couple of the exact couple (3.7.6) for r. Additionally, usually, these exact couples are in fact *bigraded* by pairs of integers (p, q) whose *total degree* $p + q$ is the degree mentioned in the definition of an exact couple, p is referred to as the *filtration degree* and q is called the *complementary degree*. In (3.7.6), i^r, j^r, k^r are bigraded homomorphisms where i^r has bidegree $(1, -1)$, j^r has bidegree $(1 - r, r - 1)$ and k has bidegree $(-1, 0)$. (Note that if the bidegree convention holds for a given r, it also holds for all the successively derived exact couples.)

Typically, we are using the spectral sequence to gain information about D^1, while (E^1, d^1) (and therefore, usually, E^2) are known. A *homological spectral sequence* is then a system of abelian groups (or objects of an abelian category)

$$E^r_{p,q}, \quad r \in \mathbb{N}, \ p, q \in \mathbb{Z}$$

and differentials

$$d^r : E^r_{p,q} \to E^r_{p-r,q+r-1}, \ d^r \circ d^r = 0$$

where

$$E^{r+1} = H(E^r, d^r).$$

We say that the spectral sequence *converges* if for every pair of integers (p, q), there exists an N such that for $r > N$, the differential d^r is 0 on $E^r_{p,q}$. Then the expression

$$\tilde{E}^\infty_{p,q} = \operatorname*{colim}_r E^r_{p,q}$$

makes sense. This happens for example when for each given total degree, the sequence

$$\cdots \overset{i}{\longrightarrow} D_{p,q} \overset{i}{\longrightarrow} D_{p+1,q-1} \overset{i}{\longrightarrow} \cdots \tag{3.7.7}$$

satisfies the Mittag-Leffler condition. If this condition holds, we have

$$E^\infty = (k^{-1} \bigcap_n Im(i^n))/(j \bigcup_n Ker(i^n)). \tag{3.7.8}$$

As already mentioned, in a *cohomological spectral sequence* the indexing is reversed, so we have abelian groups (or objects of an abelian category)

$$E^{p,q}_r, \ r \in \mathbb{N}, \ p, q \in \mathbb{Z}$$

and differentials

$$d_r : E^{p,q}_r \to E^{p+r,q-r+1}_r$$

with

$$E_{r+1} = H(E_r, d_r).$$

The definition and criterion of convergence are completely symmetrical.

3.7.1   The Spectral Sequence of a Filtered Chain Complex

An *increasing filtration* on a chain complex C is a sequence of inclusions

$$\cdots \subseteq F_n C \subseteq F_{n+1} C \subseteq \cdots$$

of subcomplexes of C. We shall call an increasing filtration *convergent* if for every $n \in \mathbb{Z}$, there exists an $N \in \mathbb{Z}$ such that $F_N C_n = 0$, and if $\bigcup_p F_p C = C$. The *associated graded* chain complex of an increasing filtration is

$$E^0 C = \bigoplus_{n \in \mathbb{Z}} F_n C / F_{n-1} C.$$

A *decreasing filtration* is a sequence of subcomplexes $(F^n C)$ where $(F_n C) = (F^{-n} C)$ is an increasing filtration. Under this correspondence, the convergence condition on decreasing and increasing filtrations are the same, and for a decreasing filtration, the associated graded chain complex is

$$E_0 C = \bigoplus_{n \in \mathbb{Z}} F^n C / F^{n+1} C.$$

With a filtration of a chain complex, there is an associated spectral sequence, which is usually written homologically when the filtration is understood as increasing and cohomologically when the filtration is understood as decreasing. We shall treat the homological case (both cases are equivalent).

The spectral sequence of an increasing filtration $F_p C$ on a chain complex C has

$$D_{p,q}^1 = H_{p+q}(F_p C),$$

$$E_{p,q}^1 = H_{p+q}(F_p C / F_{p-1} C),$$

and the homomorphisms i^1, j^1, k^1 are the homomorphisms in the long exact sequence in homology associated with the short exact sequence of chain complexes

$$0 \to F_{p-1} C \to F_p C \to F_p C / F_{p-1} C \to 0.$$

From (3.7.8), we then obtain (with the understanding that all unlabeled arrows are the canonical homomorphisms):

$$E_{p,q}^\infty = k^{-1}(0)_{p,q} / j(Ker(H_{p+q} F_p C \to H_{p+q} C)) =$$
$$j(H_{p+q} F_p C) / j(Ker(H_{p+q} F_p C \to H_{p+q} C)) =$$
$$H_{p+q} F_p C / (Ker(H_{p+q} F_p C \to H_{p+q} C) + Im(H_{p+q} F_{p-1} C \to H_{p+1} F_p C))$$
$$= Im(H_{p+q} F_p C \to H_{p+q} C) / Im(H_{p+q} F_{p-1} C \to H_{p+q} C).$$

In other words, if we introduce a filtration

$$F_p H_{p+q} C = Im(H_{p+q} F_p C \to H_{p+q} C),$$

then this filtration is convergent (by commutation of homology with colimits of a sequence) and

$$E_{p,q}^{\infty} = E^0 H_{p+q} C.$$

In general, if we have a convergent spectral sequence and a $\mathbb{Z}$-graded abelian group G with an increasing filtration $(F_p G)$ such that

$$E_{p,q}^{\infty} = (E^0 G_{p+q})_p,$$

we write

$$E_{p,q}^r \Rightarrow G_{p+q},$$

and say that the spectral sequence $E_{p,q}^r$ converges to G (with the understanding that the associated graded object is "the best information we can get from a spectral sequence," at least without further input). Thus, in the case of the spectral sequence associated with a convergent filtration on a chain complex C, we can write

$$E_{p,q}^1 = H_{p+q}(F_p C / F_{p-1} C) \Rightarrow H_{p+q} C. \qquad (3.7.9)$$

Again, the discussion for convergent decreasing filtrations is entirely equivalent, and in that case one usually writes the spectral sequence cohomologically, and C as a cochain complex, so one has

$$E_1^{p,q} = H^{p+q}(F^p C / F^{p+1} C) \Rightarrow H^{p+q} C. \qquad (3.7.10)$$

3.7.2 Proof of Theorem 3.6.3

We prove the statement for homology with coefficients in $\mathbb{Z}$. For a CW complex X, let

$$F_p C X = C(X_p)$$

be the singular chain complex of the p-skeleton. Now apply (3.7.9). By Lemma 3.6.2, we have

$$E_{p,q}^2 = H_p^{CW}(X) \text{ for } q = 0$$

and

$$E_{p,q}^2 = 0 \text{ for } q \neq 0.$$

Thus, all differentials d^r with $r > 1$ have either source or target 0, and hence are 0, and we have

$$E^2_{p,q} = E^\infty_{p,q},$$

but also the associated graded group of the convergent filtration on $H_n X$ has only one non-trivial term in filtration degree $p = n$, which is therefore canonically isomorphic to $H_n X$. The statement follows.

For homology with coefficients, the proof is the same. For cohomology, the decreasing filtration on $C^* X$ is

$$F^p C^*(X; A) = Ker(C^*(X; A) \to C^*(X_{p-1}; A))$$

(where the arrow is induced by inclusion). Alternately, one can also deduce this case from the universal coefficient theorem, but one must be careful about functoriality. □

We shall now give the example of the Grothendieck spectral sequence, which is possibly the one most used in algebraic geometry. We will state it here for right derived functors, because it is often used in the context of sheaves, which have enough injectives. However, a precisely symmetrical statement holds for left derived functors.

3.7.3 Theorem (The Grothendieck Spectral Sequence) *Let $F : \mathcal{A} \to \mathcal{B}$, $G : \mathcal{B} \to \mathcal{C}$ be additive functors between abelian categories, where $\mathcal{A}$ and $\mathcal{B}$ have enough injectives. Suppose further that for every injective object $Q \in Obj(\mathcal{A})$, $F(Q)$ is G-local. Then for every object $A \in Obj(\mathcal{A})$, there exists a convergent cohomological spectral sequence*

$$E^{p,q}_2 = R^p G \circ R^q F(A) \Rightarrow R^{p+q}(G \circ F)(A). \tag{3.7.11}$$

Proof Let Q be an injective resolution of A, and let $C^{(0)} = FQ$. Assuming we have constructed $H_n C^{(m)} = 0$ for $n > -m$, we have an embedding

$$H^* C^{(m)} \to I^{(m)} \tag{3.7.12}$$

where $I^{(m)}$ is a complex of injectives with $I^{(m)}_n = 0$ for $n > 0$. Now (3.7.12) extends to a graded homomorphism

$$f_m : C^{(m)} \to I^{(m)}$$

which is then necessarily a chain map. Let

$$C^{(m+1)} = Cf_m[-1]$$

be the mapping co-cone. We have $H_n C^{(m+1)} = 0$ for $n > -m - 1$ by Theorem 2.3.7. From (2.3.5), we obtain an inverse sequence

$$\ldots C^{(2)} \to C^{(1)} \to C^{(0)}$$

(3.7.13)

and we let C be its limit. Note then that the canonical chain map $C \to C^{(n)}$ is an isomorphism in degrees $> -n$, and thus the Mittag-Leffler condition is satisfied both for (3.7.13) and its homology, and hence

$$H_* C = \lim H_* C^{(m)} = 0.$$

Note that by construction, we also have a short exact sequence of chain complexes

$$0 \to I \to C \to FQ \to 0.$$

This gives a quasiisomorphism

$$FQ \xrightarrow{\ \sim\ } I[1].$$

(3.7.14)

By the assumption on F, and by Proposition 2.3.21, FQ and I are both G-local, so applying G to (3.7.14) induces a quasiisomorphism

$$GFQ \xrightarrow{\ \sim\ } GI[1].$$

(3.7.15)

Now the short exact sequences

$$0 \to F^p(I) \to I \to C^{(p)} \to 0$$

give a decreasing filtration on I by chain complexes which are again G-local by Proposition 2.3.21. Then

$$H^n GI[1] \cong R^n (G \circ F)(A)$$

by (3.7.15), while

$$H^* E_0 I[1] = E_0 I[1]$$

with the differential given by the connecting maps is a direct sum of injective resolutions of $R^q F(A)$, $q \geq 0$, and thus the E_2-term of the spectral sequence associated with the decreasing filtration $(GF^p I[1])$ of the chain complex $GI[1]$ is $R^p G \circ R^q F(A)$. Thus, our statement follows from the result of Sect. 3.7.1.   □

3.7.4 Corollary (The Leray Spectral Sequence) *Let $f : X \to Y$ be a continuous map of topological spaces and let $\mathcal{F}$ be a sheaf on X. Then we have a convergent spectral sequence*

$$H^p(Y, R^q f_*(\mathcal{F})) \Rightarrow H^{p+q}(X, \mathcal{F}). \qquad (3.7.16)$$

Note that the homology (in the category of abelian sheaves) of a chain complex of abelian sheaves is calculated by taking its homology on sections, and then sheafifying. Thus, $R^q f_*(\mathcal{F})$ is the sheaf associated with the presheaf on Y given by

$$U \mapsto H^q(f^{-1}(U), i^*\mathcal{F})$$

where $i : f^{-1}(U) \to X$ is the inclusion.

Proof Apply Theorem 3.7.3 with $F = f_*$, $G = \Gamma$ (global sections). This is possible, since f_* clearly preserves flasque sheaves.   □

3.8   Singular Cohomology vs. Sheaf Cohomology

A topological space X is called *locally contractible* if for every $x \in X$ and every open set $U \ni x$, there exists an open subset $V \subseteq U$ with $x \in V$ such that V is contractible. For example, it is easy to see that all smooth manifolds, as well as all CW complexes, are locally contractible.

3.8.1 Theorem *Let X be a locally contractible Hausdorff topological space. Then for an abelian group A, there is a canonical isomorphism*

$$H^n(X; A) \cong H^n(X, \underline{A})$$

(where the left hand side denotes singular cohomology and the right hand side denotes sheaf cohomology with constant coefficients).

Proof Suppose X is locally contractible. Consider the cochain complex $\mathcal{C}$ on X associated to the cochain complex of presheaves

$$U \mapsto C^*(U; A)$$

for $U \subseteq X$ open. Then, by construction, C is a cochain complex of flasque presheaves, and by the assumption that X is locally contractible, the canonical morphism of cochain complexes of abelian sheaves

$$\underline{A} \to C$$

is a quasiisomorphism. In other words, C is a flasque resolution of $\underline{A}$. On the other hand, the cohomology of the cochain complex of abelian groups ΓC is the singular cohomology $H^*(X; A)$ by Lemma 3.2.1. (We can pass from homology to cohomology using the Universal Coefficient Theorem). Thus, our statement follows. $\square$

Theorem 3.8.1 together with Theorem 2.5.10 imply that for a smooth manifold M, we have

$$H^n_{DR}(M) \cong H^n(M; \mathbb{R})$$

(3.8.1)

and similarly for a complex manifold M,

$$H^n_{DR}(M; \mathbb{C}) \cong H^n(M; \mathbb{C}),$$

(3.8.2)

where the right hand sides denote singular cohomology. This is the *de Rham Theorem*. There is a more direct approach to that result. Consider some n-simplex $K \subset \mathbb{R}^n$ and a linear homeomorphism

$$\phi = \phi_k : K \to \Delta^k$$

(where Δ^k is the standard k-simplex (3.1.1)). For a smooth manifold M and a smooth singular simplex $\sigma : \Delta^k \to M$ (i.e. one which extends to a smooth map on an open neighborhood), let ω be a k-form on M. Then we have

$$(\sigma \circ \phi)^*\omega = hdx_1 \wedge \cdots \wedge dx_n$$

for some smooth function on K. We can define a bilinear pairing

$$\langle \omega, \sigma \rangle = \int_K hdx_1 \dots dx_n.$$

(3.8.3)

By the substitution theorem for multivariable integrals, this depends on the choice of the linear homeomorphism ϕ only up to sign (note that the substitution theorem for k-dimensional integrals on subsets of $\mathbb{R}^k$ involves the *absolute value* of the determinant of the Jacobian matrix). Now obviously, (3.8.3) extends uniquely to an $\mathbb{R}$-bilinear pairing

$$\Omega^k(M) \otimes_{\mathbb{R}} C^{sm}_k(M) \to \mathbb{R}$$

(3.8.4)

were $C_k^{sm}(M)$ denotes the free abelian group on smooth singular k-simplices in M. The formula (3.1.3) defines a differential

$$d : C_k^{sm}(M) \to C_{k-1}^{sm}(M),$$

making $C^{sm}(M) \subset CM$ a chain subcomplex, and one can prove that this inclusion of chain complexes is a quasiisomorphism.

Additionally, the *Stokes theorem* states that with an appropriate choice of the maps ϕ_k, the signs work out so that (3.8.4) induces a chain map

$$\Omega^k(M) \to Hom(C_k^{sm}(M), \mathbb{R}) \tag{3.8.5}$$

(for example, ϕ_k can be chosen to be the projection to the last k coordinates). One can then prove directly that (3.8.5) is a quasiisomorphism, and it is easy to check that it coincides with the isomorphism obtained by Theorems 3.8.1 and 2.5.10.

By applying $? \otimes_\mathbb{R} \mathbb{C}$, this entire discussion is also valid with complex coefficients, which is relevant to the case of complex manifolds.

3.8.2 Sheaves Do Not Have Enough Projectives

We can now show that the category of abelian sheaves on a space X sometimes (perhaps typically) does not have enough projectives. The point is that there is such a thing as a *free sheaf*, which is a direct sum of sheaves of the form $j_!\mathbb{Z}$ where $j : U \to X$ is the inclusion of an open subset (with possibly varying U). By definition, for every abelian sheaf S on X, there is a surjection

$$\mathcal{F} \to S \to 0 \tag{3.8.6}$$

where $\mathcal{F}$ is a free sheaf.

Now let, for example, $X = \mathbb{R}^2$. If S were projective, then the morphism (3.8.6) would have to split, i.e. have a right inverse. This means that for every $x \in X$, S_x is a direct summand of the free abelian group $\mathcal{F}_x$. Suppose

$$S_x \neq 0.$$

Then S_x has a direct summand isomorphic to $\mathbb{Z}$, let it be generated by an element u. Then u comes from a section s on an open set $U \ni x$, and thus, denoting by $j : U \to X$ the inclusion, $\mathbb{Z}$ generated by s is a direct summand of j^*S. However, the constant sheaf $\mathbb{Z}$ on U cannot be projective, since then, by Theorem 3.8.1, every subset of U would have 0 singular cohomology in degrees > 0 (since sheaf cohomology is computed as morphisms from $\mathbb{Z}$ in the derived category of abelian sheaves). We see that for $X = \mathbb{R}^2$, this fails to hold for any non-empty open subset U, which is a contradiction.

4 Exercises

1. On the smooth manifold $\mathbb{R}^n \setminus \{(0, \ldots, 0)^T\}$, consider the $(n-1)$-form

$$\omega = \frac{\sum_{k=1}^{n}(-1)^k x_k dx_1 \wedge \cdots \wedge dx_{k-1} \wedge dx_{k+1} \wedge \cdots \wedge dx_n}{(x_1^2 + \cdots + x_n^2)^{n/2}}.$$

   (a) Prove that $d\omega = 0$.
   (b) Is there a differential form η such that $d\eta = \omega$? [Integrate ω over the unit sphere, more precisely a smooth singular chain complex triangulating it.]

2. A differential form ω on a smooth manifold M is said to have *compact support* if there exists a compact subset $K \subseteq M$ such that $\omega \mid_{M \setminus K} = 0$. Using the de Rham differential, forms with compact support make a cochain subcomplex $\Omega_c^* M \subseteq \Omega^* M$. Its cohomology is called *de Rham cohomology with compact support* and denoted by $H_{DR,c}^k(M)$. Prove that $H_{DR,c}^m(\mathbb{R}^n) = \mathbb{R}$ for $m = n$ and 0 for $m \neq n$. [Integrate over $\mathbb{R}^n$ to get a map $H_{DR,c}^n(\mathbb{R}^n) \to \mathbb{R}$ which is easily seen to be onto. Use induction on n and Fubini's theorem to show this is the only cohomology we get.]

3. Noting that $\Omega_c^*(U)$ is *covariantly* functorial in inclusions of open submanifolds, (while $\Omega^*(U)$ is, of course, contravariantly functorial), observe that for open submanifolds U, V of a smooth manifold M, there are short exact sequences

$$0 \to \Omega_c^*(U \cap V) \to \Omega_c^*(U) \oplus \Omega_c^*(V) \to \Omega_c^*(U \cup V) \to 0,$$

$$0 \to \Omega^*(U \cup V) \to \Omega^*(U) \oplus \Omega^*(V) \to \Omega^*(U \cap V) \to 0.$$

   Write down the corresponding long exact sequences in cohomology. These are the *Mayer-Vietoris sequences* in de Rham cohomology. (There are some sign choices. If, for example, we take the $\mathbb{R}$-vector space homomorphisms induced by inclusions $U \cap V \subseteq U, V$, then we need to reverse the sign of the homomorphism induced by one of the inclusions $U, V \subseteq U \cup V$.)

4. Let M be a compact smooth (real) n-manifold. An *orientation form* (if one exists) is an n-form ω on M which is not zero at any point. Two orientation forms ω, η are considered *equivalent* if $\omega = h\eta$ for a nowhere zero smooth function h. An equivalence class of orientation forms is called an *orientation*. A smooth n-manifold which has an orientation is called *orientable*. When an orientation is chosen we say that M is *oriented*.

   (a) Prove that an orientable connected smooth manifold has precisely two orientations. Give an example of a smooth manifold which is not orientable.
   (b) Prove that if M is an oriented connected smooth n-manifold, then $H_{DR,c}^n(M) = \mathbb{R}$. [Use induction to prove it for a union of finitely many coordinate neighborhood, then use a colimit argument. For the induction step, use the long exact sequence in cohomology associated with the first short exact sequence of Exercise 3. Observe

that on an open subset of $\mathbb{R}^n$, we have a canonical orientation form $dx_1 \wedge \cdots \wedge dx_n$ from which, at each point x, our given orientation form differs by a non-zero scalar multiple which $h(x)$. Further, the sign of $h(x)$ cannot change signs by the intermediate value theorem.]

(c) Observe that for an oriented smooth n-manifold M, we have a canonical chain map

$$\Omega_c^*(M) \to \mathbb{R}[-n],$$

given in degree n, on a coordinate neighborhood by integration, with coordinates chosen so that $dx_1 \wedge \cdots \wedge dx_n$ is equivalent to the pullback of the orientation form (and extended, say, by partition of unity).

(d) Observe that every complex smooth manifold has a canonical orientation. [If $z_1 = x_1 + iy_1, \ldots z_n = x_n + iy_n$ is a complex basis of the dual of the tangent space, then $dx_1 \wedge dy_1 \wedge \cdots \wedge dx_n \wedge dy_n$ is well defined up to positive scalar multiple.]

5. Observe that for any smooth n-manifold M, the exterior product gives a homomorphism of cochain complexes

$$\Omega^*(M) \otimes \Omega_c^*(M) \to \Omega_c^*(M).$$

When M is oriented and connected, by composing with the chain map constructed in Exercise 4 (c), we obtain a canonical chain map

$$\Omega^*(M) \otimes \Omega_c^*(M) \to \mathbb{R}[-n],$$

which can by reinterpreted by adjunction as

$$\Omega^*(M) \to Hom_\mathbb{R}(\Omega_c^*(M), \mathbb{R})[-n].$$

Prove that this chain map is a quasiisomorphism. [Show that this chain map is compatible with the Mayer-Vietoris exact sequences of Exercise 3. Then prove the quasiisomorphism on a coordinate neighborhood, use induction to extend it to a union of finitely many coordinate neighborhoods, and finally a limit argument. Be mindful of tricky signs throughout. Also, note that an intersection of coordinate neighborhoods does not have to be homeomorphic to $\mathbb{R}^n$. Probably the easiest way to get around this is to prove the statement for open subsets of $\mathbb{R}^n$ first by induction for unions of k convex open subsets, and a colimit argument.] Conclude that for a connected smooth n-manifold M, we have an isomorphism

$$H_{DR}^k(M) \cong Hom_\mathbb{R}(H_{DR,c}^{n-k}(M), \mathbb{R}).$$

This is *Poincaré duality for DeRham cohomology.*

6. Prove the uniqueness of a derived category.

7. Prove that the kernel and cokernel of a morphism f in Diagram (2.2.1) with rows exact are 0.

8. Prove directly that the chain maps (2.3.1) are quasiisomorphisms.

9. Complete the proof of Theorem 2.3.7.

10. Prove the 5-lemma (Lemma 2.3.8).

11. Prove Lemma 2.3.13.

12. For a commutative ring R, an R-module M is called *flat* when tensoring with M over R is an exact functor. Prove that the following are equivalent
    - (i) M is flat
    - (ii) For every R-module N, $Tor_i^R(M, N) = 0$ for $i > 0$
    - (iii) For any fixed R-module N, if F_N denotes the functor of tensoring with N over R, then M is F_N-colocal.

13. Prove that if M is a flat R-module, $S \subset R$ a multiplicative set (i.e. a set closed under multiplication which contains 1 and does not contain 0) then $S^{-1}M = S^{-1}R \otimes_R M$ is flat.

14. Prove that $cd_{\mathbb{Z}}(\mathbb{Q}) = 1$ (and thus, is not equal to the supremum of $cd(N)$ with $N \subseteq \mathbb{Q}$ finitely generated $\mathbb{Z}$-modules).

15. Let $R \to S$ be a morphism of commutative rings and let M be an S-module. Prove that

$$cd_R(M) \leq cd_S(M) + cd_R(S).$$

16. Let $R = \mathbb{Z}[x_1, \ldots, x_n]$. Consider the R-modules $\mathbb{Z}$, $\mathbb{Z}/p$ where x_i act by 0. Compute $Ext_R^i(\mathbb{Z}, \mathbb{Z})$ and $Ext_R^i(\mathbb{Z}/p, \mathbb{Z}/p)$ together with the product given by composition in the derived category DR.

17. Using the result of Exercise 16, prove (2.6.8) in the proof of Lemma 2.6.5.

18. Prove that the ring $S^{-1}R$ of Exercise 38 of Chap. 1, under the assumption of (c), has infinite cohomological dimension of $S^{-1}R$, while its localization at every prime ideal is regular (and thus, by our definition from Sect. 3.5 of Chap. 1, $S^{-1}R$ is regular).

19. Prove that the localization $S^{-1}R$ at a multiplicative set (containing 1 and not containing 0) of a regular local ring is regular. Then prove the analogous statement after dropping the word "local."

20. Let R be a regular ring and let M be a finitely generated R-module. Prove that M has a resolution of the form

$$P_n \to P_{n-1} \to \cdots \to P_0$$

where P_i are finitely generated projective R-modules. [Letting P_0 be a finitely generated free R-module mapping onto R, let K_0 be the kernel. K_0 is finitely generated since R is Noetherian. If K_i is a finitely generated R-module, let P_i be a finitely generated free R-module mapping onto K_i, and let K_{i+1} be the kernel. We need to prove that some K_n is projective for some n. To this end, by Lemma 2.2.4 of Chap. 4, it suffices to prove that M_p has a finite free resolution of uniformly bounded length over R_p for any prime p of R. This is true for each prime individually, and if we have such a resolution for a given prime p, a resolution also exists in its open neighborhood, i.e. over $R[a^{-1}]$ for some $a \notin p$. Then use the quasicompactness of an affine scheme.]

21. Let R be a regular ring. Recalling the concepts of Sects. 2.1 and 2.2 of Chap. 4, construct a homomorphism $\phi : G(R) \to K(R)$ by sending a finitely generated module M to

$$\sum_{i=0}^{n} (-1)^i P_i$$

where P_i are the projective modules in the resolution constructed in Exercise 20. Prove that this gives a well-defined homomorphism which is inverse to the canonical homomorphism $\psi : K(R) \to G(R)$ given by sending a finitely generated projective module to itself. [To show that the map ϕ is well-defined, i.e. independent of the resolution, first reduce to the case when we have a term-wise epimorphism of finite projective resolutions of the same module. Then the kernel is a finite projective resolution of 0, and splits as a direct sum of isomorphisms.]

22. Using Exercise 27 of Chap. 4, prove that every coherent sheaf on a Noetherian regular scheme has a resolution by finitely many locally free sheaves. Using this, extend the result of Exercise 21 to regular Noetherian schemes.

23. Prove that if R is a Noetherian ring and $G(R) = 0$ then $G(R[x]) = 0$. [On a finitely generated $R[x]$-module M, consider the filtration $M \supseteq xM \supseteq x^2M \supseteq \dots$. Using the fact that R is Noetherian, conclude that the class $[M] \in G(R[x])$ is the sum of a class $[N \otimes_R R[x]]$ for an R-module N and a class of an $R[x]$-module on which x acts nilpotently, hence, without loss of generality, trivially.]

24. Prove that every finitely generated projective $k[x_1, \dots, x_n]$-module P is *stably free* which means that there exists a finitely generated free module F such that $P \oplus F$ is free. [Prove that this is true for any ring R which satisfies $K(R) = 0$. (In fact, both properties are equivalent.) Then use Exercises 23, 21.]

25. (Quillen and Suslin) Prove that for a field k, every finitely generated projective $k[x_1, \dots, x_n]$-module is free. Therefore, every algebraic vector bundle over $\mathbb{A}_k^n$ is trivial. The following proof is due to Vaserstein. Let R be an integral domain. A column vector $(f_1, \dots, f_n)^T \in R^n$ is called *unimodular* if $(f_1, \dots, f_n)$ is the unit ideal in R. We say that R has the *unimodular extension property* if every unimodular vector over R is the column of an invertible matrix over R.

(a) Prove that if R is a commutative ring which has the unimodular extension property such that every finitely generated R-module is stably free, then every finitely generated projective R-module is free. Thus, it suffices to prove that $k[x_1, \dots, x_n]$

has the unimodular extension property. [It suffices to prove that if $P \oplus R$ is free then P is free.]

(b) (Horrocks) Let R be a local ring. Then every unimodular vector v over $R[x]$ whose one coordinate is a monic polynomial is a column of an invertible matrix. [Let $(f_1, \ldots, f_n)^T$ be a unimodular vector and let, say, f_1 be monic of degree d. Proceed by induction on d. Use the Euclidean algorithm to reduce to the case where $f_2, \ldots, f_n$ have degree $< d$. Without loss of generality, say, f_2 has a coefficient which is a unit in R. Using the method of the Buchberger algorithm, show that some linear combination of f_1, f_2 is a monic polynomial of degree $< d$.]

(c) Let R be an integral domain and let $v(x)$ be a unimodular vector in $R[x]$ which has a monic coordinate. Then $v(x)$ is equivalent to $v(0)$ over $R[x]$, meaning that there exists an invertible matrix A over $R[x]$ with $Av(x) = v(0)$. [Let J be the set of all $x \in R$ such that $v(x + cy)$ is equivalent to $v(x)$ over $R[x, y]$. Prove that J is an ideal which is not contained in any prime ideal and thus is the unit ideal. From there, $v(x + y)$ is equivalent to $v(x)$, hence $v(y)$ is equivalent to $v(0)$. To show that J is not contained in a prime p, we have $v(x) = M(x)v(0)$ for an invertible matrix M over R_p. Let $G(x, y) = M(x)M(x+y)^{-1}$, so $G(x, y)v(x+y) = v(x)$. Then $G(x, 0)$ is the identity matrix, so G is congruent to the identity modulo y. Thus, for some $c \notin p$, $G(x, cy)$ has coefficients in R.]

(d) Let v be a unimodular vector in $k[x_1, \ldots, x_n]$. By (c), we can use induction on n if v has a coordinate monic in x_n. But this can be achieved by substituting $y_n = x_n$, $y_i = x_i - x_n^{m_i}$, similarly as in the proof of the Noether normalization lemma.

26. From the definition (3.1.3), prove $d \circ d = 0$. [Composing the differential with itself, we obtain a double sum. Breaking up into the cases $i \leq j$ and $i > j$, we obtain pairs of equal summands with opposite signs, using the relation (3.1.2).]

27. (Eilenberg-Zilber Theorem) Observe that for spaces X, Y, the chain group $C_i(X) \otimes C_j(Y)$ is canonically identified with the chain group $C_{i+j}(X \times Y)$ if $i = 0$ or $j = 0$. Using this identification, prove that there exist *natural* chain maps between any two of the chain complexes $C(X) \otimes C(Y)$ and $C(X \times Y)$ which are the identity in degree 0, and that any two such chain maps are further naturally chain homotopic. By a natural chain homotopy we mean that each of the homomorphisms of abelian groups constituting the chain homotopy is natural. [This is an application of the method of *acyclic models* used in Sect. 3.3. Proceed by induction on dimensional degree. By naturality, it is necessary and sufficient to construct the value of each chain map on a *universal element*. In $C_n(X \times Y)$, the universal element is the diagonal singular n-simplex $\Delta^n \to \Delta^n \times \Delta^n$ for $X = Y = \Delta^n$. In $C_k(X) \otimes C_\ell(Y)$, the universal element is $\iota_k \otimes \iota_\ell$ where $\iota_i = Id : \Delta^i \to \Delta^i$. Now the condition on the value of the desired chain map on the universal element is that its boundary be a cycle given by induction. The obstruction for doing so therefore lies in a homology group, which is 0 by the homotopy axiom. In constructing chain maps (not homotopies), be mindful of degree 1, where the homology group in question is in degree 0 and therefore is not 0 but $\mathbb{Z}$, and thus, an additional argument is needed.]

28. Complete the proof that (3.7.2) is an exact couple, and prove formula (3.7.3).
29. Prove formulas (3.7.4), (3.7.5).
30. *Poincaré duality for singular (co)homology.* The argument of Exercises 2–5 essentially also works for singular homology, replacing the sheaf of de Rham forms by the sheafification $\mathcal{C}$ of the complex of presheaves

$$U \mapsto C^k(U, \mathbb{Z}).$$

It also works for a topological manifold M (i.e. a Hausdorff space in which every point has an open neighborhood homeomorphic to $\mathbb{R}^n$ whose every connected component has a countable basis of topology). The reader should consult [20] for details on singular homology. Here are the basic steps and technical points:

(a) Denoting by $\mathcal{C}_c(U)$ the sections of $\mathcal{C}$ over U with compact support and $H_c^k(U)$ its cohomology, prove that $H_c^k(\mathbb{R}^n)$ is $\mathbb{Z}$ for $k = n$ and 0 for $k \neq n$. The covariant functoriality of $\mathcal{C}_c$ in open subsets of a manifold and the Mayer-Vietoris sequences work the same as in the de Rham case.

(b) An *orientation* of a topological n-manifold M is a homomorphism $\omega : H_c^n(M) \to \mathbb{Z}$ which is an isomorphism when composed with the map induced by the inclusion of coordinate neighborhoods. Again, a topological manifold which has an orientation is called *orientable*, and when an orientation is chosen, we call it *oriented*. Prove that if M is an orientable connected topological n-manifolds, an orientation is an isomorphism of abelian groups. Denote, for a chain complex C, by $\tau_{\leq k}(C)$ the factor complex which in degree i is C_k for $i \leq k$, in degree $i + 1$ is B_k, and is 0 in other degrees. Thus, if M is oriented connected, we have a chain map

$$\mathcal{C}_c(M) \to \tau_{\leq -n}\mathcal{C}_c(M),$$

and the target chain complex has homology $\mathbb{Z}[-n]$.

(c) Prove that $H^i\mathcal{C}_c(M) = 0$ for $i > n$. [Use the Mayer-Vietoris sequence and a colimit argument.]

(d) The Eilenberg-Zilber map (see [20]) induces a chain map

$$\mathcal{C}_c(U) \otimes \mathcal{C}(U) \to \mathcal{C}_c(U)$$

for $U \subseteq M$ open. Composing with the chain map constructed in (b), construct a chain map

$$\mathcal{C}(U) \to RHom(\mathcal{C}_c(U), \tau_{\leq -n}\mathcal{C}_c(M))$$
$$\sim RHom(\mathcal{C}_c(U), \mathbb{Z})[-n].$$

Similarly as in the de Rham case, this chain map is compatible with Mayer-Vietoris exact sequence with and without compact support, and thus, combining induction on number of coordinate neighborhoods and a limit argument, one shows that it is a quasiisomorphism. One denotes

$$H_k^{BM}(M; \mathbb{Z}) = H_k RHom(\mathcal{C}_c(M), \mathbb{Z}),$$

and calls these groups *Borel-Moore homology*. Thus, we have proved the following version of Poincaré duality for oriented connected topological manifolds:

$$H^k(M; \mathbb{Z}) \cong H_{n-k}^{BM}(M; \mathbb{Z}).$$

31. Let M be a compact topological n-manifold.
    (a) Prove that M can be covered by finitely many coordinate neighborhoods U_i, $i = 1, \ldots m$, so that there exist continuous maps $\phi_i : M \to S^n$ so that $\phi|_{U_i}$ is a homeomorphism onto an open subset of S^n. Then $\phi = \prod \phi_i$'s is an embedding of M as a closed subset of a product of spheres T.
    (b) Prove that the image N of ϕ is a retract of an open neighborhood U. [Prove, by induction on j, that $U_1 \cup \ldots U_j$ is a retract of an open neighborhood V_j by a retraction r_j. This is true for $j = 1$. Suppose it is true for a given j. There exists an open set $W \subseteq T$ such that $U_{j+1} \subset W$ is closed. Now there exists a closed subset $Z \subseteq V_j$ whose interior contains $V_j \smallsetminus W$, and without loss of generality, $Z \cap M \subseteq U_1 \cup \ldots U_j$. By making the sets smaller if necessary, we may assume that $V_j \cap W$ is mapped, by r_j, into a set K homeomorphic to a product of compact intervals which contains U_{j+1} as an open subset. Now extend the map $Z \cup U_{j+1} \to K$ to a continuous map $\phi : W \to K$. (Use the fact if $Q \subseteq \mathbb{R}^N$ is open and $P \subseteq Q$ is closed, then a continuous map $P \to [0, 1]$ extends to Q.) Now glue ϕ with r_j, and replace its source by the open subset $\phi^{-1}(U_1 \cup \cdots \cup U_{j+1})$. Let r_{j+1} be the restriction.]
    (c) By making U smaller if necessary, U can be assumed to be homotopy equivalent to a finite CW-complex. Conclude that every homology group $H_i(M)$ is finitely generated.
32. (Poincaré duality continued) When M is a compact topological n-manifold, prove that

$$H_k^{BM}(M; \mathbb{Z}) \cong H_k(M; \mathbb{Z}).$$

Thus, in this case, when M is oriented, we obtain a Poincaré duality isomorphism

$$H^k(M; \mathbb{Z}) \cong H_{n-k}(M; \mathbb{Z}).$$

The class in $H_n(M)$ Poincaré dual to $1 \in H^0(M)$ is sometimes called the *fundamental class*. [In this case, Borel-Moore homology is the homology of the chain complex $RHom_{\mathbb{Z}}(C^*(M; \mathbb{Z}), \mathbb{Z})$. Thus, in view of Exercise 31, it suffices to show that if C

is a chain complex whose sum of homology groups is finitely generated, then the canonical map

$$C \to RHom(RHom(C; \mathbb{Z}), \mathbb{Z})$$

in the derived category of abelian groups is a quasiisomorphism.]

33. Let M be a smooth n-manifold. Prove that an orientation in the sense of Exercise 4 determines an orientation in the sense of Exercise 30. [Use the Mayer-Vietoris sequence and a limit argument, with the slogan "the ambiguity is only up to sign."]

34. Using Poincaré duality, prove that $H^*((\mathbb{P}_\mathbb{C}^n)_{an}) = \mathbb{Z}[x]/(x^{n+1})$ where $x \in H^2(\mathbb{P}_\mathbb{C}^n)_{an}) \cong \mathbb{Z}$ is a generator. [Induction on n, using the fact that the inclusion $\mathbb{P}_\mathbb{C}^{n-1} \to \mathbb{P}_\mathbb{C}^n$ induces an isomorphism in singular cohomology in dimensions $\leq 2n-2$.]

35. Due to Poincaré duality, for a continuous map between compact oriented topological manifolds $f : X \to Y$, we have a functorial homomorphism $f_* : H^k(X) \to H^k(Y)$. (All cohomology is, say, with coefficients in $\mathbb{Z}$.) Prove the projection formula

$$f_*(x \cdot f^*(y)) = f_*(x) \cdot y$$

where $f^* : H^\ell(Y) \to H^\ell(X)$ is ordinary cohomological functoriality. You may assume the existence of a natural cup product map $C^*X \otimes C^*X \to C^*X$ (which follows from the Eilenberg-Zilber theorem). [Investigate the effect of f^* on the map ω of Exercise 30.]

36. *Cohomology of non-degenerate projective quadrics* In this exercise, we shall calculate the singular cohomology ring $H^*(X_{an})$ where X is a projective quadric of dimension n, i.e. a hypersurface in $\mathbb{P}_\mathbb{C}^{n+1}$ defined by one non-degenerate quadratic form. Recall that all such quadrics are isomorphic smooth projective varieties. The answer differs depending on whether n is even or odd.

    (a) If n is even, change coordinates so the equation of the quadric is

$$x_1 x_2 + \cdots + x_{n+1} x_{n+2} = 0$$

where $x_1, \ldots x_{n+2}$ are the projective coordinates. Consider the filtration on X_{an} where

$$X_{2n-2} = Z(x_{n+2}),$$
$$X_{2n-4} = Z(x_{n+2}, x_n),$$
$$\ldots,$$
$$X_{n-2} = Z(x_{n+2}, x_n, \ldots, x_2) \cong (\mathbb{P}_\mathbb{C}^{n/2-1})_{an}.$$

If n is odd, change coordinates so the equation of the quadric is

$$x_1^2 + x_2 x_3 + \cdots + x_{n+1} x_{n+2} = 0.$$

Consider the filtration on X_{an} where

$$X_{2n-2} = Z(x_{n+2}),$$
$$X_{2n-4} = Z(x_{n+2}, x_n),$$
$$\cdots,$$
$$X_{n-1} = Z(x_{n+2}, x_n, \ldots, x_3) \cong (\mathbb{P}_{\mathbb{C}}^{(n-1)/2})_{an}.$$

Prove that these are cellular filtration on X_{an}, and compute the singular (co)homology of X_{an} by counting cells.

(b) Prove that the embedding in the analytic topology

$$f : X_{an} \to (\mathbb{P}_{\mathbb{C}}^{n+1})_{an}$$

is not onto. Using the CW-structure on $(\mathbb{P}_{\mathbb{C}}^{n+1})_{an}$, prove that f is homotopic to a map f_0 whose image is in $(\mathbb{P}_{\mathbb{C}}^{n})_{an}$. Prove that on fundamental classes, the map f_0 induces multiplication by 2. Use this and Poincaré duality to deduce the complete ring structure on $H^*(X_{an})$.

37. Prove that if $f : X \to Y$ is a birational morphism of smooth projective varieties over $\mathbb{C}$, then

$$f_* f^* = 1 : H^*(Y_{an}) \to H^*(Y_{an}).$$

Deduce that $H^k(Y)$ is a direct summand of $H^k(X)$ for any k. [By the projection formula of Exercise 35, it suffices to prove this on the class $1 \in H^0$.]

38. (Artin–Mumford)

(a) Let X be a smooth projective variety over $\mathbb{C}$, and let $Y \subseteq X$ be a smooth projective subvariety of codimension $r + 1$. Let $X' = Bl_Y(X)$ let $f : X' \to X$, be the projection, and let $i : Y \to X$ be the inclusion. Prove that in the analytic topology, $R^0 f_* \mathbb{Z} = \mathbb{Z}$, $R^q \mathbb{Z} = i_* \mathbb{Z}$ for $q = 2i$, $1 \leq i \leq r$, and $R^q \mathbb{Z} = 0$ for other values of q. (We omit the subscript $?_{an}$ in this exercise to make the notation shorter.)

(b) Deduce from the Leray spectral sequence that we have an exact sequence of the form

$$
\begin{array}{ccccccccc}
0 & & & & & & & & \\
\downarrow & & & & & & & & \\
H^0(Y) & \xrightarrow{d_3} & H^3(X) & \longrightarrow & H^3(X') & \longrightarrow & H^1(Y) & \xrightarrow{d_3} & H^4(X) \\
& & & & & & & & \downarrow \\
& & & & & & & & H^4(X').
\end{array}
$$

(All cohomology is with coefficients in $\mathbb{Z}$.)

(c) Deduce from part (b) and Exercise 37 that the torsion part of $H^3(X)$ is isomorphic to the torsion part of $H^3(X')$. [Observe from that Universal Coefficient Theorem that $H^1(Y)$ is torsion free.]

(d) Suppose X and Y are smooth projective varieties over $\mathbb{C}$ which are birationally equivalent. By a result of Hironaka (see [4] and Sect. 2.4 of Chap. 6 below), there exists a smooth projective variety Z over $\mathbb{C}$ and birational morphisms $f : Z \to X$, $g : Z \to Y$ where f is a composition of projections of blow-ups of smooth subvarieties. Deduce that the torsion part of $H^3(X_{an})$ is isomorphic to a subgroup of the torsion part of $H^3(Y_{an})$ and hence, by symmetry, the torsion subgroups of $H^3(X_{an})$, $H^3(Y_{an})$ are isomorphic. In other words, the torsion part of the third singular cohomology group is a birational invariant.

Cohomology in Algebraic Geometry

We shall now look more closely at how cohomology is used in algebraic geometry. This will include exploring additional structure present on the cohomology of algebraic varieties in the analytic topology as defined in the last chapter, as well as defining new cohomology theories which are closer to the Zariski topology.

As in the last chapter, we will begin with de Rham cohomology. We shall study in more detail what properties de Rham cohomology has when applied to complex manifolds, and, eventually, to the analytic topology on smooth projective varieties over $\mathbb{C}$. The additional structure we see in this context is called *Hodge theory*.

Next, we will observe that there is an algebraic version of de Rham cohomology obtained by plugging in a smooth algebraic variety instead of a manifold, and using regular functions instead of analytically smooth functions. Thus, we encounter, once again, the principle of imitating topological concepts algebraically. In the process, we shall study the more general concept of cohomology with coefficients in quasicoherent sheaves. Putting these methods together, we will be able to compare algebraic and geometric de Rham cohomology in the smooth projective case (Serre's GAGA Theorem), as well as Grothendieck's algebraic de Rham Theorem over $\mathbb{C}$.

Algebraic de Rham cohomology of smooth varieties can be defined over any field, or even a ring. For fields of characteristic 0, its behavior is largely as we would expect. For fields of characteristic $p > 0$, (in addition to its good properties generally weakening), one restriction of de Rham cohomology is that it produces only vector spaces over the same field. Crystalline cohomology gives a method for obtaining, among other things, non-torsion groups, and also Hodge-theory type information. We present here an approach to crystalline cohomology using the de Rham-Witt complex, which does allow the computation of some basic examples.

Crystalline cohomology of a smooth projective variety over the finite field $\mathbb{F}_p$ produces $\mathbb{Z}_p$-modules. Is there another method which would give information in characteristic

© The Author(s), under exclusive license to Springer Nature Switzerland AG 2021
I. Kriz, S. Kriz, *Introduction to Algebraic Geometry*,
https://doi.org/10.1007/978-3-030-62644-0_6

$\ell \neq p$? As it happens, a source of such information is in a concept we already studied in Chap. 3, namely *étale morphisms*. Using étale morphisms, we can obtain entirely new topological information. (Recall the concept of *étale fundamental group* in Chap. 3, Sect. 5.6.) We shall briefly explore this topic in more detail here, introducing the concept of a *Grothendieck topology*. Concretely, we shall describe the *étale topology* and the cohomology theory it leads to, called *étale cohomology*. We shall compute some basic examples, such as the étale cohomology of curves. We will see that quite strikingly, étale cohomology behaves well precisely with finite coefficients whose order is not divisible by the characteristic of the field. A guide to the foundational cornerstone of this theory, called *Zariski's Main Theorem*, is presented in the Exercises.

Can there be a cohomology theory which would combine, for varieties over any field, all the information from different characteristics, and produce cohomology groups with coefficients in $\mathbb{Z}$? In the case of smooth varieties, such a "universal" cohomology theory (in a less precise sense than used for spaces in Chap. 5) was defined by Bloch and Voevodsky using the theory of algebraic cycles. We will present the most basic definitions of this theory, called *motivic cohomology*, compute some basic examples, and state Voevodsky's main theorem comparing motivic and étale cohomology, in the final section.

The subject of cohomology theory is involved, and all the concepts introduced here can, and should, be studied in further detail. Our principal purpose is mapping out the basic directions of study, and treating the very first steps.

1 Hodge Theory

The goal of this section is to explore additional structure on the cohomology of smooth projective varieties over $\mathbb{C}$ in the analytic topology.

1.1 Dolbeault Cohomology

From the material of Sect. 1.4 of Chap. 5, we see that for a complex manifold M, we have a decreasing filtration on the de Rham complex $\Omega^*(M; \mathbb{C})$ given by

$$F^p \Omega^*(M; \mathbb{C}) = \bigoplus_{k \geq p} \Omega^{k,*}(M; \mathbb{C}),$$

(and similarly on the sheaf level). Now in Sect. 3.7.1 of Chap. 5, we have constructed a spectral sequence associated with a decreasing filtration on a cochain complex, which, in the present case, will take the form

$$E_1^{p,q} = H_{\bar{\partial}}^{p,q}(M) \Rightarrow H^{p+q}(M; \mathbb{C}). \tag{1.1.1}$$

Here by $H_{\bar{\partial}}^{p,q}(M)$, we mean the q'th cohomology of the cochain complex

$$\Omega^{p,0}(M) \xrightarrow{\bar{\partial}} \Omega^{p,1}(M) \xrightarrow{\bar{\partial}} \cdots \xrightarrow{\bar{\partial}} \Omega^{p,n}(M) \tag{1.1.2}$$

where $\Omega^{p,q}(M)$ is placed in degree q. The cohomology groups $H_{\bar{\partial}}^{p,q}(M)$ are called *Dolbeault cohomology*. The decreasing filtration on $H^k(M, \mathbb{C})$ associated with the spectral sequence (1.1.1) is called the *Hodge filtration*.

The key fact about Dolbeault cohomology is the following result:

1.1.1 Theorem (Dolbeault Theorem) *We have*

$$H_{\bar{\partial}}^{p,q}(M) \cong H^q(M, \underline{\Omega}_{Hol}^p). \tag{1.1.3}$$

Proof We already proved that the complex of sheaves

$$\underline{\Omega}^{p,0}(M) \xrightarrow{\bar{\partial}} \underline{\Omega}^{p,1}(M) \xrightarrow{\bar{\partial}} \cdots \xrightarrow{\bar{\partial}} \underline{\Omega}^{p,n}(M) \tag{1.1.4}$$

consists of soft sheaves, and it is also obvious that its 0'th cohomology is $\underline{\Omega}_{Hol}^p$. Thus, we can precisely mimic our proof of Theorem 2.5.10 of Chap. 5, if we we can prove that the cochain complex of sheaves (1.1.4) has 0 cohomology in degrees > 0. This follows from the $\bar{\partial}$-*Poincaré Lemma*, which states that if $D = \{z \in \mathbb{C} \mid ||z|| \leq 1\}$, we have

$$H_{\bar{\partial}}^{p,q}(\underbrace{D \times \cdots \times D}_{n}) = 0 \text{ for } q > 0. \tag{1.1.5}$$

(By the de Rham complex of $D \times \cdots \times D$, we mean the direct limit of the de Rham complexes of open neighborhoods of $D \times \cdots \times D$ in $\mathbb{C}^n$.) This is proved analogously to Lemma 2.5.8 of Chap. 5. Let $B_a = \{z \in \mathbb{C} \mid ||z|| < a\}$. We consider subcomplexes $C_{(m,a)}$ of (1.1.2) with $M = B_a \times \cdots \times B_a$, $a > 1$, spanned by differential forms

$$f \, d\bar{z}_{i_1} \wedge \cdots \wedge d\bar{z}_{i_k}$$

such that $i_1 < \cdots < i_k < m$, and f is holomorphic in coordinates $m, m+1, \ldots, n$. We construct, for $0 < b < a$, chain homotopies between the restriction $C_{(m,a)} \to C_{(m,b)}$ to a chain map $C_{(m,a)} \to C_{(m-1,b)} \subset C_{(m,b)}$. The homotopy is given by a formula precisely analogous to (2.5.12) of Chap. 5, where the integral is replaced by

$$\int_{B_b} \frac{g(w)}{w - z} dw \wedge d\bar{w} \tag{1.1.6}$$

with $g(w) = f(z_1, \ldots, z_{m-1}, w)$. Cauchy's integral formula in complex analysis (see for example [1]) implies that for a smooth function g on B_a, the integral (1.1.6) produces a function h on on B_b which satisfies

$$\frac{\partial h}{\partial \bar{z}} = g.$$

The result is then proved by taking direct limits over $a > 1$. (Note the subtlety of having to integrate in (1.1.6) over a disk whose closure is in the domain of definition of the smooth function g; this technical difficulty is the reason for the limit argument.)

□

1.1.2 Corollary *Let M be a complex manifold. Then we have a canonical isomorphism*

$$H^k(M; \mathbb{C}) = H^k_{DR}(M; \mathbb{C}) \cong H^k(M, \underline{\Omega}^*_{Hol}).$$

(Recall that the right hand side means hypercohomology of M with coefficients in the holomorphic de Rham complex $\underline{\Omega}^*_{Hol}$.)

Proof By the Dolbeault Theorem and the sheaf version of the spectral sequence (1.1.1), the complex de Rham complex $\underline{\Omega}^*_{M,\mathbb{C}}$ is a soft resolution of the holomorphic de Rham complex $\underline{\Omega}^*_{M,Hol}$.

□

1.2 Riemann and Hermitian Metrics

For a smooth manifold M, a *Euclidean metric* on a sheaf $\mathcal{F}$ of C^∞_M-modules M is a morphism of sheaves of C^∞_M-modules

$$g : \mathcal{F} \otimes_{C^\infty_M} \mathcal{F} \to C^\infty_M \tag{1.2.1}$$

which is symmetric (i.e. for sections $s, t \in \mathcal{F}(U)$, $g(s, t) = g(t, s)$) and is *positive definite*, which means that for a section $s \in \mathcal{F}(U)$, $g(s, s) \geq 0$ (at every point of U), with equality arising at a point x if and only if $s = 0$ at x. (By the value of a section $s \in \mathcal{G}(U)$ of a sheaf of C^∞_M-modules $\mathcal{G}$ at a point $x \in U$ we mean the image of s in the tensor product $\mathcal{G}(U) \otimes_{C^\infty_M(U)} k_x$ where k_x is the residue field of the local ring $C^\infty_{M,x}$ - clearly, a similar definition can be made for a sheaf of modules over any locally ringed space.)

For a sheaf $\mathcal{F}$ of $C^\infty_{M,\mathbb{C}}$-modules, a *Hermitian metric* on $\mathcal{F}$ is a morphism of sheaves of $C^\infty_{M,\mathbb{C}}$-modules

$$g : \mathcal{F} \otimes_{C^\infty_{M,\mathbb{C}}} \overline{\mathcal{F}} \to C^\infty_{M,\mathbb{C}} \tag{1.2.2}$$

(where $\overline{\mathcal{F}}$ is equal to $\mathcal{F}$ as a sheaf of $\mathcal{C}_M^\infty$-modules, with multiplication by i replaced by multiplication by $-i$), satisfying, for sections $s, t \in \mathcal{F}(U)$,

$$g(s, t) = \overline{g(t, s)},$$

and $g(s, s) \geq 0 \in \mathcal{C}_M^\infty$, with equality arising at a point $x \ni U$ only for $s = 0$ at x. (Note that a Hermitian metric on $\mathcal{F}$ is equivalent to a Hermitian metric on $\overline{\mathcal{F}}$.) When M is a single point, these concepts reduce to the usual notions of positive definite real symmetric (resp. Hermitian) forms (or, in another word, inner products) on a vector space over $\mathbb{R}$ (resp. $\mathbb{C}$).

The real part of a Hermitian metric is a Euclidean metric on $\mathcal{F}$ considered as a sheaf of $\mathcal{C}_M^\infty$-modules. The imaginary part of a Hermitian metric defines an *antisymmetric* form

$$\Lambda^2_{\mathcal{C}_M^\infty} \mathcal{F} \to \mathcal{C}_M^\infty.$$

(Here again we mean the obvious generalization to sheaves of modules of the exterior power construction of Exercise 16 of Chap. 4.)

Note that a $\mathcal{C}_M^\infty$-linear combination with positive real function coefficients of (finitely many) Euclidean (resp. Hermitian) metrics is a Euclidean (resp. Hermitian) metric. For a Euclidean (resp. Hermitian) metric g, and sections $s, t \in \mathcal{F}(U)$, one often writes

$$\langle s, t \rangle = g(s, t).$$

A *Riemann metric* on a smooth manifold M is a Euclidean metric on $\underline{\Omega}_M^1$, and a *Hermitian metric* on a complex manifold M is a Hermitian metric on $\underline{\Omega}_M^{1,0}$. (Note that complex conjugation gives a canonical isomorphism $\overline{\underline{\Omega}_M^{1,0}} \cong \underline{\Omega}_M^{0,1}$.) Riemann (resp. Hermitian) metrics always exist by smooth partition of unity. A *Riemann (resp. Hermitian) manifold* is a smooth (resp. complex) manifold with a Riemann (resp. Hermitian) metric.

For a complex manifold M, we have a canonical isomorphism of real bundles (i.e. $\mathcal{C}_M^\infty$-modules)

$$\underline{\Omega}_M^{1,0} \cong \underline{\Omega}_M^1 \tag{1.2.3}$$

given by

$$\kappa \mapsto \kappa + \overline{\kappa}.$$

Thus, the real part of a Hermitian metric on M defines a Riemann metric on M considered as a smooth manifold.

Now for a smooth manifold M, a *smooth vector bundle* is a finitely generated locally free sheaf of $\mathcal{C}^\infty$-modules $\mathcal{F}$ (i.e. M is covered by open sets U such that $\mathcal{F}|_U$ is isomorphic, as a $\mathcal{F}_U^\infty$-module, to a finite sum of copies of $\mathcal{F}_U^\infty$). Smooth complex vector bundles and holomorphic vector bundles over complex manifolds are defined analogously. One notes that given Euclidean metrics on smooth vector bundles $\mathcal{F}$, $\mathcal{G}$, we get canonical Euclidean metrics on $\mathcal{F} \otimes_{\mathcal{C}_M^\infty} \mathcal{G}$, $\Lambda_{\mathcal{C}_M^\infty}^k(\mathcal{F})$, $\mathcal{F}^* = \underline{Hom}_{\mathcal{C}_M^\infty}(\mathcal{F}, \mathcal{C}_M^\infty)$. If $e_1, \ldots, e_m$ is an orthonormal basis of $\mathcal{F}(U)$ (i.e. $\langle e_i, e_j \rangle = 1$ when $i = j$ and 0 otherwise), and $f_1, \ldots, f_p$ is an orthonormal basis of $\mathcal{G}(U)$, then $\langle e_i \otimes f_j, e_k \otimes f_\ell \rangle = 1$ when $i = \ell$, $j = k$, and 0 otherwise. Similarly, for $i_1 < \cdots < i_k$, $j_1 < \cdots < j_k$, $\langle e_{i_1} \wedge \cdots \wedge e_{i_k}, e_{j_1} \wedge \cdots \wedge e_{j_k} \rangle = 1$ when $i_s = j_s$ for all $s = 1, \ldots, k$, and 0 otherwise. In $\mathcal{F}^*$, the dual basis to an orthonormal basis is orthonormal. It is a standard exercise to show that this does not depend on the choice of orthonormal bases.

Completely analogous comments apply to Hermitian metrics (with the exception that complex orthonormal bases are sometimes referred to as *unitary*).

Thus, given a Riemann metric on a smooth manifold (resp. a Hermitian metric on a complex manifold) M, we automatically obtain canonical Euclidean metrics on $\underline{\Omega}_M^k$ (resp. Hermitian metrics on $\underline{\Omega}^{p,q}(M)$). We call such a manifold M a *Riemann (resp. Hermitian) manifold*.

Note that for an n-dimensional Riemann manifold M, the smooth vector bundle $\underline{\Omega}_M^n$ is 1-dimensional, i.e. locally isomorphic to $\mathcal{C}_M^\infty$. Recall from Exercise 4 of Chap. 5 that an *orientation* on M is given by an isomorphism of $\mathcal{C}_M^\infty$-modules

$$\underline{\Omega}_M^n \cong \mathcal{C}_M^\infty. \tag{1.2.4}$$

Two orientations are considered the same if one is a positive multiple of the other. If an orientation exists, then the smooth manifold M is called *orientable*. We already know from the calculation of homology in Sect. 3.6 of Chap. 5 that, for example, $\mathbb{R}P^n$ is not orientable for n even.

On the other hand, (compare with Exercise 4 of Chap. 5), a complex n-manifold M always has a canonical orientation. This is because for $U \subseteq \mathbb{C}^n$ with coordinates $x_1, \ldots, x_n$, letting $dy_k = I \cdot dx_k$,

$$dx_1 \wedge dy_1 \wedge \cdots \wedge dx_n \wedge dy_n$$

changes by a positive factor if we change coordinates holomorphically. (Here we denote the imaginary unit acting on the real cotangent space of M by I, to avoid confusion with the action of i after complexification.)

An orientation together with a Riemann metric on a smooth n-manifold M specify a unique n-form $dV \in \Omega^n(M)$ such that

$$\langle dV, dV \rangle = 1$$

and dV is positive under the isomorphism (1.2.4). The form dV is called the *volume form*.

Now it is a theorem of algebraic topology that if M is a compact connected oriented manifold, then we have

$$H_n(M; \mathbb{Z}) \cong \mathbb{Z}. \tag{1.2.5}$$

A proof can be found in the Exercises of Chap. 5, in particular Exercise 32. For more information, see [13]. We then have a unique generator $[M]$ such that

$$vol(M) = \int_{[M]} dV > 0.$$

This number is called the *volume* of M. For a compact Riemann manifold with connected components $M_1, \ldots, M_k$ (there are necessarily finitely many), one puts

$$[M] = [M_1] + \cdots + [M_k].$$

1.3 Hodge Theorem

Given real (resp. complex) smooth bundles $\mathcal{F}, \mathcal{G}$ on a smooth manifold M with a Riemann (resp. Hermitian) metric, for a morphism of bundles (i.e. morphism of sheaves of $\mathcal{C}_M^\infty$-modules, resp. $\mathcal{C}_{M,\mathbb{C}}^\infty$-modules)

$$Q : \mathcal{F} \to \mathcal{G},$$

we have a unique morphism

$$Q^* : \mathcal{G} \to \mathcal{F}$$

satisfying the identity

$$\langle Qs, t \rangle = \langle s, Q^*t \rangle$$

for sections $s \in \mathcal{F}(U)$, $t \in \mathcal{G}(U)$. We refer to Q^* as the *adjoint operator* to Q. (An analogy with this terminology is what inspired the term "adjoint functor.")

The difficulty is that we also want to talk about adjoints of certain operators

$$Q : \underline{\Omega}_M^* \to \underline{\Omega}_M^* \tag{1.3.1}$$

for a Riemann manifold M or

$$Q : \underline{\Omega}_{M,\mathbb{C}}^* \to \underline{\Omega}_{M,\mathbb{C}}^* \tag{1.3.2}$$

for a Hermitian manifold M which are morphisms of sheaves of $\mathbb{C}$-modules but *not* of C_M^∞-modules, such as d, ∂ or $\bar{\partial}$. For such operators Q, the definition of an adjoint operator Q^* requires more care. If M is a compact oriented Riemann manifold, we certainly require that for forms κ, η on M of the applicable dimensions, we have

$$\int_{[M]} \langle Q\kappa, \eta \rangle dV = \int_{[M]} \langle \kappa, Q^*\eta \rangle dV. \tag{1.3.3}$$

This formula even makes sense for morphisms of sheaves of $\mathbb{C}$-modules between any two smooth vector bundles on M. However, apart from the fact that the formula (1.3.3) applies only to global sections of bundles on a compact oriented Riemann manifold, the difficulty is that it does not suggest how to prove existence (or uniqueness) of the operator Q^*, since we are dealing with infinite-dimensional $\mathbb{C}$-vector spaces.

In the case of bundles of differential forms when Q is one of the operators d, ∂, $\bar{\partial}$, this difficulty can be circumvented by a device called the *Hodge $*$-operator*. For an oriented Riemann n-dimensional manifold M, we have a unique morphism of smooth bundles

$$* : \underline{\Omega}_M^k \to \underline{\Omega}_M^{n-k} \tag{1.3.4}$$

given by

$$\langle \kappa, \eta \rangle dV = \kappa \wedge *\eta. \tag{1.3.5}$$

For a Hermitian manifold M of complex dimension n, the same formula defines a morphism of smooth complex bundles

$$* : \underline{\Omega}_M^{p,q} \to \overline{\underline{\Omega}_M^{n-p,n-q}} \cong \underline{\Omega}_M^{n-q,n-p}. \tag{1.3.6}$$

One has

$$** = (-1)^{k(n-k)} : \underline{\Omega}_M^k \to \underline{\Omega}_M^k \tag{1.3.7}$$

in the oriented Riemann case, and

$$** = (-1)^{p(n-p)+q(n-q)} : \underline{\Omega}_M^{p,q} \to \underline{\Omega}_M^{p,q} \tag{1.3.8}$$

in the Hermitian case. (Note: For (1.3.8), we consider $*$ as a $\mathbb{C}$-linear operator landing in $\underline{\Omega}^{n-q,n-p}$. If we considered it as an antilinear operator landing in $\underline{\Omega}^{n-p,n-q}$, the sign would be $(-1)^{(p+q)(2n-p-q)} = (-1)^{p+q}$.)

Now we define

$$d^* = - * d*, \tag{1.3.9}$$

$$\partial^* = - * \overline{\partial} *, \quad \overline{\partial}^* = - * \partial *. \tag{1.3.10}$$

In (1.3.10), we use, again, the convention of treating $*$ as a $\mathbb{C}$-linear operator. It is useful to note that in the Riemann case, while we assumed that M is oriented in the definition of the $*$-operator, the definition of d^* in fact is local and does not depend on the sign of dV, and thus makes sense for any Riemann manifold without assuming orientability.

Let us verify that for an oriented compact Riemann manifold, the definition (1.3.9) satisfies (1.3.3): We have, using Stokes' theorem,

$$\int_{[M]} \langle \kappa, d^* \eta \rangle dV =$$

$$- \int_{[M]} \langle \kappa, *d * \eta \rangle dV = - \int_{[M]} \kappa \wedge d * \eta =$$

$$\int_{[M]} d\kappa \wedge *\eta = \int_{[M]} \langle d\kappa, \eta \rangle dV.$$

In the complex case, using Stokes' theorem (keeping in mind that one summand vanishes for reasons of type), we have

$$\int_{[M]} \langle \kappa, \partial^* \eta \rangle dV =$$

$$- \int_{[M]} \langle \kappa, *\overline{\partial} * \eta \rangle dV = - \int_{[M]} \kappa \wedge \overline{\overline{\partial} * \eta} = - \int_{[M]} \kappa \wedge \partial \overline{*\eta} =$$

$$\int_{[M]} \partial \kappa \wedge \overline{*\eta} = \int_{[M]} \langle \partial \kappa, \eta \rangle dV.$$

The proof for $\overline{\partial}$ is analogous.

Now one defines the *Laplacean operators* by

$$\Delta_d = dd^* + d^*d, \quad \Delta_\partial = \partial\partial^* + \partial^*\partial, \quad \Delta_{\overline{\partial}} = \overline{\partial}\,\overline{\partial}^* + \overline{\partial}^*\overline{\partial}. \tag{1.3.11}$$

One has

1.3.1 Lemma *For a differential form κ, one has*

$$\Delta_d(\kappa) = 0 \tag{1.3.12}$$

if and only if

$$d\kappa = d^*\kappa = 0.$$

(1.3.13)

Similarly for ∂ and $\overline{\partial}$.

Proof It is clear that (1.3.13) implies (1.3.12). For the converse, note that

$$\langle \Delta_d \kappa, \kappa \rangle = \langle d^*\kappa, d^*\kappa \rangle + \langle d\kappa, d\kappa \rangle,$$

and similarly in the case of ∂ and $\overline{\partial}$.

□

One usually denotes the kernel of Δ_d on $\Omega^k(M)$ for a Riemann manifold M by $\mathcal{H}_d^k$, and calls its elements *d-harmonic k-forms*. The definitions of ∂-harmonic and $\overline{\partial}$-harmonic (p,q)-forms on a Hermitian manifold M is analogous.

1.3.2 Theorem (Hodge Theorem) *For a compact Riemann manifold M, the $\mathbb{R}$-vector spaces $\mathcal{H}_d^k$ are finite-dimensional, and one has a direct sum decomposition*

$$\Omega^k(M) = \mathcal{H}_d^k \oplus Imd \oplus Imd^*.$$

For a compact Hermitian manifold M, the $\mathbb{C}$-vector spaces $\mathcal{H}_\partial^{p,q}$, $\mathcal{H}_{\overline{\partial}}^{p,q}$ are finite-dimensional, and one has a direct sum decomposition

$$\Omega^{p,q}(M) = \mathcal{H}_\partial^{p,q} \oplus Im\partial \oplus Im\partial^*,$$

$$\Omega^{p,q}(M) = \mathcal{H}_{\overline{\partial}}^{p,q} \oplus Im\overline{\partial} \oplus Im\overline{\partial}^*.$$

(By the images, we understand their summands in the given (bi)degree.)

Proving the Hodge theorem requires some advanced analysis, and exceeds the realm of this text. The case of $\overline{\partial}$ is proved in detail in [10]. The reader may wonder how come an analogous result holds for the a priori quite different operators d, ∂, $\overline{\partial}$. That is because there is a version of the Hodge theorem for a broader class of *elliptic differential operators*, to which they all belong. A treatment of that theory can be found in [26]. We will now state some consequences of the Hodge Theorem.

1.3.3 Theorem

1. For a compact Riemann manifold M, we have a canonical isomorphism

$$\mathcal{H}_d^k \cong H^k(M; \mathbb{R}).$$

For a compact Hermitian manifold M, we have

$$\mathcal{H}_{\bar{\partial}}^{p,q} \cong H^q(M, \Omega^p_{Hol,M}).$$

2. For a compact oriented Riemann n-manifold M, have a canonical isomorphism

$$H^k(M; \mathbb{R}) \cong H^{n-k}(M; \mathbb{R}).$$

3. For a compact Hermitian manifold M of complex dimension n, we have a canonical isomorphism

$$H_{\bar{\partial}}^{p,q}(M) \cong \overline{H_{\bar{\partial}}}^{n-p,n-q}(M).$$

Proof Equation (1) follows from Lemma 1.3.1 and from the Hodge Theorem. For example, for the case of d, it suffices to notice that d is injective on $Im(d^*)$, since we have

$$\langle dd^*\kappa, \kappa \rangle = \langle d^*\kappa, d^*\kappa \rangle.$$

For the case of $\bar{\partial}$, we additionally invoke Dolbeault's theorem (Theorem 1.1.1).

For (2), the version of the Hodge $*$-operator for an oriented Riemann manifold gives an isomorphism

$$\mathcal{H}_d^k \cong \mathcal{H}_{d_*}^{n-k},$$

so the result follows from part (1).

For (3), the version of the Hodge $*$-operator for a Hermitian manifold gives an isomorphism

$$\mathcal{H}_{\bar{\partial}}^{p,q} \cong \mathcal{H}_{\partial}^{n-q,n-p},$$

while the right hand side is, by definition, isomorphic to $\overline{\mathcal{H}_{\bar{\partial}}}^{n-p,n-q}$. □

It should be pointed out that part (2) of Theorem 1.3.3 has a version which does not use the Hodge theorem, or a Riemann metric. In fact, it even has a version valid in singular homology and cohomology: for an oriented compact n-manifold M, the abelian groups $H_p(M)$ are finitely generated, and

$$? \cap [M] : H^q(M; \mathbb{Z}) \to H_{n-q}(M; \mathbb{Z})$$

is an isomorphism of abelian groups. This statement is referred to as *Poincaré duality*, and it was proved in Exercise 32 of Chap. 5 (see also [13]). For de Rham cohomology, we can rephrase it to state that

$$H^k_{DR}(M) \otimes H^{n-k}_{DR}(M) \xrightarrow{\wedge} H^n_{DR}(M) \xrightarrow{\int_{[M]}} \mathbb{R}$$

defines an isomorphism

$$H^k_{DR}(M) \cong Hom_{\mathbb{R}}(H^{n-k}_{DR}(M), \mathbb{R}).$$

Note that for a compact complex manifold M, for reasons of type (i.e. examining what p, q can satisfy $p + q = 2n$), we have a canonical homomorphism of $\mathbb{C}$-vector spaces

$$H^{n,n}_{\bar{\partial}}(M) \to H^{2n}_{DR}(M; \mathbb{C}).$$

The *Kodaira-Serre duality* states that this is an isomorphism of $\mathbb{C}$-vector spaces and that the composition

$$H^{p,q}_{\bar{\partial}}(M) \otimes H^{n-p,n-q}_{\bar{\partial}}(M) \xrightarrow{\wedge} H^{n,n}_{\bar{\partial}}(M) \xrightarrow{\int_{[M]}} \mathbb{C}$$

defines an isomorphism

$$H^{p,q}_{\bar{\partial}}(M) \cong Hom_{\mathbb{C}}(H^{n-p,n-q}_{\bar{\partial}}(M), \mathbb{C}).$$

1.4   Kähler Manifolds

Consider a complex manifold M with a Hermitian metric g. As remarked in Sect. 1.2, g defines a Hermitian metric on the complex vector bundle $(\underline{\Omega}^{1,0}_M)^*$, which, in turn, can be viewed as a global section γ of

$$\underline{\Omega}^{1,0}_M \otimes_{C^\infty_{M,\mathbb{C}}} \underline{\Omega}^{0,1}_M \cong \underline{\Omega}^{1,1}_M.$$

As a matter of convention,

$$\omega = i\gamma \in \Omega^{1,1}(M)$$

is called the *Kähler form* of M. The Hermitian manifold M is called a *Kähler manifold* if we have

$$d\omega = 0. \tag{1.4.1}$$

This is called the *Kähler condition*. On a Kähler manifold M, one defines a morphism of $\mathcal{C}^\infty_{M,\mathbb{C}}$-modules

$$L = ? \wedge \omega : \underline{\Omega}^{p,q}_M \to \underline{\Omega}^{p+1,q+1}_M.$$

One also also denotes

$$\Lambda = L^*$$

the adjoint operator. Denote for two graded operators A, B, one of which is of even degree, $[A, B] = AB - BA$.

1.4.1 Lemma (Kähler Identities) *We have*

$$[L, \partial] = [L, \bar{\partial}] = [\Lambda, \partial^*] = [\Lambda, \bar{\partial}^*] = 0, \tag{1.4.2}$$

and

$$[L, \bar{\partial}^*] = -i\partial, \quad [L, \partial^*] = i\bar{\partial}, \quad [\Lambda, \bar{\partial}] = -i\partial^*, \quad [\Lambda, \partial] = i\bar{\partial}^*. \tag{1.4.3}$$

Proof The identities (1.4.2) follow immediately from the Kähler condition (1.4.1). Regarding (1.4.3), we shall prove

$$[L, \bar{\partial}^*] = -i\partial. \tag{1.4.4}$$

All the other identities follow by complex conjugation or adjunction.

To prove (1.4.4), let $\phi_1, \ldots, \phi_n$ be an orthonormal basis of $\Omega^{1,0}_M$. Then we have

$$\omega = i \sum_{s=1}^{n} \phi_s \wedge \bar{\phi}_s.$$

Moreover, putting for $I = \{i_1 < \cdots < i_p\}$, $J = \{j_1 < \cdots < j_q\}$

$$\phi_I = \phi_{i_1} \wedge \cdots \wedge \phi_{i_p}, \quad \bar{\phi}_J = \bar{\phi}_{j_1} \wedge \cdots \wedge \bar{\phi}_{j_q},$$

for a complex differential form

$$\kappa = f\phi_I \wedge \bar{\phi}_J \in \Omega^{p,q}(U),$$

we can write

$$\partial \kappa = \sum_{s \notin I} \partial_{s,I,J} f \phi_s \wedge \phi_I \wedge \overline{\phi}_J,$$

$$\overline{\partial} \kappa = \sum_{s \notin J} \overline{\partial}_{s,I,J} f \overline{\phi}_s \wedge \phi_I \wedge \overline{\phi}_J.$$

Furthermore, we have

$$\overline{\partial}_{s,I,J} \overline{f} = \overline{\partial_{s,J,I} f},$$

and the Kähler condition implies that for $k \notin I \cup J$, we have

$$\partial_{s,I\cup\{k\},J\cup\{k\}} = \partial_{s,I,J}, \quad \overline{\partial}_{s,I\cup\{k\},J\cup\{k\}} = \overline{\partial}_{s,I,J}.$$

Letting $I' = \{1,\ldots,n\} \setminus I$, $J' = \{1,\ldots,n\} \setminus J$, we note that this also implies

$$\partial_{s,I,J} = \partial_{s,J',I'}, \quad \overline{\partial}_{s,I,J} = \overline{\partial}_{s,J',I'},$$

since

$$J' \setminus I = I' \setminus J = J' \cap I',$$

$$J \setminus I' = I \setminus J' = I \cap J.$$

Using these relations, and denoting $J_{\leq j} = \{s \in J \mid s \leq j\}$ (and $J_{>j}$ similarly), one computes:

$$\overline{\partial}^*(f \phi_I \wedge \overline{\phi}_J) = -\sum_{j \in J} (-1)^{|J_{>j}|} (\partial_{j,J',I'} f) \phi_I \wedge \overline{\phi}_{J \setminus \{j\}}.$$

Thus,

$$(\overline{\partial}^* f \phi_I \wedge \overline{\phi}_J) \wedge \omega =$$

$$-i \sum_{\substack{j \in J \\ k \notin I \cup J}} (-1)^{|J_{\leq j}|+|I|} (\partial_{j,J',I'} f) \phi_I \wedge \overline{\phi}_{J \setminus \{j\}} \wedge \phi_k \wedge \overline{\phi}_k$$

$$-i \sum_{j \in J} (-1)^{|J_{\leq j}|+|I|} (\partial_{j,J',I'} f) \phi_I \wedge \overline{\phi}_{J \setminus \{j\}} \wedge \phi_j \wedge \overline{\phi}_j,$$

while

$$\overline{\partial}^{*}(f\phi_I \wedge \overline{\phi}_J \wedge \omega) =$$

$$-i \sum_{\substack{k\notin I\cup J \\ j\in J}} (-1)^{|J_{\leq j}|+|I|}(\partial_{i,J'\smallsetminus\{k\},I'\smallsetminus\{k\}}f)\phi_I \wedge \overline{\phi}_{J\smallsetminus\{j\}} \wedge \phi_k \wedge \overline{\phi}_k$$

$$-i \sum_{k\notin I\cup J} (-1)^{|I|+|J|}(\partial_{k,J',I'}f)\phi_I \wedge \overline{\phi}_J \wedge \phi_k.$$

Subtracting these expressions, we see that the first terms on the right hand side cancel, while the second terms add up to

$$-i\partial(f\phi_I \wedge \overline{\phi}_J),$$

$\square$

as claimed.

1.4.2 Corollary *On a Kähler manifold M, we have*

$$[L, \Delta_d] = 0, \ [\Lambda, \Delta_d] = 0, \tag{1.4.5}$$

$$\Delta_d = 2\Delta_\partial = 2\Delta_{\overline{\partial}}. \tag{1.4.6}$$

Proof For (1.4.5), both identities are adjoint, so we shall only treat the case of L. It is possible to simply plug in and use Lemma 1.4.1. This is basically the only proof there is, but it is possible to make it slightly more conceptual by noting the following: For three even-degree operators A, B, C, one easily checks that we have

$$[[A, B], C] + [[B, C], A] + [[C, A], B] = 0. \tag{1.4.7}$$

This is called the *Jacobi identity*. A vector space over a field of characteristic $\neq 2, 3$ with a bilinear operation $[?, ?]$ which is antisymmetric (i.e. satisfies $[A, B] = -[B, A]$), and the Jacobi identity, is called a *Lie algebra*. The operation $[?, ?]$ is then called the *Lie bracket*. Thus, in particular, an associative algebra with the operation $[a, b] = ab - ba$ is a Lie algebra.

Now for graded operators A, B, we can also put

$$[A, B] = AB - (-1)^{|A| |B|} BA$$

where $|A|$ denotes the degree of A. Note that, for example, in this sense,

$$\Delta_d = [d, d^*].$$

Then we have the *graded Jacobi identity*

$$[[A, B], C] + (-1)^{(|B|+|C|)|A|}[[B, C], A] + (-1)^{(|A|+|B|)|C|}[[C, A], B] = 0, \qquad (1.4.8)$$

which is just as easy to verify as the ungraded one. From this point of view, we have

$$[[d, d^*], L] - [[d^*, L], d] + [[L, d], d^*] = 0,$$

so it just suffices to note that $[[L, d], d^*] = 0$, which easily follows from the Kähler identities.

Accordingly, a $\mathbb{Z}/2$-graded vector space over a field of characteristic $\neq 2, 3$ which satisfies, for homogeneous elements,

$$[A, B] + (-1)^{|A|\,|B|}[B, A] = 0$$

and (1.4.8) is called a *super-Lie algebra*.

Using the graded Jacobi identity, (1.4.6) is easily proved as well. For example,

$$-[\partial, [\Lambda, \partial]] + [\Lambda, [\partial, \partial]] + [\partial, [\partial, \Lambda]] = 0$$

implies

$$[\partial, \overline{\partial}^*] = -i[\partial, [\Lambda, \partial]] = 0.$$

Similarly,

$$[\overline{\partial}, \partial^*] = 0.$$

Therefore,

$$\Delta_d = [\partial + \overline{\partial}, \partial^* + \overline{\partial}^*] = [\partial, \partial^*] + [\overline{\partial}, \overline{\partial}^*] = \Delta_\partial + \Delta_{\overline{\partial}}.$$

Now the graded Jacobi identity

$$-[\overline{\partial}, [\Lambda, \partial]] + [\Lambda, [\partial, \overline{\partial}]] + [\partial, [\overline{\partial}, \Lambda]] = 0$$

(together with the fact that $[\partial, \overline{\partial}] = 0$) imply that

$$\Delta_\partial = \Delta_{\overline{\partial}}.$$

$\square$

Now in view of the Hodge Theorem, (1.4.6) in particular implies that on a compact Kähler manifold M, we have

$$\mathcal{H}_d^{p,q} = \mathcal{H}_\partial^{p,q} = \mathcal{H}_{\bar{\partial}}^{p,q}.$$

Since, by definition,

$$\mathcal{H}_\partial^{p,q} = \overline{\mathcal{H}_{\bar{\partial}}^{q,p}},$$

we conclude that

$$\mathcal{H}_d^{p,q} = \overline{\mathcal{H}_d^{q,p}}.$$

Now recall the spectral sequence (1.1.1). By the Hodge Theorem, then, the spectral sequence (1.1.1) collapses to E_1, and the Hodge filtration on $F^p H^k(M, \mathbb{C})$ satisfies, for $p + q = k$, the isomorphism

$$F^p H^k(M; \mathbb{C}) \cap \overline{F^q H^k(M; \mathbb{C})} \cong H_{\bar{\partial}}^{p,q}(M; \mathbb{C}). \tag{1.4.9}$$

(At this point, one simply denotes the left hand side by $H^{p,q}(M)$.)

In fact, note that the spectral sequence (1.1.1), by collapsing to E_1, determines an isomorphism

$$H_{\bar{\partial}}^{p,q}(M, \mathbb{C}) \cong F^p H^k(M; \mathbb{C})/F^{p+1} H^k(M; \mathbb{C}),$$

while we also have a canonical homomorphism

$$F^p H^k(M; \mathbb{C}) \cap \overline{F^q H^k(M; \mathbb{C})} \to F^p H^k(M; \mathbb{C})/F^{p+1} H^k(M; \mathbb{C}),$$

which we have now proved is an isomorphism. Thus, the isomorphism (1.4.9) *does not depend on the choice of a Kähler metric!*

Now we can combine this with the fact that by the finiteness statement of Poincaré duality and the universal coefficient theorem, for a compact complex manifold M, we have a canonical isomorphism

$$H^k(M; \mathbb{Z}) \otimes \mathbb{C} \cong H^k(M; \mathbb{C}).$$

Define then a *Hodge structure* of weight k to consist of a finitely generated abelian group A and a direct sum decomposition

$$A \otimes \mathbb{C} = \bigoplus_{p \in \mathbb{Z}} A^{p,k-p}$$

such that

$$A^{p,q} = \overline{A^{q,p}}.$$

(Sometimes we also speak of a *weight k Hodge structure on A*.) Define a morphism of Hodge structures as a homomorphism of abelian groups $h : A \to B$ such that

$$(h \otimes \mathbb{C})(A^{p,q}) \subseteq B^{p,q}.$$

We have therefore proved the following

1.4.3 Theorem (Hodge Decomposition Theorem) *Let M be a compact complex manifold on which there exists a Kähler metric. Then we have a canonical Hodge structure of weight k on $H^k(M; \mathbb{Z})$, functorial with respect to holomorphic diffeomorphisms (i.e., in particular, independent of the choice of Kähler metric).*

$\square$

1.5 The Lefschetz Decomposition

For a Lie algebra $\mathcal{L}$ over a field F, a *representation* of $\mathcal{L}$ over F is an F-vector space V with a linear morphism

$$h : \mathcal{L} \to Hom_F(V, V)$$

such that for $x, y \in \mathcal{L}$, we have

$$[h(x), h(y)] = h([x, y])$$

(i.e. a homomorphism of Lie algebras). A representation of a super-Lie algebra on a $\mathbb{Z}/2$-graded vector space is defined analogously. The above argument really rested on the fact that the de Rham complex of a Kähler manifold is a representation of the super-Lie algebra generated by the symbols $\partial, \overline{\partial}, \partial^*, \overline{\partial}^*, L, \Lambda$ subject to relations given by the Kähler identities and the usual anticommutation relations between ∂ and $\overline{\partial}$, and ∂^* and $\overline{\partial}^*$.

It is worth mentioning that this can be pushed a little further. If one defines

$$H = [L, \Lambda], \tag{1.5.1}$$

one can show that

$$[H, L] = 2L, \quad [H, \Lambda] = -2\Lambda, \tag{1.5.2}$$

and in fact that on $\Omega^{p,q}(M)$ for a Kähler manifold M, H acts by

$$p + q - n. \tag{1.5.3}$$

The Lie algebra generated by the symbols H, L, Λ subject to the relations (1.5.1), (1.5.2) is easily seen to be isomorphic to the Lie algebra $sl_2(\mathbb{C})$ of 2×2 matrices with trace 0 with Lie bracket given by $[A, B] = AB - BA$, where L, Λ, H correspond to the matrices

$$\begin{pmatrix} 0 & 1 \\ 0 & 0 \end{pmatrix}, \begin{pmatrix} 0 & 0 \\ 1 & 0 \end{pmatrix}, \begin{pmatrix} 1 & 0 \\ 0 & -1 \end{pmatrix},$$

respectively.

For a compact Kähler manifold M, since the operators L, Λ commute with Δ, we have, for each k, a finite-dimensional representation

$$\bigoplus_{p-q=k} \mathcal{H}^{p,q}.$$

(We omit the subscripts $d, \partial, \bar{\partial}$, since they do not matter.) The numbers by which the operator H acts are called *weights*. Finite-dimensional representations of $sl_2(\mathbb{C})$ have been classified, and are, in fact, isomorphic to direct sums of representations of the form $V_\ell = Sym^\ell(V)$ where $V = \mathbb{C}^2$ is the "standard" representation given by multiplication of matrices. Here $Sym^\ell(V)$ denotes the ℓth *symmetric power*, which, for a basis $x_1, \ldots, x_m$ of a vector space V, can be identified with the vector space of homogeneous polynomials in $x_1, \ldots, x_m$ of degree ℓ. In the present case, the $\mathbb{C}$-vector space V has dimension 2, so V_ℓ has dimension $\ell + 1$.

In fact, we can pick basis elements of V of weights $-1, 1$, thus showing that the representation V_ℓ has weights

$$-\ell, -\ell + 2, \ldots, \ell - 2, \ell.$$

Thus, we know that $\mathcal{H}^{*,*}$ is a direct sum of representations of the form V_ℓ, (for varying ℓ) with bottom weight $-\ell$ in $\mathcal{H}^{p,q}$,

$$p + q = n - \ell.$$

Furthermore, one notes that the operator L does not depend on the Kähler metric (even though the operator Λ does), so if we define

$$PH^{p,q}(M) = \{\alpha \in H^{p,q}(M) \mid \alpha \cup \omega^{n+1-p-q} = 0\},$$

one obtains the following result:

1.5.1 Theorem (Hard Lefschetz Theorem) *The de Rham cohomology of a compact complex manifold on which there exists a Kähler metric decomposes canonically, and functorially with respect to holomorphic diffeomorphisms, as*

$$\bigoplus_{p,q} \bigoplus_{0 \le i \le n-p-q} PH^{p,q}(M) \cup \{\omega^i\}. \tag{1.5.4}$$

□

The decomposition (1.5.4) is referred to as the *Lefschetz decomposition*. To read about basic representation theory, including the classification of finite-dimensional representations of $sl_2(\mathbb{C})$, we recommend [8].

1.6 Examples

1.6.1 The Fubini-Study Metric

By the discussion of the beginning of Sect. 1.4, a $(1,1)$-form

$$\omega \in \Omega^{1,1}(M) \cong \Omega^{1,0}(M) \otimes_{C^\infty(M,\mathbb{C})} \overline{\Omega^{1,0}(M)}$$

$\gamma = -i\omega$ is a Hermitian metric on M if

$$\gamma^* = \gamma \tag{1.6.1}$$

where $(?)^*$ is the composition of the homorphisms of $\mathbb{R}$-vector spaces

$$\Omega^{1,0}(M) \otimes_{C^\infty(M,\mathbb{C})} \overline{\Omega^{1,0}(M)}$$

$$\downarrow T$$

$$\overline{\Omega^{1,0}(M)} \otimes_{C^\infty(M,\mathbb{C})} \Omega^{1,0}(M)$$

$$\downarrow \overline{?}$$

$$\Omega^{1,0}(M) \otimes_{C^\infty(M,\mathbb{C})} \overline{\Omega^{1,0}(M)}$$

where T is switching factors, and $\overline{?}$ is complex conjugation, and we have *positivity*, which can be expressed by requiring that for every point $x \in M$, and every non-zero complex-linear homomorphism $f : \Omega^{1,0}(x) \to \mathbb{C}$, we have

$$(f \otimes_{\mathbb{C}} \overline{f})(\gamma_x) > 0. \tag{1.6.2}$$

(Here by $\Omega^{1,0}(x)$ we mean again the finite-dimensional $\mathbb{C}$-vector space of "values at x," technically obtained from the stalk by tensoring over the stalk $\mathcal{C}^\infty_{M,\mathbb{C},x}$ with the residue field $\mathbb{C}$.)

We can use the conditions (1.6.1), (1.6.2) to construct Kähler metrics from closed $(1, 1)$-forms. Most notably, on $\mathbb{C}P^n$, we have the *Fubini-Study metric*, constructed as follows. Denote by $\pi : \mathbb{C}^{n+1} \smallsetminus \{0\} \to \mathbb{C}P^n$ the projection, consider for an open set $U \subset \mathbb{C}P^n$ a holomorphic section $s : U \to \mathbb{C}^{n+1} \smallsetminus \{0\}$ (i.e. a holomorphic map such that for $x \in U$, $\pi(s(x)) = x$). Then put

$$\omega = \frac{i}{2\pi} \partial \bar\partial \ln(\langle s, \bar s \rangle). \tag{1.6.3}$$

Since for a holomorphic function $g : U \to \mathbb{C} \smallsetminus \{0\}$,

$$\partial \bar\partial \ln(g\bar g) = 0,$$

we see that (1.6.3) does not depend on the choice of holomorphic section, and thus is a well-defined $(1, 1)$-form on $\mathbb{C}P^n$. Additionally, for $\gamma = -i\omega$, (1.6.1) is immediately obvious from the definition (the multiplication by i is only a matter of historical convention, anyway). Thus, to show that γ defines a Kähler metric on $\mathbb{C}P^n$, it remains to verify the condition (1.6.2). Since clearly applying a unitary transformation on $\mathbb{C}^{n+1}$ (i.e. one that preserves the inner product) preserves ω, it suffices to verify (1.6.2) at the point $[1 : 0 : \cdots : 0] \in \mathbb{C}P^n$. Letting

$$s([1 : z_1 : \cdots : z_n]) = (1, z_1, \ldots, z_n),$$

we have

$$\partial \bar\partial \ln(1 + z_1\bar z_1 + \cdots + z_n\bar z_n) = \partial \frac{z_1 d\bar z_1 + \cdots + z_n d\bar z_n}{1 + z_1\bar z_1 + \cdots + z_n\bar z_n}.$$

At $z_1 = \cdots = z_n = 0$, the right hand side is

$$dz_1 d\bar z_1 + \cdots + dz_n d\bar z_n,$$

which, for a non-zero linear form $f(dz_j) = a_j \in \mathbb{C}$ gives

$$(f \otimes \bar f)(dz_1 \otimes d\bar z_1 + \cdots + dz_n \otimes d\bar z_n) = a_1\bar{a_1} + \cdots + a_n\bar{a_n} > 0.$$

Thus, the Fubini-Study metric is a Kähler metric, and using restriction, we proved

1.6.2 Theorem *On every smooth projective variety over $\mathbb{C}$ with the analytic topology there exists a Kähler metric.*

□

1.6.3 The Cohomology of the Complex Projective Space

In Sect. 3.6 of Chap. 5, we proved that the singular homology of $\mathbb{C}P^n$ is $\mathbb{Z}$ in even degrees $0 \leq k \leq 2n$, and 0 else. By the universal coefficient theorem of Sect. 3.5 of Chap. 5, the same is true about singular cohomology with coefficients in $\mathbb{Z}$. Using Poincaré duality and (3.5.13) of Chap. 5, we see that there is class $u \in H^2(\mathbb{C}P^n; \mathbb{Z})$ such that

$$H^*(\mathbb{C}P^n; \mathbb{Z}) \cong \mathbb{Z}[u]/(u^{n+1})$$

(see also Exercise 34 of Chap. 5). By Theorem 1.4.3, then, there is weight $2k$ Hodge structure on

$$H^{2k}(\mathbb{C}P^n; \mathbb{Z}) \cong \mathbb{Z}$$

for $0 \leq k \leq n$. We see that the only possibility is $H^{k,k}(\mathbb{C}P^n) \cong \mathbb{C}$, $H^{k,\ell}(\mathbb{C}P^n) = 0$ for $\ell \neq k$. We also note that there is, in fact, up to isomorphism, a *unique* Hodge structure on $\mathbb{Z}$ of even weight $2k$, $k \in \mathbb{Z}$. There is no Hodge structure on $\mathbb{Z}$ of odd weight. The Hodge structure of weight $2k$ on $\mathbb{Z}$ is, by convention, denoted by $\mathbb{Z}(-k)$, and called the *Tate Hodge structure*. (In some sources, the term Tate Hodge structure is more narrowly reserved for $\mathbb{Z}(1)$.) It should be also noted that, as we shall see later, the symbol $\mathbb{Z}(k)$ is also used for objects which play, in some sense, analogous roles to the Tate Hodge structures in other categories.

1.6.4 Elliptic Curves over $\mathbb{C}$

Suppose L is a *lattice* in $\mathbb{C}$, i.e. a subgroup of $\mathbb{C}$ isomorphic to $\mathbb{Z}^2$, which is discrete in the induced topology. Lattices in $\mathbb{C}$ form a category where a morphism $L_1 \to L_2$ is a number $\lambda \in \mathbb{C}$ such that $\lambda(L_1) \subseteq L_2$. For a lattice L in $\mathbb{C}$, the *Weierstrass function* is defined by

$$\mathcal{P}_L(z) = \frac{1}{z^2} + \sum_{a \in L \smallsetminus \{0\}} \left(\frac{1}{(z+a)^2} - \frac{1}{a^2} \right).$$

The series converges absolutely locally uniformly in $z \notin L$, so this is, in fact, a function holomorphic on $\mathbb{C} \smallsetminus L$, and has, in fact, a pole of degree 2 at each point of L. Moreover, by grouping terms (which is possible due to absolute convergence), one sees that for $a \in L$, we have

$$\mathcal{P}_L(z+a) = \mathcal{P}_L(z).$$

Expanding the Weierstrass function in the neighborhood of 0, one gets

$$\mathcal{P}_L(z) = z^{-2} + \frac{1}{20}g_2 z^2 + \frac{1}{28}g_3 z^4 + \geq 6\text{'th powers of } z$$

where

$$g_2 = 60 \sum_{a \in L \smallsetminus \{0\}} a^{-4},$$

$$g_3 = 140 \sum_{a \in L \smallsetminus \{0\}} a^{-6}.$$

One has the *Weierstrass equation*

$$(\mathcal{P}_L')^2 = 4(\mathcal{P}_L)^3 - g_2 \mathcal{P}_L - g_3. \tag{1.6.4}$$

To show this, one expands

$$(\mathcal{P}_L'(z))^2 = \frac{4}{z^6} - \frac{2}{5}\frac{g_2}{z^2} - \frac{4}{7}g_3 + \text{higher powers,}$$

$$(\mathcal{P}_L(z))^3 = \frac{1}{z^6} + \frac{3}{20}\frac{g_2}{z^2} + \frac{3}{28}g_3 + \text{higher powers,}$$

so the difference of the two sides of (1.6.4) is bounded holomorphic function on $\mathbb{C}$ with zero at $z = 0$, which is therefore 0 constantly.

Thus, setting $x = \mathcal{P}_L'$, $y = \mathcal{P}_L$ turns (1.6.4) into the equation

$$x^2 = 4y^3 - g_2 y - g_3, \tag{1.6.5}$$

which is the equation of an affine part of an elliptic curve C_L over $\mathbb{C}$. One can show that all elliptic curves over $\mathbb{C}$ arise in this fashion. (See [23] for more details.) For us, it is important that the elliptic curve $(C_L)_{an}$ with the analytic topology is, in fact, holomorphically diffeomorphic to the quotient $\mathbb{C}/L$ (which has an obvious canonical structure of a complex manifold). To see this, in the neighborhood of the point at ∞, change variables to

$$u = \frac{y}{x}, \quad v = \frac{1}{x},$$

so (1.6.5) becomes

$$v = 4u^3 - g_2 u v^2 - g_3 v^3.$$

The point at ∞ then has coordinates $u = 0$, $v = 1$, and $u'(0) \neq 0$, so one can solve for z by the implicit function theorem.

Thus, we can apply Hodge theory to $(C_L)_{an} \cong \mathbb{C}/L$. In fact, picking generators a_1, a_2 of the free abelian group L, we have a homeomorphism

$$S^1 \times S^1 \cong \mathbb{C}/L$$

given by

$$(x, y) \mapsto a_1 x + a_2 y$$

(thinking of S^1 as $[0, 1]/(0 \sim 1)$ with the quotient topology). By the Künneth theorem, we then have

$$H^0(\mathbb{C}/L; \mathbb{Z}) \cong \mathbb{Z}, \ \ H^1(\mathbb{C}/L; \mathbb{Z}) \cong \mathbb{Z} \oplus \mathbb{Z}, \ \ H^2(\mathbb{C}/L; \mathbb{Z}) \cong \mathbb{Z}$$

(and $H^k(\mathbb{C}/L; \mathbb{Z}) = 0$ for $k \neq 0, 1, 2$). By the Hodge decomposition, we then know that

$$H^{1,0}(\mathbb{C}/L) \cong H^{0,1}(\mathbb{C}/L) \cong \mathbb{C}.$$

In fact, we can see that $H^{1,0}(\mathbb{C}/L)$ is generated by dz. To see this, one needs to observe that $dz \neq 0 \in H^1(\mathbb{C}/L)$. To this end, one notes that

$$\int_s dz = a_1, \ \ \int_t dz = a_2 \tag{1.6.6}$$

where s resp. t is the singular simplex mapping $[0, 1]$ linearly to $[0, a_1]$ (resp. $[0, a_2]$). These integrals are referred to as *periods*. Since s, t represent cycles in $H_1(\mathbb{C}/L; \mathbb{Z})$, if dz were a coboundary, these integrals would be 0 by Stokes' theorem.

Since the homology classes s, t, in fact, generate $H_1(\mathbb{C}/L; \mathbb{Z})$, we see that that the Hodge structure on $A = H^1((C_L)_{an}; \mathbb{Z}) \cong \mathbb{Z} \oplus \mathbb{Z}$ has

$$H^{1,0} = \mathbb{C} \cdot (a_1, a_2) \ H^{0,1} = \mathbb{C} \cdot (\overline{a_1}, \overline{a_2}). \tag{1.6.7}$$

(It follows, but is also easily verified directly, that the vectors (a_1, a_2) and $(\overline{a_1}, \overline{a_2})$ are $\mathbb{C}$-linearly independent.)

We see that an automorphism of the abelian group $\mathbb{Z} \oplus \mathbb{Z}$ replaces in (1.6.7) a_1, a_2 by a different set of generators of the lattice L. Thus, the Hodge structures of $C_L, C_{L'}$ are isomorphic if and only if the lattices L, L' are isomorphic.

Therefore, for the curves $C_L, C_{L'}$ to be isomorphic as algebraic varieties over $\mathbb{C}$, it is necessary that the lattices L, L' be isomorphic. One can show that this is sufficient as well, by displaying an algebraic isomorphism of elliptic curves given by the transformation of g_2, g_3 induced by the isomorphism of lattices (see [23] for details).

2 Algebraic de Rham Cohomology

After having studied some of the constructions by which cohomology arises in topology, it makes sense to consider the question of imitating these constructions algebraically, i.e. in the Zariski topology. We shall mostly focus on the case of a smooth variety over a field. Apart from being "more intrinsic" to algebraic geometry, such constructions allow us for example to consider cohomology of varieties over fields of characteristic > 0, or have a cohomology theory functorial with respect to automorphisms of the ground field.

2.1 The Algebraic de Rham Complex

Let us begin by noting that the first possible approach that may come to mind, namely taking the cohomology of a constant abelian sheaf, fails, in a spectacular fashion, to produce an analogous effect as in the analytic topology:

2.1.1 Proposition *Suppose X is a non-empty irreducible topological space and A is an abelian group. Then the constant sheaf $\underline{A}$ on X is flasque. Consequently, for any abelian group B,*

$$H^0(X, \underline{B}) = B, \quad H^n(X, \underline{B}) = 0 \ for \ n > 0.$$

Proof Dualizing the irreducibility condition shows that all non-empty open sets in X are connected. Given this, we see that $\underline{A}(U) = A$ for $A \neq \emptyset$, and $\underline{A}(\emptyset) = 0$. The statement follows. $\square$

2.1.2 Kähler Differentials

Given this, it may come as a pleasant surprise that an algebraic analogue of the de Rham complex in fact works much better. Let us begin with the affine case. Let $A \rightarrow B$ be a homomorphism of commutative rings (in other words, let B be an A-algebra). Then a *differentiation on B over A* consists of a B-module M and a homomorphism of A-modules

$$d : B \rightarrow M$$

such that

$$d(1) = 0$$

and for $x, y \in B$,

$$d(xy) = xdy + ydx.$$

It is easy to see that there exists a *universal* differentiation, i.e. a differentiation

$$d : B \to \Omega_{B/A}$$

such that for any differentiation $d' : B \to M$, there exists a unique homomorphism of B-modules $f : \Omega_{B/A} \to M$ with $d' = f \circ d$:

To see this, simply take the free B-module on elements db, $b \in B$, and factor by the submodule generated by da, $a \in A$, and $d(xy) - xdy - ydx$ for $x, y \in B$. The B-module $\Omega_{B/A}$ is called the *B-module of Kähler differentials of B over A.*

2.1.3 Lemma *Let A be a Noetherian ring and suppose that B is a standard smooth A-algebra of dimension k. Then $\Omega_{B/A}$ is a rank k (in particular, finitely generated) projective B-module.*

Proof By Lemma 2.2.4 of Chap. 4, it suffices to prove that $\Omega_{B/A}$ is a finitely generated locally free module. To this end, it suffices to assume that A is a local ring. If B is smooth over A, then we have

$$B = A[z_1, \ldots, z_n]/(p_1, \ldots, p_m), \quad m + k = n$$

where the ideal generated by the determinants of the $m \times m$ submatrices of the Jacobi matrix in B contains 1. Now the B-module $\Omega_{B/A}$ is, by definition, the quotient of the free B-module on elements $dz_1, \ldots, dz_n$, modulo the relations

$$\frac{\partial p_1}{\partial z_1} dz_1 + \cdots + \frac{\partial p_1}{\partial z_n} dz_n = 0,$$

$$\cdots$$

$$\frac{\partial p_m}{\partial z_1} dz_1 + \cdots + \frac{\partial p_m}{\partial z_n} dz_n = 0. \tag{2.1.1}$$

Now let p be a prime ideal of B. We claim that $\Omega_{B/A} \otimes_B B_p$ is a finitely generated free B_p-module. To this end, note that a maximal subdeterminant of the Jacobi matrix, say, on columns $i_1 < \cdots < i_m$, must be non-zero in the residue field B_p/p, which means that it is a unit in B_p. Thus, the linear equations (2.1.1) can be solved for $dz_{i_1}, \ldots, dz_{i_m}$ in terms of the other dz_j's, thus proving that $\Omega_{B/A} \otimes_B B_p$ is a free B_p-module on those generators. $\qquad\square$

2.1.4 Algebraic de Rham Cohomology

Now let X be a smooth variety of dimension n over a field k. Then X is covered by open sets $U_i = Spec(A_i)$ where each A_i is a standard smooth k-algebra. Since we clearly have, for $g \in A_i$, $\Omega_{g^{-1}A_i/k} = g^{-1}\Omega_{A_i/k}$, the $\underline{\Omega}_{A_i/k}$'s glue to a coherent sheaf $\underline{\Omega}_{X/k}$, and we have a morphism of sheaves of k-modules (although not of $\mathcal{O}_X$-modules)

$$d : \mathcal{O}_X \to \underline{\Omega}_{X/k}.$$

We may now form a cochain complex $\underline{\Omega}^*_{X/k}$ of sheaves of k-modules on X of the form

$$\mathcal{O}_X \xrightarrow{\ d\ } \underline{\Omega}^1_{X/k} \xrightarrow{\ d\ } \underline{\Omega}^2_{X/k} \xrightarrow{\ d\ } \cdots \xrightarrow{\ d\ } \underline{\Omega}^n_{X/k} \qquad (2.1.2)$$

where

$$\underline{\Omega}^\ell_{X/k} = \Lambda^\ell_{\mathcal{O}_X} \underline{\Omega}_{X/k}$$

and the differential is, as in the analytic case, uniquely determined by the conditions

$$d \circ d = 0$$

and for sections $\omega \in \underline{\Omega}^\ell_{X/k}(U), \eta \in \underline{\Omega}^m_{X/k}(U),$

$$d(\omega \wedge \eta) = (d\omega) \wedge \eta + (-1)^\ell \omega \wedge d\eta.$$

(The point is that, again, by the fact that the sheaf $\underline{\Omega}_{X/k}$ is locally free, each section can locally be written as a linear combination of sections of the form

$$f \, dz_{j_1} \wedge \cdots \wedge dz_{j_m}$$

for some basis dz_j.) The complex (2.1.2) is referred to as the *algebraic de Rham complex of the variety* X. The hypercohomology groups

$$H^m_{DR}(X) = H^m(X, \underline{\Omega}^*_{X/k})$$

are referred to as the *algebraic de Rham cohomology* of X over k. (Compare with Corollary 1.1.2.)

2.1.5 The Non-smooth Case

While we will focus on the case of a smooth variety in this section, it is important for later to note that the definition of the algebraic de Rham complex actually does not need this assumption. For any homomorphism of commutative rings $A \to B$, we may define

$$\Omega^*_{B/A} = \Lambda_B \Omega_{B/A}$$

(see Exercise 16 of Chap. 4). The differential

$$d : \Omega^\ell_{B/A} \to \Omega^{\ell+1}_{B/A}$$

is well-defined. In fact, in the case of a polynomial algebra $B_0 = A[x_i \mid i \in S]$, we see that $\Omega^*_{B_0/A}$ is the exterior algebra on B_0 on generators $dx_i, i \in S$, and when $B = B_0/I$ for an ideal I, we can describe $\Omega^*_{B/A}$ as the factor of the differential graded A-algebra $\Omega^*_{B_0/A}$ by the differential graded ideal generated by I, which is the (two-sided) ideal generated by I, dI. Here by a *differential graded A-algebra* we mean a graded-commutative graded A-algebra $R = \bigoplus_{n \in \mathbb{N}_0} R^n$ with a differential $d : R^n \to R^{n+1}$ such that for $x \in R^m$, $y \in R^n$,

$$x \cdot y = (-1)^{mn} y \cdot x, \quad d(x \cdot y) = (dx) \cdot y + (-1)^m x \cdot dy,$$

and $da = 0$ for $a \in A$. A *differential graded ideal* is an ideal J generated by homogeneous elements such that

$$dJ \subseteq J.$$

The factor of a differential graded algebra by a differential graded ideal is easily verified to be again a differential graded algebra with the induced differential. In particular,

$$\Omega^*_{B/A} = \Omega^*_{B_0/A}/J$$

where J is the ideal generated by $\{x, dx \mid x \in I\}$ (this is proved by considering the universal property of both sides).

Thus, by gluing sheaves, for a scheme X over $Spec(A)$, we also have a complex of sheaves $\underline{\Omega}^*_{X/A}$ on X.

2.2 Quasicoherent Cohomology

One of our tasks will be comparing, for a smooth variety X over $\mathbb{C}$, the algebraic de Rham cohomology of X with the analytic de Rham cohomology of X_{an}. To this end, however, it will be useful to make a few remarks about cohomology of schemes with coefficients in quasicoherent sheaves.

2.2.1 Lemma *Let R be a Noetherian ring. Suppose I is an injective R-module. Then the quasicoherent sheaf on $Spec(R)$ corresponding to I is flasque.*

Proof We need to prove that for elements $f_1, \ldots, f_n \in R$, $x_i \in f_i^{-1} I$ such that x_i, x_j map to the same element of $(f_i f_j)^{-1} I$, there exists an element $x \in I$ which maps to $x_i \in f_i^{-1} I$. This is done by induction on n. For $n = 1$, let $f = f_1$. The *annihilator* of an element $m \in M$ of an R-module M is the R-ideal

$$Ann_R(m) = \{g \in R \mid gm = 0\}.$$

Since R is Noetherian, there exists an N such that

$$Ann(f^N) = Ann(f^{N+1}) = \cdots . \tag{2.2.1}$$

Now for an element $f^{-r} z \in f^{-1} I$, define

$$(f^{N+r}) \to I \tag{2.2.2}$$

by

$$f^{N+r} \mapsto f^N z,$$

which is possible by (2.2.1). Extending (2.2.2) to a map

$$(f^N) \to I,$$

the image of f^N maps to $f^{-r} z \in f^{-1} I$, as required.

Now consider our assumption for a given $n > 1$, and suppose the statement is true with n replaced by $n - 1$. Let $y \in I$ be an element which maps to $x_n \in f_n^{-1} I$. Then for some N,

$$f_n^N (y - x_i) = 0 \in f_i^{-1} I, \ i = 1, \ldots, n - 1, \tag{2.2.3}$$

Suppose $x_i = f_i^{-r_i} z_i$, $z_i \in I$. Then (2.2.3) implies the existence of an M such that

$$f_n^N f_i^M (y f_i^{r_i} - z_i) = 0 \in I.$$

(2.2.4)

Now consider the ring $R' = R/(f_n^N)$ and the R'-module

$$I' = Hom_R(R/f_n^N, I).$$

By (2.2.4), $y - x_i$, $i = 1, \ldots, n-1$, can be considered as elements of $f_i^{-1} I'$, which, additionally, satisfy our hypothesis with n replaced by $n-1$ and R replaced by R'. Additionally, the R'-module I' is injective. Thus, by the induction hypothesis, there exists a $u \in I'$ such that

$$u \mapsto y - x_i \in f_i^{-1} I', \ i = 1, \ldots, n-1.$$

Considering u as an element of I, we then have

$$y - u \mapsto x_i \in f_i^{-1} I, \ i = 1, \ldots n,$$

as required.

□

2.2.2 Proposition *Suppose R is a Noetherian ring and $\mathcal{F}$ is a quasicoherent sheaf of R-modules on $Spec(R)$. Then*

$$H^n(Spec(R), \mathcal{F}) = 0 \text{ for } n > 0.$$

Proof Consider an injective resolution

$$I_0 \to I_1 \to \ldots$$

(2.2.5)

of the R-module M corresponding to $\mathcal{F}$. Then by Lemma 2.2.1, the corresponding exact sequence of quasicoherent sheaves

$$\mathcal{I}_0 \to \mathcal{I}_1 \ldots$$

is a flasque resolution of $\mathcal{F}$. Applying global sections, we get (2.2.5) again, and the statement follows.

□

2.2.3 Lemma *Let X be a Noetherian scheme. Then the abelian category of quasicoherent sheaves of $\mathcal{O}_X$-modules has enough injectives. Moreover, from every object, there exists a monomorphism into an injective $\mathcal{I}$ such that $\mathcal{I}$ is also flasque as an abelian sheaf. For a* *quasicoherent sheaf $\mathcal{F}$ on X, the abelian group $H^n(X, \mathcal{F})$ is canonically isomorphic to the n'th right derived functor $R^n\Gamma$ in the category of quasicoherent sheaves of $\mathcal{O}_X$-modules.*

Proof Let $U_1, \ldots, U_n$ be a cover of X by affine open sets. Then quasicoherent sheaves of the form

$$\bigoplus_{s=1}^{n} j_{s*}(\mathcal{F}_s) \qquad (2.2.6)$$

where $j_s : U_s \to X$ are the inclusions, and $\mathcal{F}_s$ correspond to injective $\mathcal{O}_{U_s}(U_s)$-modules are clearly injective in the category of quasicoherent sheaves on X, and obviously there is a monomorphism from every quasicoherent sheaf on X to a quasicoherent sheaf of the type (2.2.6) (by restriction). On the other hand, abelian sheaves of the form (2.2.6) are flasque by Lemma 2.2.1.

The last statement follows from Proposition 2.3.21 of Chap. 5. $\qquad\square$

2.2.4 Quasicoherent Čech Cohomology

Let X be a separated Noetherian scheme, let $\mathcal{F}$ be a quasicoherent sheaf on X, and let $\mathcal{U} = (U_1, \ldots U_n)$ be an open affine cover. Recall that all intersections $U_{i_1} \cap \cdots \cap U_{i_k}$ are also affine. Then the *Čech complex $\check{C}_{\mathcal{U}}(X, \mathcal{F})$ of X with coefficients in $\mathcal{F}$ with respect to the covering $\mathcal{U}$* is defined as the cochain complex

$$\bigoplus_{1 \le i_0 \le n} \mathcal{F}(U_{i_0}) \xrightarrow{\;d\;} \cdots \xrightarrow{\;d\;} \bigoplus_{1 \le i_0 < \cdots < i_k \le n} \mathcal{F}(U_{i_0} \cap \cdots \cap U_{i_k}) \xrightarrow{\;d\;} \cdots$$

$$(2.2.7)$$

where the first term is placed in cohomological degree 0 and the differential is the direct sum of maps

$$\mathcal{F}(U_{i_0} \cap \ldots \widehat{U_{i_s}} \cdots \cap U_{i_k}) \to \mathcal{F}(U_{i_0} \cap \cdots \cap U_{i_k})$$

is $(-1)^s$ times restriction. (As usual, $\widehat{?}$ denotes a missing term.) The cohomology of $\check{C}_{\mathcal{U}}(X, \mathcal{F})$ is denoted by $\check{H}_{\mathcal{U}}^k(X, \mathcal{F})$, and called the *quasicoherent Čech cohomology of X with coefficients in $\mathcal{F}$ with respect to the covering $\mathcal{U}$.*

When $\mathcal{F}$ is a chain complex of sheaves then (2.2.7) is a double chain complex, and by $\check{C}_{\mathcal{U}}(X, \mathcal{F})$, we mean its totalization. Its cohomology $\check{H}_{\mathcal{U}}^k(X, \mathcal{F})$ is then referred to as the *quasicoherent Čech hypercohomology of X with coefficients in $\mathcal{F}$ with respect to the covering $\mathcal{U}$.*

Note that $\check{H}_{\mathcal{U}}^k(X, \mathcal{F})$ makes sense for every finite open cover of a topological space X, and any abelian sheaf $\mathcal{F}$.

2.2.5 Lemma *Let $\mathcal{U} = (U_1, \ldots, U_n)$ be a finite open cover of a topological space X, and let $\mathcal{F}$ be a flasque sheaf on X. Then*

$$\check{H}^0_{\mathcal{U}}(X, \mathcal{F}) = \mathcal{F}(X)$$

and

$$\check{H}^k_{\mathcal{U}}(X, \mathcal{F}) = 0 \text{ for } k > 0.$$

Proof Induction on n. For $n = 1$, there is nothing to prove. Suppose the statement is true with n replaced by $n - 1$. Let $V = U_1 \cup \cdots \cup U_{n-1}$, $\mathcal{V} = (U_1, \ldots, U_{n-1})$, $\mathcal{W} = (U_1 \cap U_n, \ldots, U_{n-1} \cap U_n)$. Then $\check{C}_{\mathcal{U}}(X, \mathcal{F})$ is isomorphic to the mapping co-cone of the canonical chain map (induced by restriction)

$$\check{C}_{\mathcal{V}}(V, \mathcal{F}|_V) \oplus \mathcal{F}(U_n) \to \check{C}_{\mathcal{W}}(V \cap U_n, \mathcal{F}|_{V \cap U_n}).$$

By the induction hypothesis, this is quasiisomorphic to the mapping co-cone of

$$\mathcal{F}(V) \oplus \mathcal{F}(U_n) \to \mathcal{F}(V \cap U_n),$$

which has 0'th cohomology isomorphic to $\mathcal{F}(X)$ by the gluing property of sheaves, and first cohomology isomorphic to 0 by the fact that $\mathcal{F}$ is flasque. $\square$

2.2.6 Proposition *Let X be a separated Noetherian scheme, let $\mathcal{U}$ be an open cover of X by affine open sets, and let $\mathcal{F}$ be a quasicoherent sheaf (or bounded above chain complex of quasicoherent sheaves) on X. Then we have a canonical isomorphism*

$$\check{H}^k_{\mathcal{U}}(X, \mathcal{F}) \cong H^k(X, \mathcal{F}).$$

Proof Let $\mathcal{I}$ be an injective resolution of $\mathcal{F}$ in the category of quasicoherent sheaves on X by sheaves which are also flasque as abelian sheaves. Then the Čech complex

$$\check{C}_{\mathcal{U}}(X, \mathcal{I})$$

is a double chain complex bounded above in both degrees. Denote the $\mathcal{I}$-differential by $d^{\mathcal{I}}$ and the Čech differential by $\check{d}$. Then we have two spectral sequences convergent to $\check{H}^*_{\mathcal{U}}(X, \mathcal{I})$, depending on by which degree we filter.

One spectral sequence with E_1-term

$$H^*(\check{C}_{\mathcal{U}}(X, \mathcal{I}), \check{d})$$

which is isomorphic to

$$\Gamma(\mathcal{I})$$

concentrated in $q = 0$ by Lemma 2.2.5. Thus, the spectral sequence collapses to $E_2 = H^*(X, \mathcal{F})$.

The other spectral sequence has E_1-term

$$H^*(\check{C}_{\mathcal{U}}(X, \mathcal{I}), d^{\mathcal{I}}),$$

which is canonically isomorphic to $\check{C}(X, \mathcal{F})$ by Proposition 2.2.2. □

COMMENT Čech cohomology may be defined for an arbitrary abelian sheaf on a topological space X by allowing ordered infinite covers (replacing $\bigoplus$ with $\prod$ in (2.2.7)), and taking a colimit with respect to refinement of covers. (One needs to notice that up to canonical isomorphism, the Čech complex does not really depend on the ordering of the cover.) This way, we obtain a generalization of the first Čech cohomology considered in Sect. 3.1 of Chap. 4. There is also a canonical map from Čech to sheaf cohomology. However, for $k > 1$, the map is not in general an isomorphism. (For $k = 0, 1$, it is an isomorphism by the observations of Sect. 3.1 of Chap. 4, in particular Theorem 3.1.2.) On the other hand, the method of the proof of Proposition 2.2.6 clearly gives the following generalization:

2.2.7 Proposition *Let $\mathcal{F}$ be an abelian sheaf on a topological space X, let $\mathcal{I}$ be a flasque resolution of $\mathcal{F}$, and let $\mathcal{U} = (U_1, \ldots, U_n)$ be a finite open cover of X. Then we have*

$$\check{H}^p_{\mathcal{U}}(X, \mathcal{I}) = H^p(X, \mathcal{F})$$

where on the left hand side, we mean Čech hypercohomology, i.e. the cohomology of the totalization of the double cochain complex given by the Čech and $\mathcal{I}$-differentials. In particular, we have a natural spectral sequence of the form

$$E_2^{p,q} = \check{H}^p_{\mathcal{U}}(X, H^q(\mathcal{I})) \Rightarrow H^{p+q}(X, \mathcal{F})$$

where on the left hand side, $H^q(\mathcal{I})$ denotes cohomology in the abelian category of sheaves on X.

□

2.3 Algebraic de Rham Cohomology of Projective Varieties—GAGA

We begin with a calculation of the cohomology of the projective space with coefficients in the structure sheaf.

2.3.1 Theorem *For a field k, we have*

$$H^0(\mathbb{P}^n_k, \mathcal{O}_{\mathbb{P}^n_k}) = k,$$
$$H^m(\mathbb{P}^n_k, \mathcal{O}_{\mathbb{P}^n_k}) = 0 \; for \; m > 0. \tag{2.3.1}$$

Proof We will write $\mathcal{O}$ without a subscript to simplify notation when no confusion can arise. Denote the projective coordinates of $\mathbb{P}^n_k$ by $[z_0 : \cdots : z_n]$, and consider the open affine cover $\mathcal{U} = (U_0, U_1, \ldots, U_n)$ where

$$U_i = Spec(k[\frac{z_0}{z_i}, \ldots, \frac{\widehat{z_i}}{z_i}, \ldots, \frac{z_n}{z_i}].$$

Thus, by definition, for $\emptyset \neq J \subseteq \{0, \ldots, n\}$,

$$\mathcal{O}(\bigcap_{j \in J} U_j)$$

is the free k-module on monomials of the form

$$z_1^{a_1} \ldots z_n^{a_n} \tag{2.3.2}$$

where $a_i \in \mathbb{Z}$

$$\sum a_i = 0 \tag{2.3.3}$$

and $a_i \geq 0$ for $i \notin J$.

Now the Čech complex (2.2.7) is a direct sum, over $(n+1)$-tuples

$$\underline{a} = (a_0, \ldots, a_n)$$

satisfying (2.3.3), of the subcomplexes $C_{\underline{a}}$ where $C_{\underline{a}}^m$ is the free k-vector space on subsets $J \subseteq \{0, \ldots, n\}$ with $|J| = m+1$ and such that

$$a_i < 0 \Rightarrow i \in J.$$

Let $I_{\underline{a}} = \{i \in \{1, \ldots, n\} \mid a_i < 0\}$. By (2.3.3), we have $I_{\underline{a}} \neq \{1, \ldots, n\}$. Denoting by I_k the cochain complex

$$k \xrightarrow{\ d = Id\ } k$$

where the bottom term is in cohomological degree 0, we have

$$C_{\underline{a}} = \bigotimes_{J \cap I_{\underline{a}} = \emptyset} I_k[1 - |J|]$$

(the shift is counted homologically) if $I_{\underline{a}} \neq 0$, which has 0 cohomology by the Künneth theorem (see Sect. 3.5 of Chap. 5). When $J = \emptyset$, then $\underline{a} = (0, \ldots, 0)$ by (2.3.3), so $C_{\underline{a}}$ is isomorphic to the subcomplex of

$$\bigotimes_{J \subseteq \{1, \ldots, n\}} I_k[1]$$

in non-negative cohomological (i.e. non-positive homological) degrees, which has cohomology k concentrated in degree 0 by the long exact sequence in cohomology (Theorem 2.3.7 of Chap. 5). $\square$

Using the short exact sequence (3.3.1) of Chap. 4, we can use induction on n to generalize this computation to cohomology with coefficients in $\mathcal{O}(j)$:

2.3.2 Theorem *For $j \geq 0$, we have*

$$H^m(\mathbb{P}_k^n, \mathcal{O}(j)) = \begin{cases} k^{\binom{n+j}{n}} & \text{for } m = 0 \\ 0 & \text{for } m > 0. \end{cases}$$

For $j < 0$, we have

$$H^m(\mathbb{P}_k^n, \mathcal{O}(j)) = \begin{cases} k^{\binom{-1-j}{n}} & \text{for } m = n \text{ and } j \leq -n - 1 \\ 0 & \text{else.} \end{cases}$$

$\square$

Proof Exercise 1.

2.3.3 Serre's GAGA Theorem

In his famous paper [21] referred to as GAGA (Géométrie Algébrique et Géométrie Analytique), Serre proved that for a smooth projective variety X over $\mathbb{C}$, the category of coherent sheaves on X is equivalent to the category of coherent sheaves on the complex manifold X_{an}. The comparison functor from coherent sheaves on X to coherent sheaves on X_{an} is ι^* where

$$\iota : X_{an} \to X \qquad\qquad (2.3.4)$$

is the morphism of locally ringed spaces given by the identity on X.

2.3.4 Lemma *The functor ι^* is exact.*

Proof This follows from the fact that for a closed point $x \in X$, $\mathcal{O}^{an}_{X,x}$ is a flat $\mathcal{O}_{X,x}$-module. The key step in proving this is to show that the ring $\mathcal{O}^{an}_{X,x}$ is Noetherian (for then, it is clearly regular local, and the inclusion $\mathcal{O}_{X,x} \subset \mathcal{O}^{an}_{X,x}$ induces an isomorphism on associated graded rings).

To show that $\mathcal{O}^{an}_{X,x}$ is Noetherian, without loss of generality, $X = \mathbb{C}^n$, $x = 0$. In this setting, this fact is known as the *Rückert Basis Theorem*. To prove it, we proceed by induction on n. Let $0 \neq g \in \mathcal{O}^{an}_{\mathbb{C}^n,0}$. Let the standard coordinates in $\mathbb{C}^n$ be $z_1, \ldots, z_n$. Then there exists a coordinate z_i such that

$$g = \sum_{j=0}^{\infty} g_j z_i^j$$

where g_j are holomorphic functions in the remaining coordinates in some neighborhood of 0, and for some (finite) number b, $g_j(0) = 0$ for $j < b$, while $g_b(0) \neq 0$. The *Weierstrass Division Theorem* then states that any holomorphic function on $\mathbb{C}^n$ in a neighborhood of 0 can be uniquely written as

$$a_0 + a_1 z_i + \cdots + a_{b-1} z_i^{b-1} + gh$$

where h is a holomorphic function in $z_1, \ldots z_n$ and a_j are holomorphic functions in the variables z_ℓ, $\ell \neq i$, in a neighborhood of 0. Thus, $\mathcal{O}^{an}_{\mathbb{C}^n,0}/(g)$ is isomorphic to $(\mathcal{O}^{an}_{\mathbb{C}^{n-1},0})^b$ as an $\mathcal{O}^{an}_{\mathbb{C}^{n-1},0}$-module, which is Noetherian by the induction hypothesis. The details of this argument can be found in [9]. $\qquad\square$

By (2.5.4) and (2.5.3) of Chap. 5, we also obtain, for a coherent sheaf $\mathcal{F}$, a canonical homomorphism

$$H^\ell(X, \mathcal{F}) \to H^\ell(X_{an}, \iota^*\mathcal{F}). \qquad\qquad (2.3.5)$$

We will prove the following result of Serre [21]:

2.3.5 Theorem *For a smooth projective variety X over $\mathbb{C}$ and any coherent sheaf $\mathcal{F}$ (or, more generally, any chain complex of coherent sheaves non-zero in only finitely many degrees), the canonical homomorphism (2.3.5) is an isomorphism.*

COMMENTS

1. Note that Theorem 2.3.5 does not extend to general smooth varieties over $\mathbb{C}$. For example, letting $X = \mathbb{A}_{\mathbb{C}}^1$, we have

$$H^0(X, \mathcal{O}) = \mathbb{C}[x]$$

while $H^0(X_{an}, \mathcal{O}_{an})$ is the $\mathbb{C}$-vector space of holomorphic functions on $\mathbb{C}$.

2. For a smooth projective variety X over $\mathbb{C}$, Theorem 2.3.5 can be, in particular, applied to the de Rham complex $\mathcal{F} = \underline{\Omega}_X^*$, thus giving an isomorphism between the de Rham cohomology of X and its algebraic de Rham cohomology. We will show below that this isomorphism, in fact, extends to all smooth varieties over $\mathbb{C}$.

3. For a smooth projective variety X over $\mathbb{C}$, we can also apply Theorem 2.3.5 to a single sheaf $\mathcal{F} = \underline{\Omega}_X^p$, and define *algebraic Dolbeault cohomology* as

$$H_{db}^{p,q}(X) = H^q(X, \underline{\Omega}_X^p). \tag{2.3.6}$$

Thus, we also have an algebraic version of the spectral sequence (1.1.1). Theorem 2.3.5 then shows that algebraic and analytic Dolbeault cohomology groups coincide, and thus, the algebraic Dolbeault spectral sequence also collapses to E_1.

4. The concepts of algebraic Dolbeault cohomology (2.3.6), and the algebraic version of the spectral sequence (1.1.1), make sense for a smooth projective variety X over *any field*. Mumford [17] showed, however, that for fields of positive characteristic, the algebraic analogue of the spectral sequence (1.1.1) does not in general collapse to E_1. Also, an algebraic version of a Hodge-like isomorphism between $H_{db}^{p,q}(X)$ and the dual of $H_{db}^{q,p}(X)$ fails in general for smooth projective varieties X over fields of positive characteristic of dimension > 1 (even in dimension 2). There is, however, an analog of Kodaira-Serre duality, and also a version of Serre duality which holds for coherent sheaves in great generality (see Exercises 3, 4, and [11] for an even greater generality).

Proof of Theorem 2.3.5 First, using the comments in Sect. 2.5.2 of Chap. 5, we may restrict attention to the case when $X = \mathbb{P}_{\mathbb{C}}^n$. In that case, note that the statement is true for $\mathcal{F} = \mathcal{O}$ by Theorem 2.3.1 and Sect. 1.6.3 (and the easy observation that it holds in cohomological degree 0). Now we can show by induction on n that the statement is true for $\mathcal{F} = \mathcal{O}(m)$ for any $m \in \mathbb{Z}$. Indeed, for $n = 0$, there is nothing to prove. Assuming the statement is true when replacing n by $n - 1$, using the short exact sequence (3.3.1) of Chap. 4 and its twists by $m \in \mathbb{Z}$, we see from the 5-lemma that the statement is true for

$\mathcal{O}(m)$ if and only if it is true for $\mathcal{O}(m+1)$, for any $m \in \mathbb{Z}$. But since it is true for $m = 0$, it must therefore be true for all $m \in \mathbb{Z}$.

Thus, we proved that for all $n \in \mathbb{N}_0$, and all $m \in \mathbb{Z}$, the statement is true for $\mathcal{F} = \mathcal{O}(m)$. Thus, our statement is also true for a direct sum of finitely many sheaves of the form $\mathcal{O}(m)$. But any coherent sheaf on $\mathbb{P}^n_\mathbb{C}$ has a resolution of $\mathcal{O}$-modules of the form

$$\mathcal{P}_n \to \cdots \to \mathcal{P}_0 \tag{2.3.7}$$

where each $\mathcal{P}_i$ is a direct sum of coherent sheaves of the form $\mathcal{O}(m)$. This is a consequence of Lemma 2.4.3 of Chap. 4, the fact that $\mathbb{C}[x_0, \ldots, x_n]$ is Noetherian (Hilbert basis theorem), and Hilbert's syzygy theorem (Theorem 2.4.5 of Chap. 5). Recall from Sect. 2.4 of Chap. 4 that the category of quasicoherent sheaves is a reflexive subcategory of the category of graded $\mathbb{C}[x_0, \ldots, x_n]$-modules. Theorem 2.4.5 of Chap. 5 gives a resolution in the category of graded $\mathbb{C}[x_0, \ldots, x_n]$-modules which correspond to quasicoherent sheaves, since the inclusion of a reflexive subcategory preserves kernels (and therefore the kernel of each step of the resolution again corresponds to a coherent sheaf). Thus, we may invoke the spectral sequences on the algebraic and analytic side obtained by filtering the complex (2.3.7), and its ι^*, by elements of degree $\leq p$. We obtain a morphism of spectral sequences which induces an isomorphism on E^1, and thus on E^∞. A homomorphism of filtered abelian groups where the filtrations in the source and target are convergent and which induces an isomorphism on associated graded abelian groups is an isomorphism. □

It is worth noting that since the resolution (2.3.7) works over any field, using Theorem 2.3.2, it also implies the following result:

2.3.6 Proposition *Let X be a projective variety over a field k, and let $\mathcal{F}$ be a coherent sheaf over X. Then there exists an N such that the k-module*

$$H^m(X, \mathcal{F}) \tag{2.3.8}$$

is finite-dimensional for all m, and 0 for $m > N$.

□

2.3.7 Corollary *Let X be a smooth projective variety over a field k. Then the algebraic de Rham cohomology $H^i_{DR}(X)$ is a finite-dimensional k-module.*

Proof We can use the algebraic analogue of the spectral sequence (1.1.1) and Proposition 2.3.6.

□

2.4 Algebraic and Analytic de Rham Cohomology of Smooth Varieties over $\mathbb{C}$

As we remarked, there is no chance that the canonical homomorphism (2.3.5) would be an isomorphism for general smooth varieties X. This makes the following result all the more remarkable:

2.4.1 Theorem (Grothendieck) *For every smooth variety W over $\mathbb{C}$, the homomorphism (2.3.5) induces an isomorphism of hypercohomology*

$$H^\ell_{DR}(W) = H^\ell(W, \underline{\Omega}^*) \xrightarrow{\ \cong\ } H^\ell(W_{an}, \underline{\Omega}^*_{Hol}) = H^\ell_{DR}(W_{an}; \mathbb{C}) \qquad (2.4.1)$$

(i.e. an isomorphism between algebraic and analytic de Rham cohomology).

The main ingredient in the proof is *resolution of singularities*. Let X be a smooth variety over a field k. A *closed smooth subvariety of codimension ℓ* of X is a subvariety $Y \subseteq X$ such that for every closed point $y \in Y$, there exists a Zariski open neighborhood U of x in X and a smooth morphism $g : U \to \mathbb{A}^\ell_k$ over $Spec(k)$ such that

$$Y \cap U = g^{-1}(\{0\})$$

(as before, by $\cap$ we mean pullback).

Now let X be a smooth variety over a field k and let $Y_1, \ldots, Y_m$ be closed subvarieties such that for all closed points $y \in X$, there exists a Zariski open neighborhood U of y in X and a smooth morphism $g : U \to \mathbb{A}^m_k$ over $Spec(k)$ such that $Y_i \cap U$ is the inverse image of the subvariety of $\mathbb{A}^m_k$ given by vanishing of the i'th coordinate. Note that in particular, for $1 \le i_1 < \cdots < i_k \le m$, $Y_{i_1} \cap \cdots \cap Y_{i_k}$ is then a disjoint union of smooth subvarieties of X of codimension k. Then $Y_1, \ldots, Y_m$ are called *divisors with normal crossings*. The resolution of singularities theorem has many versions. We will need the following version (see for example [4]).

2.4.2 Theorem *Suppose U is a smooth affine variety over $\mathbb{C}$. Then there exists a smooth projective variety X over $\mathbb{C}$ and divisors with normal crossings $Y_1, \ldots, Y_m$ in X such that*

$$U \cong X \smallsetminus (Y_1 \cup \cdots \cup Y_m).$$

$\square$

Next, we need to prove a local result. Let $B = \{z \in \mathbb{C} \mid |z| < 1\}$. Let

$$U = U_m = B^m, \quad V = V_{m,\ell} = (B \smallsetminus \{0\})^\ell \times B^{m-\ell},$$

$1 \leq \ell \leq m$. Denote by $i : V \rightarrow U$ the inclusion. Denoting the coordinates by $z_1, \ldots, z_m$, consider the complex of sheaves of $\mathbb{C}$-modules

$$\mathcal{F} = \mathcal{F}_{m,\ell} = \text{colim}\, (\underline{\Omega}^*_{U,Hol} \xrightarrow{z_1 \ldots z_\ell} \underline{\Omega}^*_{U,Hol} \xrightarrow{z_1 \ldots z_\ell} \ldots). \tag{2.4.2}$$

(In other words, $\mathcal{F}$ is the sheaf of holomorphic differential forms with at most pole singularities at 0 in the first ℓ coordinates.)

2.4.3 Lemma *We have*

$$H^*(\mathcal{F}(U)) = \Lambda_{\mathbb{C}}(\frac{dz_1}{z_1}, \ldots \frac{dz_\ell}{z_\ell}).$$

Proof Consider, for $0 \leq j \leq m$, the subcomplex $C_j \subseteq \mathcal{F}(U)$ spanned by all differential forms

$$\omega = f(z_1, \ldots, z_m) dz_{i_1} \wedge \cdots \wedge dz_{i_s}, \tag{2.4.3}$$

$1 \leq i_1 < \cdots < i_s \leq m$ where there exists a form

$$\eta \in \mathcal{F}_{j,\min(j,\ell)}(U_j)$$

such that

$$\omega = \eta \wedge \bigwedge_{j < i_s \leq \ell} \frac{dz_{i_s}}{z_{i_s}}.$$

Then we will construct a chain homotopy between the identity on C_j and the chain map

$$g_j : C_j \rightarrow C_{j-1}$$

given on a form (2.4.3) by

$$g_j(\omega) = \frac{res_{z_j=0}(f)}{z_j} dz_{i_1} \wedge \cdots \wedge dz_{i_s}$$

when $j = i_t$, and $g_j(\omega) = 0$ when no such t exists. (To review the concept of a residue, we refer the reader to [1].) In fact, we may pick an $0 < |a| < 1$, and take, for $i_t = j$,

$$h(\omega) = (-1)^{t-1} (\int_a^{z_j} \left(f - \frac{res_{z_j=0}(f)}{z_j} \right) dz_j) dz_{i_1} \wedge \ldots \widehat{dz_{i_t}} \cdots \wedge dz_{i_s}$$

and $h(\omega) = 0$ if no such t exists. The result then follows, since the identity on $\mathcal{F}(U)$ is homotopic to the composition $g_1 \circ \cdots \circ g_m$, thus producing a chain homotopy equivalence between $\mathcal{F}(U)$ and C_0. $\qquad\square$

We have the following immediate

2.4.4 Corollary *The canonical morphism*

$$\mathcal{F}_{m,\ell} \to Ri_*\underline{\Omega}^*_{V_{m,\ell},\mathbb{C}}$$

(obtained by adjunction) is a quasiisomorphism in the category of sheaves of $\mathbb{C}$-modules on U_m.

Proof It suffices to show that we have a quasiisomorphism on stalks. This is done by induction on ℓ. For $\ell = 0$, the complexes are isomorphic. For a given ℓ, the result follows from the induction hypothesis on all stalks except at the point 0. There, we may apply Lemma 2.4.3, shrinking the ball B by factors converging to 0, and taking colimits. $\qquad\square$

Proof of Theorem 2.4.1 First, note that it is sufficient to prove the statement for W a smooth affine variety. This is because in general, we may cover W by finitely many open affine subvarieties $W_1, \ldots, W_n$ (all the intersection of which will then also be open affine). Let $i_s : W_s \to X$ be inclusions. Then for an injective resolution $\mathcal{I}$ of a bounded above chain complex of sheaves $\mathcal{F}$ on X, we may always obtain another injective resolution of $\mathcal{F}$ in the form of a cochain complex which, in degree s, will be a sum of terms of the form

$$i_*i^*(\mathcal{I}), \ i = i_{j_1} \cap \cdots \cap i_{j_s}, \ 1 \le j_1 < \cdots < j_s \le n$$

and the differentials alternating sums analogously as in (2.2.7). This construction is functorial, and thus shows, using a spectral sequence of a double chain complex, that an isomorphism in cohomology on restrictions to all the intersections of the W_j's implies isomorphism globally.

Thus, assume that W is affine. Then, by Theorem 2.4.2, W can be embedded, via an open embedding which we denote by i, into a smooth projective variety X over $\mathbb{C}$ as a complement of a finite set of divisors $Y_1, \ldots, Y_p$ with normal crossings. Now note that for each point $x \in X_{an}$, we have a holomorphic diffeomorphism ϕ of an open neighborhood $U \ni x$ to B^m which intersects only $Y_{i_1}, \ldots, Y_{i_\ell}$, $i_1 < \cdots < i_\ell$ and ϕ maps $U \cap Y_{i_s}$ holomorphically diffeomorphically to the set of points of B^m with s'th coordinate equal to 0. Thus, the complexes of sheaves of the form (2.4.2) in open neighborhoods of points of X glue to a complex of sheaves $\mathcal{F}$ on X, and by Corollary 2.4.4, the canonical morphism

$$\mathcal{F} \to Ri_*\underline{\Omega}^*_{W_{an},Hol}$$

is a quasiisomorphism of sheaves on X_{an}. However, note that $\mathcal{F}$ is a direct limit of a sequence of sheaves of the form $\iota^*\mathcal{G}_m$ where ι is as in (2.3.4), and $\mathcal{G}_m$ is the complex of coherent sheaves on X in the Zariski topology of algebraic differential forms with poles of degree $\leq m$ along the divisors Y_s. The colimit of the coherent sheaves $\mathcal{G}_m$ is, in fact, the pushforward $i_*\underline{\Omega}^*_W$ of the algebraic de Rham complex on W. Thus, by Theorem 2.3.5, we obtain an isomorphism

$$
\begin{aligned}
H^\ell(W, \underline{\Omega}^*_W) &= H^\ell(\underline{\Omega}^*_W(W)) = \\
H^\ell((i_*\underline{\Omega}^*_W)(X)) &= H^\ell(\mathcal{F}(X_{an})) = \\
H^\ell(X_{an}, Ri_*\underline{\Omega}^*_{W_{an},Hol}) &= H^\ell(W_{an}, \underline{\Omega}^*_{W_{an},Hol}).
\end{aligned}
$$

We also see that the identifying homomorphism is, in fact, induced by ι. $\qquad\square$

2.5 Examples

2.5.1 The Affine and Projective Lines

The cohomology $H^*_{DR}(\mathbb{A}^1_k)$ is the cohomology of the complex

$$
k[x] \xrightarrow{\ d\ } k[x]\{dx\}. \tag{2.5.1}
$$

Since d is a differentiation, we have

$$
d(x^n) = nx^{n-1}dx.
$$

Thus, if k is a field of characteristic 0, then $H^i_{DR}(\mathbb{A}^1_k)$ is equal to k for $i = 0$, and 0 for $i > 0$, as we expect. However, if k is a field of characteristic $p > 0$, then we get

$$
H^0_{DR}(\mathbb{A}^1_k) = k[x^p],
$$

$$
H^1_{DR}(\mathbb{A}^1_k) = k[x^p]\{x^{p-1}dx\}.
$$

In particular, these are *not finite-dimensional k-vector spaces*.
    Similarly, for $\mathbb{G}_m$, we have the complex

$$
k[x, x^{-1}] \xrightarrow{\ d\ } k[x, x^{-1}]\{dx\}, \tag{2.5.2}
$$

so if k is of characteristic 0,

$$H^0_{DR}(\mathbb{G}_{m,k}) = k,$$

$$H^1_{DR}(\mathbb{G}_{m,k}) = k\{\frac{dx}{x}\},$$

and if $char(k) = p > 0$, then

$$H^0_{DR}(\mathbb{G}_{m,k}) = k[x^p, x^{-p}],$$

$$H^1_{DR}(\mathbb{G}_{m,k}) = k[x^p, x^{-p}]\{\frac{dx}{x}\}.$$

Now as an exercise, let us calculate the algebraic de Rham cohomology of $\mathbb{P}^1_k$. We may cover $\mathbb{P}^1_k$ with two affine open sets $Spec(k[x])$ and $Spec(k[t])$ where $t = x^{-1}$. Their intersection is $\mathbb{G}_{m,k}$, so we obtain a long exact sequence

$$\cdots \to H^i_{DR}(\mathbb{P}^1_k) \to H^i_{DR}(Spec(k[x])) \oplus H^i_{DR}(Spec(k[t])) \to$$
$$\to H^i_{DR}(\mathbb{G}_{m,k}) \to H^{i+1}_{DR}(\mathbb{P}^1_k) \to \cdots$$

Of course $Spec(k[t])$ is another copy of $\mathbb{A}^1_k$, and the map into the cohomology of $\mathbb{G}_{m,k}$ is calculated by

$$t \mapsto x^{-1},$$

$$dt \mapsto -x^{-2}dx.$$

We see then that regardless of characteristic, we obtain

$$H^i_{DR}(\mathbb{P}^1_k) = \begin{cases} k \text{ for } i = 0, 2, \\ 0 \text{ otherwise,} \end{cases}$$

as expected. Recall that we know from Proposition 2.3.6 that the algebraic de Rham cohomology of the projective line is finite-dimensional, regardless of characteristic.

2.5.2 Algebraic Dolbeault and de Rham Cohomology of the Projective Space

In fact, we can prove that for a field k of any characteristic,

$$H^i(\mathbb{P}^n_k, \Omega^\ell_{\mathbb{P}^n_k/k}) = \begin{cases} k \text{ for } 0 \le i = \ell \le n \\ 0 \text{ else.} \end{cases} \tag{2.5.3}$$

Thus, the E_1-term of the algebraic version of the spectral sequence (1.1.1) is concentrated in even total degrees, and therefore must collapse, concluding that

$$H^i_{DR}(\mathbb{P}^n_k) = \begin{cases} k \text{ for } 0 \le k \le 2n \text{ even} \\ 0 \text{ else.} \end{cases} \tag{2.5.4}$$

To prove (2.5.3), note that it is another generalization of Theorem 2.3.1, which settles the case of $\ell = 0$. Letting $\mathbb{P}^n_k = Proj(k[x_0, \dots, x_n])$, in the general case, we use the same method, covering again $\mathbb{P}^n_k$ by the open affine sets

$$U_i = Spec[\frac{x_0}{x_i}, \dots, \frac{\widehat{x_i}}{x_i}, \dots, \frac{x_n}{x_i}].$$

In the higher degrees, it is convenient to write for $i \ne j$

$$d \ln(\frac{x_i}{x_j}) = \frac{d(x_i/x_j)}{x_i/x_j} = \frac{dx_i}{x_i} - \frac{dx_j}{x_j}, \tag{2.5.5}$$

and write every element of $\Omega^\ell(U_{i_0} \cap \dots \cap U_{i_m})$ as a product of elements of the form (2.5.5), and a monomial

$$p = x_1^{a_1} \dots x_n^{a_n} \tag{2.5.6}$$

with $\sum a_i = 0$. The Čech complex for calculating $H^*(\mathbb{P}^n_k, \underline{\Omega}^\ell)$ with respect to the open affine cover (U_i) then splits as a sum of chain subcomplexes $Q_{(p)}$ over the monomials (2.5.6). The non-trivial contribution is again for $p = 1$. In this case, the chain complex in question is identified with the *Eilenberg-Mac Lane complex*, which can be described as follows:

Consider the filtration on the standard simplex Δ^n by the subsets of elements $(x_0, \dots, x_n)$ with $n - i$ coordinates equal to 0. Then this gives Δ^n a structure of a CW-complex, and CW chains with respect to this structure are called *simplicial chains*, and j-cells of this filtration (which are copies of Δ^j inserted by setting a chosen set of $n - j$ coordinates to 0) are called *j-faces*. Now the Eilenberg-Mac Lane complexes C^ℓ, Z^ℓ, B^ℓ are the chain complexes obtained by assigning to each i-face σ of Δ^n, the set of all simplicial ℓ-cochains, resp. ℓ-cocycles resp. ℓ-coboundaries on σ with some fixed coefficients (in our case, k). The differentials are given by alternating sums of restrictions to faces. Then one notices that

$$H^* C^\ell = 0 \text{ for } 0 \le \ell < n.$$

Additionally, one has

$$\mathcal{B}^\ell = \mathcal{Z}^\ell \text{ for } \ell > 0$$

(since $H^\ell(\Delta^n) = 0$), and thus, we have a short exact sequence of chain complexes

$$0 \to \mathcal{Z}^\ell \to \mathcal{C}^\ell \to \mathcal{Z}^{\ell-1} \to 0.$$

By induction, this then implies, for $\ell < n$,

$$H^i(\mathcal{Z}^\ell) = \begin{cases} k \text{ for } i = \ell \\ 0 \text{ else.} \end{cases}$$

Now more or less by definition, the summand $Q_{(1)}$ is isomorphic to $\mathcal{Z}^{\ell-1}[1]$, and thus has cohomology k in degree ℓ, and 0 elsewhere. For other p as in (2.5.6), the set $I(p) = \{j \mid i_j > 0\}$ is non-empty. One then sees that $Q_{(p)}$ is isomorphic to the subcomplex of $\mathcal{Z}^{\ell-1}[1]$ on faces which contain the I-face, which has cohomology 0 by the same inductive argument.

2.5.3 The Algebraic de Rham Cohomology of an Elliptic Curve

Consider, for simplicity, a field k of characteristic 0, and an elliptic curve E which is the closure in $\mathbb{P}^2_k$ of the closed subset of $\mathbb{A}^2_k = Spec(k[x, y])$ with equation defined over k

$$x^2 = (y - a)(y - b)(y - c), \tag{2.5.7}$$

with $a \in k, b, c \in \bar{k}$ (and a, b, c are different). Let

$$R = k[x, y]/(x^2 - (y - a)(y - b)(y - c)).$$

Let

$$u = (y - a)(y - b)(y - c)' = (y - a)(y - b) + (y - a)(y - c) + (y - b)(y - c).$$

Thus, we have

$$\Omega_{R/k} = R\{dx, dy\}/\langle 2x\,dx - u\,dy \rangle.$$

Now we have

$$1 = s2x + tu \in R$$

for some $s, t \in R$. Let

$$dz = tdx + sdy.$$

Then

$$udz = tudx + sudy = (s2x + tu)dx - s2xdx + sudy = dx,$$

and similarly

$$2xdz = dy.$$

In other words, we also have

$$\Omega_{R/k} = R\{dz\}. \tag{2.5.8}$$

Thinking of R as a free $k[y]$-module on $1, x$, we compute, for $n \in \mathbb{N}_0$,

$$dy^n = ny^{n-1}dy = 2ny^{n-1}xdz,$$

$$d(xy^n) = y^n dx + 2nxy^{n-1}dz = (y^n u + 2ny^{n-1}(y-a)(y-b)(y-c))dz.$$

We compute

$$H^0_{DR}(Spec(R)) = k\{1\},$$
$$H^1_{DR}(Spec(R)) = k\{dz, ydz\}. \tag{2.5.9}$$

Now consider the ring $S = (y-a)^{-1}R$. Then

$$\Omega_{S/k} = S\{dz\}. \tag{2.5.10}$$

Now using partial fractions, every element of S can be written uniquely as a k-linear combination of elements of R and elements of the form $1/(y-a)^n, x/(y-a)^n, n \in \mathbb{N}$. We compute

$$d\left(\frac{1}{(y-a)^n}\right) = -n\frac{x}{(y-a)^{n+1}}dz,$$

$$d\left(\frac{x}{(y-a)^n}\right) = \left(\frac{u}{(y-a)^n} - 2n\frac{(y-b)(y-c)}{(y-a)^n}\right)dz.$$

Thus, we conclude that

$$H^0_{DR}(Spec(S)) = k\{1\},$$

$$H^1_{DR}(Spec(S)) = k\{dz, ydz, \frac{x}{y-a}dz\}. \tag{2.5.11}$$

Now adding the point $x = 0$, $y = a$ defines a morphism of schemes

$$E \rightarrow E$$

which sends $U = Spec(R)$ to another isomorphic open set V, with $U \cap V = Spec(S)$. The 1-form dz is preserved, and y is transformed to

$$\frac{(c-a)(b-a)}{y-a} + a.$$

Note that in $H^1_{DR}(Spec(S))$, $d(x/(y-a))$ gives a k-linear relation between

$$dz, ydz, \frac{1}{y-a}dz.$$

Thus, the canonical homomorphism

$$H^1_{DR}(U) \oplus H^1_{DR}(V) \rightarrow H^1_{DR}(U \cap V)$$

can be identified with a k-linear map

$$k\{dz, ydz\} \oplus k\{dz, \frac{1}{y-a}dz\} \rightarrow k\{dz, ydz, \frac{x}{y-a}dz\}$$

whose kernel and image have dimension 2. We conclude that

$$H^i_{DR}(E) \cong \begin{cases} k & \text{for } i = 0, 2 \\ k^2 & \text{for } i = 1 \\ 0 & \text{else,} \end{cases}$$

"as expected."

2.5.4 Non-canonicity over $\mathbb{R}$

Now consider, in the previous example, $k = \mathbb{R}$. Clearly, the computation was functorial in the field, so for an elliptic curve E over $\mathbb{R}$ as above, the algebraic $\mathbb{R}$-valued de Rham cohomology is invariant under complex conjugation in the algebraic de Rham cohomology of $E_{\mathbb{C}} = E \times_{Spec(\mathbb{R})} Spec(\mathbb{C})$ which, by Serre's GAGA theorem (Theorem 2.3.5) and the de Rham theorem, is canonically isomorphic to the singular cohomology of the the complex manifold $E_{\mathbb{C}}^{an}$ of complex points on E with the analytic topology with coefficients in $\mathbb{C}$. But now we can also consider the singular cohomology of $E_{\mathbb{C}}^{an}$ with coefficients in $\mathbb{R}$. Thus, we have two 2-dimensional $\mathbb{R}$-vector subspaces

$$H_{DR}^1(E) \text{ and } H_{DR}^1(E_{\mathbb{C}}^{an}; \mathbb{R}) = H^1(E_{\mathbb{C}}^{an}; \mathbb{R})$$

of

$$H_{DR}^1(E_{\mathbb{C}}) = H_{DR}^1(E_{\mathbb{C}}^{an}; \mathbb{C}) = H^1(E_{\mathbb{C}}^{an}; \mathbb{C}).$$

These subspaces, however, are *different*! For example, the class represented by the holomorphic (or algebraic) form dz on E, as we saw, is in $H_{DR}^1(E)$, but not in the subspace $H_{DR}^1(E_{\mathbb{C}}^{an}; \mathbb{R})$. This is because, as we saw in Sect. 1.6.4, the set of all $\langle dz, \alpha \rangle$ with $\alpha \in H_1(E; \mathbb{Z})$ is a lattice in $\mathbb{C}$, and thus cannot be contained in $\mathbb{R}$. This is the reason one sometimes says that "the (algebraic) de Rham theorem is false over $\mathbb{R}$."

3 Crystalline Cohomology

One major limitation of algebraic de Rham cohomology is that the coefficients must be a module over the field of definition. If the variety is defined over a field of characteristic $p > 0$, the cohomology will always be p-torsion. Crystalline cohomology adapts the construction in a way which gives non-torsion information.

3.1 Witt Vectors

The basic technique for producing non-torsion output from a p-torsion input is the theory of *Witt vectors*.

Let A be any commutative ring. Choose a prime number $p \in \mathbb{N}$. The set of *Witt vectors* on A is

$$W(A) = A^{\mathbb{N}_0} = \{(a_0, a_1, \dots) \mid a_i \in A\}.$$

We will define a structure of a commutative ring on $W(A)$ which is different from the product ring structure on $A^{\mathbb{N}_0}$. Consider a map

$$w : W(A) \to A^{\mathbb{N}_0},$$

$$(a_0, a_1, \dots) \mapsto (w_0, w_1, \dots)$$

where

$$w_0 = a_0,$$

$$w_1 = a_0^p + pa_1,$$

$$w_2 = a_0^{p^2} + pa_1^p + p^2 a_2,$$

$$w_n = a_0^{p^n} + pa_1^{p^{n-1}} + \cdots + p^n a_n.$$

The elements w_i are called the *ghost components* of the element $(a_0, a_1, \ldots)$.

3.1.1 Proposition *There exists a unique ring structure on $W(A)$ natural with respect to homomorphism of rings such that w is a natural homomorphism of rings with the product ring structure on the target $A^{\mathbb{N}_0}$ (or, equivalently, that each of the w_i's is a ring map into A). In particular, for $a = (a_1, a_1, \ldots)$, $b = (b_0, b_1, \ldots)$, the first coordinates of $a + b$ and $a \cdot b$ are*

$$a + b = (a_0 + b_0, a_1 + b_1 + (a_0^p + b_0^p - (a_0 + b_0)^p)/p, \ldots),$$

$$a \cdot b = (a_0 b_0, a_0^p b_1 + a_1 b_0^p + pa_1 b_1, \ldots).$$

The multiplicative unit in $W(A)$ is $(1, 0, \ldots)$.

The proof rests on the following

3.1.2 Lemma *Let $f \in \mathbb{Z}[a_1, \ldots, a_n]$. Then*

$$f(a_1, \ldots, a_n)^{p^{k+1}} \equiv f(a_1^p, \ldots, a_n^p)^{p^k} \mod (p^{k+1}).$$

Proof It clearly suffices to prove the statement of the Lemma for

$$f(a_1, \ldots, a_n) = a_1 + \cdots + a_n. \tag{3.1.1}$$

This can be done by induction on n. This is proved by induction on n. For $n = 1$, there is nothing to prove. Assuming the statement is true with n replaced by $n - 1$, compute, by the induction hypothesis, in $\mathbb{Z}/(p^{k+1})$:

$$(a_1 + \cdots + a_n)^{p^{k+1}} =$$
$$(a_1^p + \cdots + a_{n-2}^p + (a_{n-1} + a_n)^p)^{p^k} =$$
$$(a_1^p + \cdots + a_n^p + pc)^{p^k} =$$
$$(a_1^p + \cdots + a_n^p)^{p^k} + \sum_{i=1}^{p^k} \binom{p^k}{i}(a_1^p + \cdots + a_n^p)^{p^k - i} p^i c^i.$$

Now for each i in the sum on the right hand side, let j be the maximal number such that $p^j \mid i$. Then $\binom{p^k}{i}$ is divisible by p^{k-j}. On the other hand, that summand has a factor p^i where $i \geq p^j \geq j+1$, proving that each summand is divisible by p^{k+1}. $\qquad\square$

Proof of Proposition 3.1.1 First consider the case

$$A_0 = \mathbb{Z}[a_0, a_1, \ldots, b_0, b_1, \ldots].$$

Then letting $c = (c_0, c_1, \ldots)$ be $a + b$ or $a \cdot b$, clearly, the formulas for $c_0, \ldots, c_n$ are recursively determined by $w_i(c)$ being $w_i(a) + w_i(b)$ resp. $w_i(a) \cdot w_i(b)$. This proves uniqueness. For existence of the operations in this ring $W(A_0)$, we must first prove that these recursive formulas just mentioned are integral (since upon tensoring with $\mathbb{Q}$, w clearly becomes bijective, and hence the definition is consistent).

To prove integrality, in the case of addition, we must prove that

$$(a_0^{p^n} + pa_1^{p^{n-1}} + \ldots p^{n-1}a_{n-1}^p) + (b_0^{p^n} + pb_1^{p^{n-1}} + \ldots p^{n-1}b_{n-1}^p)$$
$$-(c_0^{p^n} + pc_1^{p^{n-1}} + \ldots p^{n-1}c_{n-1}^p)$$

is divisible by p^n. To this end, let

$$(\bar{c}_0, \bar{c}_1, \ldots) = (\bar{a}_0, \bar{a}_1, \ldots) + (\bar{b}_0, \bar{b}_1, \ldots)$$

where $\bar{a}_i = a_i^p, \bar{b}_i = b_i^p$. Then, by Lemma 3.1.2, modulo p^n,

$$(a_0^{p^n} + pa_1^{p^{n-1}} + \ldots p^{n-1}a_{n-1}^p) + (b_0^{p^n} + pb_1^{p^{n-1}} + \ldots p^{n-1}b_{n-1}^p)$$
$$= (\bar{a}_0^{p^{n-1}} + p\bar{a}_1^{p^{n-2}} + \ldots p^{n-2}\bar{a}_{n-2}) + (\bar{b}_0^{p^{n-1}} + p\bar{b}_1^{p^{n-2}} + \ldots p^{n-1}\bar{b}_{n-1})$$
$$= \bar{c}_0^{p^{n-1}} + p\bar{c}_1^{p^{n-2}} + \ldots p^{n-2}\bar{c}_{n-2} = c_0^{p^n} + pc_1^{p^{n-1}} + \ldots p^{n-1}c_{n-1}^p,$$

proving what we need. The multiplicative case is completely analogous.

Now in an arbitrary ring A, to add (or multiply) $\alpha = (\alpha_0, \alpha_1, \ldots)$, $\beta = (\beta_0, \beta_1, \ldots)$, read off the answer $\gamma = (\gamma_0, \gamma_1, \ldots)$ by substituting α_i, β_i for a_i, b_i in the polynomials c_i. To prove consistency, it suffices to show that in the case of addition, γ_i is in the ideal generated by $\alpha_0, \alpha_1, \ldots, \beta_0, \beta_1, \ldots$ and in the case of multiplication, γ_i is in the ideal generated by $\alpha_0, \alpha_1, \ldots$. Both statements are obvious from the definitions in the case $\alpha_i = a_i$, $\beta_i = b_i$, $\gamma_i = c_i$, and thus follow. $\qquad\square$

3.1.3 Proposition *There exists a unique natural transformation*

$$F : W(A) \to W(A)$$

in the category of commutative rings and homomorphisms (called the Frobenius*) such that*

$$F(a_0, a_1, \ldots) = (f_0, f_1, \ldots)$$

where

$$a_0^{p^{n+1}} + pa_1^{p^n} + \cdots + p^{n+1} a_{n+1} = f_0^{p^n} + pf_1^{p^{n-1}} + \cdots + p^n f_n. \qquad (3.1.2)$$

(In other words, the Frobenius "shifts the ghost components down by 1.")

Proof The proof is completely analogous to the proof of Proposition 3.1.1. First, one considers the ring

$$A = \mathbb{Z}[a_0, a_1, \ldots].$$

There, the given formula completely determines f_n inductively (and consistently, by tensoring with $\mathbb{Q}$), up to integrality. For integrality, one puts $\bar{a}_i = a_i^p$,

$$F(\bar{a}_0, \bar{a}_1, \ldots) = (\bar{f}_0, \bar{f}_1, \ldots)$$

and invokes Lemma 3.1.2 to conclude inductively that modulo p^n,

$$a_0^{p^{n+1}} + \cdots + p^n a_n^p - f_0^{p^n} - \cdots - p^{n-1} f_{n-1} =$$
$$\bar{a}_0^{p^n} + \cdots + p^n \bar{a}_n - \bar{f}_0^{p^{n-1}} - \cdots - p^{n-1} \bar{f}_{n-2} = 0,$$

and thus, $\bar{f}_n$ is integral.

The formulas on an arbitrary commutative ring are then again induced by sending a_i to chosen elements, and consistency translates to the fact that each f_i is in the ideal generated by $a_0, a_1, \ldots$, which, given integrality, is obvious from the definition. $\qquad\square$

One defines a map $A \to W(A)$ by putting, for $a \in A$,

$$\underline{a} = (a, 0, 0, \ldots). \qquad (3.1.3)$$

It is immediate from the definition that this map is multiplicative (although not additive). Now the *Verschiebung* is the map

$$V : W(A) \to W(A)$$

which sends

$$(a_0, a_1, \ldots) \mapsto (0, a_0, a_1, \ldots).$$

3.1.4 Lemma *In $W(A)$, one has*

$$F \circ V = p,$$
(3.1.4)

$$V(F(x) \cdot y) = x \cdot V(y).$$
(3.1.5)

Additionally, V preserves addition,

$$p \mid (f_n x - a_n^p),$$
(3.1.6)

and one has $f_i = a_i^p$ for all i if and only if A is an $\mathbb{F}_p$-algebra. This happens if and only if $V \circ F = p \in W(A)$.

If A is an $\mathbb{F}_p$-algebra, we additionally have

$$(V^m x)(V^n y) = V^{m+n}(F^n x, F^m y)$$
(3.1.7)

for $x, y \in W(A)$.

Proof The formula (3.1.4) only needs to be checked in $\mathbb{Z}[a_0, a_1, \ldots]$, where it follows inductively immediately from (3.1.2) (which says that the n'th ghost component of $F \circ V(a)$ is p times the n'the ghost component of a). For (3.1.5), it suffices to check in

$$A = \mathbb{Z}[a_0, a_1, \ldots, b_0, b_1, \ldots]$$

with $x = (a_0, a_1, \ldots)$, $y = (y_0, y_1, \ldots)$. In this case, we see that the n'th ghost component of both sides is p times the n'th ghost component of x times the $(n-1)$'st ghost component of y. Now since for this ring

$$A = \mathbb{Z}[a_0, a_1, \ldots, b_0, b_1, \ldots],$$

$W(A)$ is non-torsion (being a subgroup of $W(A \otimes \mathbb{Q}) = (a \otimes \mathbb{Q})^{\aleph_0}$, it follows from (3.1.4) that V is additive in A, and hence in any commutative ring. Using Lemma 3.1.2 again, one proves analogously that in $\mathbb{Z}[a_0, a_1, \ldots]$ (and hence in any commutative ring), $f_n - a_n^p$

is divisible by p, thus proving (3.1.6). Thus, $f_n = a_n^p$ if A is an $\mathbb{F}_p$-algebra. On the other hand, by definition, $f_0 = a_0^p + pa_1$, so the converse also holds.

Regarding $V \circ F$, we see from (3.1.5) that $V \circ F(x) = x \cdot V(1)$, which is equal to p if and only if

$$V(1) = p \in W(A). \tag{3.1.8}$$

However, in the ring $W(\mathbb{Z})$, writing $p = (c_0, c_1, \ldots)$, we have $p = c_0$, and hence $p = (p, \ldots) \in W(A)$ for any commutative ring A, while $V(1) = (0, 1, 0, \ldots)$. Thus, for (3.1.8) to hold, we must have $p = 0 \in A$. On the other hand, by what we already proved, in an $\mathbb{F}_p$-algebra,

$$p = F \circ V(1) = F(0, 1, 0, \ldots) = (0, 1, 0, \ldots) = V(1).$$

When A is an $\mathbb{F}_p$-algebra, then V and F commute, so we may compute from (3.1.5):

$$(V^m x)(V^n y) = V(F(V^m x) \cdot V^{n-1} y) = $$
$$V((V^m Fx) \cdot V^{n-1} y),$$

and we see that formula (3.1.7) follows by induction. □

3.1.5 Truncated Witt Vectors

By (3.1.5), $V^n W(A)$ is an ideal in $W(A)$, and one puts

$$W_n(A) = W(A)/V^n W(A).$$

The ring $W_n(A)$ is called the ring of *truncated Witt vectors* of A. By inclusion of ideals, we have a canonical surjective homomorphism of rings

$$R : W_n(A) \to W_{n-1}(A),$$

called the *restriction*. By (3.1.4), F induces a homomorphism of rings

$$F : W_n(A) \to W_{n-1}(A). \tag{3.1.9}$$

When A is an $\mathbb{F}_p$-algebra, by Lemma 3.1.4, (3.1.9) canonically factors through the homomorphism

$$(a_0, \ldots, a_{n-1}) \mapsto (a_0^p, \ldots, a_{n-1}^p)$$

which, by a somewhat unfortunate convention, is also denoted by

$$F : W_n(A) \to W_n(A).\tag{3.1.10}$$

Elements of $W_n(A)$ can be, by definition, represented by sequences of length n:

$$(a_0, \ldots, a_{n-1}), \quad a_i \in A,$$

and the definition of addition also implies that we have the (perhaps surprisingly simple) formula in $W_n(A)$:

$$\begin{aligned}(a_0, \ldots, a_{n-1}) &= \\ (a_0, 0\ldots, 0) + (0, a_1, 0, \ldots, 0) &+ \cdots + (0, \ldots, 0, a_{n-1}) = \\ \underline{a}_0 + V\underline{a}_1 + \cdots &+ V^{n-1}\underline{a}_{n-1}.\end{aligned}\tag{3.1.11}$$

3.1.6 The Witt Vectors of $\mathbb{F}_p$

For $A = \mathbb{F}_p$, by Lemma 3.1.4, the Frobenius on $W(\mathbb{F}_p)$ is the identity by Lemma 3.1.4, and thus, in $W_n(\mathbb{F}_p)$, $V^i(1) = p^i$, and thus we see that

$$W_n(\mathbb{F}_p) \cong \mathbb{Z}/(p^n).$$

By taking limits, we then have

$$W(\mathbb{F}_p) = \mathbb{Z}_p.$$

This illustrates how Witt vectors are a device for lifting from p-torsion to non-torsion information.

3.2 The de Rham-Witt Complex and Crystalline Cohomology

In this subsection, we shall assume that A is a commutative $\mathbb{F}_p$-algebra. Then consider the de Rham complex

$$\Omega^* W_n(A) = \Omega^*_{W_n(A)/W_n(\mathbb{F}_p)} = \Omega^*_{W_n(A)/\mathbb{Z}}\tag{3.2.1}$$

(see Sect. 2.1.5). One defines a differential graded ideal N_n (i.e. an ideal stable under d) in $\Omega^* W_n(A)$ inductively as follows:

$$N_1 = 0,$$

and for $n \in \mathbb{N}$, N_{n+1} is generated, as a differential graded ideal, by elements of the form

$$\sum_j V(a_j) d(V x_{1,j}) \ldots d(V x_{m,j}) \tag{3.2.2}$$

where

$$\sum_j a_j dx_{1,j} \ldots dx_{m,j} \in N_n,$$

and elements of the form

$$V(y) d\underline{x} - V(\underline{x}^{p-1} y) d(V \underline{x}) \tag{3.2.3}$$

where $x \in A$, $y \in W_n(A)$. We put

$$W_n \Omega^*(A) = \Omega^* W_n(A) / N_n.$$

Since R commutes with V on Witt vectors, we have

$$R(N_{n+1}) \subseteq N_n,$$

and thus we obtain a surjective homomorphism of commutative differential graded rings

$$R : W_{n+1} \Omega^*(A) \to W_n \Omega^*(A),$$

called the *restriction*. One puts

$$W \Omega^*(A) = \lim_n W_n \Omega^*(A)$$

and calls this the *de Rham-Witt complex* of the $\mathbb{F}_p$-algebra A. On a smooth variety X over $\mathbb{F}_p$, this construction sheafifies to produce a complex of sheaves of $\mathbb{Z}_p$-modules $\underline{W\Omega_X^*}$. Its hypercohomology is denoted by

$$H_{crys}^\ell(X) = H^\ell(X, \underline{W\Omega_X^*}),$$

and called the *crystalline cohomology* of X. More generally, for a $\mathbb{Z}_p$-module M, we can define

$$H_{crys}^\ell(X, M) = H^\ell(X, \underline{W\Omega_X^*} \otimes_{\mathbb{Z}_p} M).$$

3.2.1 Verschiebung and Frobenius

One defines homomorphisms of rings, called again the *Frobenius*

$$F : W_n\Omega^*(A) \to W_{n-1}\Omega^*(A),$$

$$F : W\Omega^*(A) \to W\Omega^*(A),$$

and homomorphisms of abelian groups

$$V : W_n\Omega^*(A) \to W_{n+1}\Omega^*(A),$$

$$V : W\Omega^*(A) \to W\Omega^*(A)$$

called again the *Verschiebung*. Clearly, it suffices to make the definitions in the truncated case, as long as they are compatible with restriction. In the case of Verschiebung, one puts simply

$$V(adx_1 \ldots dx_m) = V(a)d(Vx_1) \ldots d(Vx_m). \tag{3.2.4}$$

The relation (3.2.2) is designed precisely to ensure that this homomorphism of abelian groups is well defined. Note that, since A is an $\mathbb{F}_p$-algebra, we have in particular for $x \in W_n(A)$

$$V(dx) = V(1)dV(x) = pdV(x).$$

In fact, (3.2.4) implies

$$V(xdy) = V(x)dV(y) \tag{3.2.5}$$

for $x, y \in W_n\Omega^*(A)$ (and hence also in $W\Omega^*(A)$). In fact, when $x = adx_1 \ldots dx_m$, $y = bdy_1 \ldots dy_\ell$, both sides are equal to

$$V(a)dV(x_1) \ldots dV(x_m)dV(b)dV(y_1) \ldots dV(y_\ell).$$

To define the Frobenius, the key point is that it should coincide with the Frobenius on Witt vectors, and further, (recalling (3.1.3)), one requires

$$F d\underline{x} = \underline{x}^{p-1}d\underline{x}, \tag{3.2.6}$$

$$F dV = d. \tag{3.2.7}$$

Note that by (3.1.11), the formulas (3.2.6), (3.2.7) determine the definition of F on dx where x is a (truncated or not) Witt vector, and thus, in general by the property that it should be a ring homomorphism. In particular,

$$F d(a_0, \dots, a_{n-1}) = \underline{a_0}^{p-1} d\underline{a_0} + d\underline{a_1} + dV\underline{a_2} + \dots + dV^{n-2}\underline{a_{n-1}}.$$

To prove consistency, it suffices to verify that the Frobenius thus defined preserves the relations (3.2.2), (3.2.3). In case of (3.2.3), one computes for a Witt vector y and $x \in A$:

$$F(V(y)d\underline{x}) = F \circ V(y)F d\underline{x} =$$

$$py\underline{x}^{p-1}dx = F \circ V(\underline{x}^{p-1}y)F d(V\underline{x}).$$

In the case of the relation (3.2.2), one notes that for Witt vectors $a, x_0, \dots, x_m$, by definition,

$$F(V(a))F dV(x_1) \dots F dV(x_m) = padx_1 \dots dx_m.$$

Clearly, the definition of the Frobenius also commutes with restriction, and thus is consistent on the truncated, as well as untruncated, de Rham-Witt complex. Note also that by plugging in $adx_1 \dots dx_m$, one sees immediately that (3.2.7) holds in general on the de Rham-Witt complex. In a similar way, one also sees that (3.1.4) holds in general on the de Rham-Witt complex.

3.2.2 Proposition *The relation (3.1.5) holds for arbitrary elements x, y of the de Rham-Witt complex. Additionally, on the de Rham-Witt complex, one has*

$$dF = pFd. \tag{3.2.8}$$

Proof Note that to prove relation (3.2.8), by the Leibniz rule, we only need to prove it on elements of the form x, dx where x is a Witt vector. For the case of x, this follows from (3.2.6) (and, at least modulo torsion, essentially forces it). For the case of dx, $x = (x_0, \dots, x_{n-1})$, we compute

$$dF dx = d(\underline{x_0}^{p-1} + d\underline{x_1} + \dots + dV^{n-2}\underline{x_{n-1}}) = (p-1)\underline{x_0}^{p-2}d\underline{x_0}d\underline{x_0} = 0.$$

For relation (3.1.5), first note that multiplying y by a dy_i results in multiplying both sides by dVy_i, and thus, it suffices to assume that y is a Witt vector. Similarly, the formula is "multiplicative in the x-coordinate" in the sense that if it holds for x_1, x_2, then it holds for x_1x_2 by the calculation

$$x_1x_2Vy = x_1V((Fx_2)y) = V(F(x_1)F(x_2)y).$$

Thus, it suffices to consider the case when x is either of the form a or da where a is a Witt vector. In the case of a, it holds by Lemma 3.1.4. In the case when $a = \underline{x}$, we have, by (3.2.3) and (3.2.6):

$$V(y)d\underline{x} = V(\underline{x}^{p-1}y)d(V\underline{x}) = V(y\underline{x}^{p-1}d\underline{x}) = V(yF(d\underline{x}))$$

(Note that, again, relation (3.2.3) is forced.) Now when $a = dV^\ell\underline{x}$, $\ell \geq 1$, compute

$$V(y)dV^\ell\underline{x} = V(ydV^{\ell-1}\underline{x}) = V(yFdV^\ell\underline{x})$$

by (3.2.7).

$\square$

Thus, by Proposition 3.2.2,

$$\Phi = p^i F : W\Omega^i(A) \to W\Omega^i(A) \tag{3.2.9}$$

defines a chain map on $W\Omega^*(A)$. The induced map on crystalline cohomology is known as the *Frobenius action on crystalline cohomology*. We refer the reader to [12] for more details.

3.3 Crystalline Cohomology of $\mathbb{P}^n$

The purpose of this subsection is to work out, as an example, the de Rham-Witt complex of schemes of the form $(\mathbb{A}^i \times (\mathbb{G}_m)^j)_{Spec(k)}$, $i + j = n$ and, as an application, the crystalline cohomology of $\mathbb{P}^n_k$, where k is a perfect field of characteristic $p > 0$. Let

$$A = (T_1 \ldots T_j)^{-1}\mathbb{F}_p[T_1, \ldots, T_n], \tag{3.3.1}$$

$$C = \operatorname*{colim}_r (T_1 \ldots T_j)^{-1}\mathbb{Q}_p[T_1^{1/p^r}, \ldots, T_n^{1/p^r}]. \tag{3.3.2}$$

We shall write

$$T^k = T_1^{k_1} \ldots T_n^{k_n}, \quad k = (k_1, \ldots, k_n).$$

Denote the p-valuation on $\mathbb{Q}_p$ by v. Put

$$E = \{\sum a_k T^k \in C \mid a_k \in \mathbb{Z}_p,$$
$$v(a_k) \geq \max\{-v(k_i) \mid 1 \leq i \leq n\}, \quad k_{j+1}, \ldots, k_n \geq 0\}. \tag{3.3.3}$$

We have a $\mathbb{Z}_p$-linear homomorphism

$$V : E \to E$$

given by

$$V(T^k) = pT^{k/p}.$$

Thus, the image of V^r consists of all $\sum a_k T^k$ where all of the a_k's are divisible by p^r. Consequently, the $V^r E$ is a decreasing filtration of rings on E (i.e. $V^r E$ are ideals, and $V^r E \cdot V^s E \subseteq V^{r+s} E$).

Now we have a homomorphism of $\mathbb{Z}_p$-algebras

$$\tau : E \to W(A)$$

given by

$$p^r \tau(T^k) \mapsto V^r \underline{T^{kp^r}}.$$

3.3.1 Proposition *The map of rings τ preserves V and induces an isomorphism of rings*

$$E/V^r E \to W_r(A). \tag{3.3.4}$$

Proof The fact that τ preserves V is immediate from the definitions. The fact that (3.3.4) is an isomorphism follows from the fact that on the r'th associated graded piece of it, it is the identity on A_r, which is A, considered as an A-module where the multiplication of a scalar $a \in A$ with $x \in A_r$ is $a^{p^r} x \in A_r$ $\qquad\square$

Now put, as usual,

$$d \ln(T_i) = \frac{dT_i}{T_i}.$$

Now consider the sub-differential graded algebra

$$E^* \subset \Omega^*_{C/\mathbb{Q}_p}$$

where E^m is generated, as a $\mathbb{Z}_p$-submodule, by

$$a_k T^k d \ln(T_{i_1}) \ldots d \ln(T_{i_m}), \ 1 \le i_1 < \cdots < i_m \le n \tag{3.3.5}$$

where

$$v_p(a_k) \geq \max\{-v(k_j) \mid j \notin \{i_1, \ldots, i_m\}\}$$

and

$$k_i \geq 0 \text{ for } i > j, \tag{3.3.6}$$

where the inequality (3.3.6) is strict when $i = i_s$ for some s. Then, by definition, E^* is a quotient of $\Omega^*_{E/\mathbb{Z}_p}$. Again, we have a ring filtration $V^r E^*$ of E^*. Moreover, the relations (3.2.2), (3.2.3) hold in E^*, since they hold in $\Omega^*_{C/\mathbb{Q}_p}$ (by direct computation). Thus, we obtain a homomorphism of differential graded $\mathbb{Z}/p^n$-algebras

$$\tau^* : W_n \Omega^* A \to E^*/(V^n E^* + dV^n E^{*-1}). \tag{3.3.7}$$

3.3.2 Theorem *The homomorphism (3.3.7) is an isomorphism.*

Proof One define an inverse of τ by expressing (3.3.5) as a product of an element of E and differentials of elements in the obvious way, using the valuation conditions. The ambiguity of doing so is resolved by relation (3.2.3) (note that it happens in each variable separately).

$\square$

3.3.3 Theorem *One has*

$$H^i_{crys}(\mathbb{P}^n_{\mathbb{F}_p}) = \begin{cases} \mathbb{Z}_p \text{ for } 0 \leq i \leq 2n \text{ even} \\ 0 \quad \text{else.} \end{cases}$$

On $H^{2i}_{crys}(\mathbb{P}^n_{\mathbb{F}_p})$, the Frobenius acts by p^i.

Proof One actually proves, for each individual ℓ,

$$H^i(\mathbb{P}^n_{\mathbb{F}_p}, \underline{W\Omega^\ell}) = \begin{cases} \mathbb{Z}_p \text{ when } 0 \leq i = \ell \leq n \\ 0 \quad \text{else.} \end{cases}$$

Then one, again, applies the fact that the hypercohomology spectral sequence must collapse due to evenness of total degrees. Using Theorem 3.3.2, this is completely analogous to the proof in Sect. 2.5.2, replacing coefficients k by the appropriate $\mathbb{Z}_p$-submodule of $\mathbb{Q}_p$ for each monomial $T^k d \ln(T_{i_1}) \ldots d \ln(T_{i_m})$.

The Frobenius on the de Rham-Witt complex, by definition, acts trivially on $d \ln(T_i)$. The products of those classes represent the surviving cohomology classes. The action on cohomology then comes from formula (3.2.9). (Note that filtration degree shifts of the hypercohomology spectral sequence move the generators from cohomological degree i to cohomological degree $2i$.)   □

---

4   Étale Cohomology

If, for a variety over a field of characteristic p, we want to obtain torsion cohomology information at primes $\ell \neq p$, a different technique is required, called *étale cohomology*. This theory was developed by Grothendieck and Deligne. Deligne [5,6] used it, along with other techniques, to prove the Weil conjectures briefly mentioned in Sect. 5.7 of Chap. 3 above. In this section, we introduce étale cohomology briefly. More information can be find in [15].

4.1   Grothendieck Topology

To discuss étale cohomology, we must first generalize the concept of a topology. A *Grothendieck topology* is a category C together with a class of sets of morphisms

$$\{f_i : U_i \to U \mid i \in I\}$$

(where $U_i, U \in Obj(C)$) called *covers* such that the following axioms hold:

1. If $f : U \to V$ is an isomorphism, then $\{f : U \to V\}$ is a cover.
2. If $\{f_{ij} : U_{ij} \to U_i\}$ are covers and $\{f_i : U_i \to U\}$ is a cover, then $\{f_i \circ f_{ij} : U_{ij} \to U\}$ is a cover.
3. If $\{f_i : U_i \to U\}$ is a cover and $g : V \to U$ is any morphism, then the pullbacks $U_i \times_U V$ of f_i and g exist, and $\{f_i \times_U V : U_i \times_U V \to V\}$ is a cover.

Terminology varies, notably the words *site* and *topos* are often used. Also, axioms vary in the literature, but the definition given above is all we need. For a category D, a D-*presheaf* on a Grothendieck topology C is a functor from $C^{Op} \to D$. Analogously to (1.1.3) of Chap. 4, if the category D has limits, a presheaf $\mathcal{F}$ on a Grothendieck topology is called a *sheaf* if for every cover $\{U_i \to U \mid i \in I\}$, $\mathcal{F}(U)$ is the equalizer of

$$\prod_{i \in I} \mathcal{F}(U_i) \rightrightarrows \prod_{j,k \in I} \mathcal{F}(U_j \times_U U_k). \tag{4.1.1}$$

If the category D satisfies the Assumption of Sect. 1.1 of Chap. 4, and the Grothendieck topology is *small*, meaning that there exists a set of objects to one of which every object is isomorphic, then essentially verbatim, the entire discussion of Sect. 1.1 applies. In particular, we have an obvious analogue of the functor L on presheaves, and the obvious analogue of Proposition 1.1.1 holds, thus describing *sheafification*, i.e. a left adjoint to the forgetful functor from presheaves to sheaves. (Note: a *refinement* of a cover $\{U_i \to U \mid i \in I\}$ is a cover $\{V_j \to U \mid j \in J\}$ such that for every $j \in J$ there exists an $i \in I$ and a morphism $V_j \to U_i$ over U.)

The example of principal interest in this Section is the *étale topology* X_{et} on a locally Noetherian scheme X. The objects are étale morphisms $f : Y \to X$ of finite type, morphisms are commutative diagrams

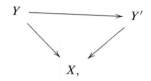

and covers are sets of morphisms $\{U_i \to U\}$ over X (we suppress the arrows to X from the notation) such that the induced morphism $\coprod U_i \to U$ is onto.

4.1.1 Lemma *For a scheme X and a scheme (resp. abelian group scheme) Z, the presheaf on X_{et} whose sections on U are morphisms of schemes from U to Z is a sheaf (resp. abelian sheaf) on X_{et}. (Usually, this sheaf is denoted by the same symbol Z.)*

It may be tempting to call such a sheaf *representable*, but note that in the present context, the representing object may not be in the category.

Proof Since we already know this statement in the Zariski topology (see Sect. 2.5 of Chap. 2), it suffices to verify the statement for an affine étale cover of an affine scheme, i.e. that if a homomorphism of commutative rings $A \to B$ is étale, and the corresponding morphism of schemes $Spec(B) \to Spec(A)$ is onto, then a morphism of schemes $Spec(A) \to Z$ is precisely specified by a morphism $Spec(B) \to Z$ which coincides when composed with the morphisms $Spec(B \otimes_A B) \to Spec(B)$ obtained from the two canonical inclusions $B \to B \otimes_A B$. Now to prove that for any scheme Z, by passing to additional Zariski open covers, we may assume that Z is also affine, i.e. one needs to prove that for any commutative ring C, a homomorphism of rings $C \to B$ factors through A uniquely if and only if it coincides when composed with the two canonical homomorphisms of rings $B \to B \otimes_A B$. This holds because B is flat over A (Exercise 20), and also $A \to B$ is an injection, since we assumed $Spec(B) \to Spec(A)$ is an étale cover.

$\square$

4.2 Geometric Points

It is possible to define geometric points in a Grothendieck topology more abstractly, but for our purposes, we may define a *geometric point* $\overline{x}$ of X_{et} as a point $x \in X$, together with a choice $k_x \to \overline{k_{\overline{x}}}$ of a separable closure of the residue field k_x. Thus, a geometric point can be equivalently specified as a morphism of schemes $\overline{x} : Spec(\overline{k}) \to X$ where $\overline{k}$ is a separable closure of the residue field of the image of $\overline{x}$. An *étale neighborhood* of $\overline{x}$ can then be defined as a diagram

$$
\begin{array}{ccc}
 & & Y \\
 & \overline{y} \nearrow & \downarrow f \\
 & \overline{x} & \\
\overline{k} & \longrightarrow & X
\end{array}
\tag{4.2.1}
$$

where f is étale. Specifying an étale neighborhood of $\overline{y}$ then by composition determines an étale neighborhood of $\overline{x}$. Thus, the diagrams (4.2.1) form a category $(X_{et})_{\overline{x}}$ of étale neighborhoods of $\overline{x}$, which, via the vertical arrow, has a canonical forgetful functor to X_{et}.

4.2.1 Lemma *The category $(X_{et})_{\overline{x}}$ of étale neighborhoods of $\overline{x}$ is directed, which means that for every two objects b, c, there is an object a with morphisms $a \to b, a \to c$, and for every two morphisms $b \rightrightarrows c$, there exists a morphism $a \to b$ with which they give equal compositions.*

Proof The first condition follows from the fact that a pullback of two étale morphisms over X is étale.

For the second statement, we need to use the fact that an equalizer of two étale morphisms over X is étale. But an equalizer can be realized by a pullback, followed by another pullback along the diagonal $Y \to Y \times_X Y$ where Y is étale over X. However, the diagonal is then an open inclusion, and hence is also étale. Our statement then follows from the fact that a composition of étale morphisms is étale. □

The *stalk* $\mathcal{F}_{\overline{x}}$ of a sheaf $\mathcal{F}$ on X_{et} at a geometric point $\overline{x}$ is defined as the colimit of sections $\mathcal{F}(U)$ where U runs through the category $(X_{et})_{\overline{x}}$ of étale neighborhoods of $\overline{x}$. Just like for sheaves over a topological space, a stalk functor preserves colimits (in fact, has a right adjoint, the étale skyscraper sheaf on the geometric point $\overline{x}$), and is exact (which is an easy general property of directed categories).

4.2.2 Lemma *For morphisms $f, g : \mathcal{F} \to \mathcal{G}$ of sheaves of sets on X_{et}, $f = g$ if and only if the induced maps on stalks $\mathcal{F}_{\overline{x}} \to \mathcal{G}_{\overline{x}}$ coincide for all geometric points $\overline{x}$. We say that the étale topology has* enough *points.*

Proof Necessity just follows from functoriality. For sufficiency, if f and g induce the same maps on stalks, then they must induce the same maps on sections for some étale neighborhood of every geometric point $\bar{x}$. Those étale neighborhoods form an étale cover of X, and thus, our statement follows from the gluing axiom for sheaves. $\qquad\square$

Consequently, for abelian sheaves (i.e. sheaves of abelian groups) on X_{et}, a sequence of morphisms is exact if and only if it is exact on stalks.

4.2.3 Stalks of the Structure Sheaf

For the structure sheaf $\mathcal{O}$ (a.k.a $\mathbb{A}^1$), the stalks at geometric points of X_{et} are obtained by taking a local ring $\mathcal{O}_{X,x}$, and successively taking étale covers and localizing at maximal ideals (taking colimits at limit ordinal steps) until one obtains a local ring R for which every étale homomorphism $R \to S$ where S is a local ring is an isomorphism. Such a ring R is called *strictly Henselian*, and the transfinite procedure just described is called *strict henselization*. For a local ring R to be strictly Henselian, it is obviously necessary that every monic polynomial $p(x)$ over R whose reduction to the residue field has no multiple roots factor to linear factors. It is also sufficient, which is a deep theorem of commutative algebra, relying on a hard result known as the *Zariski main theorem*. Zariski's main theorem (in the form one needs here) states that if $f : Y \to X$ is a separated morphism of finite type of schemes such that for every point $x \in X$, $f^{-1}(x)$ is discrete in the induced topology from Y and X is a quasicompact separated scheme (this condition can be further weakened), then we have a diagram

$$
\begin{array}{ccc}
Y & \xrightarrow{\;\;\iota\;\;} & Z \\
& {\scriptstyle f}\searrow \quad \swarrow {\scriptstyle g} & \\
& X &
\end{array}
\tag{4.2.2}
$$

such that ι is a quasicompact open immersion and g is finite. For a proof of this statement, see Exercises 16, 18, 19 below. For further details, see [18].

4.2.4 The Kummer Exact Sequence

Consider, for a natural number ℓ, the abelian group scheme

$$
\mu_\ell = Spec(\mathbb{Z}[x]/(x^\ell - 1)).
$$

The associated étale sheaf on X_{et} for a Noetherian scheme X is also denoted by $\mathbb{Z}/\ell(1)$. If X is a variety over a field *of characteristic* 0 *or prime to* ℓ, then we have a short exact sequence of sheaves on X_{et}

$$
0 \longrightarrow \mathbb{Z}/\ell(1) \longrightarrow \mathbb{G}_m \xrightarrow{\;\;\ell\;\;} \mathbb{G}_m \longrightarrow 0
\tag{4.2.3}
$$

where the morphisms are induced by the corresponding morphisms of abelian group schemes. It suffices to verify exactness on stalks, which means that for a strictly Henselian algebra R over a field of characteristic not dividing ℓ, the sequence

$$1 \to \{a \in R \mid a^\ell = 1\} \to R^\times \to R^\times \to 1$$

(written multiplicatively) is exact, where the map $R^\times \to R^\times$ is the ℓ'th power. Clearly, the only non-trivial part is verifying that the ℓ'th power map is onto, but by the assumptions on characteristic, that follows from R being strictly Henselian.

4.2.5   The Étale Topology on $Spec(k)$

The étale topology is quite complicated, and usually difficult to calculate with directly. One exception is the case of $X = Spec(k)$ for a field k. Then by reinterpreting the comments in Sect. 5.4 of Chap. 3, we have already seen that the Grothendieck topology X_{et} is simply the full subcategory of the category of schemes on disjoint unions of finite separable extensions of k. All morphisms are étale. Choosing a separable closure $\bar{k}$ of k, we see then that abelian sheaves on X_{et} are then identified with $Gal(\bar{k}/k)$-*modules*, by which we mean abelian groups M with an action of the inverse limit of $Gal(L/k)$ for Galois extensions $k \subseteq L \subset \bar{k}$ where M is the union of fixed points $M^{Gal(\bar{k}/L)}$ over finite extensions L of k contained in $\bar{k}$.

In particular, (defining tensor product of abelian sheaves by taking tensor product on sections and then sheafifying), we see that we may define a sheaf $\mathbb{Z}/\ell(m)$, $m \in \mathbb{Z}$ on $Spec(k)_{et}$ for a field k by letting $Gal(\bar{k}/k)$ act by the m'th power of its action on μ_ℓ. Additionally, we have

$$\mathbb{Z}/\ell(m) \otimes \mathbb{Z}/\ell(n) \cong \mathbb{Z}/\ell(m+n), \quad m, n \in \mathbb{Z}.$$

Consequently, by pullback, we have the same relation on X_{et} where X is any variety over a field.

4.3   Étale Cohomology

For a Noetherian scheme X, the abelian category of sheaves on X_{et} has enough injectives by the same argument as for the abelian category of sheaves on a topological space, i.e. by taking products of étale skyscraper sheaves, which are pushforwards of injective abelian groups. Thus, for a sheaf $\mathcal{F}$ on X_{et}, we can define

$$H^\ell(X_{et}, \mathcal{F}) = R\Gamma^\ell(\mathcal{F})$$

where Γ denotes global sections (i.e. sections on X). Alternately, again, this coincides with the ℓ'th Ext group from the constant sheaf $\underline{\mathbb{Z}}$ on X_{et} to $\mathcal{F}$.

4.3.1 Étale Cohomology of $Spec(k)$ for a Field k

For a group G, the free abelian group $\mathbb{Z}G$ has a unique ring structure where the product of free generators $g, h \in G$ is the free generator $gh \in G$. This is called the *group ring* of G. The ring $\mathbb{Z}G$ is not commutative when the group G is not commutative. For a non-commutative ring R, we must distinguish between *left R-modules M*, which have structure maps

$$R \otimes M \to M$$

and *right R-modules N*, which have structure maps

$$N \otimes R \to N$$

(which are required to satisfy unitality and associativity in the usual sense). For a left $\mathbb{Z}G$-module M, however, M automatically has a canonical right $\mathbb{Z}G$-module structure given by

$$mg = g^{-1}m, \ g \in G, m \in M,$$

and vice versa. Thus, left and right $\mathbb{Z}G$-modules form equivalent abelian categories, whose objects are (somewhat dangerously) often just called *G-modules*. A G-module can then be understood simply as an abelian group on which G acts by group homomorphisms. A *trivial G-module* is an abelian group on which G acts trivially (i.e. every element of G acts by the identity).

The category of left modules over any ring R has enough injectives (by the same proof as in Lemma 2.4.2), and we define *group cohomology* of G with coefficients in a module M as

$$H^{\ell}(G; M) = Ext^{\ell}_{\mathbb{Z}G}(\mathbb{Z}, M) = R^{\ell}Hom(\mathbb{Z}, M) \tag{4.3.1}$$

where $\mathbb{Z}$ is a trivial module. It should be mentioned that unlike the case of commutative rings, where Hom and Ext are canonically modules over the ring, for a non-commutative ring, they only have a canonical abelian group structure.

Now for a pro-finite group G, denote by $\widehat{G}$ the limit of the system of finite groups defining G. We can define a G-module M as a $\widehat{G}$-module which is a direct limit of submodules M_K such that the $\widehat{G}$-module structure of each M_K factors through a K-module structure, where K is one of the finite groups in the inverse system defining G (i.e. $Ker(\widehat{G} \to K)$ acts trivially on M_K). Then we define pro-finite group cohomology of G as

$$H^{\ell}(G, M) = \operatorname*{colim}_{K} H^{\ell}(K, M_K). \tag{4.3.2}$$

Now for a field k, as we remarked before, choosing a separable closure $\bar{k}$ of k, $Gal(\bar{k}/k)$ is a pro-finite group, and étale sheaves on $Spec(k)$ are an equivalent category to $Gal(\bar{k}/k)$-modules in the above sense. Thus,

$$H^\ell(Spec(k)_{et}, M) \cong H^\ell(Gal(\bar{k}/k), M). \tag{4.3.3}$$

This is one of the few cases of calculations of étale cohomology which can be done easily directly from the definition.

4.4   Étale Cohomology of Curves

Let X be a smooth curve over an algebraically closed field k, and let ℓ be a natural number *not divisible by the characteristic of k*. We will now show how to calculate the étale cohomology of X with coefficients in the constant sheaf $\mathbb{Z}/\ell$ using the Kummer exact sequence (4.2.3). Note that since k is separably closed, all the sheaves $\mathbb{Z}/\ell(n)$ on X are isomorphic.

Now let $i : Spec(L) \to X$ be the inclusion of the generic point. By taking the cokernel, we get a short exact sequence of étale sheaves on X of the form

$$0 \to \mathbb{G}_m \to i_*\mathbb{G}_m \to Div \to 0. \tag{4.4.1}$$

We claim that $Div(Y)$ is the abelian group of Weil divisors on Y for $Y \to X$ étale. In fact, defining $Div(Y)$ this way produces a short exact sequence on sections, by the equivalence of Weil and Cartier divisors. Functoriality with respect to étale maps is given by taking a closed point to the sum of the points in the fiber over it (note that we assumed the residue fields of the points to be algebraically closed). Thus, we need to prove that the étale presheaf thus defined is a sheaf. Again, since we assume the residue fields to be algebraically closed, if we let $P \in X$ be a closed point and denote by S the fiber (i.e. inverse image) of P in Y for an étale morphism $Y \to X$, then the fiber of $Y \times_X Y$ over P is $S \times S$. Thus, our claim reduces to a statement about sets: For a finite set S, the equalizer of the two maps $\mathbb{Z}S \to \mathbb{Z}S \otimes \mathbb{Z}S$ given by $x \mapsto x \otimes \sum S$, $x \mapsto \sum S \otimes x$ is $\mathbb{Z}\{\sum S\}$.

Note that the sheaf Div can also be described as the direct sum of the étale skyscraper sheaves on all Zariski closed points of X (recalling that the constant sheaf with values in $\mathbb{Z}$ on a finite set S has sections $\mathbb{Z}S$).

Now consider the diagram of étale sheaves on X

$$\begin{array}{ccccccccc}
0 & \longrightarrow & \mathbb{G}_m & \longrightarrow & i_*\mathbb{G}_m & \longrightarrow & Div & \longrightarrow & 0 \\
& & \downarrow{\scriptstyle \ell_1} & & \downarrow{\scriptstyle \ell_2} & & \downarrow{\scriptstyle \ell_3} & & \\
0 & \longrightarrow & \mathbb{G}_m & \longrightarrow & i_*\mathbb{G}_m & \longrightarrow & Div & \longrightarrow & 0
\end{array} \tag{4.4.2}$$

where ℓ_1, ℓ_2 are ℓ'th powers (thinking of the abelian groups multiplicatively), and ℓ_3 is multiplication by ℓ (thinking of Weil divisors additively).

Then the "snake lemma" (which is a special case of Theorem 2.3.7 of Chap. 5) states that there is an exact sequence

$$
\begin{array}{ccccc}
0 \longrightarrow Ker(\ell_1) \xrightarrow{\iota} & Ker(\ell_2) & \longrightarrow & Ker(\ell_3) & \\
& & & & \\
Coker(\ell_1) & \longrightarrow Coker(\ell_2) & \xrightarrow{\kappa} & Coker(\ell_3) & \longrightarrow 0.
\end{array}
$$

(4.4.3)

with the horizontal morphisms induced by the horizontal morphisms (4.4.2). However, since $Ker(\ell_3) = 0$ and $Coker(\ell_1) = 0$, ι and κ are isomorphisms. Thus, we obtain an exact sequence

$$
0 \longrightarrow \mathbb{Z}/\ell \longrightarrow i_*\mathbb{G}_m \xrightarrow{\ell} i_*\mathbb{G}_m \longrightarrow Div/\ell \longrightarrow 0. \tag{4.4.4}
$$

Now since the étale sheaf Div/ℓ is a sum of skyscraper sheaves of Zariski closed points, it is flasque and hence has no étale cohomology in degrees > 0, and in the next subsection, we will show the same about the sheaf $i_*\mathbb{G}_m$. Therefore,

$$
i_*\mathbb{G}_m \xrightarrow{\ell} i_*\mathbb{G}_m \longrightarrow Div/\ell \tag{4.4.5}
$$

is a resolution of $\mathbb{Z}/\ell$ by Γ-local étale sheaves, and $H^*(X_{et}, \mathbb{Z}/\ell)$ therefore is the cohomology of the induced complex on global sections, which is

$$
L^\times \xrightarrow{\ell} L^\times \longrightarrow Div(X)/\ell \tag{4.4.6}
$$

where $Div(X)$ denotes the group of Weil divisors on X, reduced mod ℓ. Now applying the snake lemma to the diagram of abelian groups

$$
\begin{array}{ccccccccc}
0 & \longrightarrow & L^\times/k^\times & \longrightarrow & Div(X) & \longrightarrow & Pic(X) & \longrightarrow & 0 \\
& & \downarrow{\scriptstyle \ell} & & \downarrow{\scriptstyle \ell} & & \downarrow{\scriptstyle \ell} & & \\
0 & \longrightarrow & L^\times/k^\times & \longrightarrow & Div(X) & \longrightarrow & Pic(X) & \longrightarrow & 0,
\end{array}
$$

we obtain an exact sequence

$$
0 \to {}_\ell Pic(X) \to L^\times/(L^\times)^\ell \to Div(X)/\ell \to Q \to 0,
$$

where $_\ell Pic(X)$ is the ℓ-torsion in the Picard group of X, and Q is $\mathbb{Z}/\ell$ if X is projective (because of the degree), and 0 otherwise. This is proved by generalizing the observations on the Picard group of an elliptic curve made in Sect. 3.4 of Chap. 4. One can prove that for a smooth affine curve X over an algebraically closed field k, $Pic(X)$ can be identified with the k-points of a projective group scheme, and prove the surjectivity of the ℓth power, for ℓ not divisible by the characteristic, in that context. Details would take us too far afield, but they do not depend on the foundations of Grothendieck topology. Using (4.4.5), we then conclude the following

4.4.1 Proposition *Let X be a smooth curve over an algebraically closed field k, and suppose that the characteristic of k does not divide a natural number ℓ. Then*

$$H^i(X_{et}, \mathbb{Z}/\ell) = \begin{cases} \mathbb{Z}/\ell & \textit{if } i = 0 \textit{ or } X \textit{ is projective and } i = 2 \\ _\ell Pic(X) & \textit{if } i = 1 \\ 0 & \textit{else.} \end{cases}$$

$\square$

4.4.2 Cohomology of Function Fields of Transcendence Degree 1

We begin with the following important consequence of the Nullstellensatz (Sect. 1.1.2 of Chap. 1):

4.4.3 Theorem *Let k be an algebraically closed field, let $x_1, \ldots, x_n$ be variables, and let $f_1, \ldots, f_m$ be homogeneous polynomials in $x_1, \ldots, x_n$ such that $m < n$. Then the system of equations*

$$f_1(x_1, \ldots, x_n) = 0$$
$$\cdots$$
$$f_m(x_1, \ldots, x_n) = 0$$

has at least one solution $(x_1, \ldots, x_n) \neq (0, \ldots, 0)$.

Proof By the Nullstellensatz, it suffices to prove that for a homogeneous ideal $I \subseteq k[x_1, \ldots, x_n]$ generated by $m < n$ homogeneous elements, we have

$$\sqrt{I} \neq (x_1, \ldots, x_n). \tag{4.4.7}$$

One way to of doing this is using a concept called *local cohomology*. Let R be a ring and let $I = (r_1, \ldots, r_m)$ be a finitely generated ideal. Then the *local cohomology*

$$H_I^*(R)$$

is defined as the cohomology of the totalization of the m-fold cochain complex which in m-tuple degree $(\epsilon_1, \ldots, \epsilon_m)$, $\epsilon_i \in \{0, 1\}$ is

$$(\prod_{\epsilon_i = 1} r_i)^{-1} R,$$

(and 0 otherwise), and the j'th differential from j'th degree 0 to j'th degree 1 is localization by inverting r_i, times a sign equal to (-1) to the power of

$$|\{j < i \mid \epsilon_j = 1\}|$$

(to satisfy our sign conventions regarding multiple chain complexes which is that the i'th and j'th differential should anticommute). More generally, an analogous definition can clearly be made for R-modules.

A key fact is that up to isomorphisms, the definition of local cohomology does not depend on the choice of the (finitely many) generators r_i. This is shown by showing, using the definition, that the cohomology does not change when another generator is added.

Similarly, one shows that for a polynomial ring $R = k[x_1, \ldots, x_n]$ where k is a field, the canonical map

$$H_I^*(R) \to H_{\sqrt{I}}^*(R)$$

is an isomorphism. Now if an ideal I is generated by m elements, then, by definition, $H_I^i(R) = 0$ for $i > m$. On the other hand, if $I = (x_1, \ldots, x_n)$, then by direct computation, $H_I^n(R) \neq 0$. Thus, the augmentation ideal cannot be generated by fewer than n elements.

$\square$

Now the key fact about the cohomology of function fields over an algebraically closed field k (i.e. fields generated, as fields, by k and finitely many elements) of transcendence degree 1 is the following theorem of Tsen:

4.4.4 Theorem *If K is a function field over an algebraically closed field k of transcendence degree 1, and L is a finite extension of K, then for every $x \in K^\times$, there exists a $y \in L^\times$ such that $N_{L/K}(y) = x$. (Recall that the norm $N_{L/K}(y)$ of an element $y \in L^\times$ is defined as the product of the roots of a minimal polynomial $p(y)$ of y over K, taken to the power of $[L : K]/\deg(p)$.)*

Proof Tsen (1933) argues as follows: K is isomorphic to a finite extension of the field of rational functors $k(u)$. Now choose a basis $(\alpha_i \mid i = 1, \ldots n)$ of K as a $k(u)$-vector space. By taking a common denominator which is an $[L : K] = m$'th power, we may assume, without loss of generality, that x is a linear combination of the α_i's with coefficients in $k[x]$. Choosing another basis β_j of L over K, we look for a solution to $N_{L/K}(y) = x$

where y is a linear combination of the basis elements $\alpha_i \beta_j$ of L over $k(x)$ *with coefficients in $k[x]$.* Multiplying by a common denominator and writing down those equations, one finds that they are *homogeneous polynomial equations* over k of the form

$$N_i(p_1, \ldots, p_{mn}) = p_0^m a_i, \quad i = 1 \ldots n \tag{4.4.8}$$

where p_i / p_0 are solutions to the norm equation over the $k(x)$-basis $(\alpha_i \beta_j)$ (the equations are obtained by comparing the coefficients at each power of x). If the maximum degree of the polynomials p_i, p_0 is t, and the maximum degree of the a_i's is t_0, we obtain $\leq$ $n(t_0 + mt + 1)$ homogeneous equations with $(1 + mn)(1 + t)$ unknowns. By increasing the degree of the denominator, if necessary, we can increase t and thus arrange for the number of unknowns to be greater than the number of equations, and apply Theorem 4.4.3. (The possibility of the common denominator being 0 is avoided by noting that the norm of a non-zero element is non-zero.) $\quad\square$

Now for a profinite group G, call a G-module M (in the above sense) *flasque* if for every factor K of a group in the projective system defining G, and every subgroup $H \subseteq K$, every element $x \in M^{Ker(\widehat{G} \to K) \cdot H}$, there exists an element $y \in M^{Ker(\widehat{G} \to K)}$ such that

$$x = N_{K/H}(y) = \sum_{k \in K/H} ky.$$

One has the following

4.4.5 Proposition *If G is a profinite group and M is a flasque module, then*

$$H^i(G, M) = 0 \text{ for } i > 0.$$

Proof Exactly emulates the proof for sheaves. (In fact, one can take the point of view that we are dealing with étale sheaves here.) $\quad\square$

4.4.6 Corollary *If K is a function field of transcendence degree 1 over an algebraically closed field k, then*

$$H^i(Spec(K)_{et}, \mathbb{G}_m) = 0 \text{ for } i > 0.$$

Proof Apply Proposition 4.4.5 to $G = Gal(\overline{K}/K)$ where $\overline{K}$ is the separable closure of K, and the G-module $\overline{K}^{\times}$. $\quad\square$

Now Corollary 4.4.6 is a part of what we need to finish the proof of Proposition 4.4.1. The other part is that we need to show that if $i : Spec(K) \to X$ is the inclusion of the generic point for a smooth curve over an algebraically closed field k, then

$$R^j i_* \mathbb{G}_m = 0 \text{ for } j > 0. \tag{4.4.9}$$

To this end, one notes that $R^j i_* \mathbb{G}_m$ is the sheafification of the presheaf on $Spec(L)_{et}$ which to an étale cover U of X assigns $H^j(i^{-1}(U), \mathbb{G}_m)$. However, $i^{-1}(U)$ is an étale cover of $Spec(L)$, so for $j > 0$ this group is 0 by Corollary 4.4.6.

5   Motivic Cohomology

The idea of motivic cohomology (originally due to Grothendieck) is to construct a cohomology theory of schemes which would be universal in an analogous sense as singular (co)homology is universal for topological spaces. In particular, motivic cohomology should allow coefficients in $\mathbb{Z}$. For varieties over perfect fields, such a cohomology theory was first constructed by Bloch, and reformulated by Voevodsky, who used it to prove an important conjecture of Bloch and Kato on Galois cohomology of fields. In this section, we give a brief introduction to these concepts. More information can be found in [14].

In this section, we restrict attention to the full subcategory Sm_k of the category of schemes and morphisms on smooth separated schemes X of finite type over a perfect field k.

5.1 $\mathbb{A}^1$-Homotopy Invariance

In algebraic geometry, we do not have, of course, literally the notion of homotopy as considered in Sect. 3 of Chap. 5. Nevertheless, in some sense, an analogous concept can be developed by using $\mathbb{A}^1$ as the object corresponding to the unit interval $[0, 1]$. We say that a functor F from Sm_k to another category is $\mathbb{A}^1$-*homotopy invariant* if for every object X of Sm_k, the induced morphism $F(X \times \mathbb{A}^1) \to F(X)$ is an isomorphism.

Analogously to (3.1.1) of Chap. 5, the *algebraic standard simplex* is defined as

$$\Delta^n = Spec(k[t_0, \ldots, t_n]/(\textstyle\sum t_i - 1)). \tag{5.1.1}$$

We have, again, face maps

$$\partial_i : \Delta^{n-1} \to \Delta^n, \ i = 0, \ldots, n,$$

given by setting the i'th coordinate to 0.

Now we shall discuss presheaves and Zariski sheaves on the category Sm_k of separated smooth schemes of finite type over a perfect field k, and morphisms of schemes over k. Here by the Zariski topology, we mean the Grothendieck topology on the category Sm_k

where covers are Zariski open covers. In this section, we shall always consider this Zariski topology on Sm_k, unless otherwise specified. For a chain complex $\mathcal{F}$ of abelian sheaves on Sm_k and an object of Sm_k, we define

$$H^\ell(X, \mathcal{F})$$

as the cohomology of a smooth scheme X over k with coefficients $\mathcal{F}$, restricted to the category of Zariski open sets of X, and inclusions. Denote this restriction of $\mathcal{F}$ by $\mathcal{F}_{(X)}$ (to distinguish it from the restriction to the category of Sm_k-arrows into X, which is yet a different concept).

Note that the (hyper)cohomology thus defined is contravariantly functorial with respect to morphisms in Sm_k, since for a morphism $f : X \to Y$ in Sm_k, we have a canonical morphism of complexes of abelian sheaves

$$f^{-1}\mathcal{F}_{(Y)} \to \mathcal{F}_{(X)}. \tag{5.1.2}$$

(This is because the left hand side is defined as the colimit of $\mathcal{F}_{(Y)}(V)$ over open sets V such that $f^{-1}(V) \supseteq U$, but then f restricts to a morphism of schemes $U \to V$.) Now choose quasiisomorphisms $\mathcal{F}_{(Y)} \to \mathcal{I}$, $\mathcal{F}_{(X)} \to \mathcal{K}$ with $\mathcal{I}$, $\mathcal{K}$ local. Since f^{-1} is exact, $f^{-1}(\mathcal{I})$ is quasiisomorphic to $f^{-1}\mathcal{F}_{(Y)}$, which maps (uniquely up to chain homotopy) to $\mathcal{K}$. Taking global sections and cohomology gives the required functoriality.

For an abelian presheaf F on Sm_k, define a complex of abelian presheaves C_*F on Sm_k by letting its sections on a separated smooth scheme U of finite type over k be the chain complex

$$\cdots \longrightarrow F(\Delta^2 \times U) \xrightarrow{d_2} F(\Delta^1 \times U) \xrightarrow{d_1} F(\Delta^0 \times U)$$

where, for $\alpha \in F(\Delta^n \times U)$,

$$d_n(\alpha) = \sum_{i=0}^{n} \partial_i^*(\alpha)$$

(where by f^* we mean $F(f)$ - note that F is contravariant on Sm_k). Clearly, if F is a sheaf on Sm_k, then C_*F is a chain complex of sheaves on Sm_k.

The key result is that hypercohomology with coefficients in a complex of sheaves of the form $C_*\mathcal{F}$ is $\mathbb{A}^1$-homotopy invariant:

5.1.1 Proposition *Let $\mathcal{F}$ be an abelian sheaf on Sm_k. Then for every smooth scheme $X \in Obj(Sm_k)$, the projection $X \times \mathbb{A}^1 \to X$ induces an isomorphism of hypercohomology*

$$H^\ell(X, C_*\mathcal{F}) \cong H^\ell(X \times \mathbb{A}^1, C_*\mathcal{F}). \tag{5.1.3}$$

We will prove Proposition 5.1.1 by reducing it to simpler statements. First, note that it suffices to prove that the identity on $X \times \mathbb{A}^1$ induces the same homomorphism on $C_* \mathcal{F}$-hypercohomology as the map $X \times 0$ (where $0 : Spec(\mathbb{Z}[x]) \to Spec(\mathbb{Z}[x])$ is the morphism corresponding to the homomorphism of rings given by $x \mapsto 0$). This is because then the projection $X \times \mathbb{A}^1 \to X$ and the inclusion $X \to X \times \mathbb{A}^1$ by 0 in the second coordinate induce inverse homomorphisms in $C_* \mathcal{F}$-hypercohomology.

Next, we claim that in fact, it suffices to prove that the inclusions

$$i_0, i_1 : X \to X \times \mathbb{A}^1$$

by 0 and 1 in the second coordinate induce the same homomorphism in $C_* \mathcal{F}$-cohomology. This suffices because then we can consider the composition

$$X \times \mathbb{A}^1 \rightrightarrows X \times \mathbb{A}^1 \times \mathbb{A}^1 \to X \times \mathbb{A}^1$$

where all the morphisms are identity on the X coordinate, the first two morphisms are inclusions to 0 resp. 1 in the last coordinate, and the second morphism is the product in $\mathbb{A}^1$, i.e. the morphism $Spec(\mathbb{Z}[s]) \times Spec(\mathbb{Z}[t]) \to Spec(\mathbb{Z}[u])$ given by $u \mapsto st$. If the first two morphisms induce the same on $C_* \mathcal{F}$-hypercohomology, so do their compositions with the second morphism, which are Id_X times the 0 map and the identity on $\mathbb{A}^1$, respectively.

Now the statement we just reduced Proposition 5.1.1 to will follow from the following

5.1.2 Lemma *There exists a natural (in $\mathcal{F}$) chain homotopy between the homomorphisms*

$$\Gamma(C_* \mathcal{F}_{(X \times \mathbb{A}^1)}) \rightrightarrows \Gamma(C_* \mathcal{F}_{(X)})$$

induced by i_0 and i_1.

Proof The idea here is the triangulation of a prism. (In fact, using the same argument in the topological case, we could prove the homotopy axiom for singular homology.) For the "prism"

$$\mathbb{A}^1 \times \Delta^n = Spec(k[t_0, \ldots, t_n, q]/\sum t_i = 1)$$

and

$$\Delta^{n+1} = Spec(k[s_0, \ldots, s_{n+1}]),$$

define a morphism

$$\gamma_i : \Delta^{n+1} \to \mathbb{A}^1 \times \Delta^n, \ i = 0, \ldots, n$$

by

$$t_0 = s_0, \ldots, t_{i-1} = s_{i-1},$$

$$t_i = s_i + s_{i+1},$$

$$t_{i+1} = s_{i+2}, \ldots, t_n = s_{n+1},$$

$$q = s_{i+1} + \cdots + s_{n+1}.$$

The chain homotopy is

$$\sum_{i=0}^{n} (-1)^i (\gamma_i \times Id_X)^*.$$

$\square$

5.1.3 Proof of Proposition 5.1.1

The idea is to apply Lemma 5.1.2 to a flasque resolution $\mathcal{I}$ of $\mathcal{F}$. This works, provided we can prove that $C_*\mathcal{I}$ is Γ-local. (Note that the chain complex $C_*\mathcal{I}$ is bounded below homologically, i.e. bounded above cohomologically, which is in the wrong direction!) What saves us here is that we can actually choose $\mathcal{I}$ to be non-zero in only finitely many degrees, since X is Noetherian of finite dimension, and hence its cohomology can be non-trivial in only finitely many degrees (see Hartshorne [11]). For an additive functor Γ with only finitely many non-zero right derived functors, and a double chain complex $(C_{m,n})$ whose columns $C_{(m,*)}$ are Γ-local such that $C_{m,n}$ is non-zero only for $-N \leq n \leq N$ for a fixed N, the totalization $|C|$ is Γ-local by applying $R\Gamma$ to the inverse limit

$$\lim_m (C_{m,*}/Im(\partial) \to C_{m-1,*} \to C_{m-2,*} \to \ldots).$$

$\square$

5.2 Finite Correspondences and the Definition of Motivic Cohomology

Let $X \in Obj(Sm_k)$. The *group $\mathcal{Z}^r(X)$ of algebraic cycles of codimension r on X is the free abelian group on the set of integral closed subschemes of X of codimension r* (which we will call *elementary algebraic cycles*). In particular, $\mathcal{Z}^1(X)$ is the group of Weil divisors on X.

An important feature of algebraic cycles is their *intersection product*

$$\cdot : \mathcal{Z}^r(X) \otimes \mathcal{Z}^s(X) \to \mathcal{Z}^{r+s}(X),$$

which on elementary algebraic cycles is given by

$$Z \cdot T = \sum n_j W_j$$

where W_j are the irreducible components of $Z \cap T$ and n_j is a certain number called the *multiplicity*. The multiplicity was defined by Serre [22] as follows: Let $w_j \in W_j$ be the generic point, and let $A = \mathcal{O}_{X, w_j}$ be the local ring of X at w_j. In particular, A is a regular ring, since X is regular. The dimension of A is equal to the codimension of W_j in X. Let I, J be the ideals defining Z, T in A. One notes that if we put

$$B = A/(I + J) = A/I \otimes_A A/J,$$

then

$$B/Nil(B) = K$$

where $Nil(B)$ is the nilradical of B, and K is the residue field of A. Let $N = Nil(B) = Ker(B \to K)$. Then B is filtered by powers of N, and the associated graded object is a finitely generated K-module. Its dimension is called the *length* $\ell(B)$ *of* B. More generally, a finitely generated B-module M is also filtered by powers of N ($F^i M = M \cdot N^i$), and its associated graded object is also a finitely generated K-module, whose dimension $\ell(M)$ is called the *length of* M (Compare with Sect. 5.3 of Chap. 1.). The B-modules $Tor_i^A(A/I, A/J)$ are finitely generated, so we can define

$$n_j = \sum_i (-1)^i \ell(Tor_i^A(A/I, A/J)).$$

An important point is that if the codimension of W_j is not $r + s$, then $n_j = 0$. Another important point is that the intersection product on algebraic cycles is commutative, associative and unital, thus making

$$\mathcal{Z}^*(X) = \bigoplus_r \mathcal{Z}^r(X)$$

into a graded ring. Associativity is non-trivial and is proved in [22], Chapter V.

We will focus on a particular type of algebraic cycles called finite correspondences. For $X, Y \in Obj(Sm_k)$, a *finite correspondence from X to Y* is an algebraic cycle

$$\sum a_i Z_i \in \mathcal{Z}^r(X \times Y), \; r = dim(Y)$$

(where Z_i are integral closed subschemes of $X \times Y$ of codimension r) such that the restriction of the projection $X \times Y \to X$ to each Z_i is a finite morphism of schemes. An *elementary finite correspondence* is a finite correspondence which is an elementary algebraic cycle.

Finite correspondences are thought of as a kind of "multivalued maps from X to Y with finitely many values." *Composition* of a finite correspondence F from Y to Z with a finite correspondence G from X to Y is defined as the intersection of the cycles $F \times Z$ and $X \times G$ in $X \times Y \times Z$. (The finiteness condition follows from the fact that a pullback of a finite morphism is finite.) Composition of finite correspondences is associative because intersection of cycles is.

Thus, finite correspondences on separated smooth schemes of finite type over a perfect field k form a category, which we denote by $SmCor_k$. Moreover, we have a canonical functor

$$Sm_k \to SmCor_k$$

given by sending a morphism $X \to Y$ to its graph, i.e. to the reduced closed subscheme of $X \times Y$ given by the equation $y = f(x)$ (thought of, as an algebraic cycle, as a sum of its connected components).

Now we define, for $X \in Obj(Sm_k)$, a Zariski sheaf $\mathbb{Z}_{tr}(X)$ on Sm_k by

$$\mathbb{Z}_{tr}(X)(U) = SmCor_k(U, X). \tag{5.2.1}$$

By definition, $\mathbb{Z}_{tr}$ is a covariant functor from Sm_k to abelian sheaves. In particular, applying $\mathbb{Z}_{tr}$ to the inclusion of the unit

$$Spec(k) \to \mathbb{G}_m,$$

we obtain a chain complex of Zariski sheaves on Sm_k:

$$\mathcal{G} = (\mathbb{Z}_{tr}(Spec(k)) \to \mathbb{Z}_{tr}(\mathbb{G}_m)) \tag{5.2.2}$$

(where we place $\mathbb{Z}_{tr}(Spec(k))$ in degree 0). We define, for $n \geq 0$,

$$\mathbb{Z}(n) = C_*(\underbrace{\mathcal{G} \otimes \cdots \otimes \mathcal{G}}_{n \text{ times}}). \tag{5.2.3}$$

By functoriality of C_*, then, we have a morphism of chain complexes of Zariski sheaves

$$\mathbb{Z}(m) \otimes \mathbb{Z}(n) \to \mathbb{Z}(m+n). \tag{5.2.4}$$

In fact, extending the argument of Lemma 5.1.2, one can actually show that (5.2.4) is a quasiisomorphism, but one must be mindful of the fact that in general, $\otimes$ does not preserve quasiisomorphisms. For an abelian group (more generally, abelian sheaf) A, one can also define a Zariski sheaf on Sm_k:

$$A(n) = \mathbb{Z}(n) \otimes A.$$

5.2.1 Definition (Voevodsky) *Motivic cohomology of* $X \in Obj(Sm_k)$ *is defined, for* $m, n \geq 0$, *as the Zariski cohomology*

$$H^{m,n}(X) = H^m(X, \mathbb{Z}(n)),$$

$$H^{m,n}(X; A) = H^m(X, A(n)).$$

As a direct consequence of Proposition 5.1.1, we see that *motivic cohomology is* $\mathbb{A}^1$-*homotopy invariant*. Also, by (5.2.4), we have, for a commutative ring R, a multiplication

$$H^{m_1,n_1}(X; R) \otimes H^{m_2,n_2}(X; R) \to H^{m_1+m_2,n_1+n_2}(X; R)$$

which is associative, unital and graded-commutative in the sense that for $\alpha \in H^{m_1,n_1}(X; R), \beta \in H^{m_2,n_2}(X; R)$,

$$\alpha\beta = (-1)^{m_1 m_2} \beta\alpha.$$

5.3 Some Computations of Motivic Cohomology

The sheaf $\mathbb{Z}(0)$ on Sm_k is chain-homotopy equivalent to the constant sheaf (since $\mathbb{Z}_{tr}(Spec(k))$ is). Thus, we have

$$H^{m,0}(X; A) = \begin{cases} \underline{A}(X) & \text{for } m = 0 \\ 0 & \text{else.} \end{cases} \tag{5.3.1}$$

5.3.1 Proposition *There exists a quasiisomorphism of complexes of Zariski sheaves on* Sm_k:

$$\mathbb{Z}(1) \xrightarrow{\sim} \mathbb{G}_m[-1]. \tag{5.3.2}$$

Proof ([14]) A Zariski abelian sheaf $\mathcal{M}$ is defined by letting $\mathcal{M}(X)$ be the group of rational functions on $X \times \mathbb{P}^1$ which are regular in an open neighborhood of $X \times \{0, \infty\}$ and equal to 1 on $X \times \{0, \infty\}$. One has a short exact sequence of abelian sheaves on Sm_k:

$$0 \longrightarrow \mathcal{M} \overset{\kappa}{\longrightarrow} \mathbb{Z}_{tr}(\mathbb{G}_m) \overset{\lambda}{\longrightarrow} \mathbb{Z} \oplus \mathbb{G}_m \longrightarrow 0 \qquad (5.3.3)$$

constructed as follows: The map κ is the Weil divisor of the given rational function. The first coordinate of the map λ is the degree, the second, on an elementary finite correspondence Z, is the product of the values over a given point of X. (This can be more formally defined as follows: There exists a unique rational function f on $X \times \mathbb{P}^1$ whose divisor is Z such that $f(z)/z^n$ is 1 at $z = \infty$, where $n = deg(Z)$. Then $f(z) = (z - a_1) \cdots (z - a_n)$ is a polynomial non-zero at $z = 0$, and we can define the second component of λ as $(-1)^n f(0) = a_1 \ldots a_n$.) Now by basic facts on divisors (see Sect. 3 of Chap. 4), the sequence of sheaves (5.3.3) is exact on sections, and hence is exact. Next, factor out $\mathbb{Z}_{tr}(Spec(k))$ from the last two terms (5.3.3), and apply C_*. We get a short exact sequence of complexes of Zariski sheaves on Sm_k:

$$0 \to C_* \mathcal{M} \to \mathbb{Z}(1)[1] \to C_*(\mathbb{G}_m) \to 0.$$

However, one has

$$C_*(\mathbb{G}_m) \sim \mathbb{G}_m,$$

since $\mathbb{G}_m$ is homotopy invariant (in other words, invertible regular functions on $X \times \mathbb{A}^1$ are constant on the $\mathbb{A}^1$-coordinate). Thus, (5.3.2) follows, if we can prove

$$C_* \mathcal{M} \sim 0. \qquad (5.3.4)$$

This is easy, however, since a rational function in $f \in \mathcal{M}(X)$ is obviously $\mathbb{A}^1$-homotopic to the function 1 by the $\mathbb{A}^1$-homotopy $sf + (1-s)$. Thus, (5.3.4) follows from Lemma 5.1.2. $\qquad \square$

By Proposition 5.3.1, we have

$$H^{m,1}(X) \cong H^{m-1}(X, \mathbb{G}_m).$$

The right hand side was essentially calculated in Sect. 3 of Chap. 4. On a smooth variety X over k, one considers the exact sequence of Zariski sheaves on X (4.4.1) where i is the inclusion of the generic point. The sheaves Div_X and $i_* \mathbb{G}_m$ have no Zariski cohomology in degrees > 0 (the first one is a sum of pushforwards of constant sheaves on closed subsets, the second is a constant sheaf). Thus,

$$i_* \mathbb{G}_m \to Div_X$$

can be taken as a Γ-local resolution of $\mathbb{G}_m$. Taking global sections, we get

$$K(X)^\times \to Div(X) \tag{5.3.5}$$

where $K(X)$ is the function field of X. By the discussions of Sect. 3 of Chap. 4, the kernel of (5.3.5) is $\mathbb{G}_m(X) = \mathcal{O}(X)^\times$, and the cokernel is $Pic(X)$. Thus, we have

$$H^{m,1}(X) = \begin{cases} \mathcal{O}(X)^\times & \text{for } m = 1 \\ Pic(X) & \text{for } m = 2 \\ 0 & \text{otherwise.} \end{cases} \tag{5.3.6}$$

5.3.2 Milnor K-Theory

For a field k, one defines *Milnor K-theory* $K_*^M(k)$ as the quotient of the tensor algebra (i.e. free algebra) on the set $k^\times$ (which is set in degree 1), modulo the two-sided ideal generated by

$$a \otimes (1 - a), \ a \in k \smallsetminus \{0, 1\}.$$

(This relation is called the *Steinberg relation*.) We already know from (5.3.6) that we have a canonical isomorphism

$$k^\times = K_M^1(k) \cong H^{1,1}(Spec(k)).$$

Now one can prove by explicit calculation (see [14], Lecture 5) that the Steinberg relation holds in $H^{2,2}(Spec(k))$. Thus, (using Exercise 30), we obtain a canonical homomorphism of rings

$$K_n^M(k) \to H^{n,n}(Spec(k)). \tag{5.3.7}$$

5.3.3 Theorem ([14], Theorem 5.1) *The homomorphism of rings (5.3.7) is an isomorphism. More generally, it induces an isomorphism*

$$K_n^M(k)/\ell \to H^{n,n}(Spec(k); \mathbb{Z}/\ell).$$

$\square$

5.4   Relation with étale Cohomology—Voevodsky's Theorem

Let k be a perfect field and let ℓ be a natural number not divisible by its characteristic. By the étale topology on Sm_k, we mean the Grothendieck topology on that category where covers are étale covers.

Now Proposition 5.3.1 gives a quasiisomorphism of $\mathbb{Z}/\ell(1)$ with the chain complex in homological degrees $0, -1$ of Zariski sheaves on Sm/k:

$$\mathcal{H} = (\, \mathbb{G}_m \xrightarrow{\ \ell\ } \mathbb{G}_m\,). \tag{5.4.1}$$

By the quasiisomorphism (5.2.4), we then have a canonical morphism of Zariski sheaves

$$\mathbb{Z}/\ell(n) \to \mathcal{H}^{\otimes n}. \tag{5.4.2}$$

But now $\mathcal{H}^{\otimes n}$ is a complex of étale sheaves on Sm/k which is non-zero in only finitely many cohomological degrees, and quasiisomorphic to the sheaf $\mathbb{Z}/\ell(n)$ on $(Sm_k)_{et}$. Thus, by functoriality of sheaf (hyper)cohomology, we obtain a canonical homomorphism, compatible with the product:

$$H^{m,n}(X; \mathbb{Z}/\ell) \to H^m(X_{et}, \mathbb{Z}/\ell(n)). \tag{5.4.3}$$

This homomorphism is sometimes called the *norm residue symbol*. The following theorem of Voevodsky solved the famous Bloch-Kato conjecture:

5.4.1 Theorem (Voevodsky [25]) *For a separated smooth scheme X of finite type over a perfect field k, the norm residue symbol (5.4.3) is an isomorphism for $m \leq n$, and is injective for $m = n + 1$.*

$\square$

In view of Theorem 5.3.3, we have, for example, the following

5.4.2 Corollary *For a perfect field k of characteristic not dividing a natural number ℓ, the norm residue symbol induces an isomorphism, compatible with ring structure:*

$$K_M^n(k)/\ell \cong H^n(Gal(\overline{k}/k); \mathbb{Z}/\ell(n)).$$

In particular, if k contains the ℓ'th roots of unity, then we have

$$K_M^*(k)/\ell \cong H^*(Gal(\overline{k}/k), \mathbb{Z}/\ell).$$

$\square$

This is a powerful tool for calculating Galois cohomology. For example, even the statement that for perfect fields of characteristic not dividing ℓ and containing ℓ'th roots of unity, Galois cohomology with coefficients $\mathbb{Z}/\ell$ is generated, as a ring, by H^1, is highly non-trivial.

6   Exercises

1. Prove in detail the formula of Theorem 2.3.2.
2. Let k be a field, and let $X = \mathbb{P}_k^n$.
    (a) Consider the epimorphism of coherent sheaves

$$h : \bigoplus_{i=0}^n \mathcal{O}_X(-1) \to \mathcal{O}_X$$

given by the direct sums of the morphisms corresponding to the global sections of $\mathcal{O}_X$ given by the projective coordinates $x_0, \ldots, x_n$. Prove that $Ker(h)$ is isomorphic to $\underline{\Omega}^1_{X/k}$.
    (b) Prove that $\underline{\Omega}^n_{X/k} \cong \mathcal{O}_X(-n-1)$.
3. (Serre duality I: The projective space) Prove that for a coherent sheaf $\mathcal{F}$ on $\mathbb{P}_k^n$ where k is a field, we have an isomorphism

$$H^k(X, \mathcal{F}) \cong Ext_{\mathcal{O}_X}^{n-k}(\mathcal{F}, \underline{\Omega}^n_{X/k})^\vee$$

where $V^\vee = Hom_k(V, k)$ denotes the dual of a k-vector space V. [First, note that by functoriality alone, we have a canonical map

$$Hom(\mathcal{F}, \underline{\Omega}^n_{\mathbb{P}_k^n}) \to Hom(H^n(X, \mathcal{F}), H^n(X, \underline{\Omega}^n_{\mathbb{P}_k^n}))$$
$$= H^n(X, \mathcal{F})^\vee$$

which is an isomorphism by using the resolution (2.3.7) and Exercises 2, 1. Now by using the first i stages of the resolution (2.3.7), one can pass from H^n to H^{n-i}.]
4. (Serre duality II: A smooth projective variety) Prove the formula

$$H^k(X, \mathcal{F}) \cong Ext_{\mathcal{O}_X}^{n-k}(\mathcal{F}, \underline{\Omega}^n_{X/k})^\vee$$

for any smooth projective variety X of dimension n over a field k. The line bundle $\underline{\Omega}^n_{X/k}$ is sometimes denoted by ω_X and called the *canonical line bundle on X*. [Using a very ample line bundle $\mathcal{L}$, embed $i : X \subset P = \mathbb{P}_k^N$. Prove that

$$\underline{RHom}_{\mathcal{O}_P}(i_*\mathcal{O}_X, \underline{\Omega}^N_{P/k}) \sim i_*\underline{\Omega}^n_{X/k}[N-n].$$

Note: essentially, this is a local question.]

5. Prove the *Riemann-Roch Theorem* for curves: Let X be a smooth projective curve over an algebraically closed field k. Let $\mathcal{L}$ be be a line bundle on X. Then

$$dim_k(H^0(X, \mathcal{L})) - dim_k(H^0(X, \underline{\Omega}^1_{X/k} \otimes \mathcal{L}^{-1}))$$
$$= deg(D) + 1 - g$$

where

$$g = dim_k(H^1(X, \mathcal{O}_X)).$$

The quantity g is called the *arithmetic genus* of the curve X. Note that, for $k = \mathbb{C}$, by Hodge theory, it is equal to the *geometric genus*, which is defined as $dim_k(H^0(X, \underline{\Omega}^1_{X/\mathbb{C}}))$. [Use Serre duality to show that

$$dim_k(H^0(X, \underline{\Omega}^1_{X/k} \otimes \mathcal{L}^{-1})) = dim_k(H^1(X, \mathcal{L})).$$

With this in mind, verify the statement in the case when $\mathcal{L} = \mathcal{O}_X$. Denoting by $\mathcal{L}(D)$ the line bundle corresponding to a Weil divisor D, use the long exact sequence in cohomology corresponding to the short exact sequence

$$0 \to \mathcal{L}(D) \to \mathcal{L}(D + P) \to i_*\mathcal{O}_P \to 0$$

where $i : P \to X$ is the inclusion of a closed point.]

6. *The adjunction formula*
   (a) Let X be a smooth variety over a field k and let Y be a smooth subvariety of X, defined by a sheaf of ideals $\mathcal{I}$. Let $i : Y \to X$ be the closed immersion. Prove that there is a short exact sequence

   $$0 \to i^*\mathcal{I}/\mathcal{I}^2 \to i^*\underline{\Omega}^1_{X/k} \to \underline{\Omega}^1_{Y/k} \to 0.$$

   (b) Prove that in the situation above, if the dimension of X resp. Y is n resp. m, we have

   $$\underline{\Omega}^m_{Y/k} \otimes_{\mathcal{O}_Y} \wedge^{n-m}_{\mathcal{O}_Y} i^*(\mathcal{I}/\mathcal{I}^2) \cong i^*\underline{\Omega}^n_{X/k}.$$

   (c) Prove that if D is a smooth subvariety of X of codimension 1, then $i^*\mathcal{I}/\mathcal{I}^2$ is isomorphic to $i^*\mathcal{L}(D)^{-1}$ where $\mathcal{L}(D)$ is the line bundle on X corresponding to the divisor D. Deduce that in this case,

   $$\underline{\Omega}^{n-1}_{D/k} \cong i^*(\underline{\Omega}^n_{X/k} \otimes_{\mathcal{O}_X} \mathcal{L}(D)).$$

7. Let X be a smooth curve in $\mathbb{P}^2_{\mathbb{C}}$ of degree d and genus g. Prove that $g = (d-1)(d-2)/2$. [By the adjunction formula, ω_X is the restriction of $\mathcal{O}(d-3)$ from $\mathbb{P}^2_k$ to X, and thus has degree $d(d-3)$. Apply the Riemann-Roch Theorem to $\mathcal{L} = \omega_X$.]

8. Following the method of Exercise 7, prove that a smooth closed curve in $\mathbb{P}^1_{\mathbb{C}} \times_{spec(\mathbb{C})} \mathbb{P}^1_{\mathbb{C}}$ of bidegree (d, e) (defined as the pair of degrees of compositions with the two projections to $\mathbb{P}^1$) has genus

$$g = (d-1)(e-1).$$

Conclude that there exists a smooth closed curve in $\mathbb{P}^3_{\mathbb{C}}$ of any genus. [Use the Veronese embedding.]

9. Let C be a smooth closed curve in $\mathbb{P}^3_{\mathbb{C}}$ which is the complete intersection of two smooth closed surfaces D, E of degrees d, e. (Complete intersection means that the sheaf of ideals corresponding to C is generated by the sheaves of ideals corresponding to D, E.) Using the adjunction formula, show that the degree of ω_C is $de(d+e-4)$, and consequently, using the Riemann-Roch Theorem,

$$g = de(d+e-4)/2 + 1.$$

Deduce that there exists a smooth curve in $\mathbb{P}^3_{\mathbb{C}}$ which is not the complete intersection of two smooth closed surfaces.

10. Prove the *Bott Theorem*: We have

$$H^q(\mathbb{P}^n_{\mathbb{C}}, \Omega^p_{\mathbb{P}^n_{\mathbb{C}}/\mathbb{C}}(r)) = 0$$

unless $0 \leq p = q \leq n$, $r = 0$, or $q = 0$, $r > p$, or $q = n$, $r < -n + p$. [Given the fact that we know the Hodge numbers of $\mathbb{P}^n_{\mathbb{C}}$, this can be done by mimicking precisely the method of Exercise 1, tensoring with $\Omega^p_{\mathbb{P}^n_{\mathbb{C}}/\mathbb{C}}$.]

11. Put, for a coherent sheaf $\mathcal{F}$ on a smooth projective variety X,

$$\chi(\mathcal{F}) = \sum_i (-1)^i dim(H^i(X, \mathcal{F})).$$

(Sums of this type are referred to by the term *Euler characteristic*.)

(a) Putting $\mathbb{P} = \mathbb{P}^n_{\mathbb{C}}$, prove that we have a short exact sequence of coherent sheaves on $\mathbb{P}$:

$$0 \to \underline{\Omega}^r_{\mathbb{P}/\mathbb{C}} \to \bigoplus_{\binom{n+1}{r}} \mathcal{O}_{\mathbb{P}}(-r) \to \underline{\Omega}^{r-1}_{\mathbb{P}/\mathbb{C}} \to 0.$$

(b) Prove (by induction on r) that

$$\chi(\underline{\Omega}^r_{\mathbb{P}/\mathbb{C}}(i)) = \sum_{j=0}^{r}(-1)^j\binom{n+1}{r-j}\binom{i-r+j+n}{n}.$$

12. Let X be a degree d smooth (closed) hypersurface in $\mathbb{P} = \mathbb{P}^n_{\mathbb{C}}$, and let $i : X \to \mathbb{P}$ be the inclusion.

(a) Prove that there is a short exact sequence

$$0 \to \underline{\Omega}^r_{\mathbb{P}/\mathbb{C}}(-d) \to \underline{\Omega}^r_{\mathbb{P}/\mathbb{C}} \to i_*i^*\underline{\Omega}^r_{\mathbb{P}/\mathbb{C}} \to 0.$$

[Consider the case $r = 0$ first.]

(b) Prove that for $r \geq 1$, there is a short exact sequence

$$0 \to \underline{\Omega}^{r-1}_X(-d) \to (\underline{\Omega}^r_{\mathbb{P}/\mathbb{C}})|_X \to \underline{\Omega}^r_{X/\mathbb{C}} \to 0.$$

[Consider the case $r = 1$ first.]

(c) Prove that for $p + q < n - 1$,

$$H^q(X, \underline{\Omega}^p_{X/\mathbb{C}}(-r)) = \begin{cases} H^q(\mathbb{P}, \underline{\Omega}^p_{\mathbb{P}/\mathbb{C}}) & \text{if } r = 0 \\ 0 & \text{if } r > 0. \end{cases}$$

[Use induction on p. Use (b) to prove $H^q(X, \underline{\Omega}^p_{X/\mathbb{C}}(-r)) = H^q(\underline{\Omega}^p_{\mathbb{P}/\mathbb{C}}(-r)|_X)$. Then use (a) and Exercise 10 to show that the right hand side is $H^q(\mathbb{P}, \underline{\Omega}^p_{\mathbb{P}/\mathbb{C}})$ for $r = 0$ and 0 for $r > 0$.]

13. *Cohomology of Smooth Hypersurfaces:* Let X be a degree d smooth (closed) hypersurface in $\mathbb{P} = \mathbb{P}^n_{\mathbb{C}}$. Note that in view of Exercise 12 (d), the dimension $h^{p,q}(X)$ of the $\mathbb{C}$-vector space $H^{p,q}(X)$ is determined for $p + q \neq n - 1$. Further, for $p + q = n - 1$, $h^{p,q}(X)$ is equal to

$$(-1)^{n-1-p}\chi(\underline{\Omega}^p_{X/\mathbb{C}}) + (-1)^n$$

if $p \neq q$, and to this number plus 1 if $p = q$. Thus, it is determined by $\chi(\underline{\Omega}^p_{X/\mathbb{C}})$.

(a) Use Exercise 12 (a) and Exercise 11 (b) to prove that

$$\chi(\mathcal{O}_X(i)) = \binom{i+n}{n} - \binom{i+n-d}{n}.$$

(b) Use Exercise 12 (a) and (b) to prove that

$$\chi(\Omega^p_{X/\mathbb{C}}(i)) = $$
$$\chi(\Omega^p_{\mathbb{P}/\mathbb{C}}(i)) - \chi(\Omega^p_{\mathbb{P}/\mathbb{C}}(i-d)) - \chi(\Omega^{p-1}_{X/\mathbb{C}}(i-d))$$

for $p \geq 1$.

(c) Note that (a) and (b) together with Exercise 11 (b) give a complete recursive formula for $h^{p,q}(X)$. Use this to calculate $h^{p,q}(X_{an})$ where

$$X = Proj(\mathbb{C}[x, y, z, t, u]/(x^5 + y^5 + z^5 + t^5 + u^5)).$$

14. Let X be a smooth projective variety over $\mathbb{C}$ of dimension n, and let $0 \leq i < \lfloor n/2 \rfloor$. Prove that $dim H^i(X, \mathbb{C}) \leq dim H^{i+2}(X, \mathbb{C})$. Is it necessarily true that $dim H^i(X, \mathbb{C}) \leq dim H^{i+1}(X, \mathbb{C})$?

15. *The Wonderful Compactification II (DeConcini, Procesi, Fulton, MacPherson):* Let X be a smooth variety over a field k. Assume again for simplicity that k is algebraically closed. The *ordered configuration space* $F(X, n)$ of n points in X is the open subvariety of X^n which is the complement of all loci where two coordinates of a point of X^n coincide. We describe here a generalization of the construction in Exercise 35 of Chap. 4 which gives a resolution of singularities of $F(X, n)$ in the sense of Theorem 2.4.2 in the case where X is smooth projective. Define varieties $X[n, i]$, $X[n]$ and subvarieties $\Xi^{n,i}_S \subset X[n, i]$, $\Xi^n_S \subset X[n]$ in the same way as in Exercise 35 of Chap. 4, with $\mathbb{A}^m_k$ replaced by our variety X.

The *algebraic tangent space* $T_{X,x}$ of X at a closed point x is defined as the dual of the k-vector space m/m^2 where m is the maximal ideal of $\mathcal{O}_{X,x}$.

For a closed point $x \in X$, an *S-screen* at $x \in X$ is a finite subset of the vector space $T_{X,x}$ whose elements are labeled by non-empty disjoint finite sets whose union is S. Again, screens are identified when they are related by translation and multiplication by a scalar in $k^\times$.

By an (n, i)-*tree* T in X, we mean a finite set of distinct closed points x_j of X labeled by non-empty disjoint sets whose union is $\{1, \dots, n\}$ together with a minimal collection of screens at the points x_j which satisfies the following conditions: For a subset $S \subseteq \{1, \dots, n\}$ where $|S| > 1$ and $n \notin S$, or $n \in S$ and $|S| > i$ where T contains either a point x_j labeled S or a point of a screen at x_j labeled by S, there is an S-screen at x_j. If $n \in S$ and $2 < |S| \leq i$ and T contains a point x_j labeled S or a point of a screen at x_j labeled S, there is an $(S \setminus \{n\})$-screen at x_j.

(a) Defining again D_S, for $S \subseteq \{1, \dots, n\}$ with $|S| > 1$, as the closed subvariety of X^n of n-tuples of points whose S-coordinates coincide, prove statements analogous to parts (a) and (b) of Exercise 35 of Chap. 4.

(b) Prove that in $X[n]$, Ξ^n_S for $|S| > 1$ are divisors with normal crossings. Therefore, if X is smooth projective, $X[n]$ is a resolution of singularities of $F(X, n)$ in the sense of Theorem 2.4.2.

16. Prove the following version of *Zariski's Main Theorem*: Let A be a commutative ring and B be a finitely generate A-algebra. Let $P \subset B$ be a prime which is both maximal and minimal among primes whose pullback in A is a given prime p (in other words, in its Zariski fiber over p). We also say that the prime P is *isolated in its fiber over* p. Then there exists an $f \in A \setminus p$ such that the unit map induces an isomorphism of rings $f^{-1}A \cong f^{-1}B$. (The assumptions are stronger than necessary to simplify the proof.)

(a) Suppose the A-algebra B is generated by n elements $x_1, \ldots, x_n$. Reduce to the case $n = 1$ by using induction on n: Suppose the theorem holds with $n > 1$ replaced by a smaller number. Let C be the A-algebra generated by $x_1, \ldots, x_{n-1}$. Let Q be the pullback of the prime P to C. Then P is both minimal and maximal over Q. Thus, it suffices to show that Q is both minimal and maximal over p. For minimality, use that by the $n = 1$ case, $u^{-1}B \cong u^{-1}C$ for some $u \in C \setminus Q$, so primes contained in Q and P are in a bijective correspondence. For maximality, let, as usual, the rings B_p, C_p be formed by inverting $A \setminus p$ in each of them respectively. Then, by maximality, B_p/P is a field which is finitely generated as an algebra (hence finite) over the field A_p/p. Now C_p/Q is a finitely generated A_p/p-algebra, over which the field B_p/P is a finitely generated algebra. By the Nullstellensatz, C_p/Q is a field, and hence Q is maximal.

(b) Suppose the A-algebra B is generated by a single element x. Without loss of generality, $A = A_p$ is local and $B = B_p$. Then B_p/p is generated as an algebra by the image $\overline{x}$ of x over the field A_p/p. But it cannot be a polynomial algebra on one generator (then no prime over p in B_p would be isolated in its fiber). Thus, the element $\overline{x}$ is algebraic. In other words, there is a polynomial relation

$$a_d x^d + a_{d-1} x^{d-1} + \cdots + a_0 = 0$$

with $a_i \in A_p$, and not all $a_i \in p$. Let d be minimal possible. Then $d > 0$, $a_d x$ is integral over A_p and hence in A_p, and thus, we can rewrite the equation as

$$(a_d x + a_{d-1})x^{d-1} + a_{d-2}x^{d-2} + \cdots + a_0 = 0.$$

By the assumption, $a_0, \ldots, a_{d-2}, a_d x + a_{d-1} \in p$. If $a_d x \in p$, then a_{d-1} is not in p while $a_d x + a_{d-1} \in p$, which is a contradiction. Thus, $a_d \notin p$, and hence $x \in A_p$, completing the argument.

17. Prove the following form of Zariski's Main Theorem: Let $f : X \to Y$ be a morphism of varieties over an algebraically closed field k where f is bijective on closed points, Y is normal, and the function field $K(Y)$ is a separable extension of $K(X)$ (note that f is dominant because it is bijective on closed points; by separable extension here we mean that every algebraic sub-extension is separable). Then f is an isomorphism. (The same argument also gives a version of the theorem for k a perfect field if we additionally assume that the bijection on closed points preserves residue fields.)

(a) Give examples to show that the assumptions (normal and separable) are both necessary. [For separability, consider the affine line over a field of characteristic $p > 0$, and the Frobenius map.]

(b) Reduce to the affine case, i.e. $Y = Spec(A)$, $X = Spec(B)$ for a homomorphism of k-algebras $A \to B$.

(c) Deduce from the results of Sect. 5.5 of Chap. 1 that the transcendence degree of $K(X)$ over $K(Y)$ is 0, and hence $K(X)$ is finite over $K(Y)$.

(d) Prove that for a closed point p in a non-empty open set $U \subseteq Spec(A)$, the dimension of the A/p-vector space B/Bp is equal to $[K(X) : K(Y)]$ (i.e. the dimension of the $K(Y)$-vector space $K(X)$). [B is finitely generated over A, and after inverting an element, becomes integral, hence a finite module. Since $K(X)$ is separable, it is generated by a single element $t \in B$; localize away from the denominators of its coefficients, its discriminant, and the quotient of B by the subalgebra generated by t, which is finitely generated and torsion.] Conclude from the assumptions that $K(X) = K(Y)$.

(e) We have $A \subseteq B$, and we need to prove that the inclusion is onto. If it isn't onto, then for some maximal ideal $m \subset A$, $(B/A)_m = B_m/A_m \neq 0$, thus, $B_m \neq A_m$. Since B_m and A_m have the same field of fractions, by assumption, A_m is integrally closed in B_m, and by assumption, there is only one prime in the fiber of the maximal ideal in B_m, and thus is isolated. Thus, by Exercise 16, there is an $f \in A_m$ not in the maximal ideal such that $f^{-1}A_m = f^{-1}B_m$. But since A_m is local, f is a unit.

18. Prove the following version of the Zariski Main Theorem: Let A be a commutative ring, and let B be a finitely generated A-algebra such that every prime in B is isolated in its fiber. Then there exists a finite subalgebra C of the integral closure A' of A in B and an element $f \in C$ such that the unit morphism induces an isomorphism $f^{-1}C \cong f^{-1}B$. Further, for any prime P of B we may choose C, f so that $f \notin P$. [Let A' be the integral closure of A in B. Use Exercise 16 to conclude that there exists an element $f \in A'$ such that $f^{-1}A' = f^{-1}B$. In particular, $f^{-1}A'$ is a finitely generated A-algebra, say, by elements $x_1/f^{r_1}, \ldots x_n/f^{r_n}$ with $x_i, f \in A'$. Let C be the subalgebra of A' generated by $x_1, \ldots, x_n, f$.]

19. Using the result of Exercise 18, prove that diagram (4.2.2) can be completed with ι a quasi-compact open immersion and g a finite morphism when X is a quasi-compact separated scheme, for every point $x \in X$, $f^{-1}(x)$ is discrete in the induced topology from Y, and the morphism f is separated and of finite type. [Use gluing.]

20. A morphism $A \to B$ of commutative rings is called *flat* if it makes B a flat A-module. It is called *unramified* if it is of finite type (i.e. B is a finitely generated A-algebra) and $\Omega_{B/A} = 0$. (Note: both are clearly local concepts, and hence readily generalize to schemes.) Prove that the following are equivalent:

(i) f is étale

(ii) f is flat and unramified

(iii) For every prime $q \subset B$ over a prime $p \subset A$, there exists an $g \in A \smallsetminus p$ and $h \in B \smallsetminus q$ such that $h \mid g$ and $B' = h^{-1}B$ is *standard étale* over $A' = g^{-1}A$ (i.e. $B' \cong A'[x]/(\phi(x))$ where ϕ is a monic polynomial whose discriminant is a unit).

Clearly, (iii) implies (i).

(a) Assuming (i), assume further f is standard smooth of dimension 0. $\Omega_{B/A}$ is the dual of the tangent space, which is the cokernel of the Jacobi matrix of the constraints. Thus, it is 0. Recalling what we proved about étale extensions of fields, prove that the Zariski fiber of every prime $p \subset A$ is finite and discrete. By Exercise 18, f is locally finite. Without loss of generality, then, we may assume f is finite. Choose a prime $p \subset A$. Localizing further, we may assume that B_p/pB_p is a finite separable extension of A_p/pA_p. A set of elements of B_p whose projections form a basis of B_p/pB_p over A_p/pA_p also generates B_p as a module over A_p by Nakayama's lemma. In particular, an element $x \in B_p$ which projects to a primitive element of B_p/pB_p over A_p/pA_p is the root of a monic polynomial with coefficients in A_p of the same degree, whose discriminant is a unit. This proves (iii). It also proves flatness (hence (ii)), since it shows that B_p is a free A_p-module.

(b) Assume f is flat and unramified. Let again $p \subset A$ be a prime. Then B_p/pB_p is unramified over A_p/pA_p. Verify that this, again, means that B_p/pB_p is a product of finite separable extensions of A_p/pA_p. Again, localizing further, we may assume it is just a finite separable extension. By Exercise 18, again, B is locally finite over A, and thus, by Nakayama's lemma, the set of powers of an element $x \in B_p$ which projects to a primitive element of B_p/pB_p below the degree of the minimal polynomial of its projection to B_p/pB_p forms a set of generators of B_p as an A_p-module, which is a basis since it projects to a basis of B_p/pB_p over A_p/pA_p. Thus, B_p is standard étale over A_p. Thus, (ii) implies (iii).

21. Prove that if S is a Noetherian scheme and $f : X \to S$ is a morphism of finite type, then f is smooth of dimension n if and only if for every $x \in S$ there exists an open neighborhood U, an open cover (V_i) of $f^{-1}(U)$ and étale morphisms $g_i : V_i \to \mathbb{A}^n_U$ over S.

22. (Deligne) Let k be a perfect field of characteristic $p > 0$ and let $A = k[T_1, \ldots, T_n]$. Using the computations of Sect. 3.3, prove that the de Rham-Witt complex $W_N\Omega^* A$ contains the de Rham complex $\Omega^*_{A/W_N(k)}$ as a direct summand, where the complementary summand has cohomology 0.

23. Let A be an $\mathbb{F}_p$-algebra, and let B be an étale A-algebra. Prove that there is an isomorphism

$$W_n\Omega^* B \cong W_n B \otimes_{W_n A} W_n\Omega^* A.$$

24. Let X be a smooth scheme over Wk where k is a perfect field of characteristic $p > 0$. Using the results of Exercises 21–23, prove that

$$H^*_{crys}(X \times_{Spec(Wk)} Spec(k)) \cong H^*_{DR}(X).$$

25. Proof the statement of the Comment under Proposition 2.4.4 of Chap. 4.

26. Prove an analogue of Corollary 3.7.4 of Chap. 5 in the étale topology, i.e., for a morphism of schemes $f : X \to Y$, and a sheaf $\mathcal{F}$ on X_{et}, a convergent spectral sequence of the form

$$H^p(Y_{et}, R^q f_*(\mathcal{F})) \Rightarrow H^{p+q}(X_{et}, \mathcal{F}).$$

27. Using the result of Exercise 26, prove that for any algebraic variety X over a perfect field k, the projection $p : \mathbb{A}^1_X \to X$, and, more generally, the projection $p : U_{\mathcal{M}} \to X$ for any algebraic vector bundle $\mathcal{M}$ on X induces an isomorphism in étale cohomology with coefficients in $\mathbb{Z}/\ell$ where ℓ is not divisible by the characteristic of k. (In other words, étale cohomology is $\mathbb{A}^1$-homotopy invariant.)

28. Let X be an algebraic variety over an algebraically closed field k. Let $f : (X \times \mathbb{G}_m)_{et} \to X_{et}$ be the projection and let $\ell \in \mathbb{Z}$ not be divisible by the characteristic of k. Prove that $R^i f_* \mathbb{Z}/\ell$ is equal to $\mathbb{Z}/\ell$ for $i = 0, 1$ and to 0 otherwise. Using the result of Exercise 26 together with the fact that f has a right inverse, prove that

$$H^i((X \times \mathbb{G}_m)_{et}, \mathbb{Z}/\ell) = H^i(X_{et}, \mathbb{Z}/\ell) \oplus H^{i-1}(X_{et}, \mathbb{Z}/\ell).$$

29. Let k be an algebraically closed field and let let $\ell \in \mathbb{Z}$ not be divisible by the characteristic of k. Using the results of Exercises 27, 28, calculate $H^*((\mathbb{P}^n_k)_{et}, \mathbb{Z}/\ell)$.

30. Let G be a group and let M, N be $\mathbb{Z}[G]$-modules (recall that left and right modules form equivalent categories by setting $gm = mg^{-1}$ for $g \in G$). Observe that $M \otimes_{\mathbb{Z}} N$ is a $\mathbb{Z}[G]$-module by setting $g(m \otimes n) = g(m) \otimes g(n)$ for $m \in M, n \in N, g \in G$. By using injective resolutions, construct a bilinear product

$$H^i(G; M) \otimes_{\mathbb{Z}} H^j(G, N) \to H^{i+j}(G; M \otimes_{\mathbb{Z}} N)$$

and prove that it is unital, associative and graded-commutative in the obvious sense. (This product is sometimes denoted by $\cup$ and called the *cup product*.)

31. Suppose $H \subseteq G$ is an inclusion of groups. Then by forgetting structure (which is an exact functor), we obtain an additive functor $D\mathbb{Z}[G]\text{-}Mod \to D\mathbb{Z}[H]\text{-}Mod$ from the derived category of G-modules to the derived category of H-modules, and an induced map called *restriction* $res^G_H : H^i(G; M) \to H^i(H; M)$ for a $\mathbb{Z}[G]$-module M. The forgetful functor from $\mathbb{Z}[G]$-modules to $\mathbb{Z}[H]$-modules has both a right and a left adjoint (the left and right Kan extension) which are both exact, and hence induce additive functors on the derived categories.

(a) Prove that if the index $[G : H]$ is finite, then the left and right Kan extension are canonically isomorphic. If M is a $\mathbb{Z}[G]$-module, and N is the Kan extension of itself considered as a $\mathbb{Z}[H]$-module, then the counit of the left Kan extension gives a homomorphism of G-modules $\alpha : N \to M$. The homomorphism on cohomology $H^i(H; M) \to H^i(G; M)$ given by α composed with the derived Kan extension is called the *corestriction* and denoted by cor_G^H.

(b) Prove that res_H^G is a homomorphism of rings, and cor_G^H is thus a homomorphism of modules. Explicitly,

$$cor_G^H(res_H^G(\alpha) \cdot \beta) = \alpha \cdot cor_G^H(\beta)$$

for $\alpha \in H^i(G; M)$ and $\beta \in H^j(H; M)$ where M is a $\mathbb{Z}[G]$-modules. This relation is sometimes referred to as the *projection formula*.

32. Using Hilbert's 90 Theorem and the Kummer exact sequence, prove that if ℓ is not divisible by the characteristic of a field k, then

$$H^1(k, \mathbb{Z}/\ell(1)) \cong k^\times/(k^\times)^\ell.$$

Prove that if K is a finite separable extension of k, then under this identification, inclusion $k^\times \subseteq K^\times$ induces restriction in Galois cohomology, while $N_{K/k}$ induces corestriction. (This involves an obvious extension of Exercise 31 to profinite groups.)

33. Using the results of Exercises 30–32, prove directly that for a field k, a number ℓ not divisible by its characteristic, and elements $a \neq 0, 1 \in k^\times$, the *Steinberg relation*

$$a \cup (1 - a) = 0 \in H^2(k; \mathbb{Z}/\ell(2))$$

holds. [Factor $x^\ell - a = \prod p_i$ into irreducible polynomials $p_i \in k[x]$. Let $K_i = k[x]/(p_i)$, an let α_i be the image of x in K_i. First observe that

$$1 - a = \prod_i N_{K_i/k}(1 - \alpha_i).$$

Using the projection formula and the result of Exercise 32, prove that

$$a \cup (1 - a) = \sum_i cor_i(res_i(a) \cup (1 - \alpha_i))$$

where cor_i, res_i denote corestriction and restriction between $Gal(k)$ and $Gal(K_i)$. Now observe that $res_i(a) = 0$ since $a = \alpha_i^\ell$.]

34. Referring to the definitions of Sect. 5.2, prove that for smooth schemes X, Y over a field k and for the projection $f : X \times \mathbb{G}_m \to X$, we have

$$Rf_*(\mathbb{Z}_{tr}(Y)(1)) \sim \mathbb{Z}_{tr}(Y) \oplus \mathbb{Z}_{tr}(Y)(1),$$

where f_* is applied on categories of Zariski abelian sheaves. [Consider $Y = Spec(k)$ first.] Conclude that

$$H^{m,n}(X \times \mathbb{G}_m) \cong H^{m,n}(X) \oplus H^{m-1,n-1}(X).$$

35. Using the result of Exercise 34 and $\mathbb{A}^1$-homotopy invariance of motivic cohomology, prove that

$$H^{\ell,m}(\mathbb{P}^n_k) = \bigoplus_{i=0}^{n} H^{\ell-2i,m-i}(Spec(k)).$$

[Use the spectral sequence arising from the standard affine cover of $\mathbb{P}^n$.]

36. Let k be a perfect field. Let $a \in k^\times$ not be a square. Prove that there is a short exact sequence of Zariski sheaves of Sm_k

$$0 \to \underline{\mathbb{Z}/2} \to \mathbb{Z}/2_{tr}(Spec(k[\sqrt{a}])) \to \underline{\mathbb{Z}/2} \to 0$$

where both arrows are compositions with the correspondence given by $Spec(k[\sqrt{a}])$. Write down the resulting long exact sequence in motivic and Galois cohomology. (Note: the sequence in Galois cohomology can also be obtained directly from the Leray spectral sequence of the projection $Spec(k[\sqrt{a}])_{et} \to Spec(k)_{et}$.)

Bibliography

1. L. Ahlfors, *Complex Analysis: An Introduction to The Theory of Analytic Functions of One Complex Variable* (McGraw-Hill, New York, Toronto, London, 1979)
2. M.F. Atiyah, I.G. MacDonald, *Introduction to Commutative Algebra* (Addison-Wesley, Westview Press, Boulder, 1994)
3. H. Cartan, S. Eilenberg, *Homological Algebra* (Princeton University Press, Princeton, 1999)
4. S.D. Cutkosky, *Resolution of Singularities*. Graduate Studies in Mathematics, vol. 63 (American Mathematical Society, Providence, 2004)
5. P. Deligne, La conjecture de Weil I. Inst. Hautes Études Sci. Publ. Math. **43**, 273–307 (1974)
6. P. Deligne, La conjecture de Weil II. Inst. Hautes Études Sci. Publ. Math. **52**, 137–252 (1980)
7. W. Fulton, *Intersection Theory* (Springer, New York, 1998)
8. W. Fulton, J. Harris, *Representation Theory: A First Course* (Springer, New York, 1999)
9. H. Grauert, R. Remmert, *Coherent Analytic Sheaves* (Springer, New York, 1984)
10. P. Griffiths, J. Harris, *Principles of Algebraic Geometry* (Wiley-Interscience, John Wiley and Sons, New York, 1994)
11. R. Hartshorne, *Algebraic Geometry* (Springer, New York, 1997)
12. L. Illusie, Complexe de de Rham-Witt et cohomologie cristalline. Ann. Sci. ENS **12**, 501–661 (1979)
13. J.P. May, *A Concise Course in Algebraic Topology* (University of Chicago Press, Chicago, 1999)
14. C. Mazza, V. Voevodsky, C. Weibel, Lecture Notes on Motivic Cohomology (AMS/Clay Math. Institute, Providence/Cambridge, 2011)
15. J.S. Milne, *Étale Cohomology* (Princeton University Press, Princeton, 2017)
16. J.W. Milnor, *Topology from the Differentiable Viewpoint* (Princeton University Press, Princeton, 1997)
17. D. Mumford, Pathologies of modular algebraic surfaces. Am. J. Math. **83**, 339–342 (1961)
18. D. Mumford, *The Red Book of Varieties and Schemes*. Lecture Notes in Mathematics, vol. 1358 (Springer, New York, 1999)
19. J.R. Munkres, *Analysis on Manifolds* (Taylor and Francis, Boca Raton, 2018)
20. J.R. Munkres, *Elements of Algebraic Topology* (Taylor and Francis, Boca Raton, 1996)
21. J.P. Serre, Géométrie algébrique et géométrie analytique. Annales de l'institut Fourier **6**, 1–42 (1956)
22. J.P. Serre, *Algèbre locale*. Multiplicités (Springer, New York, 1965)
23. J.H. Silverman, *The Arithmetic of Elliptic Curves* (Springer, New York, 2016)
24. V. Srinivas, *Algebraic K-Theory* (Springer, New York, 1996)
25. V. Voevodsky, On motivic cohomology with $\mathbb{Z}/\ell$-coefficients. Ann. Math. **174**, 410–438 (2011)
26. R. Wells, *Differential Analysis on Complex Manifolds*, 3rd edn. (Springer, New York, 2008)

© The Author(s), under exclusive license to Springer Nature Switzerland AG 2021
I. Kriz, S. Kriz, *Introduction to Algebraic Geometry*,
https://doi.org/10.1007/978-3-030-62644-0

Index

© The Author(s), under exclusive license to Springer Nature Switzerland AG 2021
I. Kriz, S. Kriz, *Introduction to Algebraic Geometry*,
https://doi.org/10.1007/978-3-030-62644-0

nted in the United States
Baker & Taylor Publisher Services